BEARS
Page 420

MUSTELIDS
Page 428

NORTHERN RACCOON
Page 484

EVEN-TOED UNGULATES
Page 492

BIRDS
Page 602

AMPHIBIANS
Page 725

REPTILES
Page 729

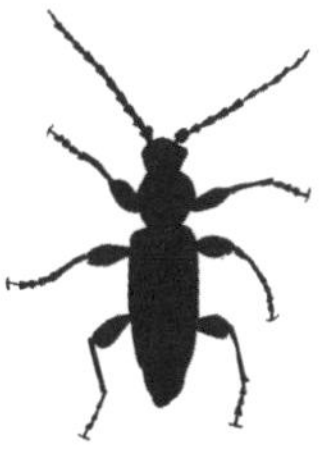

INSECTS
Page 742

CONTENTS

INSECTS AND OTHER INVERTEBRATES 742

APPENDICES 784

FOREWORD

Wildlife tracking skills are real and can be learned by anyone with patience and persistence, especially if they hold a good handbook to help them get started. In short, tracking describes the skills of identifying and interpreting the physical signs animals leave in their wake. For those of you who invest the time to be able to interpret footprints and other signs, you will both become aware of and be able to find the animals that surround you. Through tracking, you will engage in real relationships with real animals in a real world.

I know of no better way than studying tracks and other signs to see and experience the relationships between wildlife species, flora and fauna, and the biotic and abiotic components of ecosystems. Tracking grounds us in the natural history of a place, while simultaneously highlighting the individual personalities and tendencies of different animals. I encourage you to find a purpose for learning tracking skills, as purpose will push your skills to new levels. Photography, search and rescue, education, wildlife monitoring, wildlife research and conservation are but some of their varied applications.

Handbooks like this one that aid us in interpreting tracks and signs, are vital contributions to human communities – they kindle love of place, of wildlife, and make us aware of the ecological health of our local environment. Guidebooks that unlock new worlds of experience for their readers are powerful conservation tools when placed in the right hands. I own 123 tracking books from around the world. Each teaches me something, but I'm drawn to those that provide deep and comprehensive information about the diversity of signs made by wildlife – the heavy books, filled with details the casual natural historian might find uninteresting or even intimidating.

My own inspiration came from Preben Bang and Preben Dalhstrom's landmark guide *Collin's Guide to Animal Tracks and Signs*, first translated into English in 1974. At the time, it was the most comprehensive guide to animal signs ever written. Since then, numerous authors have strived to contribute something to the field, to expand to what was shared by Bang and Dahlstrom. However, it was decades before another book came close to their comprehensive work. It was a full 50 years, until this very moment in fact, that a guide to the wildlife tracks and signs of Europe became available that truly surpassed them.

The paramount skill of the tracker, and indeed any field scientist, is humility. Know your limits. You will always find signs you are unable to interpret and you will always continue to make mistakes in your interpretation, regardless of your level of expertise. But take solace in knowing that the best trackers in the world often make mistakes. Tracking requires deliberate concentration and an attention to detail. The act of creating handbooks to help others reliably differentiate between similar footprints and other signs requires even more focus, intensity and tenacity than tracking itself – and a touch of stubbornness as well. Few people can maintain the effort of will required to produce something truly comprehensive.

Let me be the first to introduce you to Joscha Grolms. Joscha is all of these things – focused, intense and more. He is a passionate, patient and humble human being. His new guide to the wildlife signs of Europe sets a new standard and one to which all subsequent books will be compared. His illustrations are both beautiful and accurate. He reveals details born of careful observation to help you differentiate among the signs of very similar species, and a greater diversity of animal footprints and signs than any other guide to the region. He has gathered the data needed to describe and illustrate the varied gaits used by wildlife. He has created thoughtful keys to lead readers to relevant sections, so as to make his comprehensive work more accessible, and more useful. He selected around 1,000 images of wildlife tracks and other signs to provide visual aid in their interpretation. Joscha's book betrays both his obsession with and love for the subject. His work is a gift not just to Europe but the entire world interested in natural history, wildlife, animal tracking and conservation.

In your hands you hold the most comprehensive guide to wildlife tracks and signs of Europe ever produced. Appreciate the effort that moulded it. Appreciate the man. Thank you, Joscha, for your contribution to the field.

Mark Elbroch, Master Tracker US
Author of the book *Mammal Tracks and Signs:
A Guide to North American Species*

More about Mark Elbroch:
https://markelbroch.com

PREFACE

'Every contact leaves a trace.'

Edmond Locard, criminologist and forensics pioneer

Wild animals can be shy and difficult to spot, so we are often unaware of their presence. But even though we rarely set eyes on them, they are always leaving traces. Traces we can read like stories to give us an insight into the hidden animal kingdom.

I've been fascinated by stories and animals since early childhood. When I discovered tracking at the age of 20, I was excited by the possibility of seeing animals' stories come alive. I was struck by the wealth of information contained in a single footprint and by the fact that even the tiniest trace can enable us to identify a species. I learned how to follow animal tracks. The fact that we can learn more about an animal and its life literally 'step by step' never fails to amaze me. Many moments and impressions, especially encounters with wild animals, have stayed with me for years.

For many early societies, the art of tracking was essential for finding food, clothing and tools. Trails left by animals indicated the location of water, safe routes for travel or possible dangers and helped people find their way. Nowadays, animal tracks aren't just interesting to hunters. Scientists, conservationists and other nature lovers are also fascinated by the many possibilities of tracking. Tracking is scientifically applied to monitor large predators such as bears, lynx and wolves. It is also becoming increasingly popular among more casual wildlife-watchers. Every year, people of different ages and from different backgrounds come together to learn to read animal tracks on our tracking courses. Once they learn what to look out for, most of these people are surprised and delighted by just how much more there is to discover on a normal walk through the forest. Their enthusiasm for tracking doesn't wane even after the course is over. But where does this fascination come from? Perhaps it's the joy of working something out or finding a new way of experiencing nature, which encourages us to slow down. Or maybe it's the incredible knowledge of a place that you gain from reading tracks in the countryside. For me, it's the tangible relationship between the land and the animals, as well as the familiar feeling of belonging that comes from it. In my teaching work, I have seen time and again how adults become like curious children, filled with a spirit of discovery and eagerness to learn more.

Tracking is essentially about a connection to nature and heightened perception. We experience nature with our senses and this experience leaves traces in us – traces that have a lasting influence on our behaviour. Tracking is a way of consciously being part of nature and feeling connected with the world.

This book is the result of my endless enthusiasm for European fauna. I hope that this reference work will help you, whether you are a professional or amateur tracker. I also hope that this book will inspire you to spend time in nature and that it will contribute to a growing fascination for animal tracks and signs.

ACKNOWLEDGEMENTS

I couldn't have written this book alone. It is based on the knowledge of other naturalists and trackers that has been gathered for generations. I have taken up and continued to follow their work, but any mistakes are my own. The scope of this project only became clear to me over the years of writing and I am grateful for all the support I've received. I would like to take this opportunity to thank everybody who contributed to the success of this book. I would like to thank the publisher for the offer to write this book. I am especially grateful to Ulf Müller for proofreading and his wife, Ina Vetter, for her friendly and professional support. Ulf's work is like the artistic final polish that a goldsmith gives a piece of jewellery, while Ina's management skills are comparable to a circus performer juggling 5,000 jigsaw pieces. I would like to thank Patrick Rösen, director of Natural Park Centre Uhlenkolk in Mölln, for contacting Verlag Eugen Ulmer, and David Moskowitz, who supported me during the first steps with his sample copies of the outline and letters to publishers. My thanks also go to the photographers for their generous contributions to the extensive images of the tracks and signs covered. With a few exceptions, the photographs have been published here for the first time. I am grateful to the administrative staff of the Oberlausitzer Heide- und Teichlandschaft Biosphere Reserve for granting me access rights and for pointing out suitable places for recording animal tracks, as well as to Jörn Lies, who drew up the track pattern image overviews. I would also like to thank Stefanie Huck from Retscheider Hof e.V. for contributing the ink prints of mustelids, as well as Simon Capt from the Swiss Cartographic Centre on Fauna (CSCF), who undertook the professional correction of these animal portraits. Thanks also go to Moritz Krämer for his expert examination of the texts on the *Microtus* voles, field mice and *Clethrionomys* voles. For the exchange of data and collaboration on species descriptions from their specialist fields, I would like to thank Jennifer Hatlauf from the Golden Jackal Project, Austria and the University of Natural Resources and Life Sciences, Vienna; Brown Bear researcher Michaela Skuban from Slovakia and Helene Möslinger from LUPUS, the Institute for Wolf Monitoring and Research in Germany. I would like to express my gratitude to Hielke Chaudron from the Musk Ox Centre in Tännäs, Sweden, as well as insect specialist Patrick Urban from AG Westfälischer Entomologen, Toni Romani from CyberTracker Italia, Immo Meyer for contributing his wild cat expertise and Markus Schwaiger from the lynx project in Bavaria. My thanks go to Dr Christine and Dr Stefan Resch from the apodemus Institut in Austria for their extensive small mammal research and helpful feedback. I would also like to thank shepherd Raphael Fuchs for contributing his 20 years of experience with domestic sheep and goats and their hoofprints. Heartfelt thanks also go to Sygmund Komancza for the care of European Bison herds in Bieszczady and for his trust in allowing me to collect tracks there. Thanks also to *Kriminaloberkommissar* Aaron Tiedemann, whose job it is to collect evidence at crime scenes and who is also passionate about animal tracks in his spare time. He created precisely scaled image collages for size comparison and contributed his professional knowledge of the feeding patterns of bark beetles. I am grateful to the team from spurenjagd.de, especially Ulrike Quartier, Antje Beneken and the Head of Biodata Recording at CyberTracker Germany, Holger Röhle, as well as Simone Roters from Erdwissen e.V., for examining and preselecting the datasets used. Thanks to Dr Wolfgang Schmidt for his extensive input into the chapter on foot morphology and gait. Its precision is down to his thoroughness and persistence. My special thanks also go to Dr Andreas Wenger, who was by my side as an exceptional ornithologist and good friend. His generous contribution was invaluable to the bird section.

I would like to thank the participants of the Wildniswissen tracking courses whose many interested questions inspired me to keep observing animal tracks, keep learning and to put my knowledge into a structured, comprehensible form.

Special thanks to my colleagues from CyberTracker Conservation, especially Louis Liebenberg, Mark Elbroch, René Nauta, John Rhyder, José Galan, Casey McFarland, George Leoniak and Nate Harvey. Thank you for your fascination and your dedication to wild animals and their tracks. You are passionately interested in the details that few people look for so closely. Along with expert criticism, you also make me feel that I belong to a group of likeminded people. I find our shared enthusiasm to be a never-ending source of inspiration and motivation.

Ulrike Quartier introduced me to the art of Indian ink dot drawing, which makes her the mother of my drawings. She accompanied me through each of my texts, provided valuable personal support and voluntarily contributed all the animal silhouettes, as well as the drawings of shrew and bird feet. The findings that enable us to differentiate between the feet of *Sorex* and *Crocidura* shrews are down to her fascination with shrew feet. Her expertise as a graphic artist, author, artist and tracker, as well as her eye for what I like to call track aesthetic, were significant contributions.

I would like to thank my mother, Marlis Grolms, from the bottom of my heart. She was the first to guide me down this path and stuck by me, even when my adventures occasionally caused her trouble. Special thanks also go to my mentors and role models: Wolfgang Peham, Tamarack Song, Abel Bean, Tony Kemnitz and Jon Young. Your personal

commitment to my professional and private development sowed the seeds from which this book grew. Tony's grandson once said of his grandpa in relation to the art of tracking that Tony had a 'wealth of useless knowledge'. Wolfgang Peham showed me that the things I'm interested in are worthwhile and he was the first to continuously strengthen my resolve to keep pursuing these interests. Wolfgang paved the way for this book by founding the Wildniswissen wilderness school and through his 25 years of pioneering work. I'd personally like to thank him for two decades of sincere interest, many good and difficult questions that kept making me think and for his enduring faith in me. Wolfgang helped me to see tracking not just as a 'wealth of useless knowledge', but as an extremely effective antidote to the widespread problem of nature-deficit disorder (Richard Louv 2005).

Of all the people who helped make this book a reality, one was there for me at every moment: my partner, Laura Gärtner. I can't begin to express how valuable our shared passion for animal tracks and her critical gaze and eloquence were to the quality of this book. She accompanied me on many of my trips through Europe, created a significant part of the animal track key, did all the first revisions and contributed important visual material. Most importantly, she was always there for me with a listening ear, time and understanding, even though this was by no means always mutual. Her sheer breadth of skills, from track reading to comma placement, to the comfort and emotional support that she continuously offered me, are what made this book possible in the first place.

Last but not least, I would like to thank all the landscapes, the lush riverbanks and the species-rich pond landscapes, the arid semi-deserts, the awe-inspiring mountains and the wonderfully peaceful forests. For me, they are sources of peace, clarity and strength. I'm grateful to all the animals: the insects and other invertebrates, the reptiles and amphibians, the birds and the mammals. They are the main characters and I dedicate this book to them.

INTRODUCTION

INTRODUCTION

'Read all the signs that tell you what happened.
Don't just look at the individual footprint, but
the universe that surrounds it.'
Tony Kemnitz, Special Investigations Tracker, Wisconsin.

European Bison in Bieszczady

Almost every year since 2005, I have travelled in and around Bieszczady National Park in south-eastern Poland. The region where Poland, Slovakia and Ukraine meet is one of Europe's most unspoilt landscapes and always worth a visit. The largely untouched mixed beech forests of the Carpathian foothills are home to animals such as wolves, lynx, bears and bison. This time, my visit was in spring, shortly after the snow had melted and while the beech trees were coming into leaf. It was a fresh morning in April when Matthias excitedly told me that, just 20 minutes earlier, he had spotted a female European Bison with two young by the river. I quickly packed the essentials into my daypack and set off.

Matthias had been sitting by a shallow part of the river that flows from the southern mountains into the valley. The thick, damp layer of leaves made it easy to find the tracks and I began to read them. The mother bison gave a warning as soon as she sensed a human, whereupon the two calves abruptly changed direction and fled uphill in rapid bounds. She was on the other side of the river and hastily trotted up the first hill, parallel to her calves. I followed the tracks and soon found the place where the young animals had crossed the river before heading further uphill with their mother at a more relaxed pace.

I was relieved that the animals had calmed down again so quickly and I decided to keep following the tracks, at sufficient distance from the family. The bison had now slowed to a walk and their path kept crossing older bison tracks, which made the fresh tracks harder to spot. Twice I lost the trail and it was a while before I was able to find the fresh tracks between the many older ones. The further I followed them, the more clearly I was able to discern the frequent and distinctive signs of the animals. Wide trails with small trees damaged by browsing animals and old, cowpat-like piles of dung told me that the bison had returned here often. I suddenly found myself standing in front of a head-high root plate, the ground under which the animals had been using as a sand bath. The distinctive underwool was scattered on the ground and their sharp aroma still hung in the air.

I then discovered a fresh dung pile with a gathering of flies. In the next valley, in a smaller meadow, the path branched out in different directions like a river. I suspected this must be a sign of grazing behaviour and I was right: I found a willow tree with freshly stripped bark. The bison had to be nearby! I looked around and, in the distance, I spotted a dense forest of young beech trees whose soft green leaves provided a screen. The wind was blowing in a favourable direction, so I began to creep towards the thicket. The smell became stronger and I suddenly saw a movement out of the corner of my eye. The first thing I spotted was a distinctive swish of a tail. Then I saw the horns and a head movement through the bushes and my heart began to beat faster. How much did I really know about European Bison and how reliable was my information? Would the animals attack or pretend to? I immediately looked around for a suitable tree and quietly climbed up it. I was tense and my breathing shallow. Now I could see them: a whole herd of European Bison standing just 20 metres away! The older animals were eating, while the calves played and sometimes ran under the adults' bellies. The large, powerful animals radiated calm and contentment but they were watchful at the same time. The young ones were more agile. They seemed playful and full of joie de vivre. For a fleeting moment, I forgot that I was hanging several metres above ground in a tree, until my weakening arms reminded me.

Observing large mammals in an undisturbed state is very special and it's something that stays with you. It's moments like these we remember our whole life.

You can find animal tracks wherever there are animals. Almost every walker will be familiar with seeing horses' hoofprints along a path and if you walk with a dog, you will probably have seen it leave paw prints on the edge of a puddle. Many people have some idea of what the tracks of a deer or the prints of a hare look like, but few know how much more there is to discover. For example, what do typical Northern Raccoon tracks look like? How do the tracks of a Eurasian Beaver differ from those of a Coypu? What marks does a lizard make in the sand when it runs away? Has a badger or a woodpecker been working on this dead tree? Most people could tell whether a house is inhabited just by looking at it, but how do we know whether the woods outside our front door are mainly populated by Western Roe Deer, Western Red Deer or Common Fallow Deer? Most people think of tracking as following the footprints of mammals, but it also involves signs like droppings and traces of eating, marking and building, which are important parts of a bigger overall picture. If you think about the many thousands of traces of insects, you will begin to grasp how enormous the list of tracks and signs of our fauna really is.

'The book of nature has no beginning, as it has no end.'
Jim Corbett, famous hunter, naturalist and author.

This book looks at the tracks and signs of mammals, birds, amphibians, reptiles, insects and other invertebrates in Europe. The extensive and detailed section on mammals forms the main part of the book. The sections on birds, amphibians and reptiles, and insects and other invertebrates are to be understood as supporting guidance. They aim to give a basic understanding of the tracks and signs of other wild animals that we can use to help us interpret mammal tracks. Insect tracks can help to put events in chronological order. Bird tracks can be confused with those of mammals. The tracks and signs of the other species described give basic information, help to avoid identification errors and assist in interpreting complex facts.

The art of tracking takes us step by step, deep into the mystery of nature and even experienced trackers will be able to find new, fascinating tracks they hadn't noticed before. Was this Western Roe Deer alone or were there several animals? How were they moving? Where were they going and why? The wealth of information that can be read from animal tracks seems inexhaustible but most of us regularly walk right past these stories without noticing them. Tracking means paying attention and asking questions. A key question is: Which tracks do you notice?

'Tracking reveals a cosmos in which every being and every event holds a unique meaning and significance.'
Martin Derrez, tracking course, 2017.

CREATION OF THE BOOK

The first email in my archive mentioning this project is dated 13 February 2012. I dedicated the next four years to photographing animal tracks and collecting data. I travelled through various European countries; most of the images come from Germany, England, the Netherlands, Spain, Poland and Sweden. I took a total of 25,000 photographs, 950 of which made it into this book. I also collected data in Switzerland, Denmark, Scotland, Norway, Austria, the Czech Republic, Slovenia, Slovakia, Belarus, Italy and France.

The dimensions given in the species accounts normally refer to adult animals. The focus was on measuring the footprints, also referred to here as 'tracks'. The minimum and maximum values for the track dimensions were determined through at least 120 individual measurements (samples) of different prints. For data to be included in the data collection, it was 'trimmed', i.e., the ten lowest and the ten highest values were deleted. The samples came from different individuals from at least three different regions of Europe. If a species description is based on fewer data points, this is indicated at the corresponding point in the text. Much more data was collected for common species or species of particular research interest. For example, the information about stride length and trail width of Western Roe Deer, Western Red Deer and Wild Boar are each based on more than 1,000 trimmed individual measurements of different tracks and gaits.

The track drawings form the heart of this book. They have been drawn by hand and are the result of many years of observation and research and around 200 days of drawing. Like Mark Elbroch, I decided to use the intricate Indian ink dot technique. Footprints of dead animals in modelling clay do not depict the dynamics of movement; they show complete footprints, which are rarely seen in the field, or, if they are, tend to be distorted. Good track drawings should depict a perfectly printed foot with all the morphological features and, at the same time, show which areas of the track are normally less easily recognisable, i.e., indicate the dynamics of movement. They should depict the most likely of the large number of possible manifestations. Every track drawing in this book shows a combination of the significant features, the typical weight distribution and the resulting emphasis in the footprint, and how the track will probably look in the field. The drawings created for this book will rarely precisely correspond to a find in the field, but at the same time, you should be able to recognise most of the prints you find in the drawings.

The work for this book became more focused as the result of a casual chat with Mark Elbroch, who tapped me on the shoulder in September 2013 and said that somebody had to research the tracks of Europe's mice and voles. I spent two winters examining the differences between the ink footprints of voles (Arvicolinae) and mice (Muridae). I concluded that, if the tracks are made under perfect conditions, it's possible to differentiate between all genera and, in some cases, even between species. This was previously unpublished knowledge; I hadn't been able to find information about it anywhere. I began actually writing the book when I signed the contract with the publisher in 2016.

The graduates of a one-year Wildniswissen tracking course were trained in collecting and correctly documenting animal tracks in the field. To collate this data and make it accessible to the public, the Wildniswissen wilderness school published the website spurenjagd.de in 2012. It provided a platform for collecting track documentation. In 2013, a team of four experts from the course came together to examine the track documentation before the data could be included in a secure data collection. More than 1,200 track documentations with more than 10,000 voluntarily checked datasets have been collected through the website since 2012. They complemented my own 15-year data collection.

More than 2,000 participants of the Wildniswissen wilderness school, who have been documenting animal tracks since 1995, have collected data and formulated countless astute questions. It was these questions that motivated me to keep researching and observing more closely so I could finally write down the knowledge that I had gathered so far.

HOW TO USE THIS BOOK

This book is made up of four parts that contain basic information about biology and detailed knowledge about the tracks and signs of European wild animals.

I start by giving an insight into the art of tracking, and tips about how to take your first steps in exploring the diverse and fascinating world of animal tracks.

The following section, Mammals, essentially consists of species portraits of adult animals. First, you will learn important principles of the foot morphology of mammals and the structure of tracks and their most important features. This is followed by an explanation of gaits and track patterns, as well as instructions for measuring tracks and track patterns. Then comes a detailed guide to identifying mammal tracks. Lifesize or precisely scaled drawings of the front foot and hind foot prints of individual species enable rapid identification. The subchapter Identifying and Interpreting Signs gives an introduction to the vast and varied world of signs. The identification guide that follows will help you to quickly assign a sign you have found to a specific species or group of animals.

The following species portraits, which make up the main part of the mammal section, describe the size and distinguishing features of the species and species groups covered, their distribution and habitat, as well as diet and reproduction. Pages with a coloured background provide information about the tracks, with descriptions of the front foot and hind foot prints and preferred gaits, as well as an overview of the most common track patterns and important measurements. It also covers typical signs of a species.

Knowledge about European fauna varies greatly. As a result, information about the biology and tracks of some species is fragmented. Species like those from the shrew family (Soricidae), where current knowledge of tracks does not allow precise identification of the species, are dealt with together as a genus. Other species, such as the Polar Bear (*Ursus maritimus*) or Axis Deer (*Axis axis*), that occur rarely or only in small, isolated populations, are only mentioned in passing.

Parts 2, 3 and 4 summarise basic information about the tracks and signs of birds (part 2), amphibians and reptiles (part 3) and insects and other invertebrates (part 4). I had to choose a slightly different approach in these sections because of the large number of different species covered and the limited knowledge about their tracks. While the tracks are, on the whole, still dealt with in – albeit shorter – species portraits, the signs are organised according to appearance, i.e., in the case of birds, by nests, feeding traces or excrement, that are then assigned distinctive features in the case of certain species or species groups. The book refers to specialist literature that provides further detail.

The animal silhouettes on the inside of the front cover and the adjacent page numbers refer to the description of the respective animal group in the text. An overview of measuring track patterns, as well as a centimetre scale, can be found on the inside of the back cover.

In the mammal section, a colour coding system is used in which each colour stands for one of the families covered. Colour coding also identifies the respective major group, i.e., birds, amphibians and reptiles, or insects and other invertebrates, in subsequent sections.

Structure of species descriptions in the mammal section

English and scientific name of the species or species group.

Head-trunk length, tail length and weight: The measurements given in the margin column cover the entire range of variation in the area covered. Most of these measurements come from Dr Eckhard Grimmberger (2009); additional sources are used for some species. Size differences between the sexes (sexual dimorphism) are also mentioned where present.

Distinguishing features: Typical build and characteristic features.

Distribution and habitat: In the case of species with very similar tracks and signs, it can occasionally be possible to distinguish between them based on their different distribution or the different habitats they prefer. This information should be understood as a complementary aid for species identification and for guidance. The same species can live in different geographical areas and different habitats.

Diet: Basic knowledge about the dietary habits of a species helps trackers to recognise ecological connections. The occurrence of a certain species allows conclusions to be drawn about the biotope or the presence of other species. The focus is on interpreting feeding traces.

Reproduction: Information about reproduction for interpreting track findings in the field. Are they the tracks of a mother animal with her young or the tracks of a bachelor group?

Notes: Various miscellaneous comments can be found in the margin column.

Tracks: Detailed track descriptions of front foot and hind foot prints, together with track drawings, corresponding photographs and measurements.

Similar tracks: Footprints of different species that can be confused with the species described, as well as features for differentiation.

Gaits: Descriptions of preferred gaits, track patterns and measurements.

Signs: Typical and distinctive signs of the species described, for example excrement, markings and feeding traces.

Track formula for mammals

The track formula gives trackers important information. It allows taxonomic order and often even family membership to be determined at a glance.

XF × Yh + C

XF   Number of front foot toes normally imprinted (front foot track)
Yh   Number of hind foot toes normally imprinted (hind foot track)
C   Claw prints are distinguishable in both tracks

The capitalisation of the letters 'F' and 'h' indicates the relative size of the front foot and hind foot track.

For example: **4F × 4h + C**

This track formula describes a front foot track with four toe prints, and a smaller hind foot track, also with four toe prints. Claw prints are distinguishable in both tracks. The number of toe prints and their relative size in this example are a match for both cats and dogs. Since the claw prints of cats are rarely visible, this is a dog track.

Abbreviations and symbols used

General

D   Diameter
W   Weight
HTL   Head-trunk length
Max.   Maximum
Min.   Minimum
Tl   Tail length
sp.   species (singular)
spp.   species (plural); example '*Tringa* spp.' means 'several species of the genus *Tringa*'
♂   male
♀   female

Tracks and track patterns

W   Width
GL   Group length
Hf   Hind foot
L   Length
LH   Left hind
LF   Left front
RH   Right hind
RF   Right front
TW   Trail width
SL   Stride length
Ff   Front foot
IGL   Inter-group length

LEARNING TRACKING

Anybody can learn to read tracks. One of my teachers emphasised that, strictly speaking, there isn't actually much to learn because we already have basic tracking skills – we are born with them. We perceive things with our senses and we are able to interpret them and draw conclusions. We can also go out in nature and discover tracks at any time. We can experience firsthand how the land is shaped and moulded by its inhabitants by asking questions, observing and listening closely. Our curiosity, joy and enthusiasm drive us to get to the bottom of our questions and to expand our knowledge. This type of learning is innate and an excellent way of deepening our connection to nature.

Trackers have a saying: 'If you read tracks by yourself then you will always be right.' Talking to other trackers, gathering different opinions about a question and scrutinising your assumptions can help you broaden your horizons and consider different possibilities. We often learn more effectively when we surround ourselves with people who share our interests. Groups can promote a culture of learning where each individual's knowledge grows rapidly. That's why I would recommend that anyone who wants to learn tracking makes contact with people who share their enthusiasm.

Charles Worsham, former FBI and HRT trainer, US

We will never know everything there is to know about nature and our lifespan isn't long enough to learn everything about animal tracks. This can be discouraging or reassuring. I'd like to encourage you to adopt the positive attitude of a childlike, enquiring mind. Dare to form hypotheses and laugh at yourself when new information proves you wrong. Keep an open mind and consider every possibility instead of jumping to conclusions.

Tracking requires close observation, deductive and logical reasoning, accurate

measurements and precise documentation. The tracker's powers of imagination are another important aspect. Mark Elbroch writes: 'A skilled tracker is a scientist and a storyteller. They must observe critically, collect good data and avoid jumping to conclusions. They must also use their imagination to interpret the signs they find.'

Extensive research into an animal's biology and behaviour helps us to interpret tracks. The better we know a species, the easier it is to understand its behaviour. Suitable specialist literature is valuable for learning this basic knowledge. At the same time, you can't learn tracking just by reading books; it takes time and practical experience in nature. You need a great deal of experience to be able to confidently interpret a track or print, and especially to be able to incorporate the many other factors, such as surface, speed and weather, into your interpretation.

'To interpret tracks and signs, trackers must project themselves into the position of the animal in order to create a hypothetical explanation of what the animal was doing. Tracking is not strictly empirical, since it also involves the tracker's imagination.'

Louis Liebenberg, Founder and Executive Director of CyberTracker Conservation and author of several books on tracking

THE SIX QUESTIONS OF TRACKING

In his books, American tracker Jon Young describes the six questions of tracking. The answer to each one of these questions can be understood as its own branch, art or discipline and all the questions are an excellent way of learning how to track.

- **Which animal has left the track?** Identification of tracks and signs, species identification.
- **What has happened here?** Interpretation of tracks and signs, the ability to use track patterns to reconstruct gaits and other animal behaviour.
- **Where is the animal now?** Trailing, i.e., tracking down animals in real time.
- **When was this track made?** Determining age and chronology of events.
- **Why was the animal here?** Ecological context, the ability to read and predict animal behaviour.
- **What sort of state was the animal in?** Imagination, the art of putting yourself in the animal's place.

TRADITIONAL TRACKING

The San Bushmen of southern Africa are among the best trackers in the world. Their knowledge of the flora and fauna in their region is exceptional. The fact that the Bushmen live so close to nature is a crucial factor in their outstanding tracking skills. They share their habitat with wild animals, so they experience many of the same things – for example, they are both directly affected by drought, rain or cold. This sensory experience of the natural environment enables the Bushmen to anticipate animal behaviour.

Along with the many advantages of technological progress, modern life is also accompanied by a kind of estrangement from our natural environment. Most of us live in houses in towns and cities and our survival rarely directly depends on our tracking skills. There are a few things we can learn from the approach of traditional trackers from

hunter-gatherer societies. San huntsman !Nqate Xqamxebe (1998) said: 'When you follow the tracks of an animal, you need to become that animal.'

Many indigenous trackers have an impressive ability to imitate how an animal moves, even if they aren't familiar with theories on gait research – the practice of the repeated imitation of movement sequences resulting in internalised muscle memory. Try it out for yourself by attentively imitating the movements of, for example, your dog. The participants of our tracking courses regularly imitate animals and recreate the different gaits with their own body. As strange as it might seem at first, the tracking success of people who do this speaks for itself.

'With growing awareness of what's really happening in nature, we begin to grasp and comprehend countless things we hadn't even noticed before.'

Hugh Falkus, British wildlife filmmaker and author

To learn tracking in depth, you must use all your senses. Feel the moisture in the air in the morning and the wind against your skin. Smell the scents and listen to the natural sounds around you. The more you understand a situation with your senses, the easier you will find it to draw appropriate conclusions and to develop an immediate relationship to the landscape and its inhabitants. Over time, your perception of your environment will change, so forests, meadows, wind and weather, and animals themselves, will become familiar to you. You will also get to know yourself much better because 'we can't increase our perception of our environment without simultaneously becoming aware of our own role within it.' (Mark Elbroch, 2003)

'Tracks may take us back to the Earth's beginning or give us a complete autobiography of a living animal from day to day.'

Ellsworth Jaeger, naturalist and author of *Wildwood Wisdom*

FIVE ROUTINES FOR TRACKERS

The following five routines have proven to be useful tools for trackers and can be used to help you develop your skills.

Get to know your own territory

Find a suitable and convenient area and make it your tracking territory. Visit it regularly and take an interest in its habitats, plants and animals. Find an observation place where you feel comfortable. Spend time there by yourself without interruption and listen to the world around you. Observe the weather, the temperature, the wind direction and the animals at your usual spot. Pay attention to behaviour of birds and insects. Record your observations. Regularly spend between 20 to 60 minutes in this place so you get to know it at different times of day and in different weathers and seasons. Over time, you will build up important basic knowledge that will help you with tracking.

Draw up a species list

Once you have got to know the different habitats in your territory, draw up a species list. Browse nature guides, study distribution maps and preferred habitats of animals, and then note which animals you would expect to find in your area. By spending time at your observation site and by exploring your territory, you will increasingly often find and identify animal tracks and signs. On your species list, make a note of which animals you were able to detect based on which information. Over time, the list of animals you might find will become a personally verified species list. You'll be surprised how close to human habitation some wild animals live without people knowing about them. Over time, you may find evidence of animals that you wouldn't have expected to find in your area, and you might even be able to prove the occurrence or return of a threatened species.

Keep animal journals

A good way of increasing your knowledge of animals is to create animal journals. Take a close look at a species that really interests you and research its way of life, diet and reproductive behaviour. Use different sources and record the key information. Try to put into your own words any knowledge that seems particularly interesting to you. You will end up with your own personal animal journals; you get to choose their scope, contents and creative design yourself. You can use the structure of the species descriptions in this book for guidance, or find inspiration from other templates.

Draw the animal and its skull. This will give you a basic understanding of the physical features of different species, for example the elongated foot bones of typical prey species or the enlarged eye sockets of species who primarily rely on their sight. Dedicate another part of your animal journal to drawings of the front foot and hind foot prints, in actual size if possible. Leave enough space to add more measurements later. Over time, you will create your own local reference work.

Mapping

Get hold of a topographical map of your territory; walking maps in 1:25 000 scale are ideal. Study your territory on the map and look at where areas of woodland, buildings or settlements, fields and meadows are, and where the footpaths and roads run. Pay attention to stretches of water, differences in elevation and exposure according to the points of the compass, as well as border areas. Look for distinctive contrasts in the landscape within a section of the map, for example hedge lines between fields, a gravel pit in the forest or a small copse surrounded by fields and farmland.

Then draw your own map of the area and mark your own animal sightings and track findings on it (see template on page 801). Draw in game trails that you find on walks and label the resting places, markings, dwellings, feeding traces or droppings. Pay attention to the direction and course of animal paths and look on the topological map to see which interesting points they might connect. Ask yourself: 'Why was a mark specifically made there? Why is this deer bed here and not over there?' This will draw your attention to the ecological connections within your territory and you will learn how they influence the animals' behaviour. If you keep looking, you will sooner or later notice changes or regularities that are influenced by the weather, the season or even by human intervention in the landscape.

Documenting tracks

Drawing up track documentation (see template on page 800) is an excellent way of recording your experiences and sharing them with other trackers. Draw or photograph a front foot and hind foot print when you are recording a track in the field. Measure the footprints, as well as the track pattern, and take note of the measurements. Stick to the guidelines for measuring track pattern (see pages 73–75, 92–95), so you are consistent when gathering data and you can later share and compare this data with other trackers.

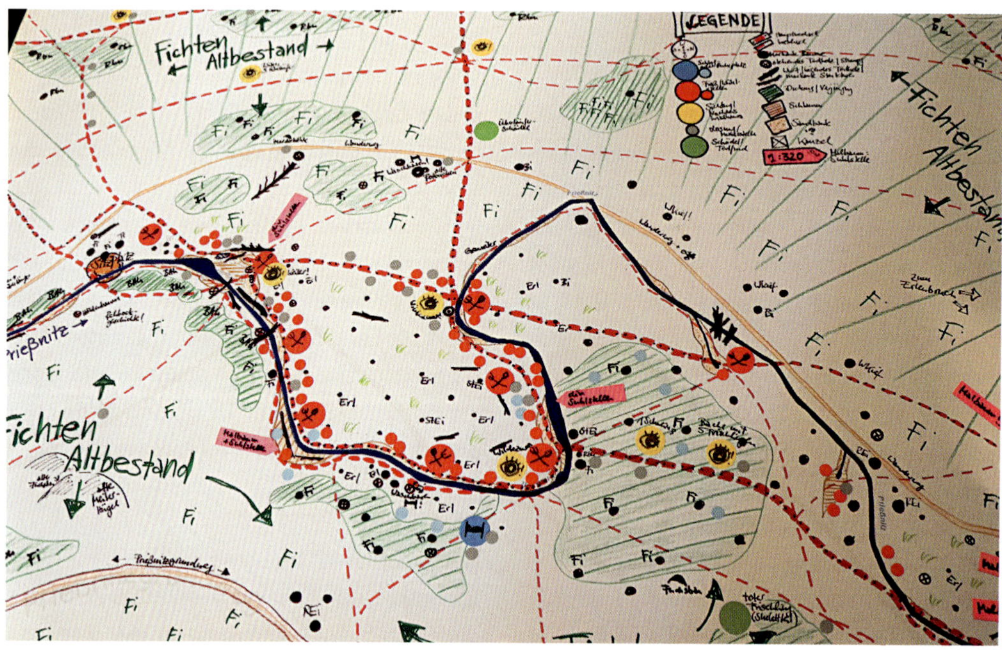

Photographing tracks

Photographing animal tracks is an excellent way of documenting findings. Tracking is highly aesthetic and, occasionally, you can capture this aesthetic in an impressive wildlife photograph. The quality of images from modern smartphone cameras is more than sufficient for everyday purposes. If you want to publish images regularly, I recommend a camera with manual setting options for shutter speed and aperture, as well as a lens with a focal length of 17–70mm.

When photographing tracks, make sure that the footprint takes up approximately two thirds of the entire image, is positioned centrally and that the photograph is taken from directly above. Use an aperture setting between F8 and F16. The shutter speed should be 1/60s or shorter because longer shutter times tend to lead to blurry images. A tripod is very useful for photographing tracks, especially for small subjects, because it enables you to take clear photographs with slower shutter speeds. The area you are photographing should be completely in the shade, because half shade can distort the appearance of the track, and areas with light and shadow are usually difficult to photograph.

▲ Beech Marten in 4 × 4 bound (transverse).

'Long before Scotland Yard or the F.B.I., Nature fingerprinted her numerous family wherever they went. She took their prints in sand, dust, mud and snow.'
Ellsworth Jaeger

EFFECTS OF THE GROUND ON TRACKS AND ANIMAL BEHAVIOUR

Trackers often ask about ground conditions. This is important because it influences the visibility, quality and appearance of tracks, as well as the size, shape, depth and ageing behaviour of a footprint. Many people will have seen animal tracks in the snow. Winter is an excellent time for tracking because the first fall of snow means we can suddenly discover animal tracks we wouldn't have noticed before. Animals obviously leave tracks all year round, but prints are much easier to spot on snow. However, not all snows are equal, so tracking in the snow isn't necessarily easy. The footprints left by a Red Fox when it crosses a surface covered with a fine layer of fresh snow are usually clear and detailed, while in deep snow, the same Red Fox will only leave behind holes with no detail.

Even if the material of a surface is identical, the conditions can differ greatly. The dampness or dryness of the ground, its depth and density all have an impact on the track. Clay soil keeps its shape so an accurately detailed print can be preserved in it for months; a footprint on dry sand, however, will quickly lose its details. An animal will leave shallower prints on a compressed surface than on a loose surface, and tracks will often appear larger on a soft surface. There are countless types of sand, dust, mud and different soil types, as well as types of snow, fallen leaves, and grass. This variation expands the range of possible appearances of tracks enormously. Always take ground conditions into consideration when interpreting tracks, as they also have a direct influence on animal behaviour.

▲ *Detailed Red Fox tracks.*
Laura Gärtner, Sweden.

▲ *Snow-covered Red Fox tracks.*
Laura Gärtner, Sweden.

Surface and substrate depth influence gait

A Red Fox will often trot across a frozen open space covered with 1cm of fresh snow. But if this same area was covered in 80cm of snow following a heavy fall, this would change its gait. The stride length would normally become shorter, while the trail width would increase. Put yourself in the position of this fox and imagine you're walking in the forest on 1cm of fresh snow. Then imagine doing the same walk in deep snow. It's very likely that when walking in the deep snow, your stride would become shorter while the width of your tracks would increase. This happens when you sink deep into a surface and is related to the amount of energy you need to move forward. It takes the same amount of energy to cover a short distance in deep snow than it does to cover a longer distance on a firm, flat surface, and so you move more slowly.

To conserve energy, animals often adapt their gait to the ground conditions. Red Foxes usually trot but if they sink deeper into the surface, they often adopt a slower walk. However, animals don't necessarily shift down into a slower gait when navigating a softer surface. They often cross short distances of soft ground, such as a marshy ditch, quickly, usually with a kind of bound. The preferred gait doesn't just depend on the species, but also on the ground conditions.

DETERMINING THE AGE OF TRACKS

Determining the age of a track is perhaps the area of tracking that takes the most time and practice to be able to judge reliably. The number of overlapping variables is virtually endless. Deer droppings that have been well preserved over winter can appear fresh again in light spring rain. Weather is a significant factor in ageing, so knowing when the last rain fell or when it stopped raining is essential for determining the age of tracks. The same applies to the last snowfall or frost. Two other variables that influence ageing behaviour are the degree of moisture in the ground, and the soil material. Sun and wind will dry out the surface of the ground more quickly than the deeper soil beneath. You can usually tell this by the difference in colour between the pale, dry surface and the dark, moist substrate. When an animal leaves a footprint, the movement transports deeper, moist substrate to the surface and the contrast to the surrounding ground is easy to see. This difference will disappear again over time, depending on the weather conditions. Make a mark of a similar depth with the toe of your shoe next to the track you're assessing. In doing so, you can compare it with a fresh track. Ask the following questions:

- Are there raindrops, snowflakes or droplets of dew in the track?
- What is the colour of the track like in comparison to the surface around it?
- Are the outlines of the track sharp or blurred?
- Have sand, leaves or other loose material accumulated in the track?
- Have insects left traces in the track?

It is often only the chronological order of events that tells me about the possible drama of a situation. For example, a few years ago in Wisconsin, we found gallop tracks of two White-tailed Deer in the snow. They had both come from the cedar swamp in the north and quickly crossed a logging road to the south. On the other side of the road, they had galloped back into the swamp at high speed.

It was a warm, sunny February day with a brilliant blue sky. The individual tracks were frozen and clearly visible. The day before had been warm and the tracks on the road were exposed to direct sunlight. In February, the sun in Wisconsin is already strong enough that the tracks would have become less clear after more than two days. Consequently, it seemed probable that the tracks had been made early the previous evening, i.e., when the snow was soft from the warmth of the day but not yet frozen by the cold of the night. However, footprints can last for an astonishingly long time in these weather conditions. Approximately 50m further west and at the level of the road, another small group of our trackers found Grey Wolf tracks. The wolves had also been moving faster than normal.

I have often caught myself creating a story in my mind without paying attention to information about the age of the track. I have imagined gripping pursuits between predators and prey, even though the animals had only chosen a similar route and the events took place at completely different times. But this time it was different. The wolf tracks were also clearly visible, frozen and similar in appearance to the deer tracks. The wolves must have been in this place at a similar time. We eagerly began to follow the tracks...

'Rain! Whose soft architectural hands have power to cut stones, and chisel to shapes of grandeur the very mountains.'
Henry Ward Beecher, American theologist.

Setting up a track box

Spread a 3cm deep layer of playground sand on an area of ground measuring approximately 50 x 50cm. This area should be easily accessible for you and exposed to the weather. Leave an approximately 1cm deep handprint and observe how it changes daily. Take note of the weather every day and place this note, with the date, next to the handprint. Now record the print and note by taking a photograph. Take note of how the handprint changes every day. After a week, put your photographs together in chronological order and, using your notes, observe how the handprint has changed according to the weather conditions. Repeat this exercise in different weather conditions and at different times of year so you get to know the effects of the weather and seasonal factors in your region. Your ability to determine the age of tracks will become more precise over time.

▲ *Observe and take notes of how your handprint changes every day.*

TRACKER ETIQUETTE

When we follow an animal, we see more than just the composition of its footprints and tracks. We discover where it spends its time, what it has eaten and where it has slept. We might observe how a young roebuck leaves more marks during mating season. We can find out whether he finds a female and, if we follow the tracks of the female for long enough, we will see at some point that she has had young. The following year, we will keep coming across the larger tracks of the mother and the smaller tracks of the fawn following in her wake and we will also see how the roebuck spends more time alone again. We begin to understand more and more about the network that these deer are part of. We learn which plants they prefer and where they seek refuge. Our relationship to the deer and their habitat becomes firmly established. We learn about the success and failure of the animals and, if we observe them for long enough, we will eventually also experience their death. Tracking is about real relationships with the real world.

It was a crystal-clear winter morning and I decided to use the first daylight for tracking. It had snowed lightly overnight and the older, frozen snow was covered in a fine layer of powder, so the conditions were ideal. I had crossed an open field and was approaching a forest of Douglas Fir near where I lived when a tiny track suddenly caught my eye. The size of the trapezoid track pattern indicated it was made by a small rodent. On closer inspection, I saw the clear footprints. There was a print of each pad and the outlines were sharp enough to allow me to identify the species: it was the track of a Wood Mouse.

Dawn soon gave way to a brighter, blue light that gave the footprints a special shine. It had only stopped snowing around half an hour ago and there was no snow on the small track. This track was less than 30 minutes old. Fascinated, I followed the track, imagining I was the tiny mouse leaping over this open space. I was overcome by a sense of danger when I realised how risky this crossing was for the mouse.

We had almost made it under the shelter of the Douglas Firs when I noticed a larger disturbance in the tracks with a hole in the snow. Two clear wing prints were clearly visible to the left and right of the hole. I peered anxiously into the hollow in the snow and at the base I saw the sharp track of an owl. The journey of my Wood Mouse had come to an end.

Be respectful and considerate when tracking, both towards other people and towards nature. When you are following a track, be careful not to disturb the animals. You should follow tracks backwards rather than forwards so you can learn about the animal's life and behaviour without disturbing it. Be especially careful during breeding season, at inhabited dwellings or at other places that are important to animals, such as feeding and resting places. Make sure that you leave behind few traces yourself, be committed to the wellbeing of our fauna and contribute to protecting and preserving our environment. Many of the trackers we have trained enjoy good relationships with landowners and conservation authorities in their region. They often get special permission to enter an area and gather valuable data that helps with projects such as otter monitoring.

Reading animal tracks can provide information about protected species. Use this knowledge responsibly. Do not make the location of these animals public and make an active contribution to their privacy. Remember: 'The first track is the end of a string. At the far end, a being is moving; a mystery, dropping a hint about itself every so many feet, telling you more about itself until you can almost see it, even before you come to it.' (Tom Brown Jr., author of numerous books on tracking.)

Committed to conservation

Observe the applicable conservation regulations, as well as breeding times, and find out about the protected species of a region. Respect this protected status and adhere to any restrictions on where you can go. If necessary, contact the relevant authorities to discuss your plans.

APPLIED TRACKING

Tracking is used in different ways across the world. The archaeology project Tracking in Caves drew the attention of international media in 2013 and was the subject of a documentary by broadcaster ARTE TV. Three indigenous San trackers from Namibia, along with a group of scientists, visited the cave of Tuc d'Audoubert in the French Pyrenees to research tracks. The project focused on reading and understanding human footprints in the caves. It combined the modern scientific methods of the archaeologists with the traditional skills of trackers from one of the last hunter-gatherer societies in the world. The successful ongoing project is an example of the many different applications that the art of tracking has to offer.

Tracking also has a firm place in modern wild animal monitoring. For example, wildlife biology office LUPUS uses it to provide information about wolves in Germany. LUPUS contributed to the development of monitoring standards for large carnivores, as a result of which it has been possible to use tracks as evidence of the presence of a species since 2009.

▲ *Specialist in tracks and signs Casey McFarland explains the features of different digging tracks. Lausitz, Germany.*

International standard for trackers

Louis Liebenberg, Co-Founder and Executive Director of CyberTracker Conservation from South Africa, initiated a certification system for trackers in 1994. The aim was to revive the traditional reading of animal tracks, which was at risk of dying out, and to contribute to developing the art of tracking into a modern profession. The evaluation system was recognised as an official standard by the FGASA (Field Guides Association of Southern Africa). In 2005, Mark Elbroch transferred this system to the US where it is also now established. These evaluations have also been used in Europe since 2011. As a result, an international standard for trackers has been created. This evaluation system allows the reliability of trackers in the field to be assessed and certified. This has even led to higher wages for professional trackers in Namibia, Botswana and South Africa. Working towards certification constitutes first-class training for beginners and professionals. More than 5,000 tracker certificates had been issued in Africa, North America and Europe by 2018.

▲ *Trailling specialist Antje Beneken with the CyberTracker software. Dresden, Germany. Arvid Müller.*

As well as providing information about threatened species, tracking is also used in Europe as a way of estimating the population size of wild land mammals. A project to determine deer populations in northern Italy employed three different methods: counting, use of trail cameras, and tracking. While all three methods were comparably effective and yielded similar results, tracking was by far the most economical method in forested areas and approximately 28 per cent cheaper than using a trail camera (Romani, Giannone 2018).

Biomapping with CyberTracker software

The CyberTracker app from CyberTracker Conservation can be used on smartphones and other GPS-enabled mobile devices to record the required data on site and assign it a GPS coordinate. It gives trackers an effective way of gathering biodiversity data and making it available for further investigations. Based on this software, CyberTracker Germany has developed applications for monitoring large predators and for bird mapping in Germany.

Trackers are also used by countries and federal states. In the Doñana National Park in Andalusia, trackers are tackling poaching, and when Poland joined the EU and became the external border of the community, the Polish government engaged members of the Navajo to train its own border officials in the art of tracking. In Germany, trackers are often commissioned to draw up reports on signs of wolves, to be used within the context of legal matters.

'The potential applications of tracking are limited only by our creativity. Deer pellet counts have been used for decades in deer density counts, but this is only the beginning of what could be done with further research and training.'
Mark Elbroch

MAMMALS

MAMMALS

OCCURRENCE, BIOLOGY AND ECOLOGY

We currently know of more than 6,000 mammal species (Mammalia) in the world. In comparison to the animal kingdom as a whole, they make up less than half a percent of all species. Nevertheless, people often form the most intimate relationships with mammals. This could be because we are mammals ourselves and and share many characteristics with them.

Mammals are a class of vertebrates (Vertebrata) that evolved from reptiles more than 200 million years ago. The first mammals were small, shrew-like creatures in both appearance and behaviour. Fauna at that time was largely dominated by dinosaurs and the rapid development of mammals only began around 65 million years ago when the non-avian dinosaurs died out.

Europe's mammals differ greatly in terms of size and appearance. With a body weight of only around 2g, the Etruscan Shrew (*Suncus etruscus*) is, along with Kitti's Hog-nosed Bat (*Craseonycteris thonglongyai*) of East Asia, the smallest mammal in the world. The heaviest and largest land mammal in Europe is the European Bison; wild bulls can reach weights of more than 800kg.

The term *mammalia* comes from the Latin *mamma*: 'breast', 'udder' or 'teat', and describes the defining feature of mammals: mammary glands that they use to feed their young. Another distinguishing feature of mammals is that they give birth to live young; the only known exception are the monotremes, found in Australia and New Guinea.

Many mammals have a relatively large brain. Other features are the three ossicles (malleus, incus and stapes bones) in the middle ear, a lower jaw consisting of a single bone element and a specialised set of teeth adapted to a species-specific diet. Another feature of most mammals is a heat-insulating fur coat and, as a result, a more even body temperature. Many mammals can therefore remain active in a wide range of climactic conditions, although they also have a relatively high metabolic rate and high energy (food) requirements.

DISTRIBUTION AND HABITAT

The mammalian brain's capacity for learning, combined with their ability to maintain a constant body temperature, has enabled mammals to live in every climatic zone on Earth and adapt to nearly any habitat. The Red Fox is an example of a generalist species that can live in a variety of habitats. This species is found in the Sahara-like climate of southern Spain as well as north of the Arctic Circle and in densely populated cities. Generalists can adapt their behaviour and social structure to their habitat. By contrast, other mammals have specialised in specific ecological niches. Moles are an example of this. They live almost exclusively underground (subterranean) and have developed corresponding adaptations in their physiology and behaviour. Other mammals have specialised in spending most of their active life under water (otters, seals and whales), in trees (squirrels) or in the air (bats).

The area where an individual mammal lives is called its **home range**. The home range encompasses all the places used by the animal throughout the year. All regular activities, such as searching for shelter and food, mating, and raising young, take place here. The size of a home range varies greatly and depends on the species, sex, the individual animal and various environmental factors. Some voles (Arvicolinae) need less than $5m^2$, while a Brown Bear can have a home range measuring several hundred square kilometres. Some areas within a home range are often only used at certain times of year. Some even-toed ungulates, such as Western Red Deer, generally spend the summer at a higher altitude, moving to lower-lying areas for winter. The home range of an individual will usually overlap to some extent with the home range of several other members of the same species, but the home ranges of animals of the same sex rarely overlap. Among carnivores, it is much more common for the home range of a male to overlap with that of several females. This increases the chances of successful reproduction and, at the same time, reduces intra-species competition for resources such as food or suitable building options. The area that an animal will defend against other members of the same species is called its **territory** and is normally much smaller than the home range. An animal will demarcate and defend its territory with territorial behaviour such as scent marking, visual signs or even fighting. The territorial behaviour of animals of the same sex often increases before mating season. During this period, animals show other typical behaviours to attract the opposite sex.

REPRODUCTION AND SEXUAL DIMORPHISM

In most species, either males or both sexes mate with several partners (mating systems known respectively as polygyny and promiscuity). Pairs normally only come together during mating season. Only a few species form pair bonds that last for several years. In most species, reproduction takes place at a time that enables the female to rear the young when plenty of food is available. Pregnancy and rearing young are energy- and time-consuming and, in most cases, the male does not assist with care of offspring. As a result, it is often the case that a small number of males will mate with many females and many males have no chance to reproduce. Fights to earn the privilege to mate are therefore common and many mammals have developed physiological features, such as antlers in deer, or other strategies to assert themselves against rivals.

These and other sex-specific differences within a species are called sexual dimorphism.

Another example is the high-contrast and often brightly coloured plumage of many male birds. Sexual dimorphism is normally much less pronounced in mammals than in birds. Examples in mammals are found in the mustelid family, where the males are approximately one third larger and stronger than the females, and the deer family (cervids), where only the males have antlers and the females tend to have a smaller build – an exception is Reindeer, in which the females also have antlers but they are significantly smaller and finer than the males' antlers. Sexual dimorphism can be recognised in some tracks and signs. Details about this are covered in the corresponding group or species descriptions.

In Europe, most mammals give birth to their young during spring or early summer. For mating, gestation and birth all to occur at the most optimal times, in some species fertilised egg cells do not immediately develop into embryos. This phenomenon of extended gestation is called embryonic diapause, or delayed implantation, and occurs in Western Roe Deer, mustelids and bears. Reproductive behaviour usually begins in late winter or early spring and normally ends in autumn. Smaller species, such as small rodents and lagomorphs, will have several litters per year if circumstances are favourable. The number of offspring varies greatly depending on the species. European Rabbits can have more than 10 young per year, while European Bison usually only give birth to one calf per year. Herbivores normally have more offspring than carnivores. Species that produce fewer young usually have longer lifespans.

Offspring are either **altricial** or **precocial**. Altricial animals are usually born naked, sightless and with limited ability to move. The offspring of small mammals with a high reproduction rate and several litters per year (for example, mice) tend to be altricial. Precocial animals are more fully developed at the time of birth. They are usually born with fur and can walk independently shortly after birth. Large mammals with a slower reproduction rate often have precocial offspring because giving birth to them takes a lot of energy, but ungulates (hoofed animals) in Europe are precocial.

DIET

An animal's diet has a major influence on its development, distribution and interaction with other animals. Mammals can be roughly divided by diet into herbivores, carnivores, omnivores and insectivores. **Herbivores** are animals that mainly eat plants. Many species of animal have specialised in eating certain parts of plants; for example, granivores prefer to eat seeds. Herbivores have teeth that are designed to cut and grind plants; wide, flat molars are typical.

Carnivores mainly or exclusively eat vertebrate animal tissue. The term 'carnivore'

can be misleading because some members of the taxonomic order Carnivora, such as the Brown Bear, are actually omnivores, while species like hedgehogs and dolphins are not in Carnivora but do take vertebrate prey. Carnivores that specialise in eating dead animals that they have not killed themselves may be known as scavengers or carrion-eaters. The teeth of a carnivore are used to kill prey and to cut and tear meat. Chisel-shaped incisors, dagger-like canine teeth and relatively sharp molars are characteristic of carnivores.

Omnivores are generalists that can eat a wide range of food. They are also able to adapt their diet to what's available depending on the season and the habitat. They normally have a varied diet that can consist of plants, fungi and animals. Members of this species include Northern Raccoon, Grey Squirrel, Brown Bear and Red Fox. They are examples of the high level of adaptability, intelligence and opportunistic nature that are common to many omnivores. Omnivores' teeth are a combination of the types described above. The incisors are similar to those of carnivores but the molars are more like those of a herbivore because they grind plant-based foods.

Insectivores like bats, moles and shrews mainly eat insects and their larvae. All the teeth of animals in this category are very sharp and pointed.

Nutritional requirements vary greatly from species to species. Wolves can go hungry for weeks on end without suffering any serious damage, while some shrews will die if they don't eat every two hours or so.

Some mammals have developed different strategies for surviving periods when there is a shortage of food, for example by storing body fat as an energy reserve. This allows them to eat large quantities when there is a surplus of food, to help them survive periods when food is scarcer. In many areas, these animals have periods of winter inactivity (bears) or hibernation (bats). Their activity levels greatly reduce, their respiratory rate, heart rate and body temperature drop, and their metabolism sometimes slows down by more than 75 per cent. This enables them to minimise their energy consumption during months when food is scarce.

SENSES

The sensory perception of different mammal species is often very different from that of human. Humans and some other primates are among the few mammals that can see the colour red. No other mammals can perceive this colour. By contrast, many nocturnal species have black and white mesopic vision (twilight vision) that is far superior to ours, and rats and hamsters can see in the ultraviolet spectrum. Bears and dogs have a much stronger sense of smell than we do and Northern Raccoons have such a powerful sense of touch that we say that they 'see with their hands'.

Other mammals have a highly developed sense of hearing that enables them to perceive frequencies beyond the human range. Bats navigate by using ultrasonic sounds and their echoes to create a kind of 'aural map' of their surroundings. There is even evidence that blind mole-rats are guided by the Earth's magnetic field. Research consistently leads to findings about the sensory abilities of other mammals that we can scarcely begin to imagine. Research into the senses of some mammals has been rather patchy, so facets of their perception remain a mystery to us.

As humans, vision is our dominant sense and our behaviour greatly depends on our response to optical stimuli. Tracking therefore involves us using our imagination to attempt to comprehend the very different sensory worlds of other species.

The behaviour of a species that perceives its environment primarily with its sense of smell differs greatly from our human behaviour and these differences can be seen in track patterns. Consequently, knowledge about the sensory perception of different species can give important indications for interpreting track patterns. The description of the dominant sense is given in each species portrait.

A few examples are enough to show just how impressively diverse the world of mammals is. Their fascinating lives are as varied as they are mysterious. Other mammals often live among us without us noticing them, and the tracks and signs they leave behind are often the only indication of their presence.

Three questions

It was a warm day in September and I was staring intently at an area of ground measuring approximately 10 x 10cm. Mark Elbroch had marked this area with a stick, drawn my attention to an approximately 6cm-long and 4.5cm-wide footprint in it and asked me three questions:

'Which animal was it?'
'Which foot made the print?'
'What sex was the animal?'

My heart began to beat faster when he asked the last question. I knew it was the left front foot print of a Beech Marten but how could Mark tell the sex? What could he see that I couldn't? These were the questions that went through my mind as I continued to stare at the track. 20 minutes later, I admitted defeat. I didn't know how to interpret the information in front of me and another 20 minutes of staring wasn't going to change that. I told Mark what I had guessed and waited eagerly for the answer. His explanation would forever change how I looked at individual footprints...

▼ *A fascinating footprint that I will always remember for its impressive wealth of information. Lausitz, Germany.*

FOOT MORPHOLOGY AND TRACKS

Foot morphology is the study of the shape and structure of feet. It is also important for the structure and features of tracks.

STRUCTURE OF THE FOOT

The foot of a land mammal consists of claws, toes and pads, among other features. The oldest land mammals had five toes with five claws and several pads on all four feet. The feet of many insectivores, such as shrews, are the closest equivalent to this original foot structure. The feet of all terrestrial mammals still share a similar bone structure today: each foot consists of toes made up of several phalanges, a midfoot with metacarpal bones in the front feet or metatarsal bones in the hind feet, and a back portion with carpal bones in the front feet, called the carpal region, or tarsal bones in the hind feet, known as the heel.

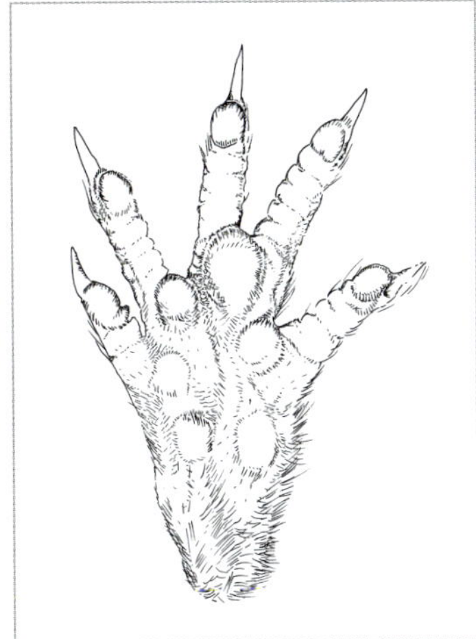

◄ *Shrew, left front.*

Proximal and distal

The anatomical terms 'proximal' and 'distal' are used to describe the position of a body part. Along an anatomical structure, such as a foot, these terms refer to the relative distance of the substructure, e.g., a bone in the foot, from the centre of the body. Both terms are independent of the position of the body. For example, the midfoot is closer to the centre of the body than the toes, but further away from the body than the heel, no matter how the body is standing or lying.

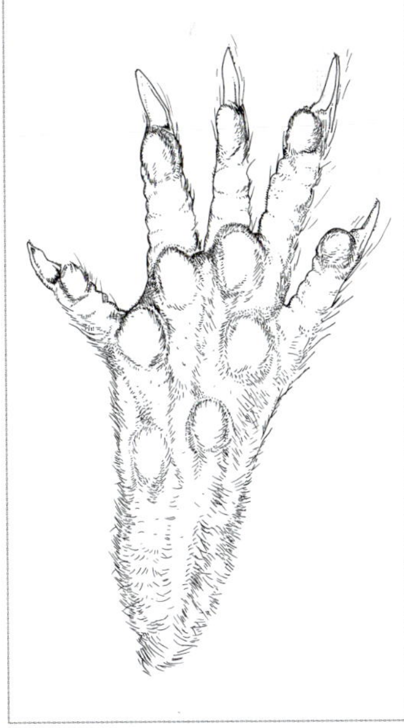

◄ *Shrew, left hind.*

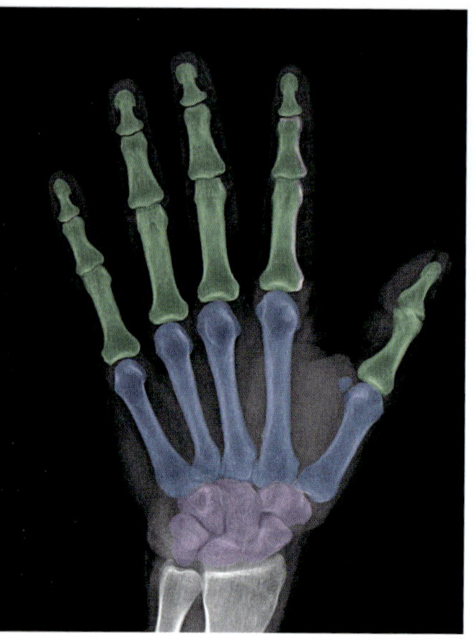

▲ *Left hand of a human, consisting of phalanges or finger bones (green), palm (metacarpus, blue) and wrist (carpus, purple).*

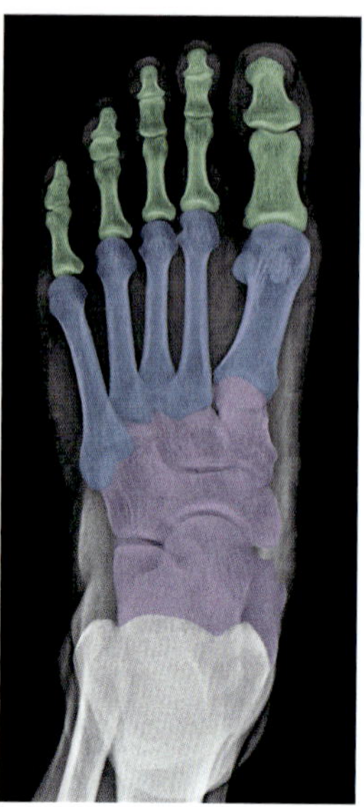

▲ *Left foot of a human, consisting of phalanges or toe bones (green), midfoot (metatarsus, blue) and heel (tarsus, purple).*

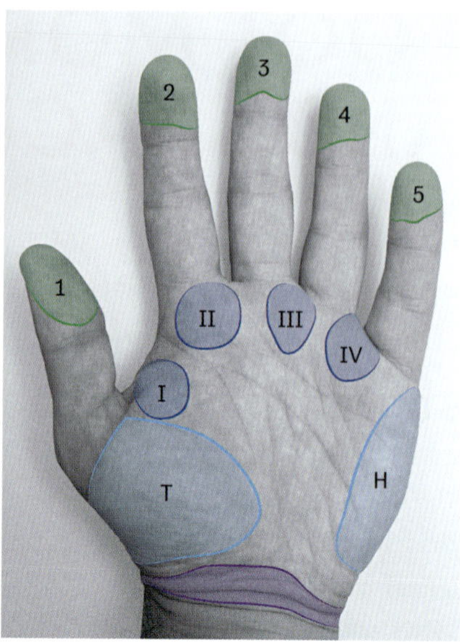

▲ *Left hand of a human. Positions of pads (see opposite for key).*

CLAWS

Most mammals have a horny covering on the tip of their most distal phalanges. In most of today's mammals, as well as in birds and reptiles, this horny structure is called a claw. Claws project over the tip of the toe; they are tapered, relatively narrow and curved in a sickle shape. They can also develop into nails, for example in primates such as humans. Nails do not project over the tip of the toe or project very little; they have a blunt end and are wide, flat and rarely curved.

TOES

Toes are the digits of the foot and each is made up of several phalanges. They are counted from the ones closest to the inside of the body outwards, and numbered accordingly.

Toe 1 is also known as the hallux on the hind foot and pollex on the front foot and, in most mammals, is the toe closest to the body. The numbering can be illustrated by comparing it with our human hand: Toe 1 corresponds to our thumb. Toe 2 corresponds to our index finger. It is shorter than toe 3, our middle finger, and usually also shorter than toe 4. Toe 3 of the human hand, the middle finger, is the longest, and is the longest toe in most mammals. Toe 4, our ring finger, is normally the second longest toe and only shorter than toe 3. Toe 5 is longer than toe 1, but shorter than toes 2–4. It is the outermost toe and corresponds to the little finger of the human hand.

In some animals, toe 1 is greatly reduced. In these species, the first toe print on the inside of the body is normally made by toe 2.

PADS

The ends of some foot bones and other parts of the foot that carry a lot of weight are cushioned by a thickened, and often raised, padded area of subcutis. These are known as pads. Pads are malleable and elastic, absorb shock during movement and protect the underlying structures (Biegert 1961, Brown & Yalden 1973, Schlaginhaufen 1905). Another important function of pads is to perceive touch sensations (Ziekur 2006). The shape and number of pads in various mammal groups has changed during the course of evolution. Scientists have been in agreement for more than 100 years about the arrangement of pads in the basic plan of the original mammal foot, but the names occasionally vary. Again, this terminology can be illustrated by comparing it with our human hand (see the annotated photograph on page 62): there is a **digital pad** on the tip of each toe. In the midfoot region (metacarpus and metatarsus), up to six pads can be found on the plantar surface (sole) of the foot: four in the distal (front) section and two in the proximal (back) section. All six are classified as **metacarpal** or **metatarsal pads**.

The four **distal pads** are located between the toes and are therefore also known as interdigital pads. Like the digital pads, they are numbered from the inside out, using Roman numerals I–IV. Over the course of evolution, the distal metacarpal or metatarsal pads in many animals have fused into a single, large pad. This structure is called the metacarpal or metatarsal pad and is commonly known as the 'palm', or 'palm pad'.

The **two proximal metacarpal/metatarsal pads** are called the thenar and hypothenar pads, or T and H pads for short. In humans, the T pad corresponds to the mound at the base of the thumb (thenar eminence) and the H pad to the fleshy part at the base of the little finger. In quadrupeds, the hypothenar pads of the front feet are closely connected to the carpus and are also referred to as carpal pads. However, Kimura (1999) presents a perspective that deviates from this traditional understanding, contributing to an ongoing discussion in scientific literature.

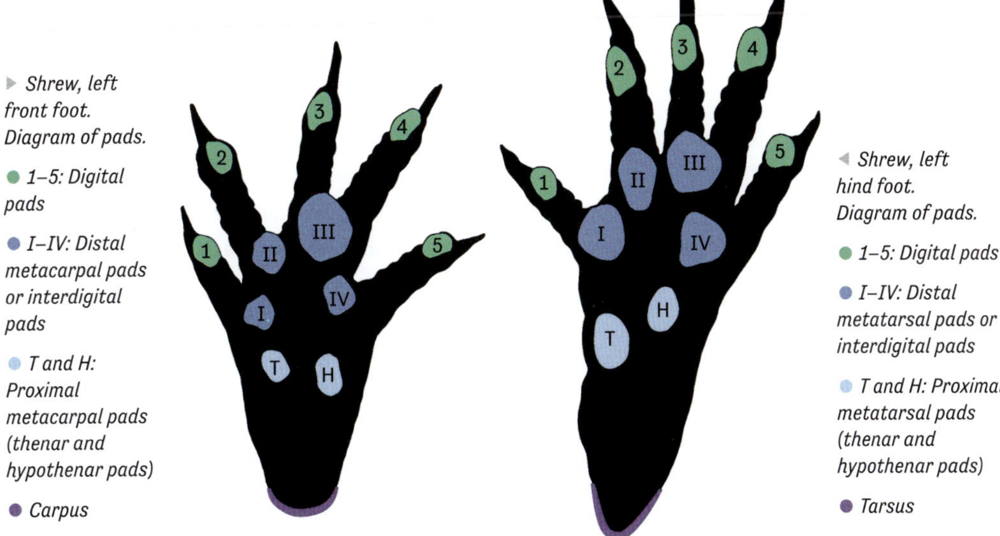

▶ *Shrew, left front foot. Diagram of pads.*

● *1–5: Digital pads*

● *I–IV: Distal metacarpal pads or interdigital pads*

● *T and H: Proximal metacarpal pads (thenar and hypothenar pads)*

● *Carpus*

◀ *Shrew, left hind foot. Diagram of pads.*

● *1–5: Digital pads*

● *I–IV: Distal metatarsal pads or interdigital pads*

● *T and H: Proximal metatarsal pads (thenar and hypothenar pads)*

● *Tarsus*

Terminology

When discussing humans and other primates, we distinguish between the upper extremities (arms) and the lower extremities (legs). The distal part of the upper limbs, also known as the autopodium, is called the hand, while the distal part of the lower limbs is referred to as the foot.

For other terrestrial vertebrates, we use the terms front limbs and hind limbs, or front legs and hind legs. In this book, we adopt the latter. The distal parts of the front legs and hind legs are referred to as the front foot and hind foot, respectively.

Some species, such as squirrels and raccoons, have front feet that resemble human hands in both structure and function, and these are often colloquially referred to as 'hands.' However, in this book, the corresponding limbs of all quadrupeds are consistently referred to as front feet rather than hands. Consequently, we use terms like toes and avoid anthropomorphic terms such as fingers, which are specific to the forelimbs of primates.

Some species, including humans, have a specialised pad on the hind foot known as the **heel pad**, which supports the tarsal region. In quadrupeds, a similar specialised pad exists on the fore foot, referred to as the **carpal pad**, located near the carpus.

The development of specific pad features is closely linked to certain types of locomotion and other factors. The palm pad, for instance, can exhibit combinations of features that reflect a species' locomotion or behaviour (Ziekur 2006). Ground-dwelling species often have fewer distal metacarpal/metatarsal pads, while climbing species typically have larger, more prominent pads to increase frictional resistance. The number, shape and size of the pads can therefore provide indications about the ecological niche of a species.

The 'protomammal foot' described has evolved into the various types of feet and limbs seen in the different mammal groups. The anatomical structures of modern mammals vary greatly. Some parts of the proto-foot have decreased in size or disappeared altogether. Other parts have changed shape dramatically. Specialisation has resulted in features that are beneficial for a frequently used function. For example, depending on whether the main function of the foot is running, grasping, swimming or digging, species have developed opposable digits (as in humans), webbing or claws. A European Badger that digs down into the ground needs a different shape of foot to a Eurasian Beaver that spends a lot of time in water.

Feet also act as lever systems for locomotion. When a land mammal moves, it leaves footprints or tracks on the ground. When examining a track, we look at the indentation in the ground caused by the sole. It provides information about the species, lifestyle and behaviour of the animal that made it. Even the smallest footprints can contain a lot of information, enabling a tracker to ascertain the relative size of the animal, its weight and its sex. In principle, it can even be possible to identify an individual animal, if it moves in a characteristic way or has certain other habits that distinguish it from other individuals. In 1997, Stander *et al.* investigated the tracking skills of four San Bushmen in Namibia. They were able to respond correctly to 557 out of 569 questions to determine individual animals by their footprints (Elbroch 2012).

Assigning individual parts of a track to the corresponding parts of the foot helps us to make detailed observations and to create precise search grids. Knowing which patterns to look for and being able to recognise these patterns are two essential tracking skills. These skills require a basic understanding of different foot structures and the resulting foot strike. In the case of mammals, we differentiate between three different foot structures and types of foot strike, which are described in more detail below.

PLANTIGRADE

The first land mammals were plantigrades. In plantigrade locomotion, the animal often puts its entire foot on the ground, from the carpus/tarsus to the tips of the toes. The term 'plantigrade' comes from the Latin *planta* 'sole' and *gradi* 'walk'. Typical examples of plantigrades are humans, bears, mustelids and most rodents.

◄ Beech Marten, left front. Beech Martens often walk on the entire sole of their foot.

◄ Beech Marten, left front.

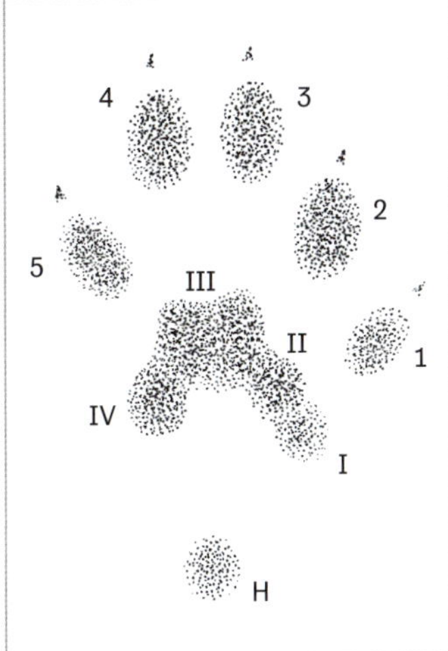

◄ Beech Marten, left front.

DIGITIGRADE

The foot structure of digitigrades evolved from the anatomy of plantigrades. It is characterised by an altered foot position where only the toes are in contact with the ground. The term 'digitigrade' comes from the Latin *digitus* 'toe' and *gradi* 'walk'. Some bones, particularly the metacarpal/ metatarsal bones, are extended, resulting in limbs that are longer overall. Longer limbs enable a longer stride length. In many digitigrades, toe 1 is much shorter so only four toes are normally in contact with the ground. Bang and Dahlström (1974) note that many species that can run fast have long legs and a relatively small ground contact area. Both these features are achieved through the development of a digitigrade foot. Dogs and cats are typical digitigrades. Note: many mammals can move both as plantigrades and digitigrades.

▲ *Domestic Cat, left front. The distal metacarpal pads have fused into one large midfoot pad that is useful for moving stealthily.*

▶ *Domestic Cat, left front. Toe 1 (not visible here) is further up the leg and rarely comes into contact with the ground.*

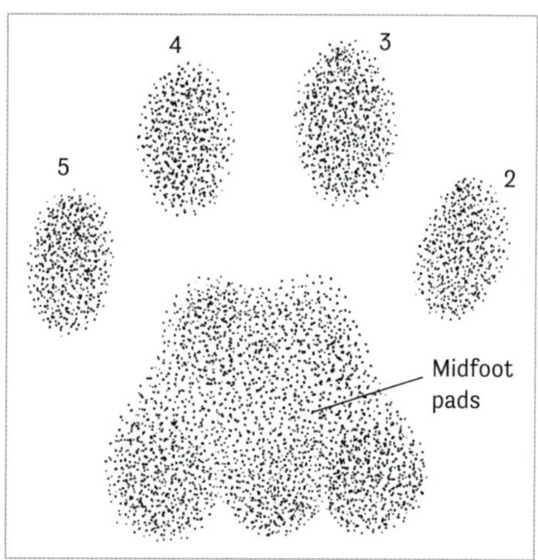

▶ *Domestic Cat, left front.*

UNGULIGRADE

The foot structure of unguligrade (Lat. *ungula* 'hoof' and *gradi* 'walk') animals is characterised by a foot position where only the tips of the toes are in contact with the ground. The metacarpal/metatarsal bones and toes have evolved to be much longer than in digitigrades. As well as an altered foot position, unguligrades also have longer limbs than digitigrades and, correspondingly, a much longer stride. It is hypothesized that the development of carnivorous digitigrades that could run faster than plantigrades exerted evolutionary pressure on unguligrades to achieve even greater speeds. Supporting this, all living unguligrades are prey species. The greatly reduced size of some of the toes is likely to also enable unguligrades to run faster. In even-toed ungulates (artiodactyls), toes 2 and 5 have shrunk so much that only toes 3 and 4 are weight-bearing, but relatively large. The greatly reduced toes 2 and 5 are positioned higher up on the foot and referred to as dew claws. Unlike most plantigrade and digitigrade animals, toe 4 is longer than toe 3 and toe 1 is entirely absent. Equines are even more specialised and are missing all toes except toe 3. They are Europe's only odd-toed ungulates (perissodactyls).

The distal toe bones of unguligrades are encased in a firm yet flexible and regenerative horn cap. Even-toed ungulates stand on two cloven hooves per foot, while odd-toed ungulates such as horses have a single hoof per leg. This adaptation is analogous to humans standing on their fingernails.

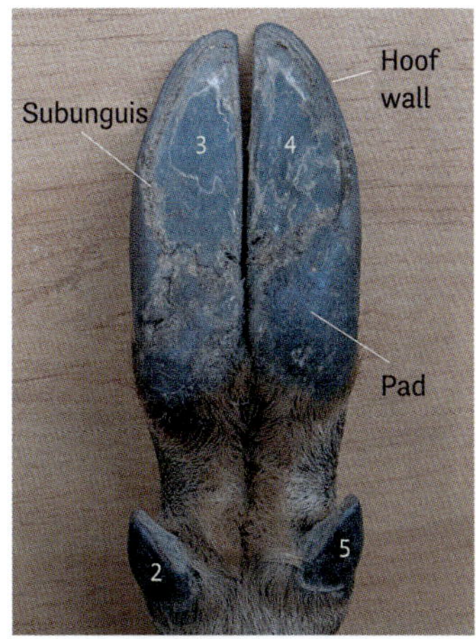

◀ *Western Red Deer, left front. Toe 4 is longer than toe 3.*

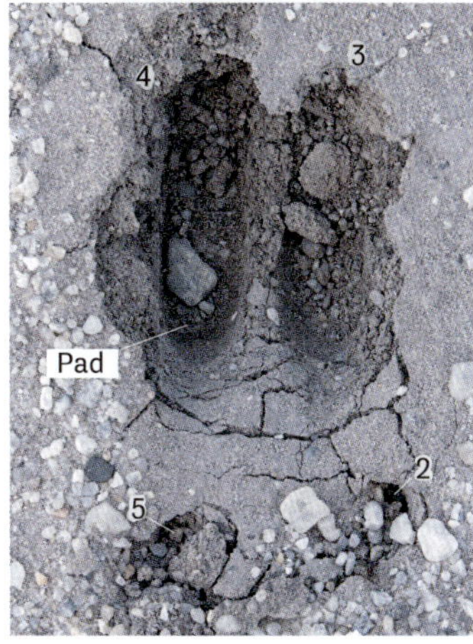

◀ *Western Red Deer, left front.*

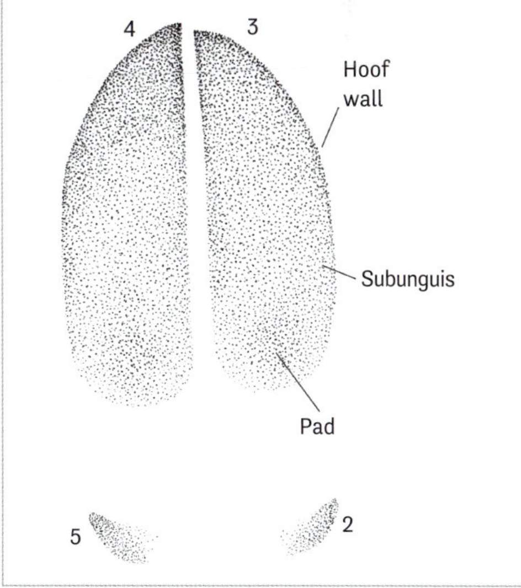

▲ *Western Red Deer, left front.*

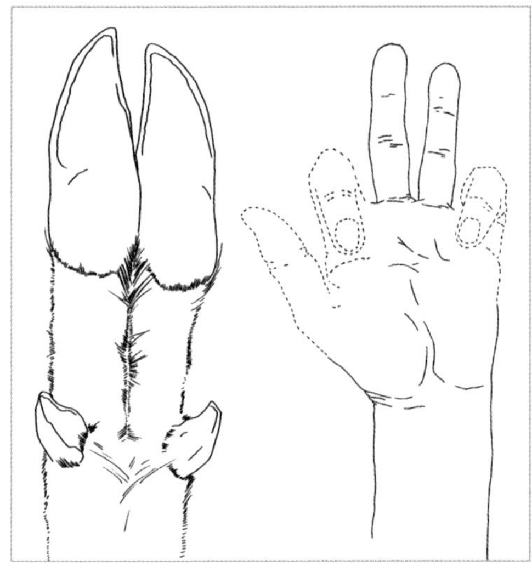

◄ *The weight-bearing toes of even-toed ungulates correspond to our human middle finger and ring finger.*

▼ *Comparison of foot position and structure of a plantigrade (left), a digitigrade (centre) and an unguligrade (right).*

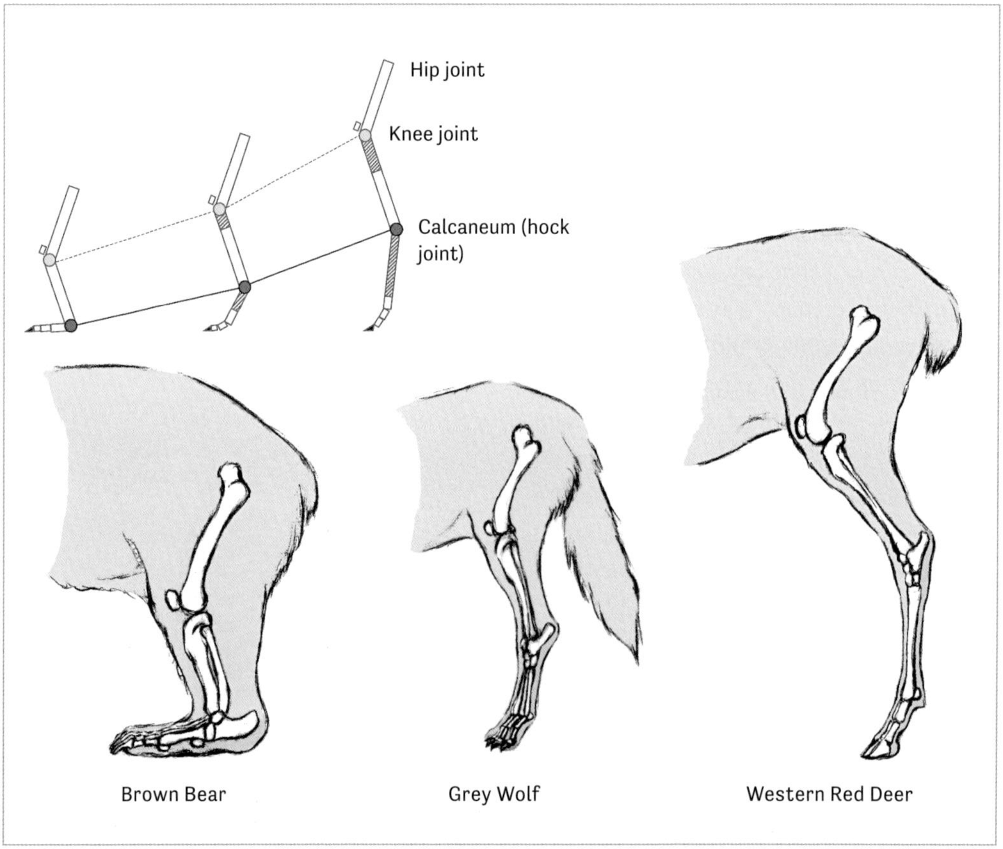

STRUCTURE AND FEATURES OF TRACKS

Most mammal tracks are made up of prints of the claws, toes and metacarpal/metatarsal pads. The negative space between the print of the digital pads and the metacarpal/metatarsal pads is also an important distinguishing feature. Even-toed ungulates might leave dew claw prints and some species might occasionally leave webbing prints. Other important features include the size and symmetry of the track.

CLAWS

The presence or absence of claw prints in the track and the type of claw prints are important features for species identification. Ask yourself: are the claws sharp or blunt, stout or fine, short or long, straight or curved? Some animals can partially or fully retract their claws but in others you can reliably identify them in the footprint.

Right or left track?

If a track consists of the five toe prints of a foot and the shortest toe, closest to the body, is on the right-hand side, then the print is of a left foot; if it is on the left-hand side, the print is of a right foot. Be aware, however, that toe 1 often doesn't make a print and this can lead to toe 5 being incorrectly identified as toe 1. If we can only make out four toe prints, they are usually toes 2–5. It can be helpful to look at how toes 2–5 are arranged and compare them with how your fingers are arranged on your own hand. Look at the back of your right hand: the little finger is on the outside right of your hand and, apart from the thumb, it is the shortest. The middle finger is the longest and sticks out the furthest. If you can make out this arrangement in a track, then it is a right footprint. Try to work out the side of the track in the diagram opposite. You'll find the solution on page 71.

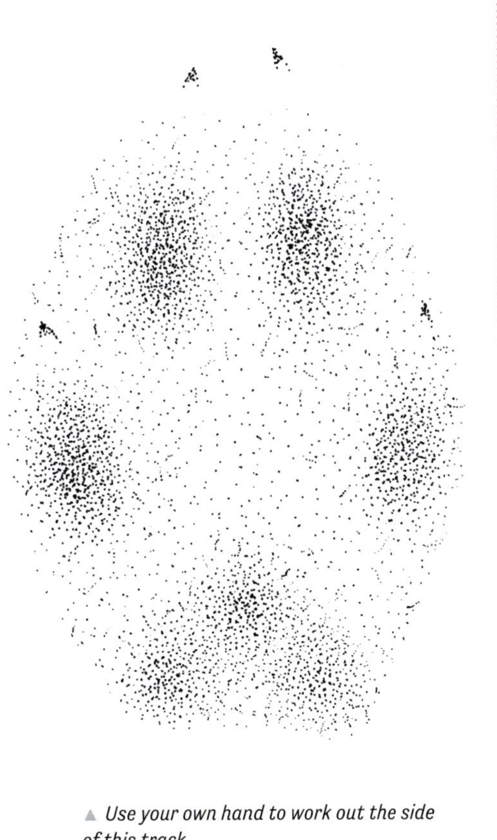

▲ *Use your own hand to work out the side of this track.*

TOES

The number, shape, arrangement and length of toe prints are important identifying criteria. For example, Northern Raccoons have quite long, finger-like toes and, in a track, they will appear to be directly connected with the fused front distal metacarpal/metatarsal pads. By contrast, the toes of an Edible Dormouse appear relatively short and, in the track, are clearly separate from the print of the midfoot pad.

Tracks with relatively small digital pads and relatively large midfoot pads are typical of cats; robust, strong digital pads with a relatively small midfoot pad are a feature of dogs.

INTERDIGITAL PADS

The arrangement, shape and number of interdigital pads (also known as distal metacarpal/metatarsal pads) are important features for species identification. Has the midfoot pad printed completely as a coherent surface or can you make out several individual interdigital pad prints? Have the individual interdigital pads printed faintly or strongly? What is the size ratio between the digital pads and the midfoot pads, and which proportion of the total surface of the track is taken up by the midfoot pads?

THENAR AND HYPOTHENAR PADS

Plantigrades often leave impressions of one or two additional pads in their tracks, corresponding to the thenar and hypothenar pads (or proximal metacarpal/metatarsal pads). These pads can appear in the posterior part of the track. In digitigrades, the prints of these pads are rarely visible unless the animal is moving quickly, walking on a deep substrate, or resting in a lying position.

The presence and position of thenar and hypothenar pad impressions in a track are critical clues for identifying the foot and the species that made it. For certain species, these pads are uniquely positioned. For example, in mustelids, the hypothenar pad is slightly outward relative to the centreline of the frontfoot, making this a distinctive characteristic of the family.

Following the scent of a track

Feet have sweat and scent glands that give every footprint the scent of the animal that made it. If you have a dog, you will notice that it sniffs other animal tracks on a walk. Dogs have an excellent sense of smell and, once they have picked up the scent trail, can follow a track for a long distance without needing to see any footprints. Dogs' sensitive noses can pick up scent trails several weeks after they were left.

NEGATIVE SPACE

The negative space of a mammal track is the area between the digital pads and midfoot pads. It is the part of the foot that doesn't leave a print. The negative space has a characteristic shape depending on the species and can help to identify an animal group or species. An example of this is the 'canid cross' seen in dog footprints.

Looking at the negative space also shows us how the digital pads are arranged in relation to the midfoot pads. The footprints of some species can seem compact (small negative space), while others are 'open' and extended (larger negative space).

▼ *Northern Raccoon, left front.*

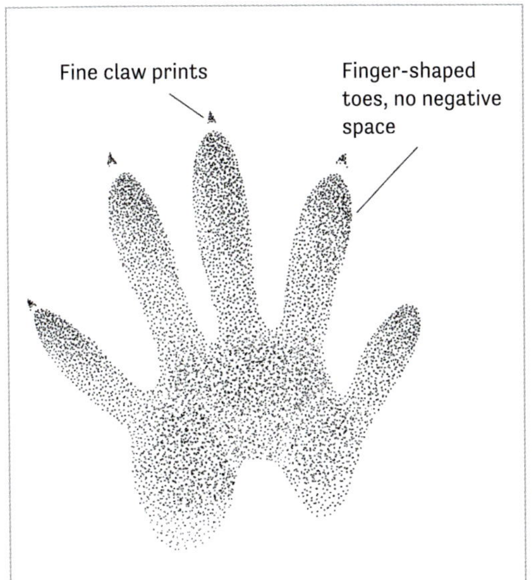

▼ *Edible Dormouse, left front.*

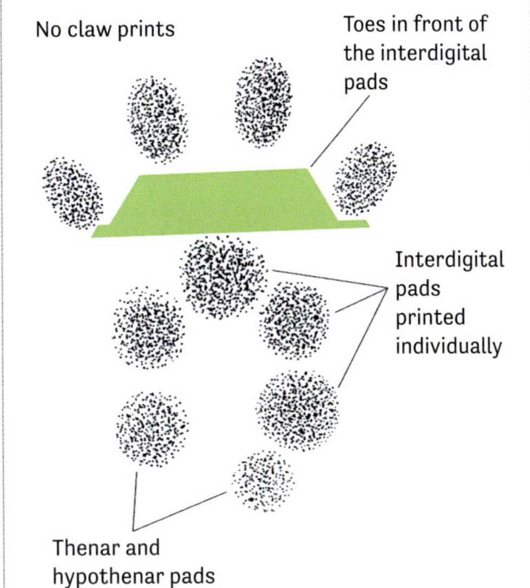

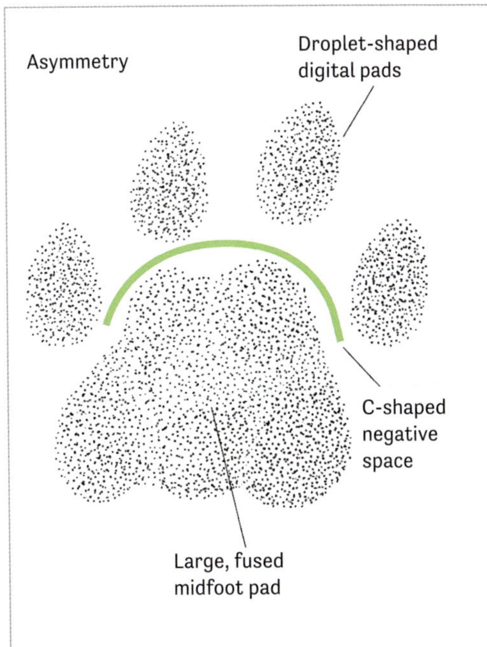

▲ *Lynx, left front.*

▲ *Grey Wolf, left front.*

Solution from page 69:

This is the left front foot print of an Arctic Fox. Were you able to work out the correct side?

DEW CLAWS

The truncated toes 2 and 5 of even-toed ungulates are higher up on the foot than toes 3 and 4 and are referred to as dew claws. Dew claw prints are often recognisable in Wild Boar but, in most other even-toed ungulates, they are normally only made in deeper surfaces such as snow or sand, or when the animal is moving at high speed. The size, shape and position of dew claws can tell us whether the print is that left by a front foot or a hind foot (see page 495).

WEBBING

Mammals that mainly live in water have flaps of skin – webbing – between or on the side of the toes. Webbing is an adaptation to the species' habitat. It enables the animal to move effectively in water and can be found between two or more toes. If the webbing spans the entire length of the toes, it is referred to as distal. Webbing that only reaches halfway up the toes is called medial. Proximal webbing is rudimentary webbing that is only found between the base of the toes. These prints are not always discernible, even in the case of pronounced distal webbing.

SIZE

Looking at the size of a track can often enable us to rule out a number of species. In every mammal portrait in this book, size classes are given at the beginning of the track description to enable initial classification. The classes are: minute, very small, small, medium, large and very large.

SYMMETRY

Looking at the symmetry of a footprint can help identify a species, or at least assign the track to a group of animals. Symmetry is also often what allows us to differentiate between the front feet and the hind feet. We roughly categorise prints as symmetrical, slightly asymmetrical and asymmetrical. In some land mammals, such as cats, the hind foot is more symmetrical than the front foot.

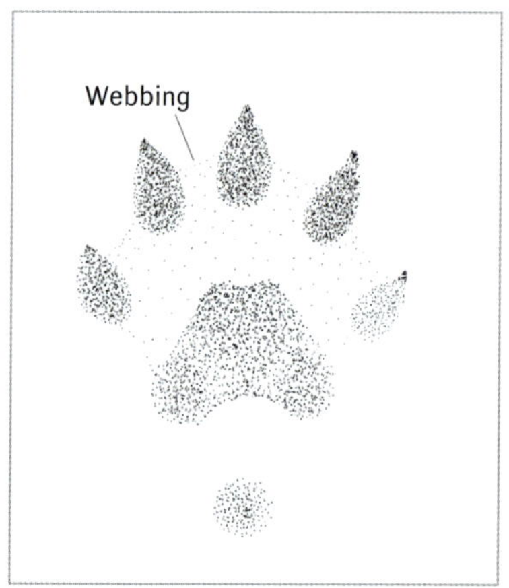

▲ *Eurasian Otter, left front.*

Measuring tracks

I keep coming across two types of people in my life: those who love taking measurements, are passionate about creating diagrams and evaluating data and feel very comfortable doing so, and those who avoid dealing with figures and measurements and prefer to work things out in a different way.

Personally, I've always counted myself among the second group and I could never have dreamt that I would meticulously measure the smallest distances as I did during my research for this book. I spent months using callipers to examine the difference between the ink footprints of voles (Arvicolinae) and mice (Muridae). I was surprised by how much I enjoyed it! In my work teaching tracking, I have come to learn how much track dimensions and the activity of measuring them can teach us. Regular measuring trains our eye. It forces us to look closely and to develop an eye for the details. Both are essential characteristics of a tracker.

I have had times when I didn't measure tracks and times when I hardly did anything else. Now I take measurements if a track seems to fall outside the pattern I'm familiar with. Taking lots of measurements has helped me to be able to spot special features. I have recorded so many strides that often just looking at chains of numbers is enough for me to be able to tell how fast the animal was going. Looking back, I wasn't initially as interested in measuring tracks and strides as I was in learning about the life of the animal. I now realise that measuring and evaluating data has brought me closer to the animals and given me important insight into their behaviour.

We measure the length and width of tracks. These values help us to identify species and can highlight characteristic features of individuals and give information about sex. Taking many measurements allows us to cooperate with other trackers. Comparing and exchanging data often paves the way for subsequent research. By scientifically documenting this data, we can reach conclusions about the species in a certain area. In each animal portrait, I give minimum and maximum values for the dimensions of tracks, gaits and signs, rather than an average value. This is much closer to reality as individual sizes can sometimes vary considerably within a species. In addition to the size of the animal, speed and surface conditions can also cause the dimensions of a track to vary greatly.

To obtain the most accurate dimensions when measuring tracks, we measure the minimum outside edge of an individual footprint (see page 74). Drag marks or material that is scattered when the foot touches or leaves the ground should be ignored because they distort the size information.

Try to measure the weight-bearing area and take into consideration any inaccuracies in your measurements due to difficult surfaces when evaluating your data.

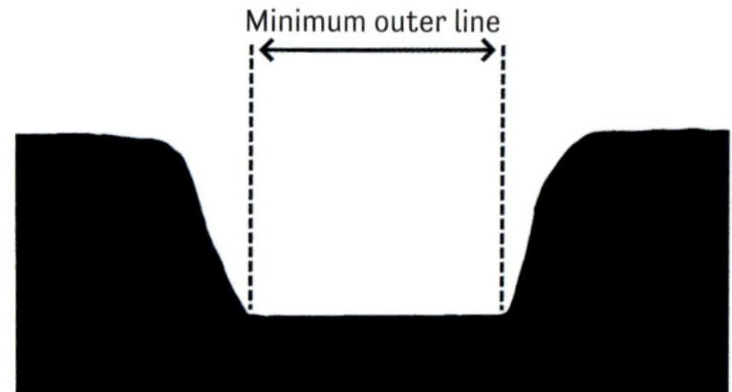

▶ *Measure the minimum outer line.*

Limitations of data collection

This method of data collection is intended to give representative minimum-maximum values that take into consideration small and large individuals of a species and, at the same time, reflect the most likely size range. For that reason, I have left out extreme values of exceptionally small or large individuals.

When comparing your measurements from the field, always remember that exceptions can occur and shouldn't be categorically ruled out of a finding even if they fall outside the specified data range.

Individuals of a species can vary greatly in their physical features in an extensive area like Europe because they are influenced by local conditions such as climate and food availability. You may also come across prints of young animals and these are not normally taken into consideration in the stated measurements.

For example, if a track has the clear features of an otter footprint but is slightly smaller than the minimum given here, it is probably the track of either a very small or young otter.

LENGTH

The length of a track is measured from the foremost to the hindmost outside edge of the footprint, in parallel to the longitudinal axis of the track. Like Rezendes (1999) and Elbroch (2003), we measure the length including the claws. Some trackers leave out the claws, but this makes it impossible to measure tracks that are often exclusively made up of claw prints (for example, moles). In this book, length information includes dimensions with and without claws.

For data to be comparable, clear information should be given as to how it was measured. If the thenar and hypothenar pads are regularly discernible (for example in the front foot of a squirrel), they are included in the measurements. In animals like bears and badgers, whose thenar and hypothenar pads make irregular prints, the minimum and maximum data include measurements with and without the thenar and hypothenar pads. As a result, the range of dimensions is generally larger in these species: the minimum values describe the footprints of smaller animals excluding the thenar and hypothenar pads and claws; the maximum values, by contrast, refer to the footprints of larger animals that are measured including thenar and hypothenar pads and including claws.

Canids and cats are different. The thenar and hypothenar pads in these animals very rarely leave prints and if they do, it is only in deep surfaces or if the animal is moving at high speed. They are therefore not included in the length information of the relevant species portraits.

Ungulate prints are measured without dew claws because they do not regularly create prints in many species, so size comparison of the tracks of different species is possible.

WIDTH

The width of a track is measured at a right angle to the length, at the widest point.

The size of the track can vary greatly even within an individual. For this reason, you should always take several measurements of the front foot and hind foot prints. Remember, we are not measuring the actual foot but the print of a moving animal, and the shape of the prints is influenced by several factors, so they won't always be the same. For setting up your own data collection, we recommend taking at least 12 measurements and deleting the lowest and highest value. This should prevent your data collection from being distorted by outliers.

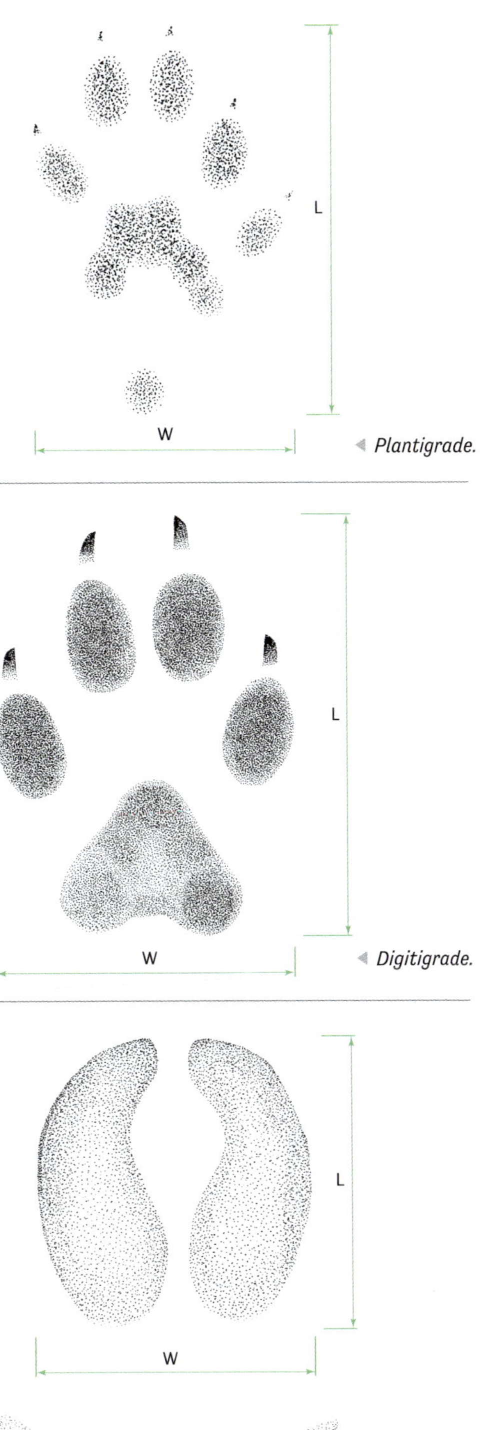

◀ Plantigrade.

◀ Digitigrade.

◀ Unguligrade.

GAITS AND TRACK PATTERNS

Mammals can move in different ways on land: quickly or slowly, walking or running, at a consistent pace or interrupted by stops. The regular sequence in which they move their legs is called the gait. The choice of gait depends on factors such as energy expenditure, speed, agility, stability and the size and structure of body (Hildebrand 2004). Different species favour different gaits and some gaits are characteristic of certain animals.

Locomotion leaves a track pattern that allows us to draw conclusions about the gait used. However, specific gaits can produce different track patterns depending on the shape of the animal's body, its speed and the sequence of its steps. For example, a trot is generally faster than a walk but both gaits can be fast or slow depending on the situation. Correctly interpreting a track pattern not only helps us to identify the chosen gait but also provides information about the specific behaviour of the animal, its condition and the context in which it was moving. Was the animal moving fast or was it more relaxed? Why did it stop and then suddenly begin to stalk? We can infer these and many other details from the track pattern.

Zoologists have been researching gaits for many years. Eadweard Muybridge and Milton Hildebrand were two pioneers in this field of research. Today, scientific investigations into gaits are conducted using modern technology such as super slow-motion recordings, blood and respiratory gas analyses, force measuring plates and other methods in the laboratory, which couldn't be applied to tracking out in the field. This book focuses on track patterns that can be linked to a specific gait using their appearance and measurements. The scope is limited to the gaits of wild animals. As a result of domestication and training by humans, domestic and farm animals have gaits that do not usually occur in animals living in the wild. Wild mammals in Europe primarily hunt, stalk, explore and flee using the gaits presented here.

All the mammals described here have four legs and four feet. Because humans only have two legs, it can be hard to visualise – especially as a beginner – the complex movement sequences of quadrupedal animals. Communication on this topic is further complicated by the variation in names for different gaits and track patterns in different languages and lack of set terminology in gait research. However, trackers began using an internationally standardised system of terms two decades ago. Louis Liebenberg, James C. Lowery and Mark Elbroch made important contributions to this that we apply and build upon in this book. At the end of this chapter, you will find an overview of the internationally standardised terms.

Understanding gaits and interpreting track patterns might seem difficult at first but, with practice, we can quickly make progress and be rewarded with a more complex understanding of events unfolding in the natural world around us. We will be able to read and understand more and more stories written in the tracks of animals.

WALKING AND RUNNING

The ways animals move using their legs can be divided into two main categories: **walking** and **running**. The primary difference between walking and running is the underlying mechanics on which the movement is based. When walking, the animal's body moves forward in a 'swinging' manner on relatively stiff legs.

This type of biomechanics is known as the **mechanics of an inverted pendulum** (see the top diagram) and is a defining feature of all forms of walking. When running, by contrast, the legs act as a suspension system, propelling the body forward over the point of foot touchdown. This type of biomechanics is called **spring-mass mechanics** (see the bottom diagram). It is characteristic of all types of running.

Walking and running encompass different gaits that leave a large variety of different

▶ *Mechanics of an inverted pendulum.*

Preferred gait

The gait an animal mainly uses under normal circumstances is called its 'preferred gait'. For example, humans primarily walk. Although we may change gait and start running if we are in a hurry or for exercise, walking remains our preferred gait.

Different animals have distinct preferred gaits depending on species and body type. Knowing which gaits are mainly used by different animals helps us to interpret track patterns. The species descriptions of mammals contain information about the preferred gait and the corresponding track dimensions.

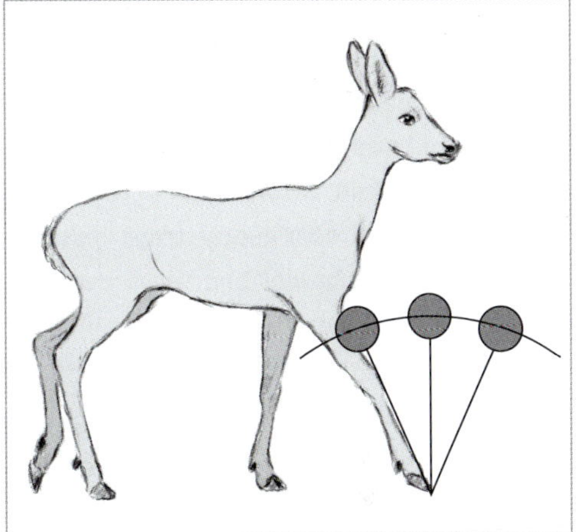

▲ *Spring-mass mechanics.*

track patterns. We divide gaits and track patterns into two main categories: **symmetrical gaits**, which result in **continuous track patterns** and **asymmetrical gaits**, which produce **group-forming track patterns**. All walking gaits are symmetrical with continuous track patterns. Running gaits can leave continuous or group-forming track patterns because they involve both symmetrical and asymmetrical movement sequences.

SYMMETRICAL GAITS, CONTINUOUS TRACK PATTERNS

In symmetrical gaits, an animal's legs on the left and right sides of its body move in identical patterns. While the movement sequence and pace are the same on both sides, they are staggered, resulting in a continuous track pattern. A continuous track pattern consists of two parallel and evenly spaced rows of successive footprints. Both rows are offset in the direction of travel and each row contains all the tracks of the two feet on the same side of the body. There are almost always alternating pairs of tracks made by a front foot and a hind foot. This basic pattern is found in all symmetrical gaits, such as **walk**, **trot**, **amble** or **grounded running** (Schmidt 2017; Biknevicius & Reilly 2006).

Walk

The walk is a slow and energy-efficient gait used by many animals. When undisturbed, most animals with a compact body shape, such as hedgehogs and bears, prefer to walk; this also applies to all deer and most cats. The walk is a symmetrical gait and the only one of the gaits presented here that uses the mechanism of an inverted pendulum. Two or three feet are usually on the ground at any one time, resulting in regularly alternating phases where the animal's weight is on two or three legs. In this case, the maximum **stride** length is limited by the animal's hip-shoulder length. The **trail width** is relatively large in relation to the stride length, often resulting in the impression of a pronounced zigzag pattern. If you look along a track like this, it is relatively easy to assign the right and left footprints to one side of the body because they fall clearly to the right or left of an imaginary centreline that runs parallel to the direction of travel.

There are different sequences of steps, where each leg normally moves separately. A hind foot usually moves first, followed by the front foot of the same side. A common sequence is as follows: the right hind foot is lifted, moves forward and, just before it touches the ground, the right front foot lifts. For a brief moment, both feet are off the ground and the animal is supported on two legs, before the right hind hits the ground again and there are three feet on the ground. The right front foot moves further forward and then makes contact with the ground. Just before the right front foot touches the ground, the left hind foot is lifted and moves forward. As before on the right side of the body, the left front foot lifts just before the left hind foot touches the ground. This is the second point when just two feet (at this point, the two right feet) are in contact with the ground. Next, the left hind foot touches the ground, the left front foot moves further forwards and, shortly before it touches the ground, the right hind foot is raised. A full cycle of footfalls is now complete and the movement cycle begins again with the right hind foot.

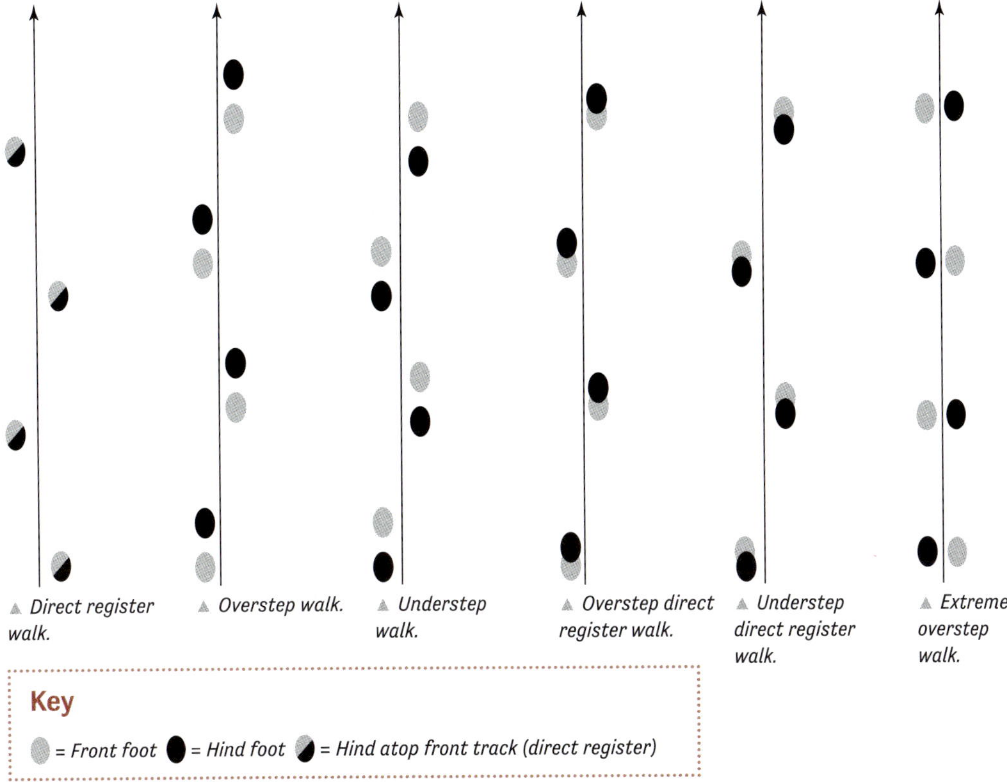

Key

The hind foot's position relative to the front foot of the same side of the body can vary in the track pattern. As the front foot is raised before the hind foot touches the ground, the hind foot can step directly into the footprint of the front foot on the same side. This creates a double print and the corresponding track pattern is called a **direct register walk**. Many animals use the direct register walk while looking for food or when moving in a relaxed way.

In some cases, the hind foot oversteps the front foot print, touching the ground slightly ahead of where the front foot was. This creates a track pattern called an **overstep**.

A third option is that the hind foot 'understeps' the corresponding front foot. Understepping indicates a slow movement, for example when stalking. Here, a hind leg moves forward and steps on the ground before the following front leg comes off the ground, which creates an **understep track pattern**.

Understepping and overstepping are also possible in a direct register walk.

The **extreme overstep walk** track pattern indicates a special gait known as an **amble**. In the amble, the legs on each side of the body move almost simultaneously; the front leg lifts off the ground earlier than in the walking sequences described above. The hind foot oversteps the front foot print of the same side of the body so much that it lands more or less next to the front foot of the opposite side. The hind foot track can be seen slightly behind, parallel to, or slightly in front of the front foot print of the opposite side. The typical position of the hind foot track at a moderate pace varies from species to species. If you know this position, you can determine whether the animal was moving in a slower or faster amble. The amble is a preferred gait of Northern Raccoons (page 487).

Note: an extreme overstep track pattern doesn't necessarily indicate that the animal was moving at high speed. For example, the relatively short back of a Northern Raccoon, which becomes hunched in this type of movement, is largely responsible for the extreme overstep of the hind foot prints.

Trot

A trot is a dynamic, faster type of movement. It is classified as running because of the spring-mass biomechanics. All forms of trot are symmetrical gaits and only leave continuous track patterns. While at least one foot, and usually two or three feet, is in contact with the ground when walking, there is almost always a moment of suspension in a trot. The trot is the preferred gait of canids and some shrews and small voles. Many other animals, such as deer, cats, badgers and bears, often use a form of trot over longer distances on flat surfaces. In contrast to walking, stride length in a trot is normally longer than the hip-shoulder length of the animal. The trail width is relatively narrow in comparison to the stride length. The right and left footprints generally fall closer to the imaginary centre line of the body that runs parallel to the direction of travel. An extreme form is shown by the Red Fox, whose trail width can be so narrow that the tracks appear 'linearised', as if forming a single line.

Unlike walking, in a trot only diagonally opposite legs move forward simultaneously. A typical movement sequence is: the right hind foot leaves the ground at the same time as the left front foot, while the opposite diagonal pair of legs is already in the air and moving forward. At this moment (apart from in slow trots) there is a brief moment of suspension before the left hind and the right front foot simultaneously (or almost simultaneously) land. They then push off together and there is a second moment of suspension before the right hind foot and the left front foot land together. The full cycle of footfalls is now complete and the movement cycle begins again.

As with a walk, the hind foot can step directly into the footprint of the front foot on the same side, creating a double print. This track pattern is called a **direct register trot**. The direct register trot is commonly observed in the tracks of Grey Wolves and other canids, as well as in small voles, European Badgers, Wild Boar and deer.

Differentiating between walk and trot

In the field, track patterns can be used to differentiate between walk and trot by comparing their stride length to the hip-shoulder length of the animal. To estimate this, draw a line at a right angle to the direction of travel from one hind foot track and then from the next hind foot track on the same side of the body. Then, roughly sketch the body of the animal between these two lines. If the sketch is much longer (over 10 per cent) than the actual hip-shoulder length of the animal, the gait is a trot. There are also species-typical trail widths for walk and trot that are given in the animal portraits. Trail widths generally decrease when the animal changes from walk to trot (see Gaits and speed, page 83). The side trot and straddle trot are exceptions to this rule. In these gaits, the trail width is slightly greater because the hind feet land slightly further out to the side. The relationship between stride length and straddle/trail width is another important feature for differentiating between walk and trot. In a walk, the stride length is relatively short, and the trail width is relatively wide. In a trot, the stride length is relatively long, and the trail width is relatively narrow.

▲ *In a walk, stride length corresponds approximately to the hip-shoulder length of the animal.*

▲ *In a trot, stride length is normally much longer than the hip-shoulder length of the animal.*

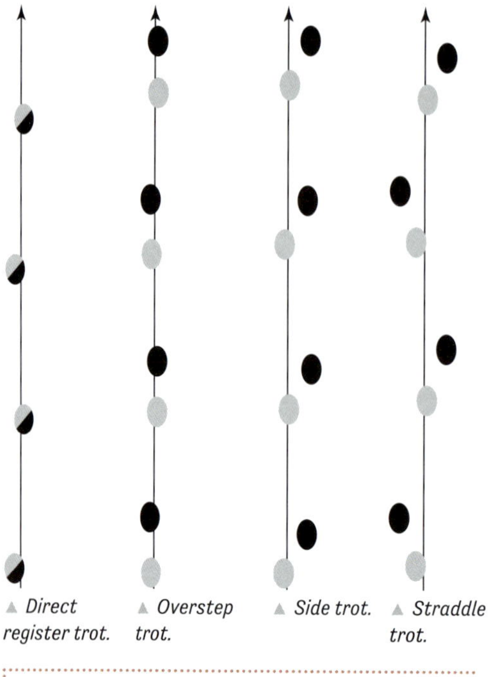

▲ *Direct register trot.* ▲ *Overstep trot.* ▲ *Side trot.* ▲ *Straddle trot.*

Key

○ = *Front foot* ● = *Hind foot*

◐ = *Hind atop front track (direct register)*

The hind feet can also overstep or understep in different ways in a trot, creating similar track patterns to those seen in a walk. They can be distinguished from a walk because of the longer strides and narrower trail widths. An **understep trot** is rare, while **side trot** and **straddle trot** are two special forms of an overstep trot.

In side trot, both hind feet consistently overstep the front feet on the same side of the body. The animal's body is held at a slight angle to the direction of travel. One hind foot oversteps the corresponding front foot on the inside of the body, while the other hind foot oversteps the corresponding front foot on the outside. The hind feet prints are overstepped and offset to one side. Side trot is typical of Domestic Dog, Grey Wolf, Golden Jackal and Red Fox.

In a straddle trot, the animal widens the stance of its hind legs so each hind leg oversteps the corresponding front leg on the outside of the animal's body. The track pattern shows the hind foot prints overstepping the front foot prints on the outside. This gait is used by many animals as a transition gait. Western Red and Common Fallow Deer, as well as Reindeer and Elk, also use the straddle trot over longer distances.

The side trot and straddle trot are exceptions to the rule that trail widths decrease when the animal changes from walk to trot. Because the hind feet are slightly offset to the outside, the trail width may, in fact, increase.

Pace

As in the amble, the pair of legs on one side of the body moves forward almost simultaneously when the animal is running. This is known as pacing. However, in contrast to the amble, pacing typically includes a moment of suspension. This allows the stride length to be much greater than the animal's hip-shoulder length, while the trail width is proportionately narrower compared to the amble. Camels and giraffes are known for pacing. European wild animals rarely use this gait, but it is occasionally observed in some domestic dog and horse breeds.

Grounded running

Grounded running is a symmetrical gait, like walk and trot, and leaves only continuous track patterns. It employs the spring-mass mechanics typical of all running gaits but is

Gaits and speed

Each gait can be performed at different speeds. Gallop is generally faster than walk. However, we need to consider different sizes within a species. The rapid direct register trot of a Western Red Deer hind, for example, is much faster than the gallop of the calf following her. In the field, we often see how young animals have to choose a faster gait to keep up with their parents. As speed increases, stride length generally becomes longer, while trail width typically narrows.

Observe how your own stride length changes in relation to your straddle/trail width. Find a place with an even surface, such as damp sand, and walk along it barefoot or wearing shoes. Start out in a relaxed walk and then slowly accelerate after a few metres. Change to an easy jog and then increase your speed to a sprint after a few more metres. Slow down again after sprinting for a few strides and then come back to a walk. Examine your tracks and note how your stride length increases as you move faster, while your trail width likely becomes narrower.

However, this rule of thumb should be applied with caution. Changes in stride length and trail width indicate acceleration within the same gait but can be misleading when an animal transitions between gaits. For instance, the rule applies when an animal shifts from a direct register walk to a direct register trot but may not hold if the animal transitions from a direct register trot to a side or straddle trot. In these cases, even though speed and stride length increase, the trail width might remain wider.

The same principle applies to gallop. The trail width of a gallop's track group is generally wider than that of a direct register trot, even when the trot is slower. However, within the gallop gait, the rule of thumb remains valid: the greater the group length and the proportionately smaller the inter-group length, the higher the animal's speed. For bounding gaits, increasing inter-group length indicates higher speed. Stride length can generally be used as a near-direct measure of speed. In trot, an increase in overstepping can also indicate higher speed.

characterised by the fact that a moment of suspension is almost entirely absent, giving it the name 'grounded.' The **tölt**, for which Icelandic horses are renowned, is one of the better-known types of grounded running and is extremely comfortable for riders. Although the cycle of footfalls is the same as in a classic walk (RH, RF, LH, LF), there is a significant difference in ground contact: while the animal regularly alternates between having two and three feet in contact with the ground when walking, it alternates between having one and two feet on the ground in tölt. This means that much longer stride lengths can be achieved than when walking, as well as speeds as fast

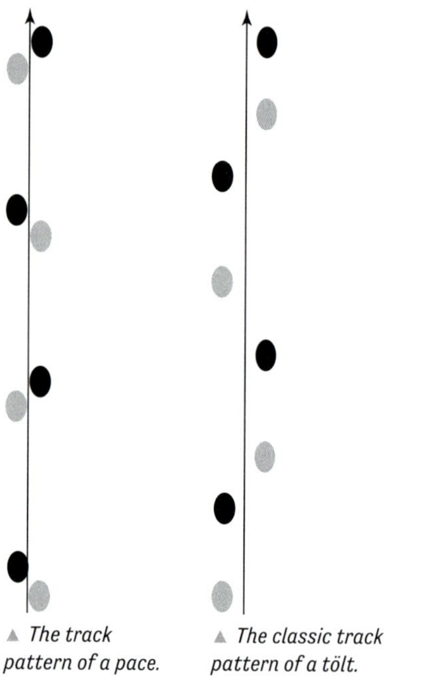

▲ *The track pattern of a pace.* ▲ *The classic track pattern of a tölt.*

Key

⬤ = *Front foot* ⬤ = *Hind foot*

as, and faster than, trot. Grounded running is rare in larger quadrupeds but occurs in elephants and nonhuman primates, as well as in some domestic horse breeds. It is also a key component of the movement repertoire of small mammals and is used, for example, in confined spaces such as narrow passageways underground (Bertram 2016).

ASYMMETRICAL GAITS, GROUP-FORMING TRACK PATTERNS

In asymmetrical gaits, the hind leg and front leg pairs come into contact with the ground one after the other at irregular intervals. These gaits leave group-forming track patterns. A group-forming track pattern consists of groups of four tracks that are each made up of the footprints of the two front feet and two hind feet. This basic pattern is found in all asymmetrical gaits and points to a form of **gallop** or **bound**. All asymmetrical gaits are classified as types of running because they are based on spring-mass biomechanics.

Gallop

Gallop is the fastest gait and used by many animals while hunting or fleeing. This 'top gear' is usually only used for as long as necessary and animals will switch into a lower intensity gait as soon as the situation allows. It's important to note that each gait can be performed at varying intensities and that there are clear differences between a slow gallop and a high-intensity gallop. A gallop is generally faster than a bound but its speed comes at the price of reduced flexibility.

In gallop, the front leg and hind leg pair move forward alternately. Each foot strikes the ground individually and in quick succession, with all four feet landing in different positions relative to the body's centre of gravity. A defining feature of gallop is the additional acceleration forces generated by the front legs, which enhance the propulsion initiated by the hind legs and enable high speeds to be achieved. The animal begins its gallop by pushing off and accelerating with the hind legs. This

▲ *Grey Wolf in gathered suspension.*
Sebastian Körner, Germany.

▲ *Grey Wolf in extended suspension.*
Sebastian Körner, Germany.

is followed by a powerful forward thrust generated by the front legs, as the hind legs come far underneath the body during contraction. Apart from a few exceptions, there is a moment of **gathered suspension** when all the legs are off the ground and come together. The basic cycle of footfalls is: hind foot, hind foot, front foot, front foot, followed by the moment of gathered suspension (see photo, top left). At higher intensities, there can also be a moment of **extended suspension** (see photo, top right), after the two hind legs have pushed off the ground. The cycle of footfalls in this case is: hind foot, hind foot, brief extended suspension, front foot, front foot, gathered suspension. Regardless of whether there are one or two moments of suspension in the gallop, the significant moment of suspension is the gathered suspension.

Except in slow forms of gallop, the two hind foot tracks typically overstep both front foot tracks. Different track patterns can result, depending on the sequence in which the feet strike the ground.

The track pattern of the **rotary gallop** shows a rotary cycle of footfalls (RF, LF, LH, RH or LF, RF, RH, LH), where the first front foot and hind foot to strike the ground are diagonally opposite each other. Depending on whether the right or left front foot touches the ground first, the track pattern resembles either a C shape or an inverted C shape. In this agile gait, the belly of the C curves outward.

The track pattern of the **transverse gallop** shows a transverse cycle of footfalls (RF, LF, RH, LH or LF, RF, LH, RH), where the first front foot and hind foot to strike the ground are on the same side. Depending on whether the right or left front foot touches the ground first, the track pattern resembles either a Z shape or an inverted Z shape. The transverse gallop is more stable than the rotary gallop because it creates a diagonal supporting line during the cycle of footfalls, directly below the body's centre of gravity. However, the increased stability of this gait comes at the expense of flexibility.

The **lope** is a relaxed gallop of low to medium intensity. The lope can reach similar speeds to a fast trot and some species use it as an energy-efficient alternative. This relaxed and relatively slow gallop leaves a track pattern where the two hind feet do not

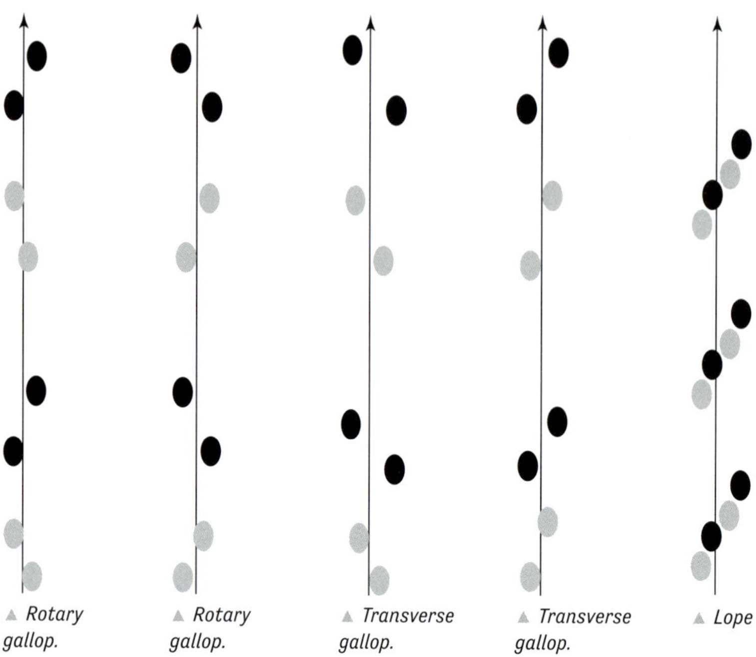

▲ *Rotary gallop.*　　▲ *Rotary gallop.*　　▲ *Transverse gallop.*　　▲ *Transverse gallop.*　　▲ *Lope*

Key

⬤ = *Front foot*　⬤ = *Hind foot*

◐ = *Hind atop front track (direct register)*

overstep the two front feet. Instead, track patterns resemble those of the 3 × 4 bound (rotary) or 4 × 4 bound (transverse) (for differentiation, see page 89).

Bound

The bound is a gait characterised by a series of individual jumps strung together. This makes it an extremely agile, nimble gait. The direction, length and height of bounds can change suddenly, making this gait much more flexible than a gallop. Mustelids (except European Badgers), whose bodies are long and tubular with relatively short legs, typically move by bounding. Many rodents are also bounders, as are hares and rabbits.

A key difference between bound and gallop lies in the propulsion generated by the hind legs. In a bound, the hind legs push off almost simultaneously, creating a large amount of propulsion. This results in a moment of extended suspension after the hind legs leave the ground, which serves as the main moment of suspension in the bound. This extended suspension gives the bound its arc-like motion. The basic cycle of footfalls is: hind foot, hind foot, extended suspension, front foot, front foot. The two front feet essentially have a stabilising function. At higher intensities, there can be a second, very short moment of gathered suspension after the two front feet have left the ground. In this case, the cycle of footfalls is: hind foot, hind foot, extended suspension, front foot, front foot, brief gathered suspension. Regardless of whether there are one or two moments of suspension,

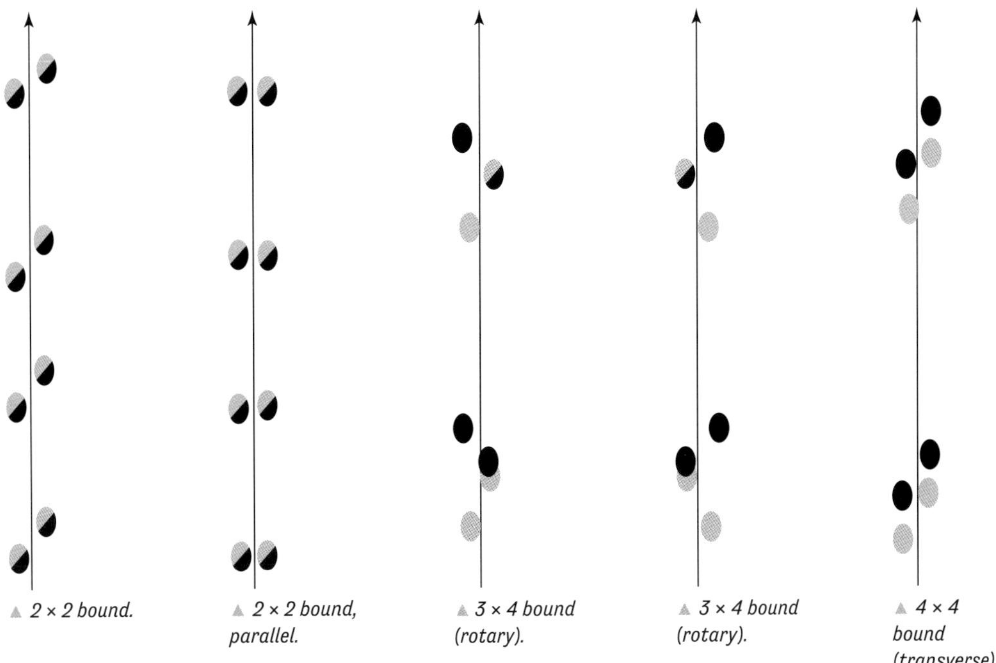

▲ *2 × 2 bound.* ▲ *2 × 2 bound, parallel.* ▲ *3 × 4 bound (rotary).* ▲ *3 × 4 bound (rotary).* ▲ *4 × 4 bound (transverse).*

the extended suspension remains the defining feature of the bound.

As in the gallop, bounding leaves group-forming track patterns consisting of groups of four tracks. These groups of four are separated by inter-group lengths. Different track patterns can result depending on the sequence in which the feet strike the ground.

In the **2 × 2 bound**, both hind feet step into the tracks of the two front feet. This is a 'direct register bound'. Be aware that the two superimposed double prints of a 2 × 2 bound can look like an individual group of two tracks, although it is a group consisting of four tracks. A variation called the **parallel 2 × 2 bound** occurs when both double prints in a cycle of footfalls are placed parallel to each other.

The **3 × 4 bound** has a rotary cycle of footfalls (LF, RF, RH, LH or RF, LF, LH, RH), where the first front foot and hind foot to strike the ground are diagonally opposite each other. The first hind foot to strike the ground can fully or partly cover the track of the second front foot, creating a double print. This overlapping can cause the track pattern to appear as a group of three tracks, giving the gait its name.

The **4 × 4 bound** has a transverse cycle of footfalls (LF, RF, LH, RH or RF, LF, RH, LH), where the first front foot and hind foot to strike the ground are on the same side. Four footprints are distinctly visible in the 4 × 4 bound.

The **parallel bound** can leave many different track patterns. As with all other bounds, the hind legs generally move at almost the same time. Both hind foot tracks of a cycle of footfalls are more or less parallel to each other. If the front leg pair moves at the same time, this is described as a full/paired bound. The full/paired bound is characteristic of tree-dwelling rodents but can also be used by other rodents and lagomorphs. Along with gallop, the parallel bound is one of the few gaits where both hind feet can overstep both front feet. It can combine high speed with extreme agility.

The **hop** is a type of slow bound, technically classified as an 'understep parallel bound'. As in the parallel bound, both hind legs move simultaneously, so the two hind foot tracks of a cycle of footfalls are nearly parallel. If the front leg pair move at the same time, this is known as a full/paired hop. In the hop, both hind foot tracks remain behind both front foot tracks. As the hind feet land further forward in the direction of travel relative to the previous prints, they approach the position of the front foot prints, until all four prints are nearly aligned in a single line. This 'direct register variation' is the fastest possible form of hop or a slow, **gathered parallel bound** and is typical of small rodents moving in the snow. If both hind feet land even further forward so they overstep the front foot tracks, the track pattern transitions into that of a parallel bound.

The **offset bound** is a distinctive gait. It can be confused with the track pattern of gallop, although the two are clearly distinguishable. In the offset bound, the hind feet and front feet are offset to the front, although the primary propulsion still comes from the hind legs. Consequently, the hind foot tracks are more defined in the track pattern. Furthermore, the tracks of the hind feet are much less offset in the direction of travel than in a typical gallop. The other differences

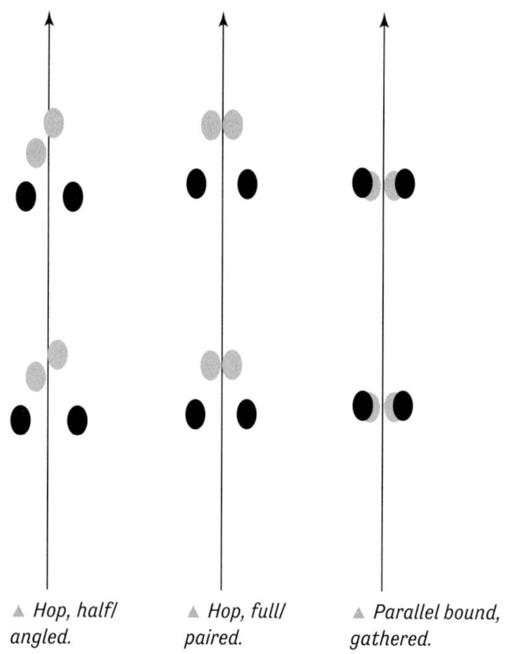

▲ *Hop, half/angled.* ▲ *Hop, full/paired.* ▲ *Parallel bound, gathered.*

to a gallop track pattern are similar to those that distinguish between the 3 × 4 bound (rotary) and the 4 × 4 bound (transverse) and the lope (see page 89). It should be noted that the offset bound can also be executed with a transverse or rotary cycle of footfalls.

The **pronk** is an unusual form of bound, where the animal pushes itself off the ground with all four feet simultaneously and lands again on all four feet at the same time. As a result, it can achieve a considerable height. The additional height may provide a better vantage point or offer greater flexibility in changing direction (Leonard Lee Rue 2013). The pronk is typical of Common Fallow Deer. Along with the hop, it's one of the only other bounds where the two hind feet stay behind the two front feet.

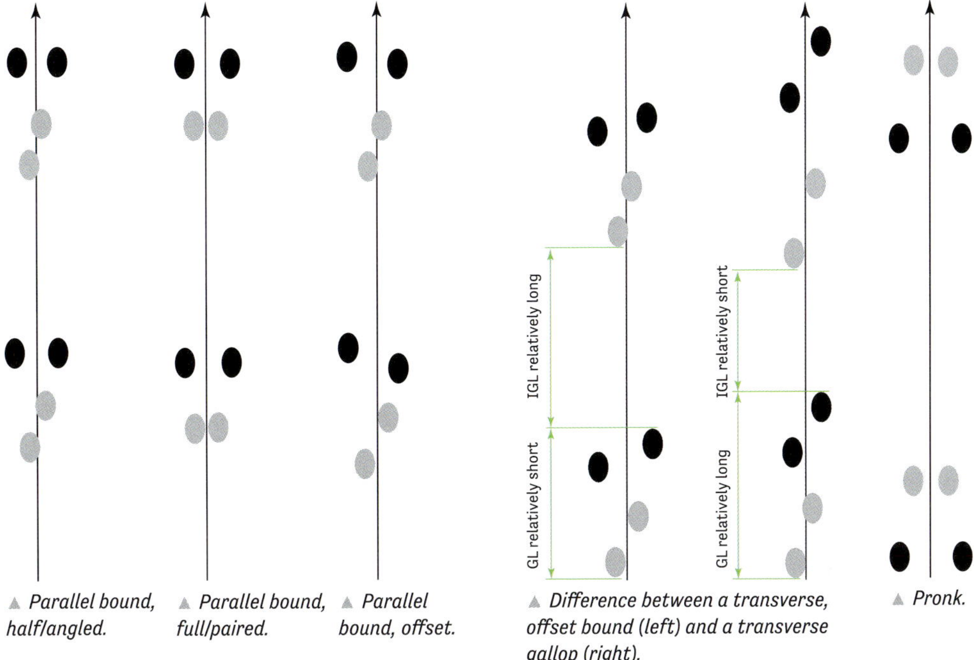

▲ *Parallel bound, half/angled.*

▲ *Parallel bound, full/paired.*

▲ *Parallel bound, offset.*

▲ *Difference between a transverse, offset bound (left) and a transverse gallop (right).*

▲ *Pronk.*

Differentiating between 3 × 4 bound (rotary), 4 × 4 bound (transverse) and lope

The boundaries between the 3 × 4 bound, 4 × 4 bound and the lope can be unclear in the track pattern. For this reason, it isn't always easy to distinguish between a lope and a bound in the field. Nevertheless, there are certain key indicators that can help:

- **Group lengths**: In a lope, group lengths are relatively large due to the additional propulsion provided by the front legs. In contrast, bounds typically exhibit shorter group lengths.
- **Track pronunciation**: The track pattern of a lope often features more pronounced front foot prints, reflecting the extra force generated by the front legs. Conversely, in a bound, the hind foot prints are usually more defined.
- **Inter-group lengths**: The inter-group lengths in a lope are comparatively short relative to the group lengths and are shorter than those found in bounds.

These differences, while subtle, can aid in identifying the gait when observing track patterns in the field.

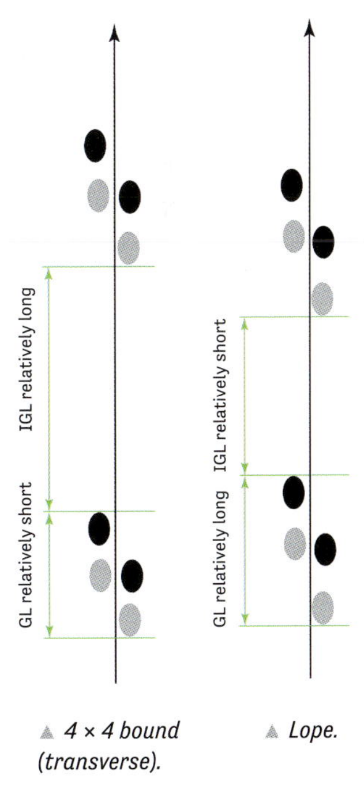

▲ *4 × 4 bound (transverse).*

▲ *Lope.*

Personal experiences and comment

The many different gaits and their names can cause confusion. Throughout my many years of tracking, I've learned that animals place their feet in a wide range of ways and ignore the categories people have made for them. As much as I value precise language and conceptual organisation, nature has taught me that it can't always be neatly classified. Every schematic system must deal with a continuum of variations, including in the gaits of quadrupeds (Hildebrand 2004). Experienced observers can differentiate between more than 60 gaits by eye and specialists describe even more. In this book, I've focused on the gaits that can be clearly recognised and described by their track patterns. One of South Africa's most respected and well-known trackers, Oom Vet Piet († 27 March 2003), famously preferred to imitate gaits with hand gestures, rather than try to describe them with words.

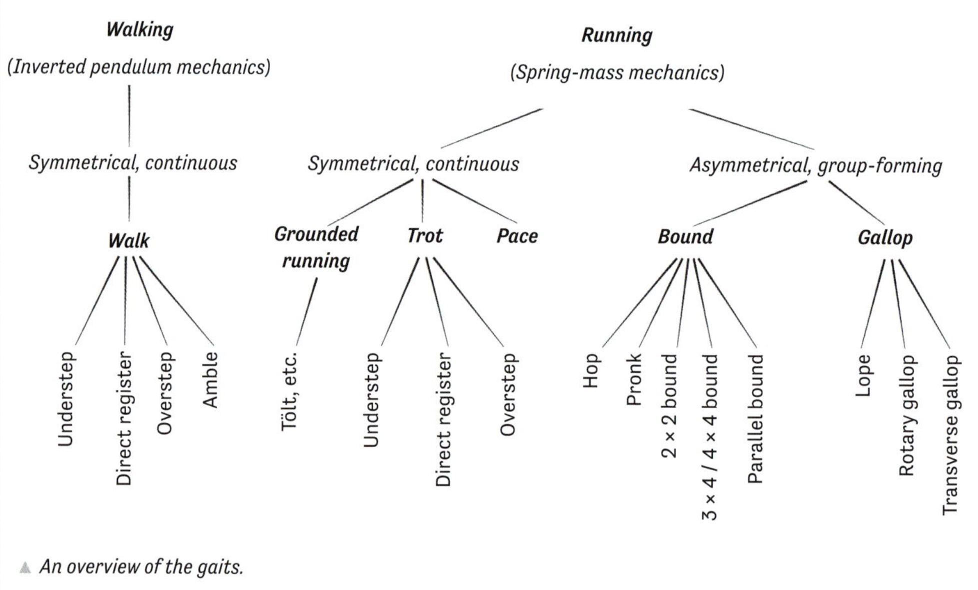

▲ An overview of the gaits.

IRREGULARITIES IN THE TRACK PATTERN

If we follow tracks over longer distances, we will most likely come across an irregularity in the track pattern. Even animals that cover very large distances will sooner or later need to stop for some reason. At first glance, this can make interpretation more difficult, but there are some characteristic irregularities that indicate a specific behaviour.

T stop

An animal moving in a walk or trot might make temporary stops, often leaving a T-shaped track pattern. During such a pause, the animal briefly hesitates when moving a front leg forward and puts it down next to the standing front leg, in a similar manner to how we put one foot next to the other when we stand still. The vertical line of the 'T' is part of the normal track pattern; the horizontal line is created

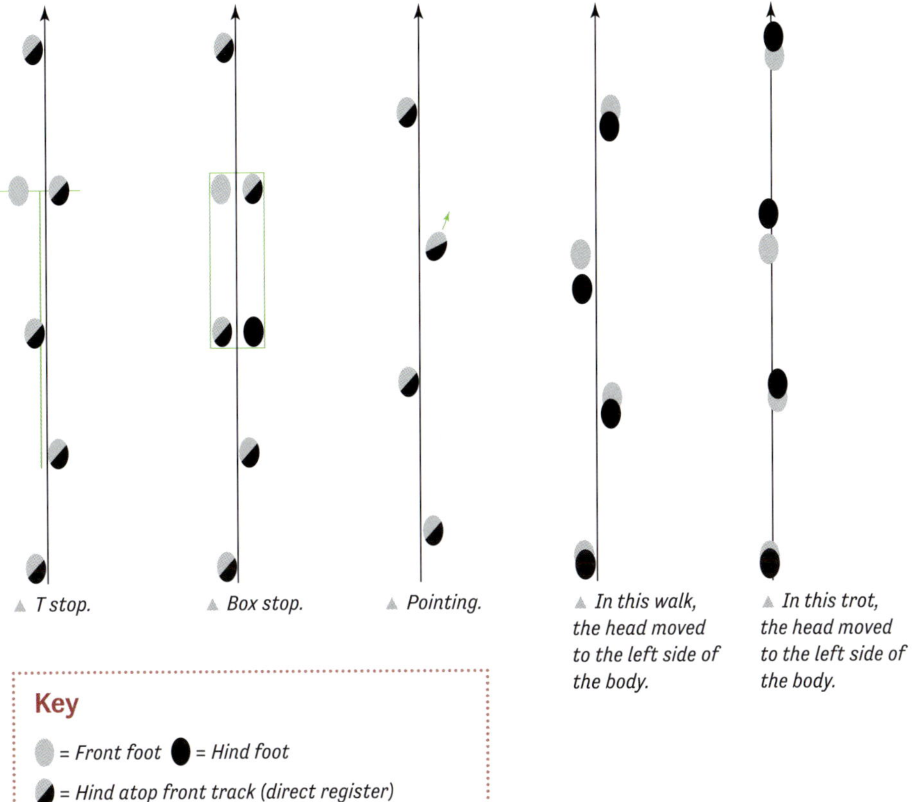

▲ *T stop.*   ▲ *Box stop.*   ▲ *Pointing.*   ▲ *In this walk, the head moved to the left side of the body.*   ▲ *In this trot, the head moved to the left side of the body.*

Key

= *Front foot*   = *Hind foot*

= *Hind atop front track (direct register)*

by the two adjacent front foot tracks. This pattern is often found before it crosses a path or open space. After the brief interruption, the normal track pattern will resume or there will be a change in gait and/or direction.

Box stop

The box stop is a variation of the T stop. In contrast to the T stop, the animal puts down its two hind feet, in addition to the front feet. The two individual prints are normally fainter than the two diagonally opposite double prints. A deep and clearly visible, single hind foot print indicates a longer stop, while a more superficial single hind foot print indicates a brief pause.

Pointing

In a walk, a front foot print might stand out clearly because its position is diagonal to the direction of travel. This diagonal placement often 'points' to an object of interest, for example an edible plant. As with the T stop, the animal might also put down the second front foot. Following Western Roe Deer walking tracks often provides ample opportunities to observe this kind of pointing. Look in the direction in which the track is pointing and you will often be able to find signs of feeding at approximately the height of the animal's head. Pointing prints can be single or double prints.

Determining the head position

In a direct register walk and trot, certain irregularities in the track pattern can point to more pronounced head movements. When an animal looks over its shoulder when walking, it will often leave three consecutive irregular double prints. The hind foot on the side to which the head was turned lags behind (see diagram on page 91). The reverse is true in a trot, where the same head movement often leads to slight overstepping. The front foot on the side to which the head was turned lands slightly further back to compensate for the resulting shift in weight, causing the corresponding hind foot to overstep (see diagram on page 91).

The position of the head can also be determined by a deeper individual front foot track (Brown 1999). This occurs because the additional weight of the head causes the front foot print on the side to which the head was turned to press deeper into the ground. The T stop is useful for trying out this method of evaluation because the individual front tracks are easy to make out.

MEASURING TRACK PATTERNS

Measuring track patterns enables us to assign certain gaits to them. This, in turn, is an important key to understanding animal behaviour. To measure track patterns, we measure stride length, trail width, group length and inter-group length.

STRIDE LENGTH (SL)

Stride length is recorded in all gaits. Stride length in symmetrical gaits is the distance between two consecutive touchdowns of the same foot. To determine the stride length, pick a clearly visible point in the track and measure to the same point in the following track made by the same foot. For example, you could either measure from the front outside edge to the front outside edge, or from the back outside edge to the back outside edge. It is important to work consistently and to always use consecutive prints of the same foot as a measuring point. Stride length in all asymmetrical gaits is calculated by adding together the group length and the inter-group length. Note: In 'man tracking', the step is also measured. Some animal trackers do this too. The data is still comparable provided you mention how they were measured.

'The stride is the distance between two consecutive touchdowns of the same foot.'

TRAIL WIDTH (TW)

The trail width is the widest point of a track pattern. To measure the trail width in continuous track patterns (all symmetrical gaits), place a measuring tape or ruler on the outside edge of two consecutive tracks made by the same foot. Then measure to the outside edge of the next footprint on the opposite side, ensuring the measurement is taken at a right angle to the direction of travel. In the case of group-forming track patterns (all types of bound and gallop), measure the distance between the outside edges of the outermost tracks within a group length, also at a right angle to the direction of travel.

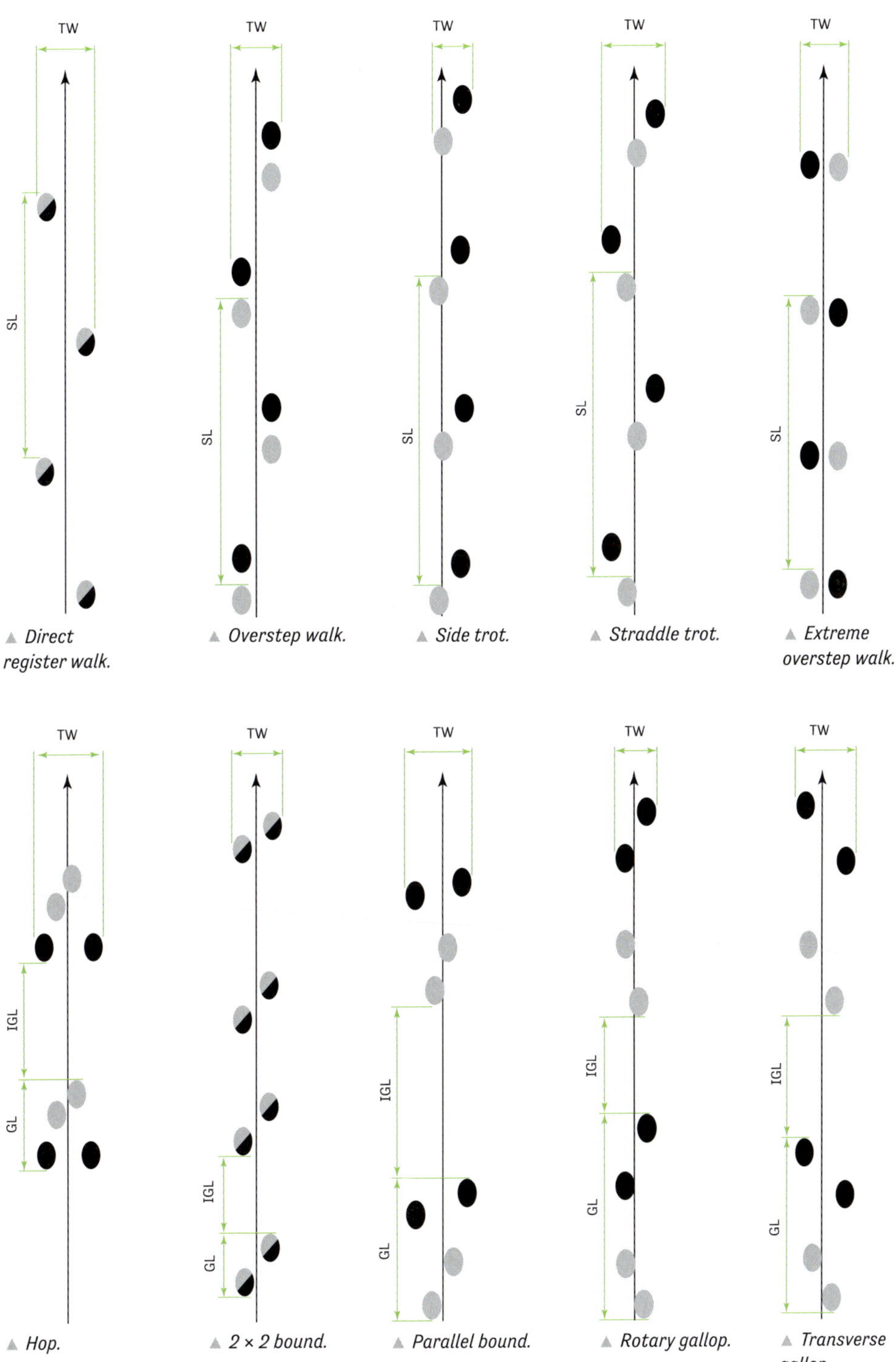

▲ Direct register walk.

▲ Overstep walk.

▲ Side trot.

▲ Straddle trot.

▲ Extreme overstep walk.

▲ Hop.

▲ 2 × 2 bound.

▲ Parallel bound.

▲ Rotary gallop.

▲ Transverse gallop.

Gait terminology

Walk
- Direct register walk
- Understep walk
- Overstep walk

Trot
- Direct register trot
- Overstep trot
- Side trot
- Straddle trot

Bound
- Half/angled bound
- Full/paired bound
- Half/angled hop
- Full/paired hop
- Pronk
- 2 × 2 bound
- 3 × 4 bound (rotary)
- 4 × 4 bound (transverse)

Gallop
- Rotary gallop
- Transverse gallop

Lope
- 4 × 4 lope (rotary)
- 4 × 4 lope (transverse)

GROUP LENGTH (GL)

The group length of a group of four tracks is measured from the back outside edge of the first footprint to the front outside edge of the last footprint of the same group, parallel to the direction of travel. This measurement encompasses all four tracks in the group.

INTER-GROUP LENGTH (IGL)

The inter-group length measures the distance between two group lengths of consecutive groups of four tracks. It is measured from the front outside edge of the last footprint of one group length to the back outside edge of the first footprint of the next group length. The tracks are not included in this measurement.

GUIDE TO IDENTIFYING MAMMAL TRACKS

Below is a guide to identifying individual mammal tracks, organised from small to large. Almost all track drawings are life-size. The exceptions are reduced to 50 or 80 per cent of their actual size and labelled accordingly.

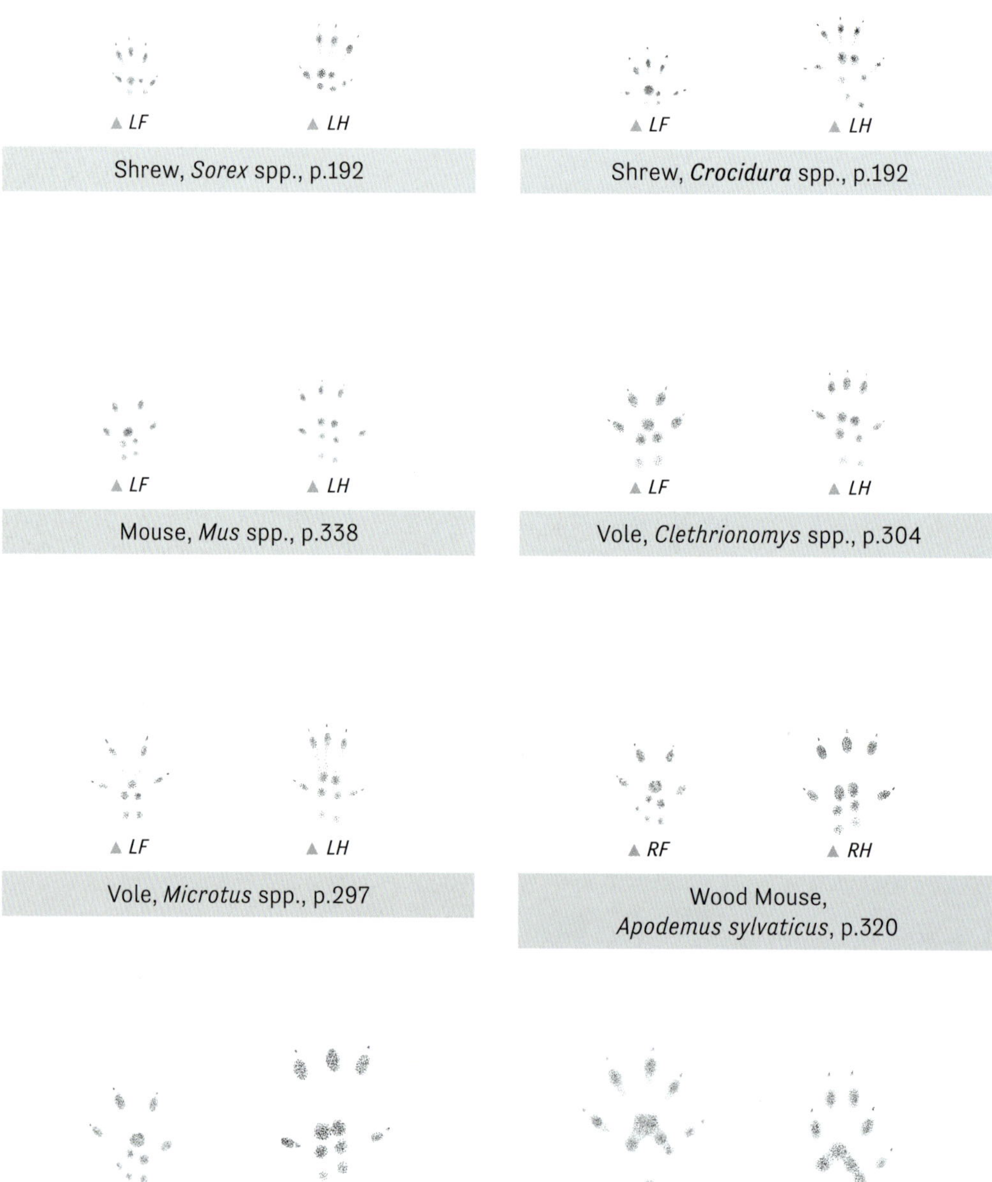

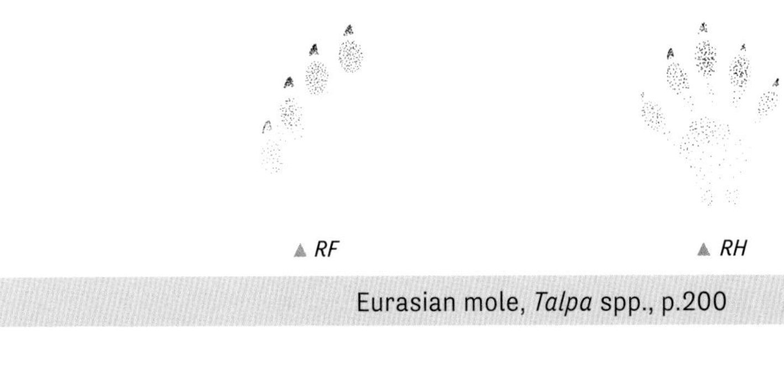

Eurasian mole, *Talpa* spp., p.200

Water vole, *Arvicola* spp., p.286

Black Rat, *Rattus rattus*, p.344

Brown Rat, *Rattus norvegicus*, p.344

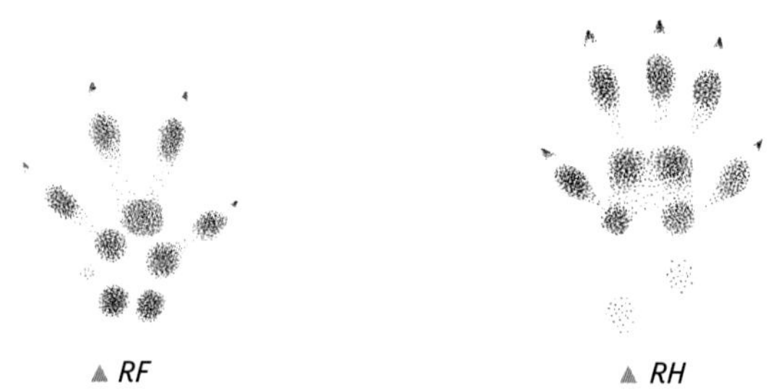

Common Hamster, *Cricetus cricetus*, p.312

Souslik, *Spermophilus* spp., p.251

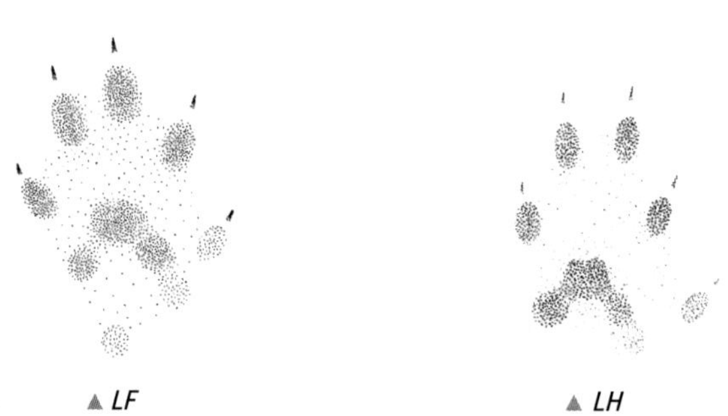

Stoat, *Mustela erminea*, p.469

Edible Dormouse, *Glis glis*, p.256

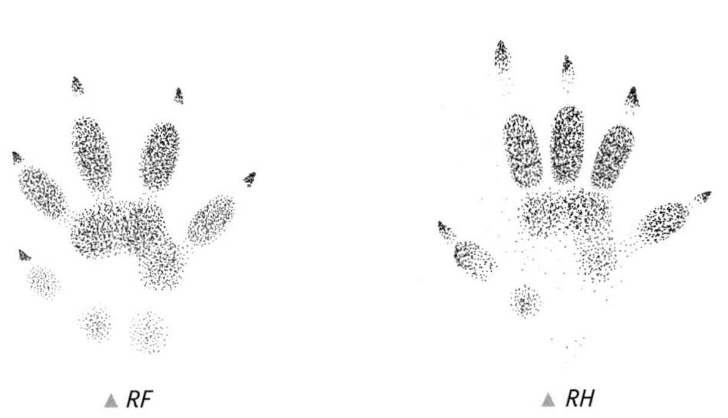

West European Hedgehog, *Erinaceus europaeus*, p.188

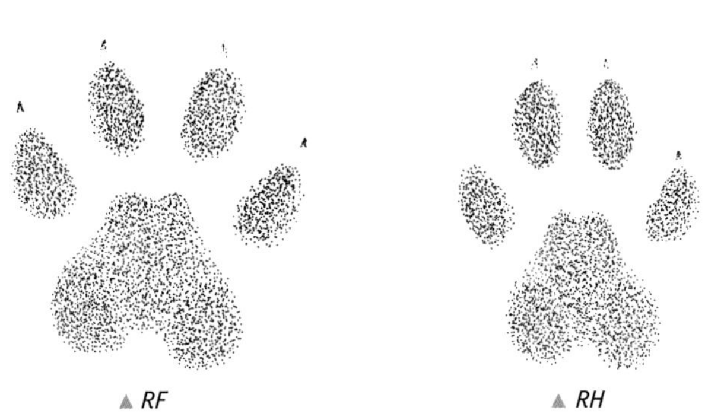

Common Genet, *Genetta genetta*, p.376

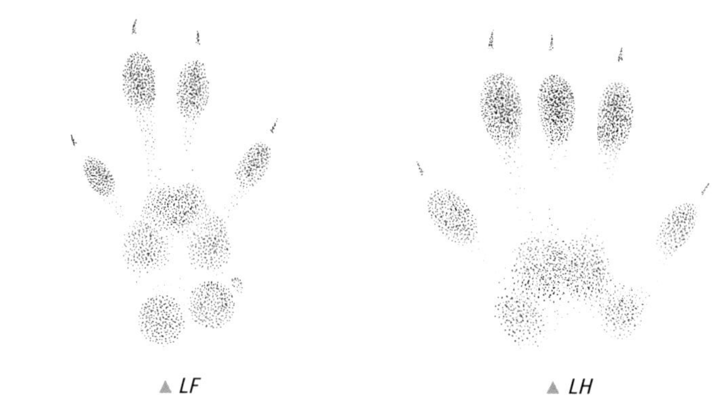

▲ *LF* ▲ *LH*

Squirrel, *Sciurus* spp., p.237

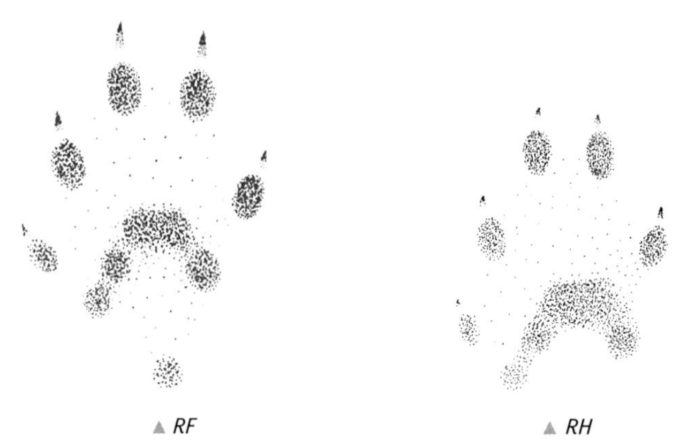

▲ *RF* ▲ *RH*

European Polecat, *Mustela putorius*, p.464

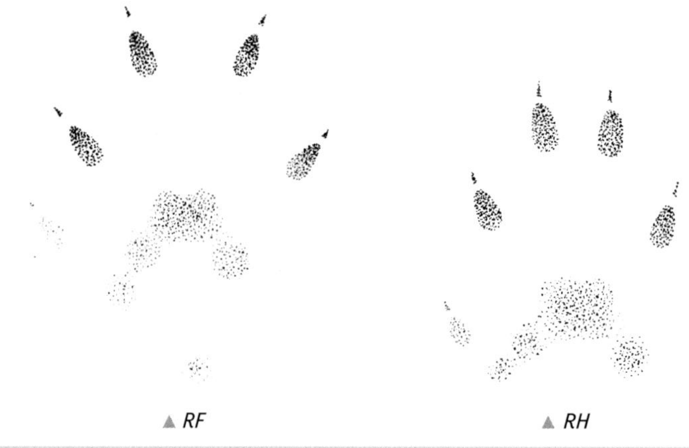

▲ *RF* ▲ *RH*

American Mink, *Neovison vison*, p.476

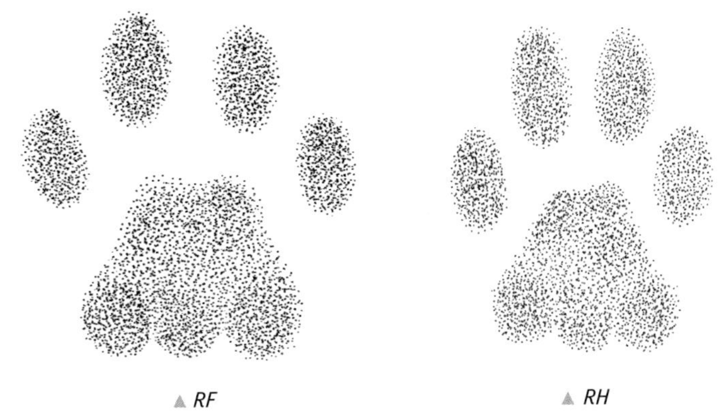

▲ *RF*　　　　　▲ *RH*

European Wildcat, *Felis silvestris*, p.371

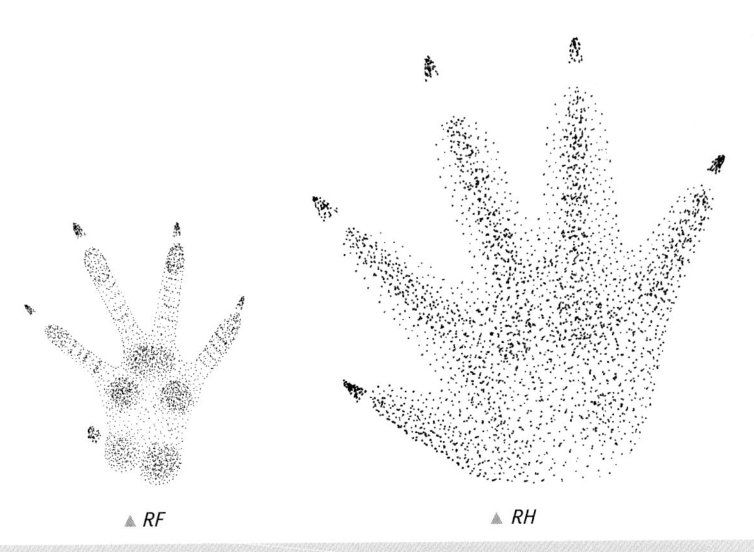

▲ *RF*　　　　　▲ *RH*

Muskrat, *Ondatra zibethicus*, p.280

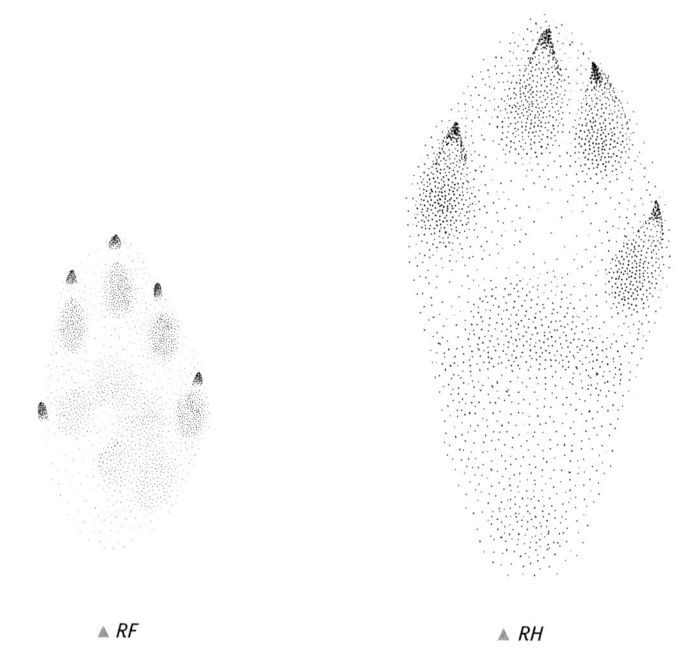

▲ *RF*   ▲ *RH*

European Rabbit, *Oryctolagus cuniculus*, p.226

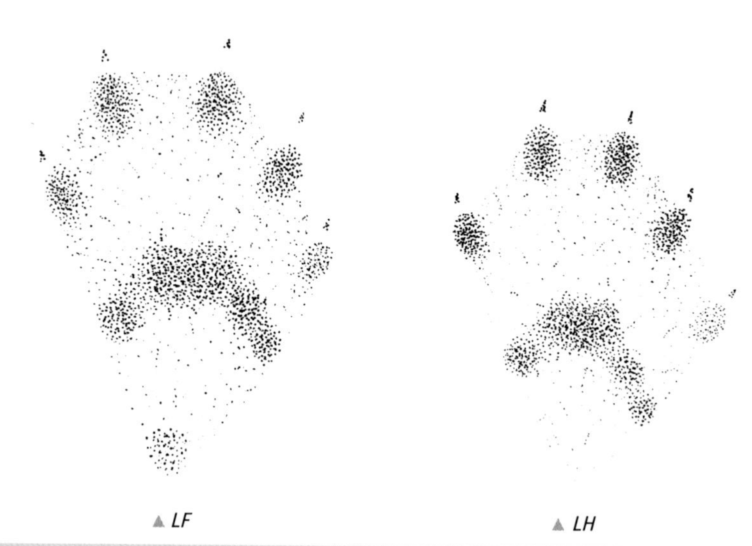

▲ *LF*   ▲ *LH*

Pine Marten, *Martes martes*, p.452

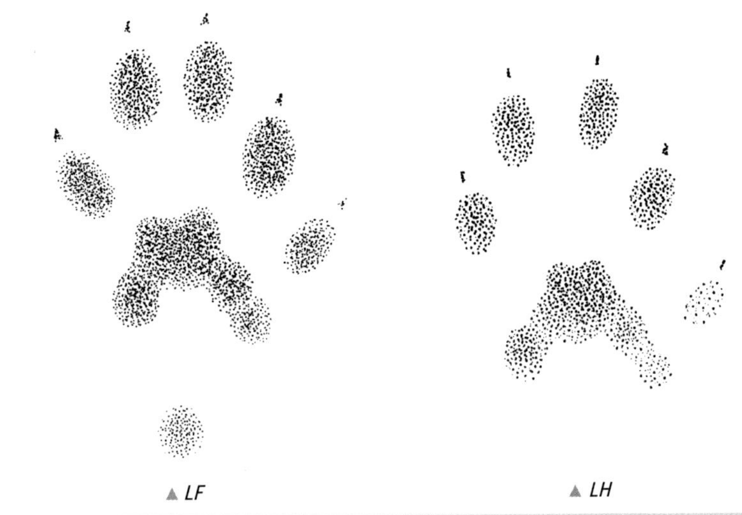

▲ LF   ▲ LH

Beech Marten, *Martes foina*, p.457

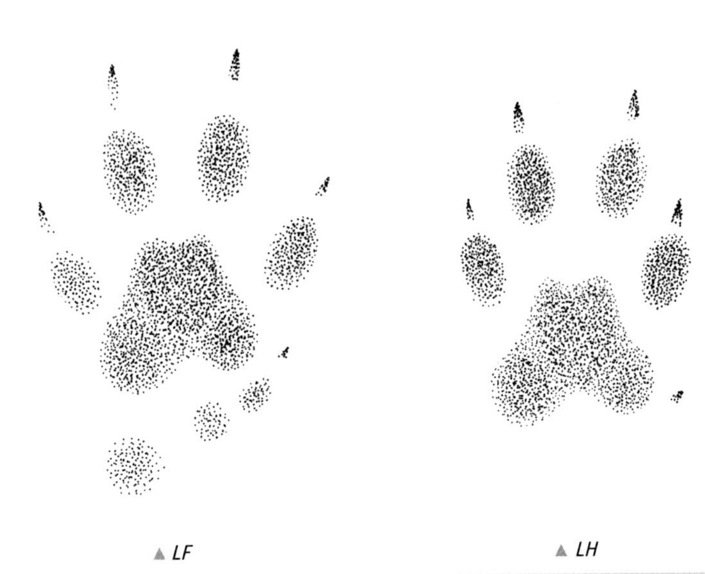

▲ LF   ▲ LH

Egyptian Mongoose, *Herpestes ichneumon*, p.380

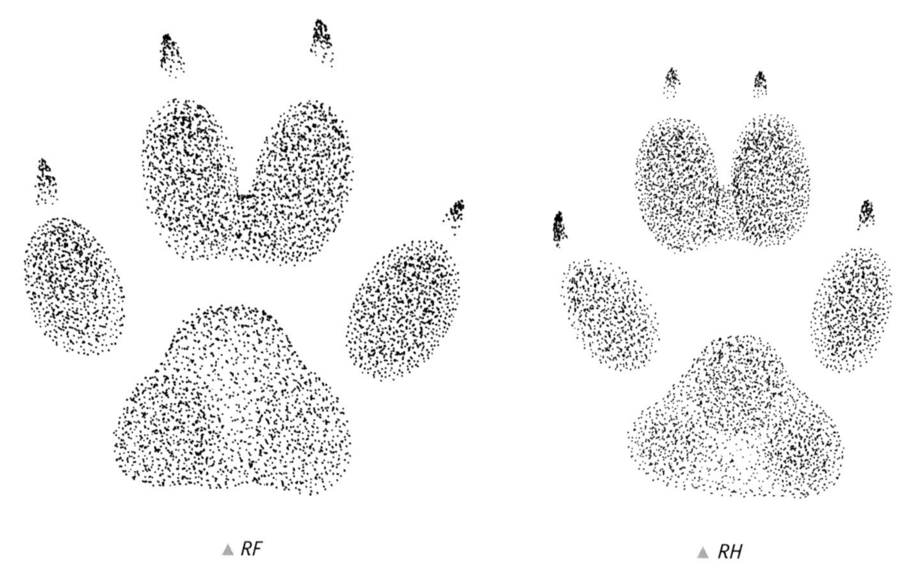

▲ *RF*　　　　　　▲ *RH*

Common Raccoon Dog, *Nyctereutes procyonoides*, p.402

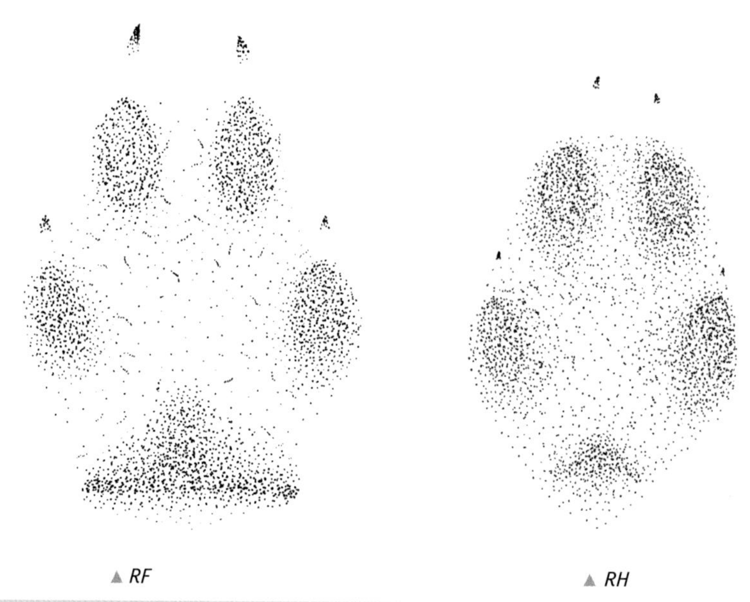

▲ *RF*　　　　　　▲ *RH*

Red Fox, *Vulpes vulpes*, p.411

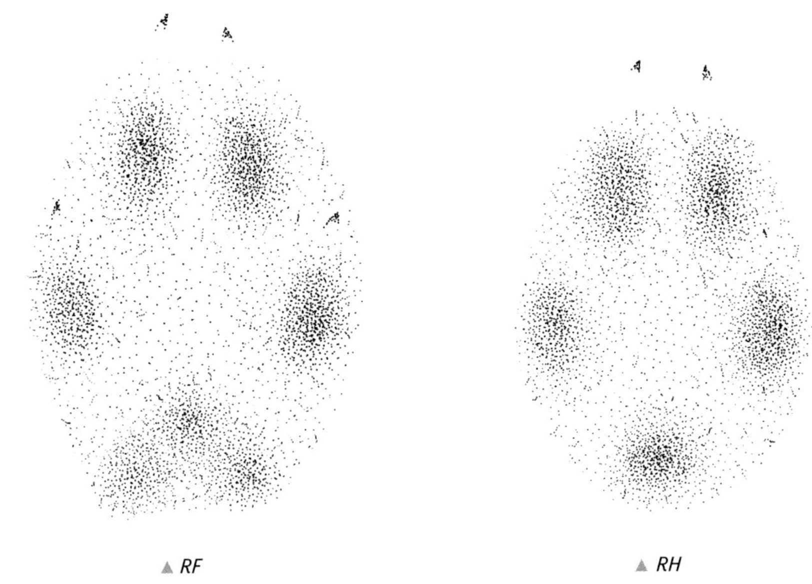

Arctic Fox, *Vulpes lagopus*, p.406

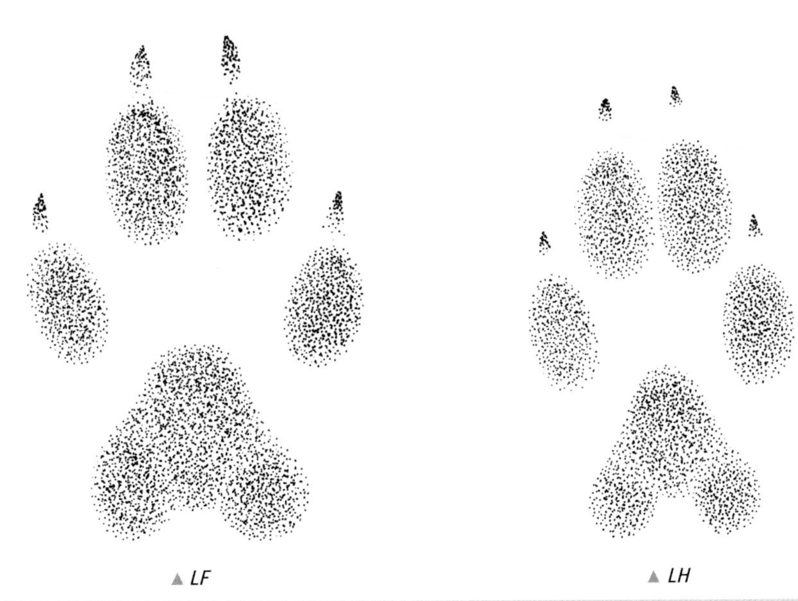

Golden Jackal, *Canis aureus*, p.386

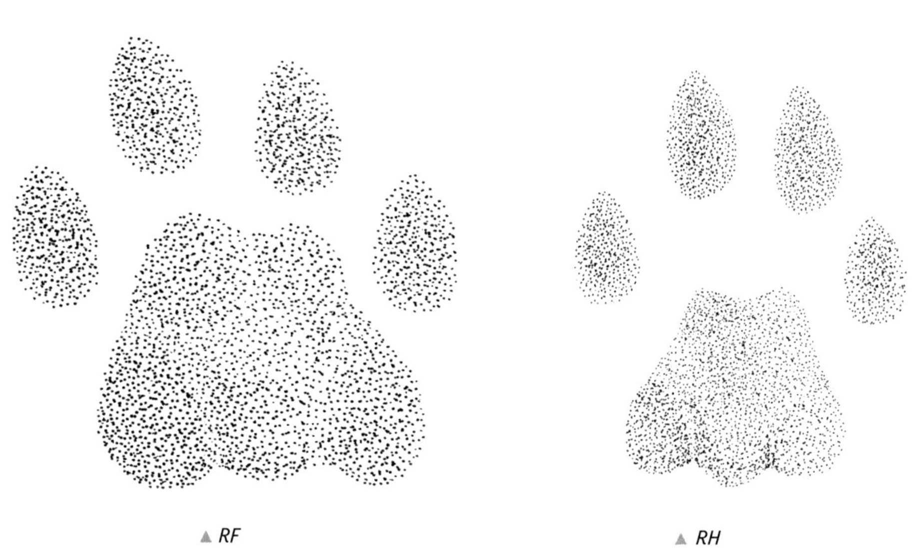

▲ RF　　　▲ RH

Iberian Lynx, *Lynx pardinus*, p.364

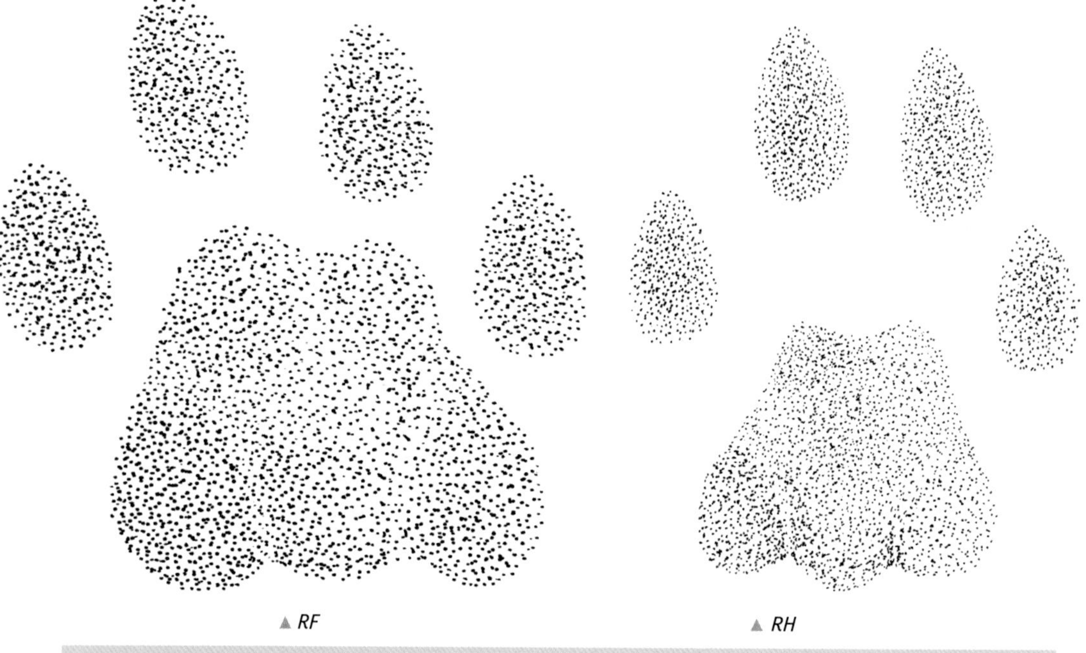

▲ RF　　　▲ RH

Eurasian Lynx, *Lynx lynx*, p.364

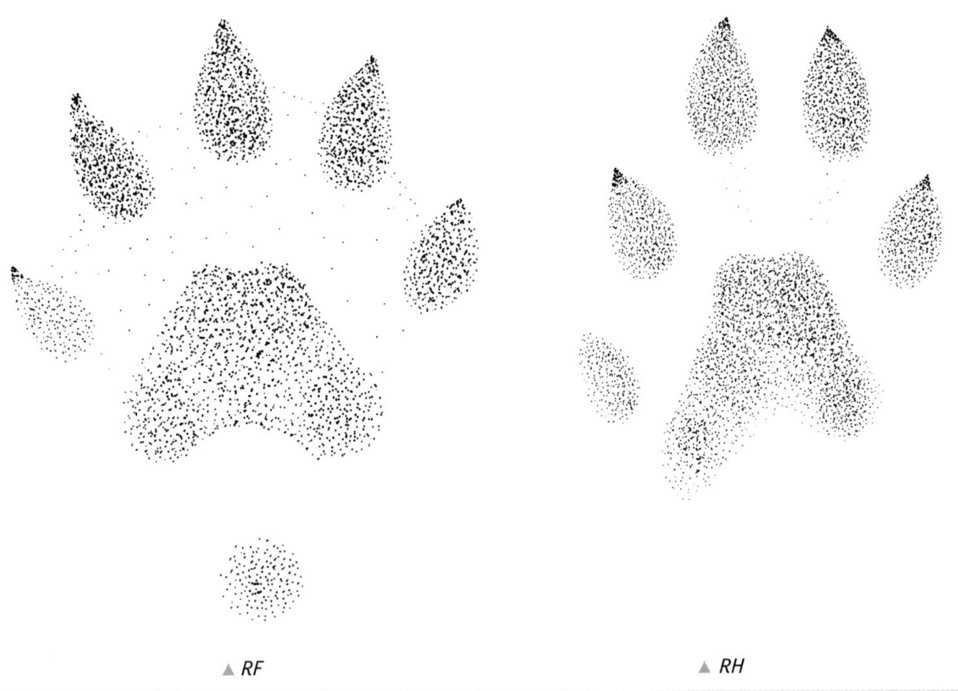

▲ RF ▲ RH

Eurasian Otter, *Lutra lutra*, p.430

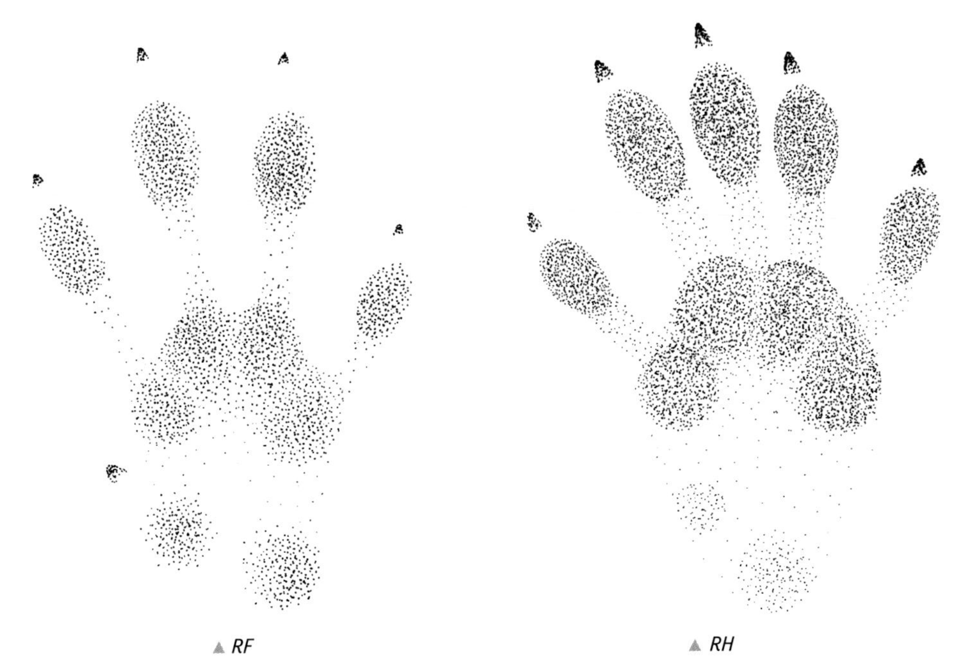

▲ RF ▲ RH

Alpine Marmot, *Marmota marmota*, p.246

Crested Porcupine, *Hystrix cristata*, p.356

▲ *LF*

▲ *LH*

European Badger, *Meles meles*, p.444

▲ *RF*

▲ *RH*

Northern Raccoon, *Procyon lotor*, p.484

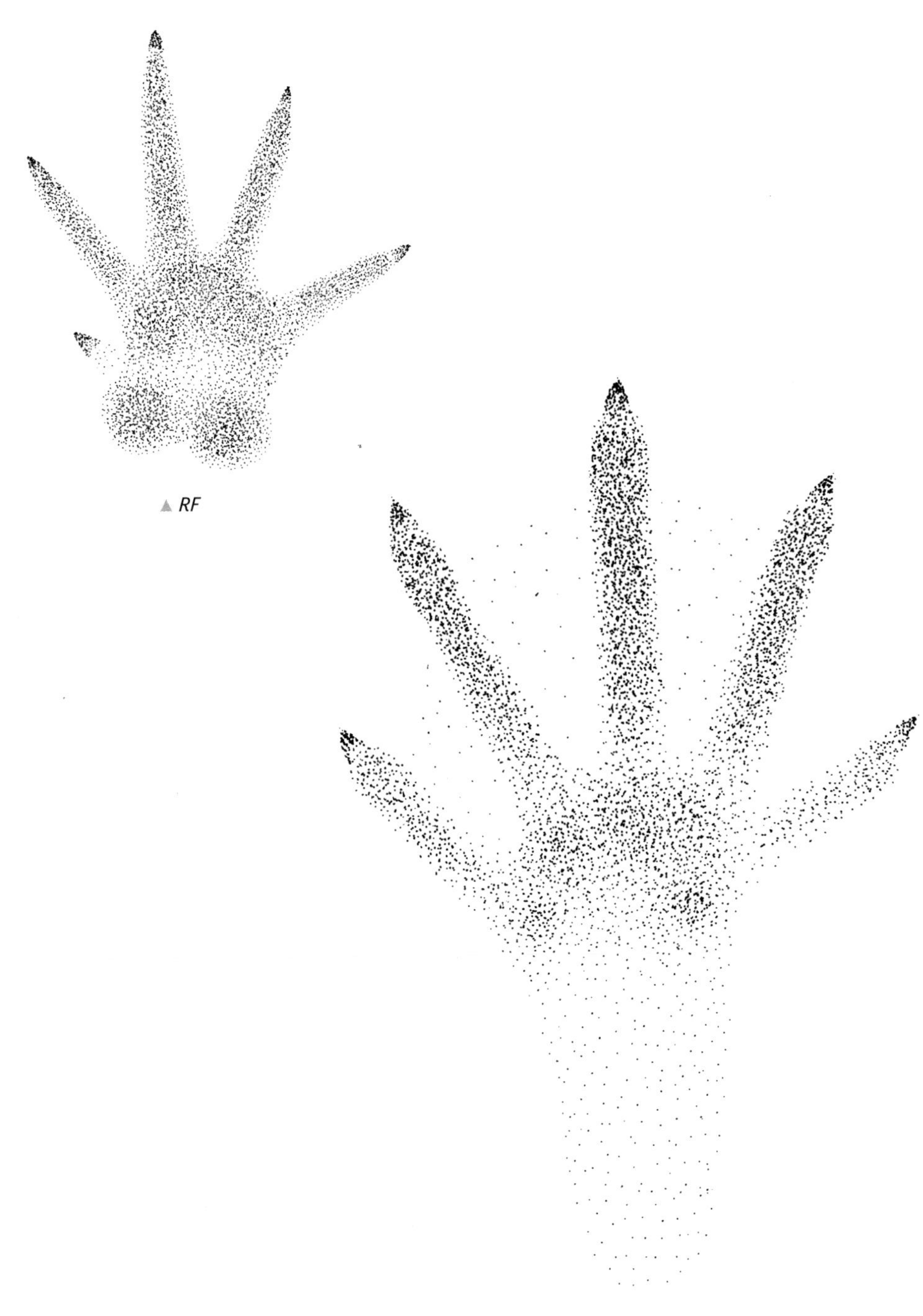

Coypu, *Myocastor coypus*, p.270

Brown Hare, *Lepus europaeus*, p.214

Mountain Hare, *Lepus timidus*, p.220

Domestic Dog, *Canis lupus familiaris*, p.390

Grey Wolf, *Canis lupus*, p.390

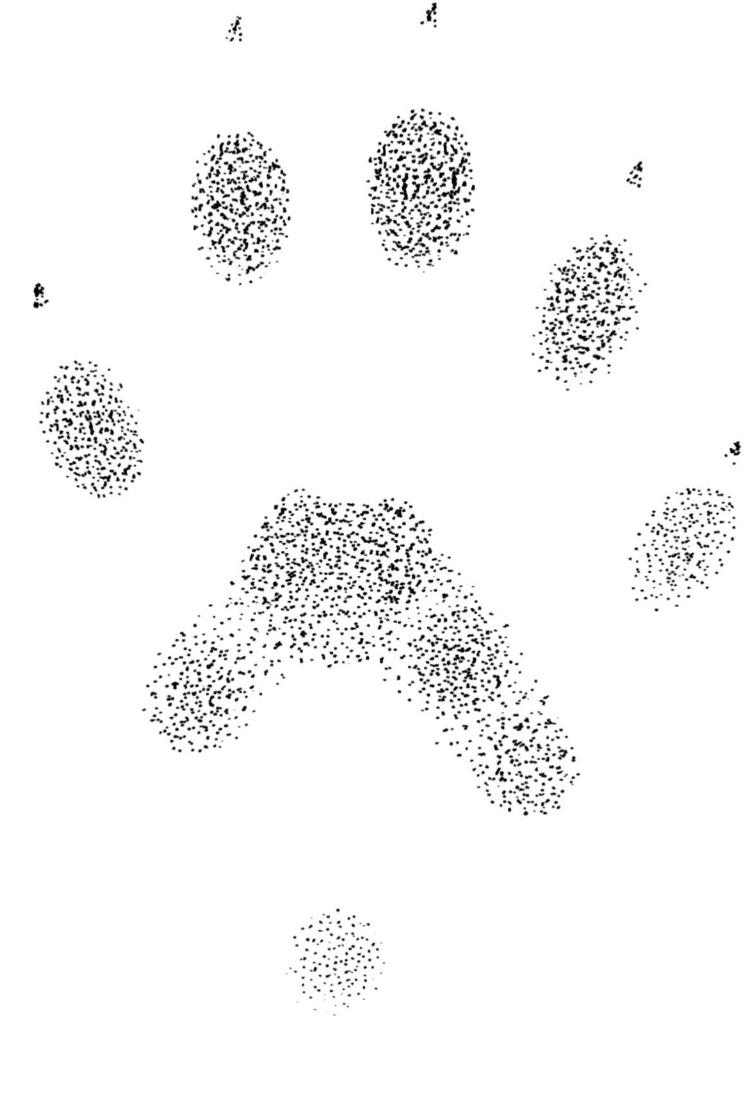

Wolverine, *Gulo gulo*, p.438

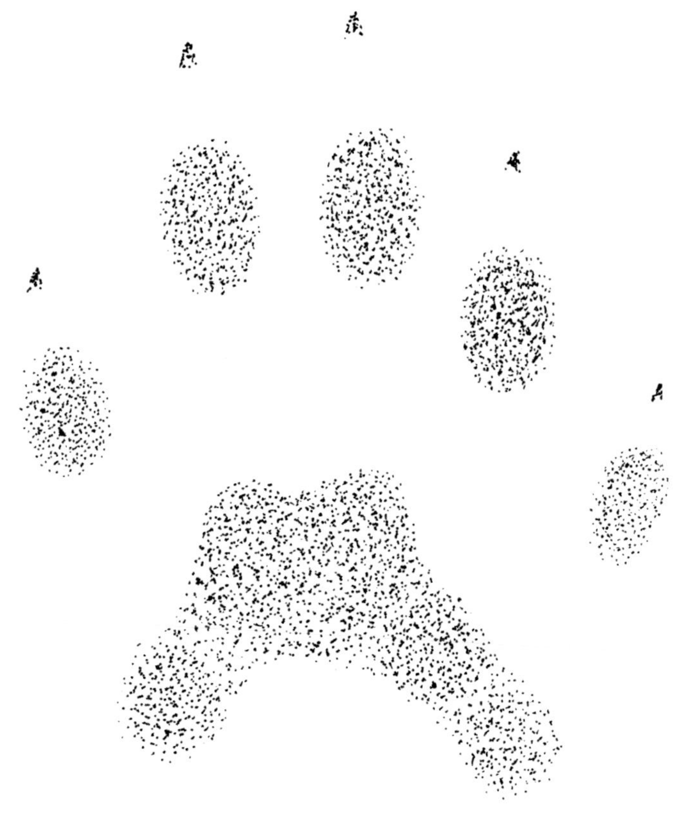

▲ *LH*

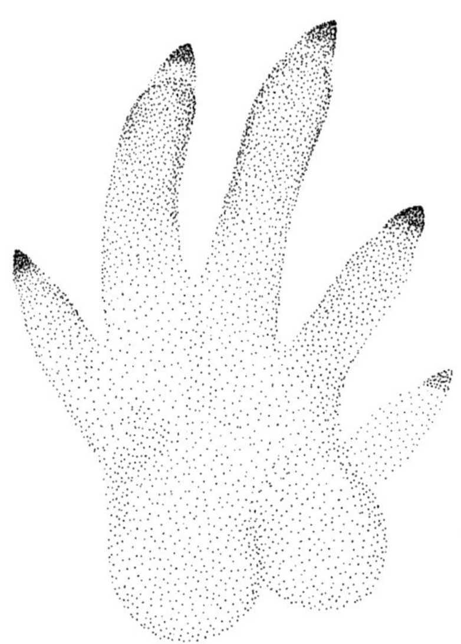

Eurasian Beaver, *Castor fiber*, p.262

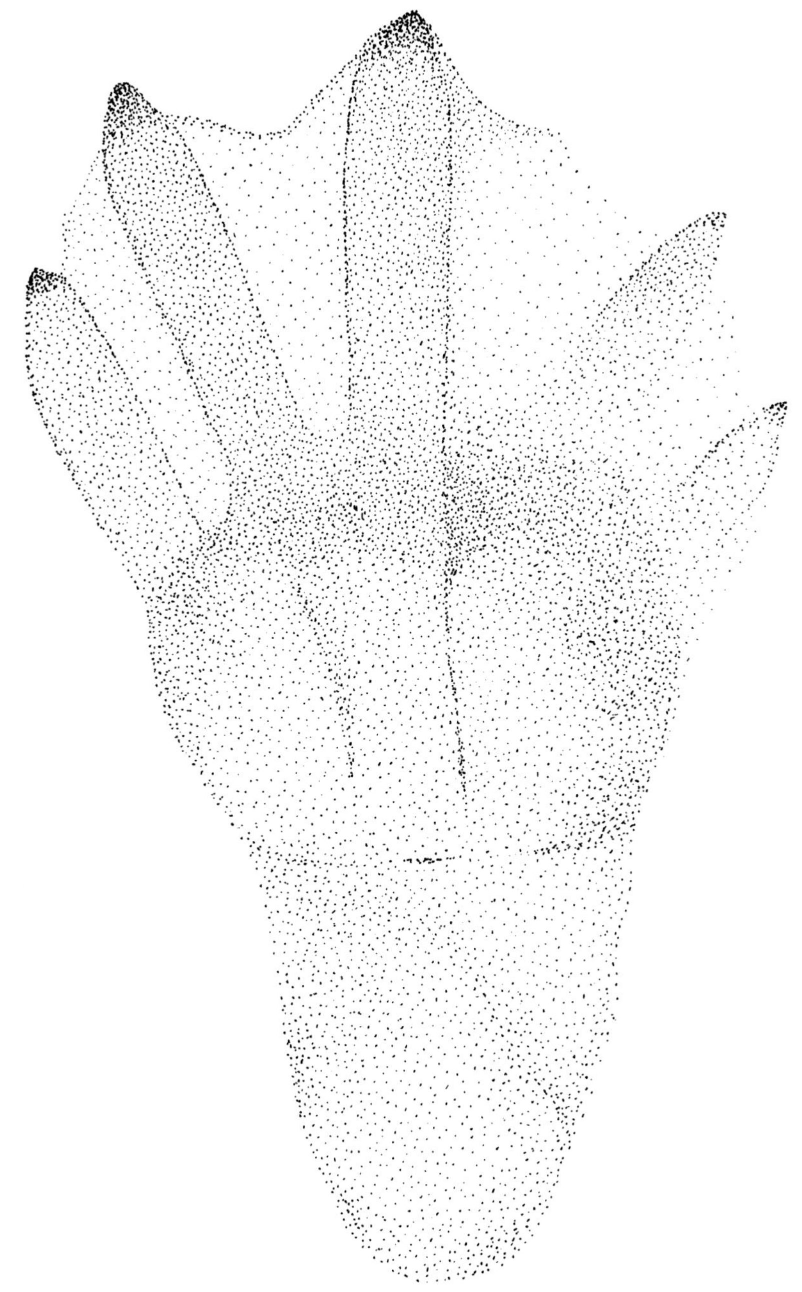

▲ LH

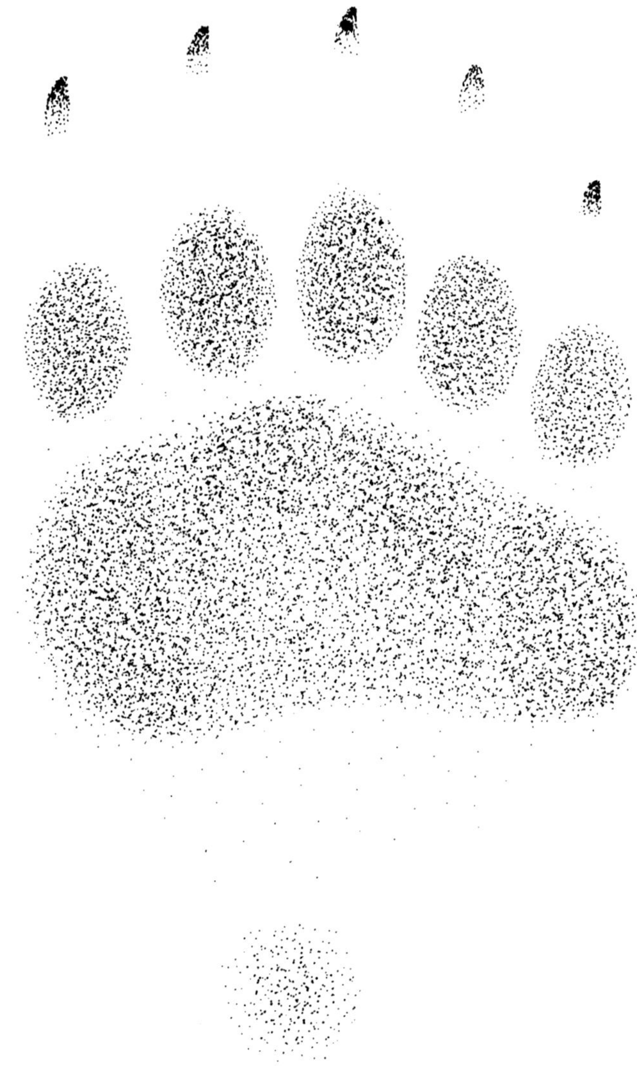

▲ LF – 50 per cent of original size

Brown Bear, *Ursus arctos*, p.422

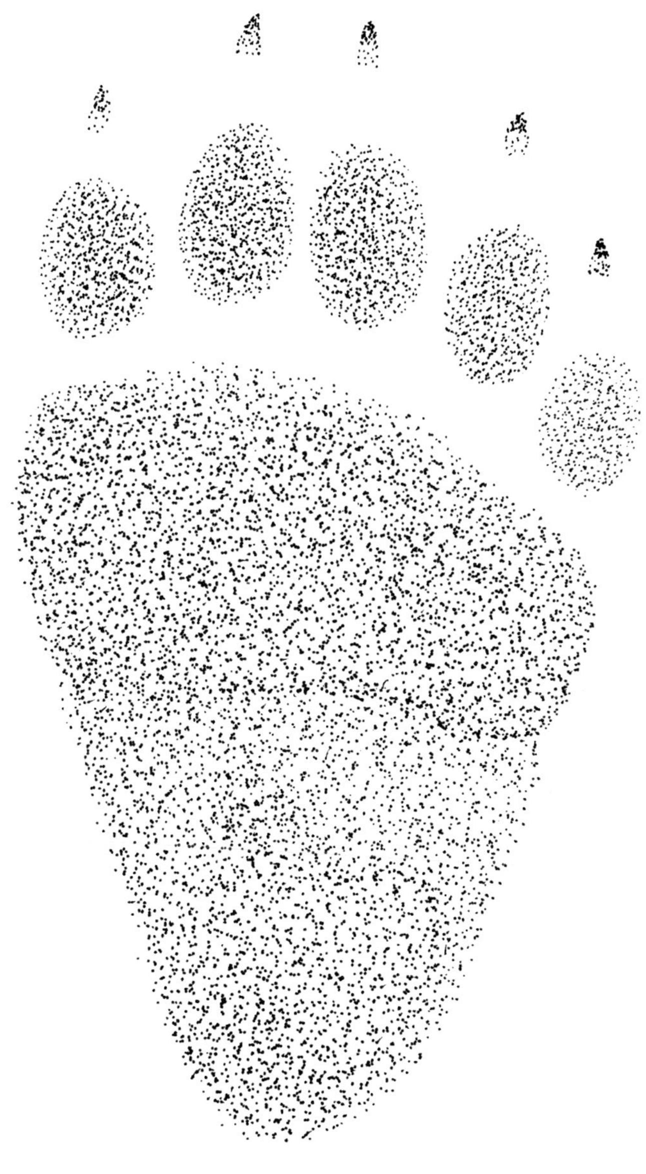

▲ *LH – 50 per cent of original size*

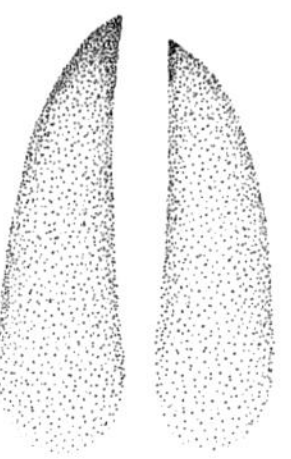

▲ *LF*

Reeves's Muntjac, *Muntiacus reevesi*, p.558

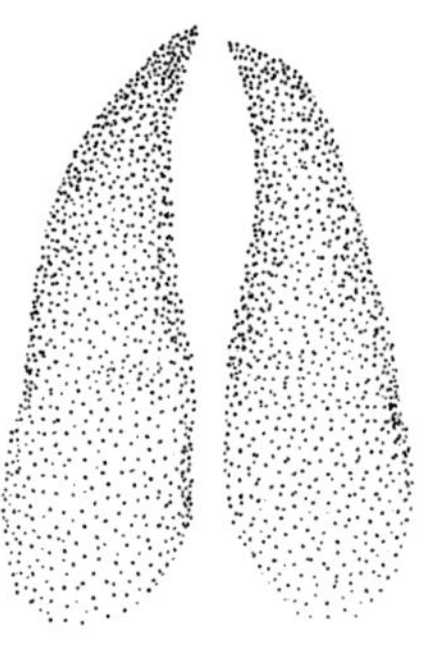

▲ *LH*

Water Deer, *Hydropotes inermis*, p.564

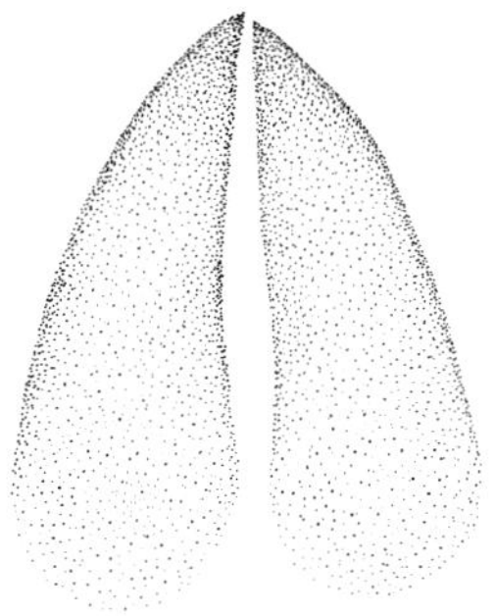

▲ *LH*

Western Roe Deer, *Capreolus capreolus*, p.518

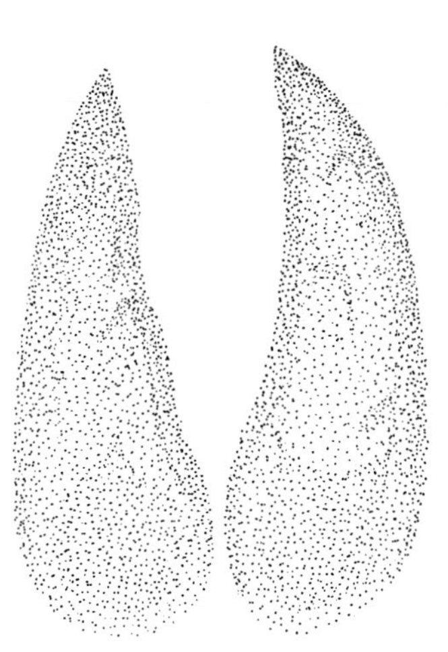

▲ *RF*

Mouflon, *Ovis gmelini musimon*, p.592

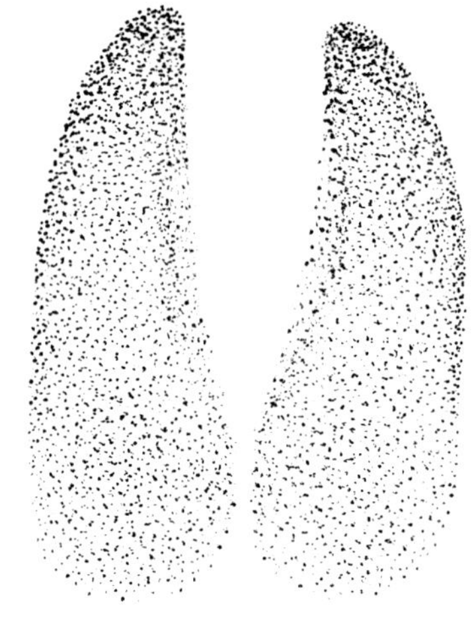

Domestic Goat, *Capra aegagrus hircus*, p.596

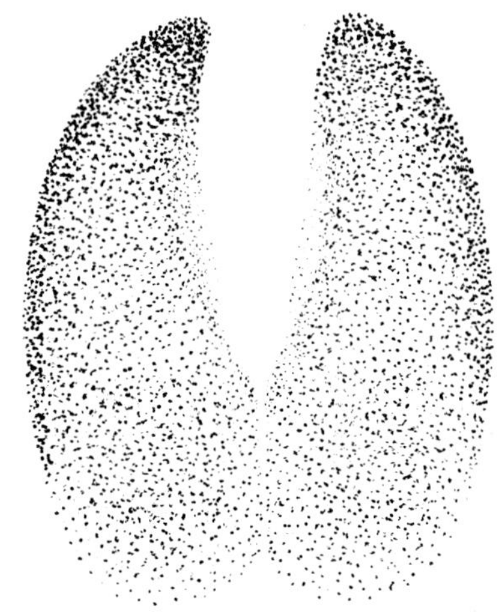

Domestic Sheep, *Ovis gmelini aries*, p.596

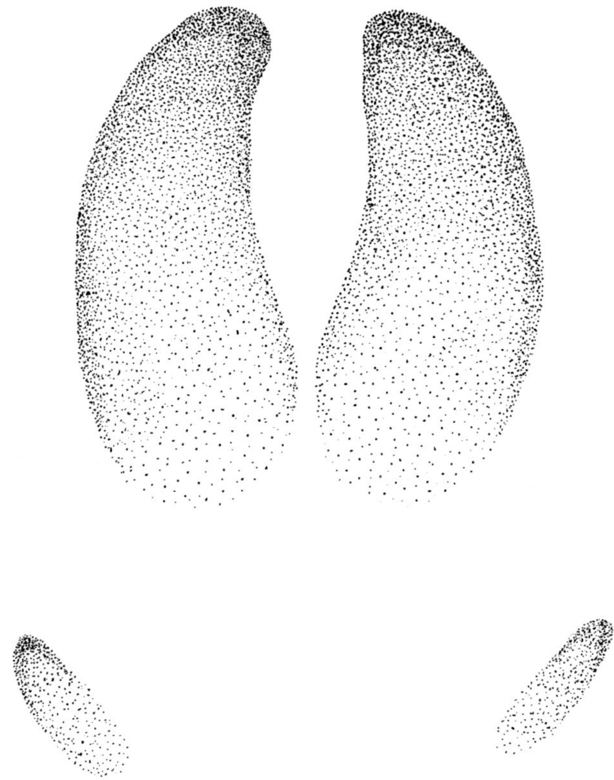

Wild Boar, *Sus scrofa*, p.498

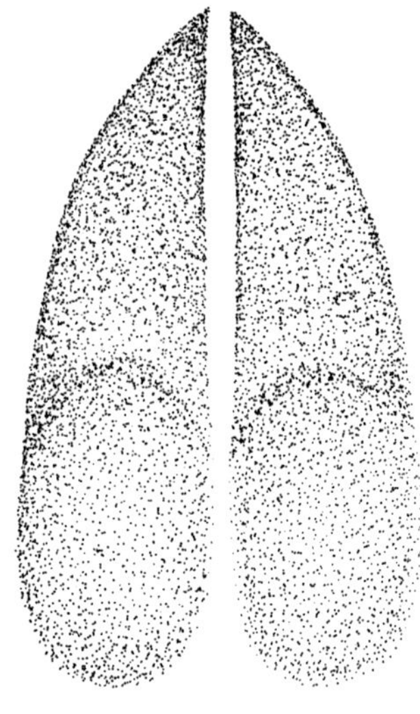

▲ *LF*

Common Fallow Deer, *Dama dama*, p.532

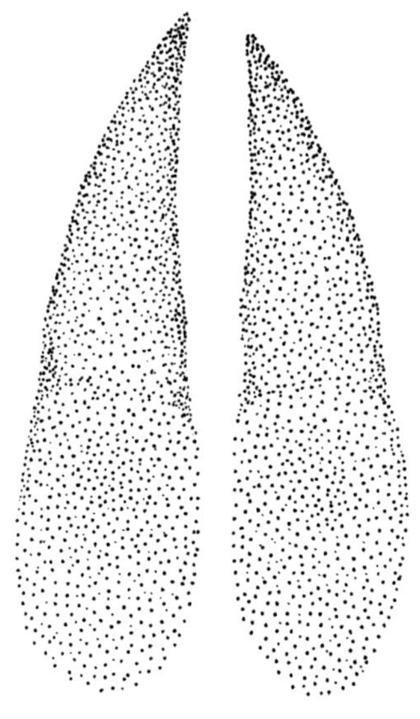

▲ *LH*

Sika Deer, *Cervus nippon*, p.552

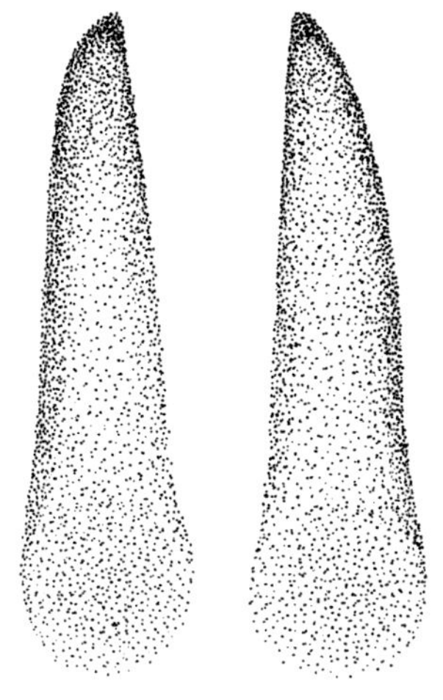

Chamois, *Rupicapra* spp., p.586

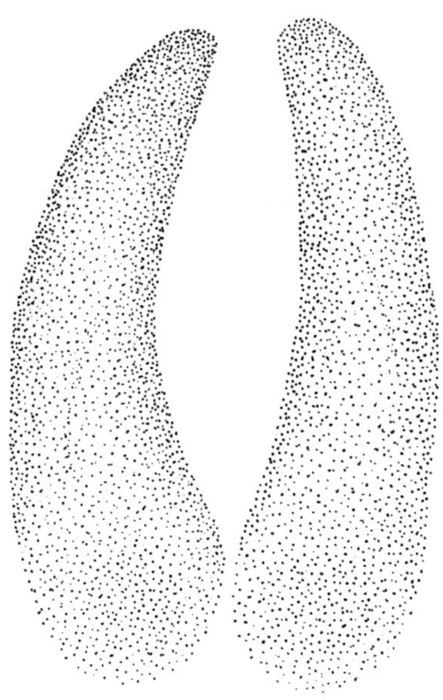

Alpine Ibex, *Capra ibex*, p.581

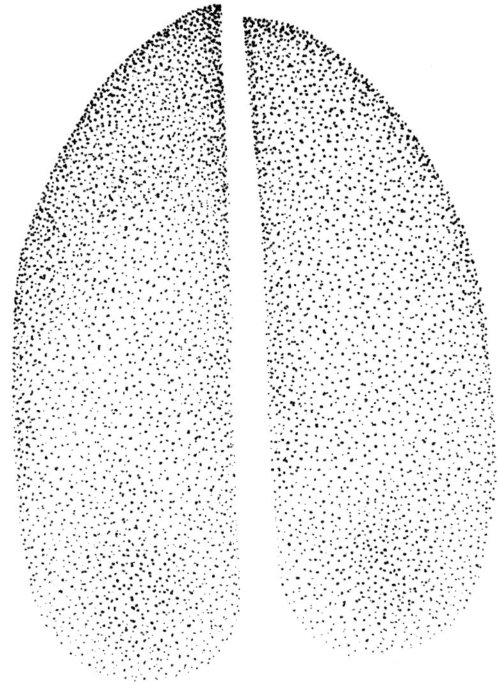

Western Red Deer, *Cervus elaphus*, p.542

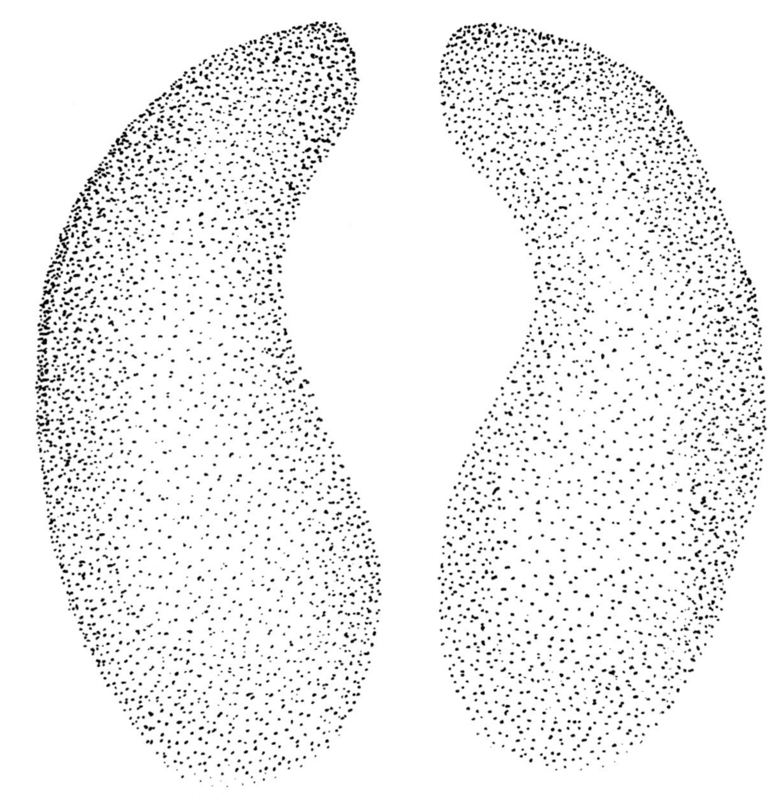

Reindeer, *Rangifer tarandus*, p.526

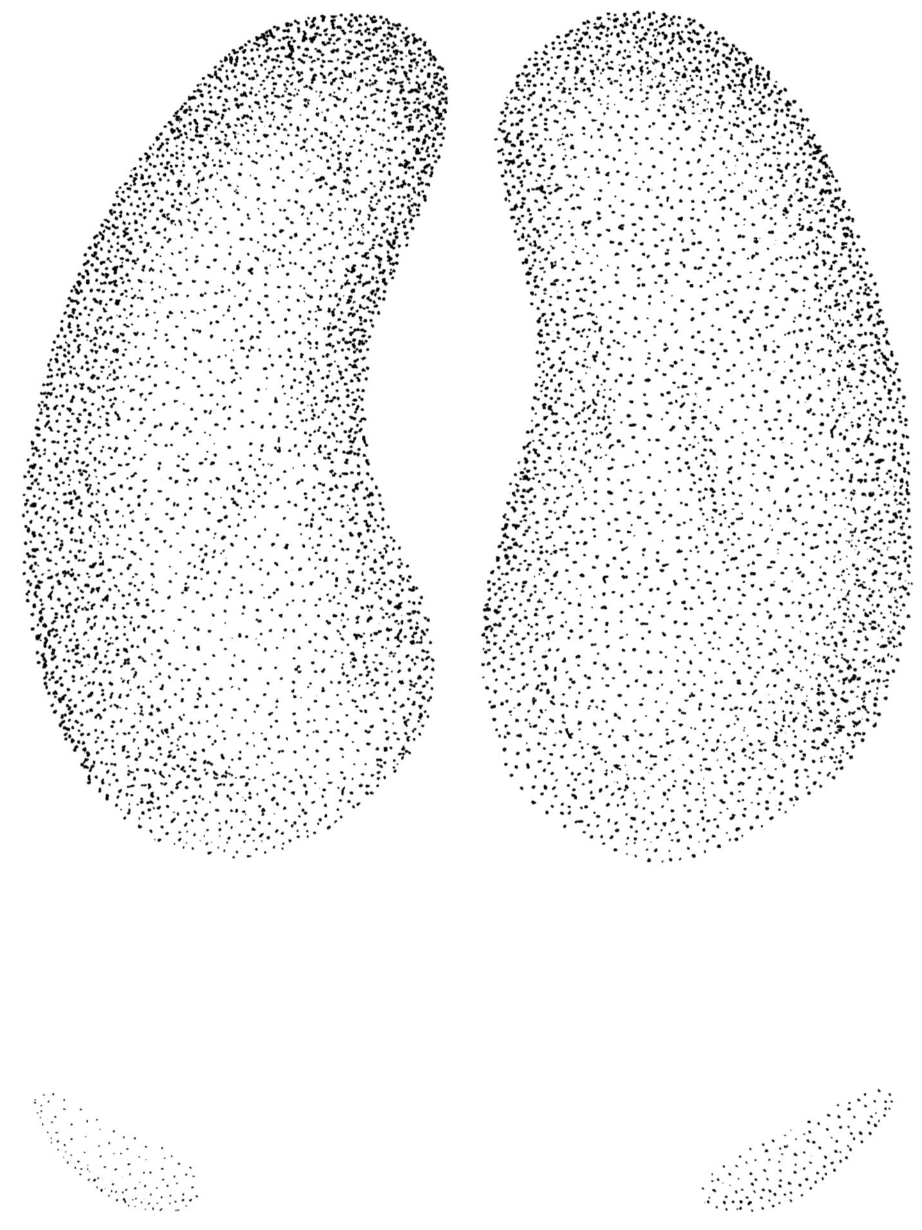

Musk Ox, *Ovibos moschatus*, p.576

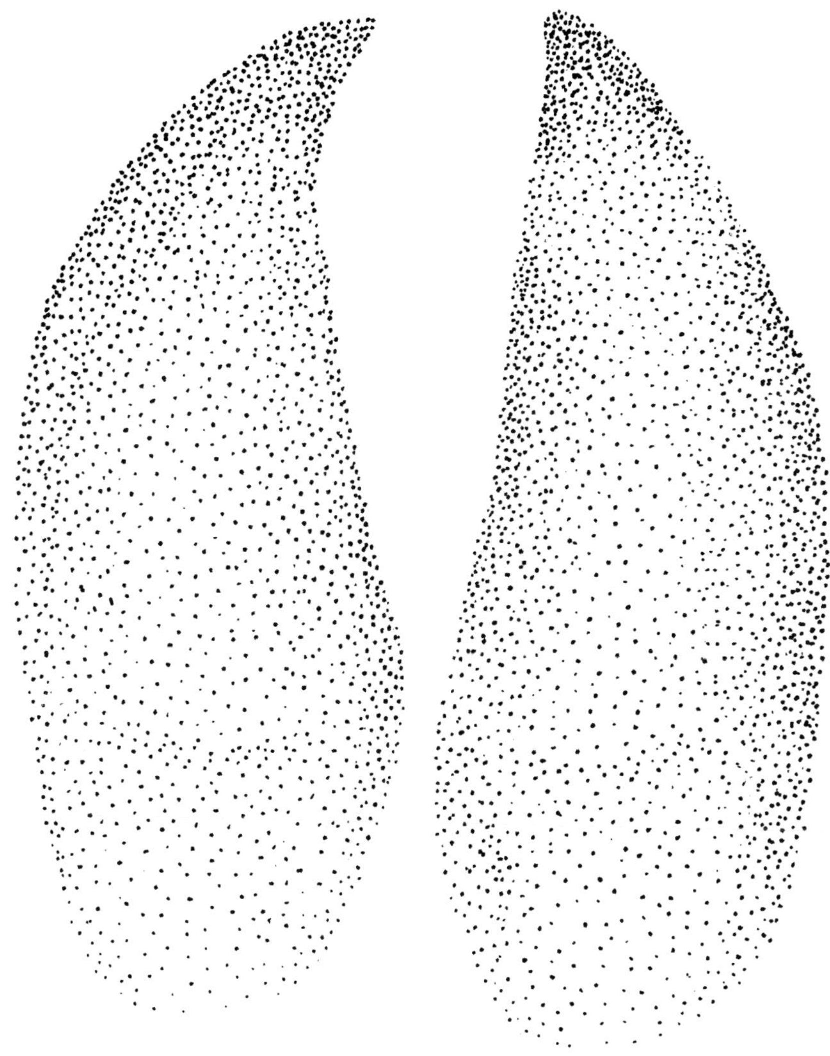

▲ *RF*

Elk, *Alces alces*, p.510

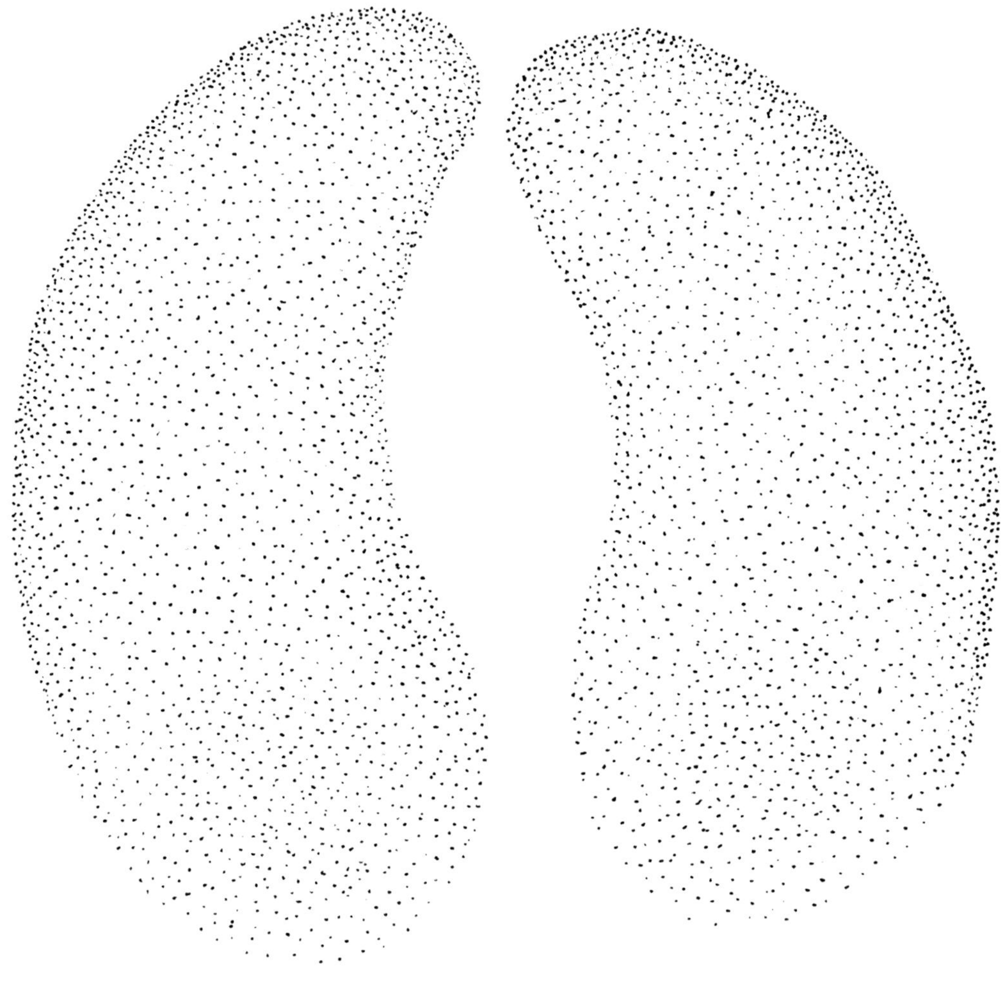

▲ LF – 80 per cent of original size

European Bison, *Bos bonasus*, p.570

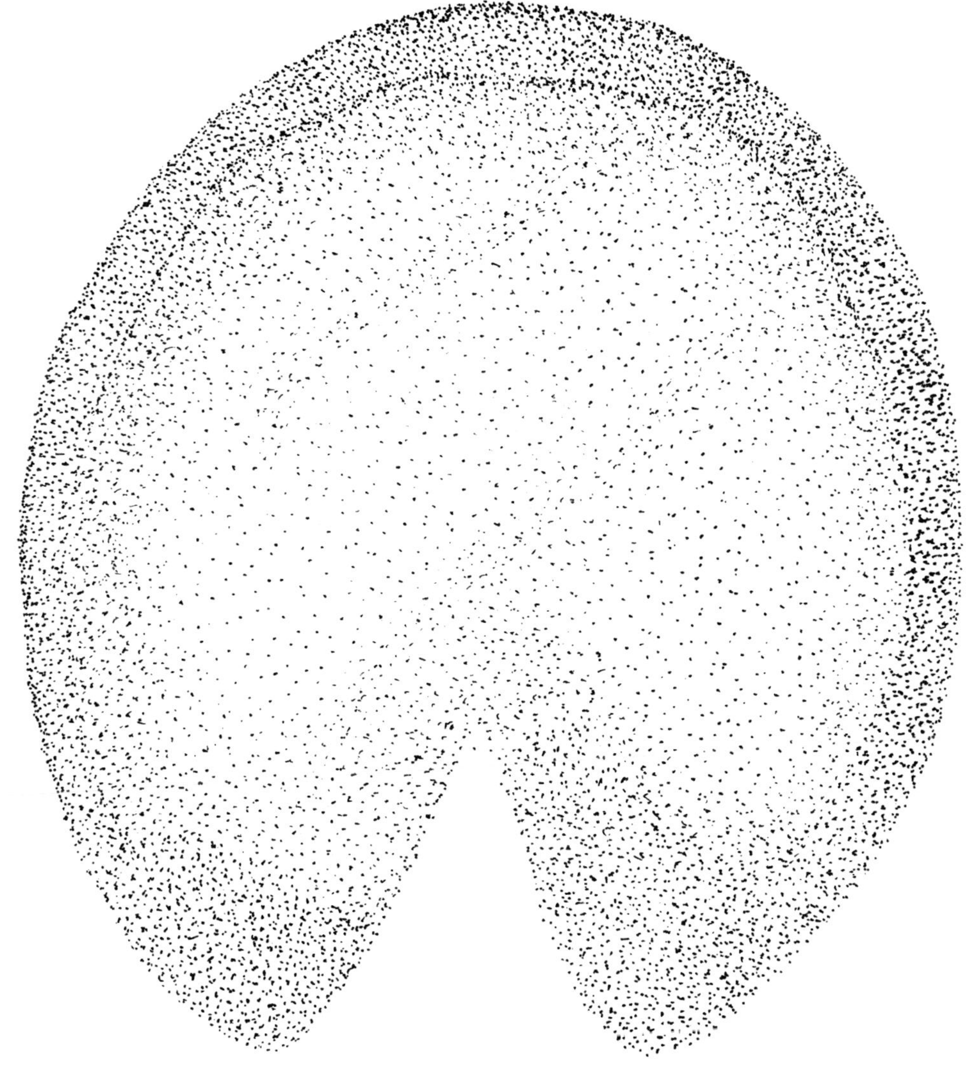

Horse, *Equus caballus*

IDENTIFYING AND INTERPRETING SIGNS

'The art of tracking involves every sign in nature that points to the presence of an animal: signs such as scent, urine and faeces, saliva, pellets, feeding signs, vocal and other auditory signs, visual signs, incidental signs, circumstantial signs, blood spoor, skeletal signs, paths, homes and shelters.'

Louis Liebenberg

Identifying and interpreting footprints is a key aspect of tracking, but recognising and interpreting signs is equally important. The term 'sign' refers to all other evidence of an animal's presence, other than its footprints. Tracks enable us to identify species. Track patterns provide information about an animal's gait and the route it was taking. Signs provide clues about an animal's behaviour and purpose for being in a specific location. We can use them to help determine exactly which behaviours an animal has engaged in while it was in a particular area.

All animals need **food**, **shelter or safety** and a **means of communicating** with each other. When meeting these needs, animals leave signs that we can identify and interpret. When they search for food and eat, they leave feeding marks. Some animals make bedding sites or create dens or nests to satisfy their need for protection. Animals also use various forms of marking for intraspecies communication. Tracks often go hand in hand with signs, and reliable identification of footprints is sometimes only possible if we can identify the surrounding signs. Even in a forest with a thick layer of leaves on the ground, where clear tracks are hard to find, signs are usually visible. Disturbed leaves on the floor of a beech forest, traces of mud on a blackberry bush, or a large oak tree with half its trunk gnawed through – all these signs tell a story even if no footprints have been left.

FOOD Animals leave a virtually endless variety of feeding marks on **herbaceous plants**, **shrubs**, **fungi**, **trees**, **nuts**, **fruit** and **carcasses**. The occurrence of a certain species is largely influenced by the food available. Knowledge about feeding behaviours and about the corresponding availability of food in an area are important clues for narrowing down the list of species that occur in a territory and for gaining an impression of the predator-prey relationships in a region. Some feeding marks are characteristic of a particular species, while others are harder to identify or can be confused with territorial

markings. Appearance and locations of feeding marks are crucial for species identification. The illustration above shows where some common feeding marks are typically found.

SHELTER Animals need somewhere safe to rest and sleep. Many animals constantly change their resting place and only use a fixed place for the breeding season. Others prefer a more permanent shelter. Some animals' shelters resemble bird nests, such as the spherical nest of the Eurasian Harvest Mouse, which is attached to grasses, or the roundish drey of a squirrel, which builds its nest from loose branches and foliage close to the trunk at the top of a tree. Many animals use simple holes in the ground or complex subterranean tunnel systems for protection. Some will use existing holes while others always dig their own burrows. Permanently occupied shelters, such as a large badger sett, can be obvious places to find a wide variety of signs such as paths, droppings, feeding marks and bones.

COMMUNICATION Animals communicate their presence with species-typical behaviour. Visual markings and scent markings play an important role because they indicate the boundaries of a territory and contain important information about health, fertility and dominance. Many male deer leave visible markings on vegetation by striking it with their antlers. Otters scrape together mounds on which they then release a pungent anal secretion; the elevated position on a mound allows the odour to spread better. Along with droppings, urine and other excretions, other common forms of marking include scratches and bite marks, rubbing on trees, and mud and dust baths. Animals intensify the effect of markings by leaving them in conspicuous places. In your area, observe which species mark which elements of the landscape. This will enable you to quickly discover signs of certain species in other areas.

To make the fascinating and diverse world of signs more easily accessible, the rest of this section is divided into six basic identification guides:

- Droppings, urine and other excretions
- Feeding marks on herbaceous plants, shrubs and fungi
- Feeding marks and other signs on trees
- Feeding marks on fruit and nuts
- Feeding marks on animal carcasses
- Signs on the ground: Sleeping and resting places, dust and mud baths, scratches, digging marks, mounds of earth and holes in the ground.

Each identification guide gives an introduction to the topic, as well as example images for initial classification of the sign you have found. Species-specific sign descriptions are given in the respective species accounts and some easily confused signs are compared in the identification guides.

IDENTIFICATION GUIDE TO DROPPINGS, URINE AND OTHER BODILY WASTE

Biologists have been collecting carnivore droppings for the purposes of studying predator ecology for decades (Elton 1927, Litvaitis 2000, Murie 1944). While carnivores are generally difficult to observe directly, it is often possible to find large amounts of their droppings, which can be useful for species identification. Droppings contain lots of information, for example about an animal's diet and health, population sizes in a specific area or the extent of a territory. DNA analyses are expensive and time-consuming so trained trackers who can identify droppings provide a reliable and cost-effective alternative.

The enormous variety of animal droppings poses a challenge when identifying a species. Appearance, shape and size are largely determined by the remains of undigested food. Since availability of food is subject to seasonal fluctuations, the droppings of certain species can look very different throughout the year. In ruminants (such as deer and bovines), for example, young protein- and energy-rich grass in spring leads to soft or, at times, cowpat-like droppings, while droppings later in the season are dry and firm as the food the animals eat becomes increasingly woody. This is why we talk about summer and winter droppings. The size of the droppings relates to the size of the animal and allows us to draw rough conclusions. However, size alone usually isn't enough, because there are often overlaps with the droppings of other species and considerable variations within a species. **Contents**, **shape**, **placement** and

odour are other important indicators for species identification. Classifying droppings into herbivore, carnivore and omnivore is the first step.

The **droppings of pure herbivores** mainly consists of the remains of undigestible plant fibres. The usually densely compressed, cylindrical, short faecal pellets are normally found in large piles. Because plant-based diets have low nutritional value compared with meat-based diets, animals must eat large quantities and consequently produce more droppings. Droppings that mainly consist of plant-based food have a neutral odour and are less unpleasant than carnivore droppings.

Carnivore droppings usually contain undigestible parts of prey, such as feathers, fur or bones. Droppings that primarily consist of meat or blood are very dark to black. Because of the high nutrient content of meat, carnivores produce much smaller quantities of excrement than herbivores.

Omnivore droppings normally contain components of more varied food sources than are found in those of pure herbivores or carnivores. For example, the remains of fur or meat are often mixed with plant seeds.

In addition to the relative size and content of droppings, you should also assess the shape and odour for precise identification. Three basic shapes are **faecal pellets**, **sausage-shaped or cylindrical droppings**, or **cowpats or formless piles**.

Faecal pellets can vary from almost spherical to cylindrical and have a rough or smooth surface. Rodents, even-toed ungulates and lagomorphs usually leave faecal pellets. Sausage-shaped or cylindrical droppings can come in different lengths and thicknesses and can be kinked or segmented. The ends can be blunt or pointed and may be drawn out or end abruptly. Many carnivores, but also wildfowl (Anatidae), leave sausage-shaped or cylindrical droppings. Cowpats and formless mounds can be caused by eating fruits with a high water content, young energy- and protein-rich plants, or offal. These soft and occasionally formless droppings are often left by omnivores, but can also be left by even-toed ungulates and carnivores, depending on their diet. Typical examples of this type of excrement are the pats left by European Bison (page 146) or the droppings of most land birds (page 703).

Many species produce droppings with a unique smell that can be useful for identification. Odour is difficult to describe but, with experience, it can become an important feature for identification. Wolf droppings give off an almost unmistakable odour, fox droppings are normally acrid-smelling and otter droppings often smell like fish.

Some animals appear to leave droppings at random whereas others repeatedly seek out

Why do carnivore droppings smell so bad?

Carnivore excrement has an odour that we humans find very unpleasant. It is the result of the bacterial breakdown of ingested proteins which produces strongly smelling skatole (3-methylindole) or foul-smelling, gas-forming hydrogen sulphide in the intestine. Our strong aversion to the odour could be evolutionary because it warns us of the presence of harmful bacteria.

the same places. Regularly used places are called **latrines**, and can have species-typical features. For example, Northern Raccoons prefer to site their latrines at the base of large, distinctive trees near water, while European Badgers dig holes in the earth. Certain species prefer raised sites, such as tree stumps or stones. The raised height helps the odour to spread effectively. Many animals use their droppings as a form of marking. Latrines can be used over long periods of time and can contain large quantities of droppings.

When an animal **urinates**, the position of the patch of urine relative to the footprints can give an indication of the animal's gender (Bang & Dahlström 1972). Female even-toed ungulates usually stand with their hind legs apart when they urinate, so the patch of urine is normally between or behind the two hind foot tracks. In the case of male even-toed ungulates, the urine patch is normally in the middle of the body between the prints of the front feet and hind feet. This method of sex identification can also be applied to many other animals, but you must be able to make out the position of the urine in relation to the footprints.

Apart from a few exceptions, the excrement of amphibians, reptiles, insects and other invertebrates is easily distinguishable from mammal droppings, and most mammals can be ruled out by the small size. Photographs and descriptions of the droppings of selected species are provided in the species descriptions of the respective sections. To enable better comparison, the following identification guides compare examples from other animal groups with mammal droppings.

Owls, raptors and many other birds produce gastric pellets that can be confused with mammal droppings. Some of these pellets are described in greater detail in Part Two on page 694.

OVERVIEW OF TYPICAL FORMS OF DROPPINGS

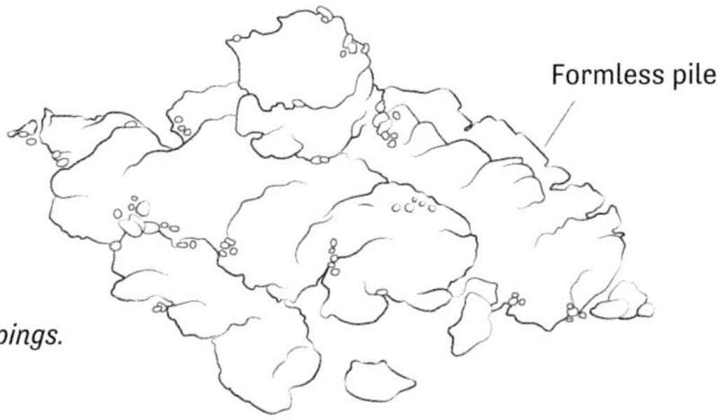

Formless pile

▶ *Brown Bear, omnivore droppings.*

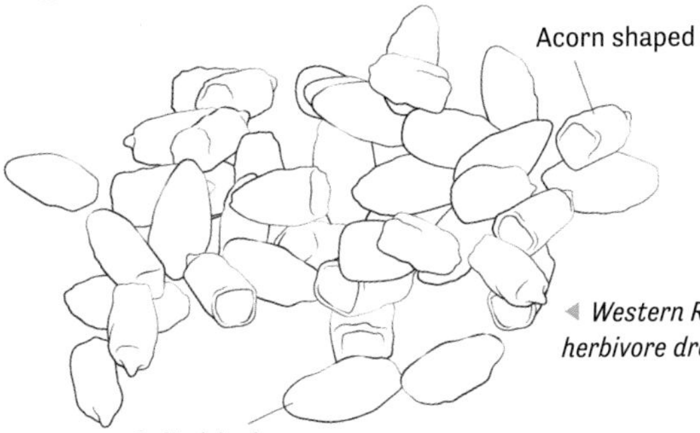

Acorn shaped

◀ *Western Red Deer, herbivore droppings.*

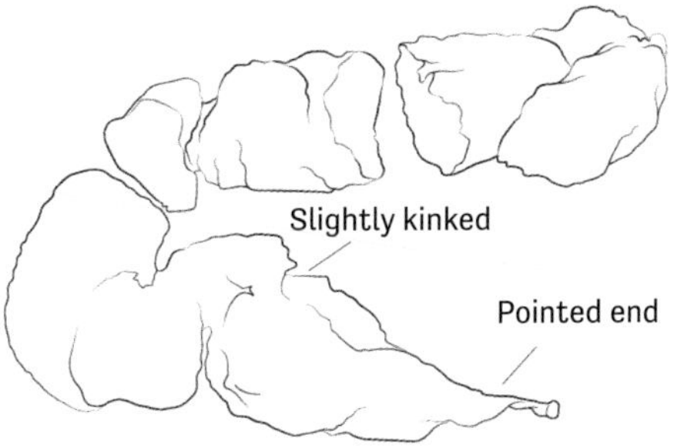

Cylindrical

Slightly kinked

Pointed end

▲ *Dog droppings.*

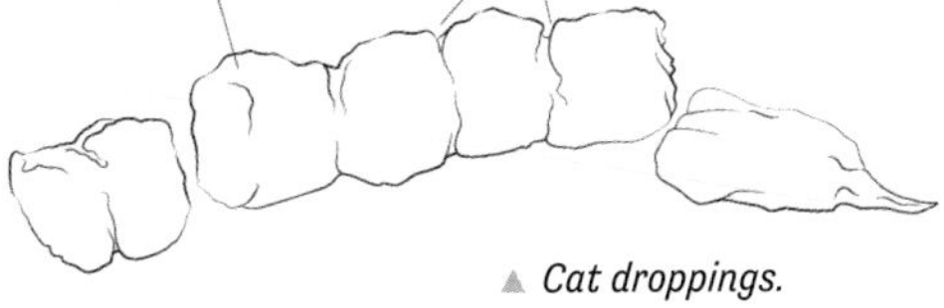

Smooth surface

Clear segmentation

▲ *Cat droppings.*

Very kinked

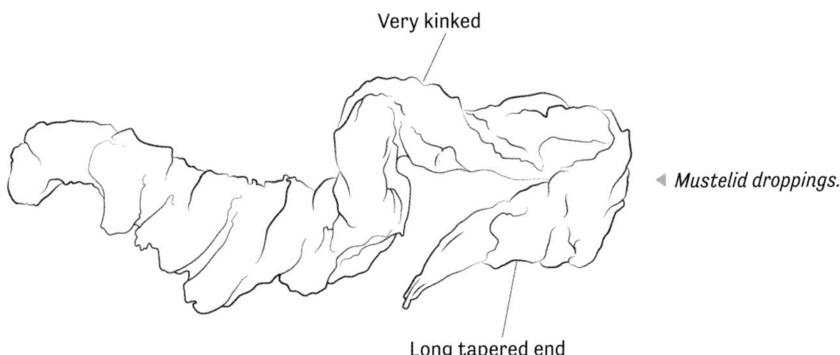

◀ *Mustelid droppings.*

Long tapered end

Oval to spherical pellets, normally slightly flattened. Rough fibrous surface ▶ **Lagomorphs** (page 210) and **Eurasian Beaver** (page 262).

▲ *The spherical pellets of a Brown Hare. The surface is much rougher than the pellets of ruminants.*

▲ *The oval pellets of a Eurasian Beaver mainly consist of waste wood. Vledder, Netherlands. René Nauta.*

Elongated, cylindrical faecal pellets. Pointed, rounded or blunt ends. Smooth or rough fibrous surface ▶ **Rodents** (page 234).

▲ *The elongated cylindrical pellets of a Coypu. The lateral furrows on the surface are characteristic. Hitzacker, Germany.*

▲ *Elongatted cylindrical faecal pellets with rounded ends are typical of voles. Extertal, Germany.*

Various forms of excrement, most with a white coating of urine ▶ **Birds** (page 702).

▲ *No white urine coating can be seen on these winter droppings from a Western Capercaillie. Jämtland, Sweden. Laura Gärtner.*

▼ *The droppings of a Eurasian Blackbird are as varied as its diet. Vledder, Netherlands. René Nauta.*

▲ *Various types of bird droppings with characteristic white urine coating. Biebergemünd, Germany.*

Elongated cylindrical pellets. Pointed, rounded or blunt ends. As with birds, one end has a white coating of urine. Rough surface and coarse texture ▶ **Lizards** and **geckos** (page 740).

▲ *The pellets of lizards and geckos often have a urine coating on one end. Matalascañas, Spain.*

Elongated cylindrical pellets. Pointed, rounded or blunt ends. Extremely rough surface and porous texture ▶ **Insectivores** (page 186), **bats** (page 206).

▲ *The porous pellets of a bat. Extertal, Germany.*

Cylindrical pellets with a bulbous, compact shape. Pointed, rounded or blunt ends. Rough surface and porous texture ▶ **Frogs** and **toads** (page 740).

▲ *Toad pellets are often surprisingly large, especially in terms of diameter. Lüneburger Heide, Germany.*

Pellets ranging in size from tiny to small. Cylindrical shape with blunt ends. Normally identifiable by a distinctively symmetrical structure. Smooth or rough surface ▶ **Caterpillars** (page 783).

▲ *The six lateral grooves are a distinguishing feature of caterpillar pellets or frass. Spessart, Germany.*

Short, cylindrical pellets. Shape can vary greatly even within a species. Often with one pointed and one blunt or indented end, but also acorn-shaped with two rounded ends. In rare cases, almost spherical. Normally smooth surface
▶ **Even-toed ungulates** (excluding Wild Boar), winter droppings (page 492).

▶ *50 per cent of original size, from left to right: droppings of Reeves's Muntjac, Western Roe, Common Fallow and Western Red Deer and Elk. Spessart, Germany.*

Short pellets stuck together in a sausage shape. Some compressed together in a disc-like form. Rough fibrous surface ▶ **Wild Boar** (page 498).

▲ *The short pellets of a Wild Boar, stuck together in a sausage shape.*
Märkische Schweiz, Germany.

▲ *Disc-like pellets of a Wild Boar, stuck together in a sausage shape. Because Wild Boar are omnivorous, the surface of their droppings is normally rougher and more fibrous than ruminant droppings.*
Märkische Schweiz, Germany.

Sausage-shaped droppings with blunt or pointed ends. Vary from almost straight to very kinked, with a smooth or rough surface ▶ **Carnivores** (page 360).

▶ *The sausage-shaped, slightly kinked droppings of a Grey Wolf. Bieszczady, Poland.*

Look at the degree of **kinking**, the amount of **segmentation** and the **tapering ends** to more precisely identify this large category of sausage-shaped droppings.

Sausage-shaped droppings with at least one pointed end. Usually no clear segmentation, but often broken into three separate parts. From mostly straight to distinctly kinked with a smooth or rough surface ▶ **Canids** (page 384).

▲ *The sausage-shaped droppings of a Red Fox. Spessart, Germany.*

▲ *Dog excrement is often dropped in three parts. Spessart, Germany.*

Sausage-shaped droppings with at least one blunt end. Heavily segmented and slightly kinked, with a relatively smooth surface ▶ **Cats** (page 361).

Sausage-shaped droppings with pointed ends that taper markedly. Often very kinked and noticeably thin in relation to the length, rope-like ▶ **Mustelids** (page 428).

▲ *Eurasian Lynx. Sausage-shaped, very segmented droppings with a relatively smooth surface is typical of cats. West Sussex, England.*

▲ *European Pine Marten. The strong kinking and long, tapered ends are characteristic of the droppings of many mustelids. Spessart, Germany.*

The appearance of droppings produced by omnivores such as Beech Martens, Red Foxes, Brown Bears and Northern Raccoons varies enormously depending on the food they have eaten so, in rare cases, the features can even be the opposite of those described above. Brown Bear and Northern Raccoon excrement is normally not kinked and usually has blunt ends.

Cylindrical excrement with blunt ends, often with a white urine cap on one end. Hardly kinked, mainly consisting of plant remains and usually with a rough, fibrous surface ▶ **Herbivorous wildfowl and land birds** (page 702).

▲ *Mute Swan droppings are very large and can even be mistaken for Grey Wolf droppings. Lausitz, Germany.*

▲ *Canada Goose droppings. Duck droppings often look identical but are much smaller. Drehna, Germany.*

Tiny to very small, thin, long, sausage-shaped droppings with blunt or pointed ends. Mostly coiled up like rope with a smooth or rough surface ▶ **Snails** (page 783).

▲ *The thin, sausage-shaped excrement string of a snail (Helicidae). Offenbach, Germany. Simone Roters.*

Cowpat-shaped excrement with a smooth or rough surface ▶ **Even-toed ungulates** (excluding Wild Boar), summer droppings (page 492).

▲ *The cowpat-like droppings of a European Bison. Bieszczady, Poland.*

Formless piles with a smooth or rough surface ▶ **Carnivores** (page 360).

▲ *Brown Bear droppings. The contents of omnivore droppings are normally more diverse than the excrement of ruminants or carnivores. Bieszczady, Poland.*

Many different species of **bird** leave gastric pellets that can be confused with mammal droppings. They are examined in greater detail on page 694.

▲ *Owl pellets can be confused with mammal droppings. Spessart, Germany.*

IDENTIFICATION GUIDE TO FEEDING MARKS ON HERBACEOUS PLANTS, SHRUBS AND FUNGI

Herbivores and omnivores, as well as insects and other invertebrates, leave a variety of feeding marks on herbaceous plants, shrubs and fungi. It often isn't possible to identify exactly which species left these marks, but you can usually roughly classify them.

HERBACEOUS PLANTS AND SHRUBS

Hares and rodents usually leave clean-cut edges at a 45° angle to the stem. Deer, by contrast, tend to leave ragged, torn edges because they have a bony dental pad in their upper jaw instead of upper incisors (see page 496). The feeding marks of small rodents, such as voles and field mice, are normally noticeably fine. Insects and their larvae often leave characteristic feeding marks on leaves. On closer inspection, we can make out countless signs of this class of animals, which contains many different species, although identification to species level may not be possible. Context is generally crucial for identifying feeding marks. For example, cleanly gnawed blades of grass alongside a body of water, together with well-defined paths and latrines, are characteristic of water voles. Ask yourself: How big would an animal need to be to make a feeding mark at this height? Which animals of this size occur in this habitat? Are there additional signs nearby, such as droppings or paths?

Torn or roughly cut buds or shoots at a 90° angle to the stem
▶ **Even-toed ungulates** (page 492).

▲ *Irregular, frayed cuts at a 90° angle are an indicator of even-toed ungulates. Buds and leaves have been eaten here. Coto de Doñana, Spain.*

▲ *The uneven, frayed feeding marks of a Western Red Deer. Bavarian forest, Germany.*

▲ *The fine, clean feeding marks of a vole* (Microtus *sp.*).
Hitzacker, Germany.

▲ *The clean feeding marks of a Brown Hare.*
The 45° angle is characteristic.
Vledder, Netherlands.

▲ *Deer bite mark on a bramble. Even-toed ungulates*
leave roughly cut or torn feeding edges on leaves.
Vledder, Netherlands.

Fine marks on the leaves of shrubs and herbaceous plants, cleanly cut, sawn or rasped ▶ **Insects** such as caterpillars, beetles, leaf-mining moths, sawflies, crickets, grasshoppers, leafcutter bees and leafcutter ants (page 767).

▲ *Leaf feeding marks made by insects are usually much smaller and finer than feeding marks of mammals. Spessart, Germany.*

▲ *Beetles (middle) and leaf miners (right) have eaten the same leaf. Spessart, Germany.*

Small to large toothmarks on the roots of herbaceous plants and shrubs underground, as well as cambium feeding marks on shrubs at ground level ▶ **Lagomorphs** (page 210), **rodents** (page 234), **even-toed ungulates** (page 492).

◀ *Voles leave fine toothmarks in the garden, for example on this carrot. Spessart, Germany.*

FUNGI

Insects, small rodents such as voles, field mice and squirrels, as well as deer, boar and birds, all eat fungi. The beak marks of birds are usually easily distinguishable from the parallel incisor marks of mammals. The feeding marks of snails are also highly characteristic. Small rodents can occasionally be identified by additional signs such as digging marks on the ground or droppings on the fungus. Wild Boar and deer feeding marks on fungi are usually distinguishable because deer do not have upper incisors.

Distinctive, often triangular beak marks on the top of a fungus
▶ **Birds** (page 711).

▲ *The characteristic beak marks of birds are usually easily distinguishable from the feeding marks on fungi made by other animals. Germany. Simone Roters.*

Many adjacent, roundish feeding marks that look as though they have been rasped ▶ **Snails** (page 782).
Fine toothmarks (up to 4mm wide) that cut into the fungus from above
▶ **Mice** (page 318).

Strong toothmarks (0.5–2cm wide) of large lower jaw incisors that have cut into the fungi from underneath
▶ **Even-toed ungulates** except Wild Boar (page 492).

▲ *Western Red Deer fungi bite mark. Spremberg, Germany.*

Strong toothmarks (0.5–2cm wide) of incisors that have bitten into the fungi from above and below (not illustrated)
Wild Boar (page 498).

▲ *Feeding marks of a mouse (left) and snail (right) for comparison. Spessart, Germany.*

IDENTIFICATION GUIDE TO FEEDING MARKS AND OTHER SIGNS ON TREES

Many different species of animals leave signs on trees. We differentiate between four categories of signs:

- Strike, rub and scratch marks
- Toothmarks, feeding marks on cambium and bark-peeling
- Hollows in standing or fallen trees
- Other signs.

STRIKE, RUB AND SCRATCH MARKS

Almost all species of deer make distinctive **strike marks** by hitting shrubs, saplings and trees with their antlers and rubbing their forehead and cheeks against them. This often results in bark being removed at the height of the animal's head, leaving behind a smooth, pale, shiny surface. Repeated striking of surrounding vegetation is a clear form of visual marking. The glands on the animal's forehead and other parts of its head also leave scent marks. Antler strike marks are usually found along the animal's paths at highly frequented crossing points, where their effect is intensified. Striking occurs more often just before mating season. This behaviour in Western Roe Deer is also known as 'fraying'.

Rubbing is a typical behaviour of Wild Boar, but Western Red Deer, Sika Deer and European Bison also rub. As well as grooming, this behaviour also serves the purpose of marking because it leaves behind both visual marks and scent marks. Animals tend to repeatedly select preferred trees for rubbing and so rub off large areas of bark over time. Animal hair is often found on the edges of the area where the bark has been rubbed off. The height of the rubbed patches can provide information about the species. Trees used for rubbing are often found near muddy pools.

Many mammals and some birds leave **scratch marks** on trees. Scratch marks left by cats sharpening their claws are a familiar form of this type of sign. Brown Bears and European Badgers scratch trees to leave visual and scent marks as a form of territory marking. The height and size of the scratch marks can provide information about the species and size of the individual.

Canids and even-toed ungulates leave scratch marks on the wood and bark when they climb over fallen trees, especially if these trees are located on a popular path. Northern Raccoons, martens and rodents leave scratch marks on trees when they climb them. Woodpeckers make similar scratch marks when they hold onto the bark while searching for food or chiselling out a cavity in a tree.

When identifying and interpreting scratch marks on trees, ask yourself the following questions: Do these scratch marks look as if they have been made intentionally in a conspicuous place, or do they seem more like signs of use that have occurred incidentally? Have the scratch marks been made by claws or hooves? Were the claws sharp or blunt? Can you make out individual scratches? Are the scratches parallel, perpendicular or diagonal to the trunk? Are the scratches visible around the entire circumference of the tree or only on one side? Do the scratches occur in conjunction with other signs such as holes or droppings?

20–200cm (or more). Frayed bark, hanging off in threads, no clean-cut edges. Often found in conjunction with smaller branches that have been broken by antlers ▶ **Striking** ▶ **Even-toed ungulates** (page 492).

▲ *Roebuck strike mark.*
Märkische Schweiz, Germany.

▲ *Western Red Deer antler strike mark.*
Bavarian forest, Germany.

20–160cm high. Area on a tree where the bark has been removed by rubbing, often in conjunction with resin discharge and usually with hairs that have become stuck to the resin or bark. Normally smooth transitions to the remaining bark ▶ **Rubbing** ▶ **Wild Boar** (page 498), **Elk** (page 510), **Western Roe Deer** (page 518), **Western Red Deer** (page 542) and **European Bison** (page 570).

▲ *Tree where a Wild Boar has rubbed. Märkische Schweiz, Germany.*

▲ *Tree where a Western Red Deer has rubbed. Midhurst, England.*

Ground level up to a height of 2.5m. More or less parallel scratch marks in wood or bark, made by claws ▶ **Rodents** (page 234), **cats** (page 361), **canids** (page 384), **Brown Bears** (page 422), **mustelids** (page 428), **Northern Raccoons** (page 484) and **woodpeckers** (page 713).

◀ *A European Badger has left claw marks when climbing over a trunk. Bieszczady, Poland.*

TOOTHMARKS, FEEDING MARKS ON CAMBIUM AND BARK-PEELING

Toothmarks on trees may be a form of deliberate marking, but they can also be feeding marks or signs of use. Squirrels mark popular trees by making closely spaced gnawing marks with their incisors. Small rodents, hares and rabbits leave feeding marks in the form of toothmarks. Muskrat, Coypu and Eurasian Beaver toothmarks can be feeding marks or signs of use, for example made when building a dam or landscaping. Rodents, such as squirrels, Coypu and Eurasian Beaver, as well as hares and rabbits and many even-toed ungulates, eat the relatively nutrient-rich cambium layer of trees.

Cambium feeding marks and bark-peeling are common signs. Bark-peeling is a special kind of cambium eating that is characteristic of even-toed ungulates. Two clear lower incisor toothmarks are usually visible. They leave a relatively sharp incision at the bottom end of the peeled area and then scrape from the bottom up. The animal pulls off the top part of the peeled bark so the often frayed remaining bark hangs down.

In contrast to markings, which tend to be placed deliberately and individually, bark is often peeled off of several trees because animals usually eat it in large amounts. The width of the individual toothmarks and the distance between them are important features for species identification. The maximum height of a peeling mark, the type of tree and the location of the toothmark on the tree can also be useful indicators. For example, voles prefer to eat the cambium layer of tree roots and hares usually eat cambium at a height typical of their species. When interpreting finds, bear in mind that seasonal phenomena, such as floods and snow, can affect the accessibility of certain parts of a plant because of the changing height ratios.

Differentiating between striking and peeling

Strike and peeling marks can easily be confused at first glance. To tell the difference between them, pay attention to the following features:

Striking

- Many other small areas of damage to the bark and broken branches are visible in addition to the main area of damage.
- Bark is uneven and frayed at the top and bottom edge of the strike mark.
- Toothmarks are not normally visible.

Peeling

- Clear main area of damage but no or hardly any other small areas of damage to the bark or branches.
- Clean horizontal cut edge on the bottom end of the peeling mark. The top edge is uneven and the remaining bark is frayed.
- Toothmarks can often be seen.

Ground level up to a height of above 200cm. Toothmarks of large incisors that run from bottom to top. Smooth cut edge at the bottom end, frayed remaining bark in the top area. Can also be at a slight angle. Can be very localised and isolated or, more commonly, on large areas and on many different trees
▶ **Bark peeling** ▶ **Even-toed ungulates** (page 492).

▲ *Bark peeling by Western Red Deer, the toothmarks are an important feature for distinguishing from strike marks. Bieszczady, Poland.*

▲ *European Bison have peeled this hazel tree. Bieszczady, Poland.*

At all heights. Small to medium toothmarks, local damage to bark caused by browsing ▶ **Marking** or **cambium feeding marks** up to maximum of 1.1m high ▶ **Hares and rabbits** (page 210), all heights ▶ **Rodents** (page 234).

▲ *Squirrels make visible territory markings with their incisors. Midhurst, England.*

Ground level up to a height of 1.5m. Small to medium toothmarks. Small branches cut at a 45° angle ▶ **Feeding mark** ▶ **Hares and rabbits** (page 210), **rodents** (page 234).

▲ *Brown Hare feeding mark, the clean 45° cut is typical of hares and rabbits. Bavarian forest, Germany.*

Ground level and underground. Small to large toothmarks on the roots of trees ▶ **Feeding mark** ▶ **Hares and rabbits** (page 210), **rodents** (page 234), **even-toed ungulates** (page 492).

▲ *Western Roe Deer have left their toothmarks on this root. Lake District, England.*

Ground level up to a height of 80cm. Large incisor toothmarks, often in conjunction with piles of large wood shavings ▶ **Cambium feeding marks, dam building** or **landscaping** ▶ **Eurasian Beavers** (page 262) and ▶ **Coypu** (page 270).

◀ *Incisor toothmarks of a Eurasian Beaver. Elbufer Lüchow-Dannenberg, Germany.*

▶ *Eurasian Beaver feeding place. Hoher Fläming, Germany.*

HOLLOWS IN STANDING OR FALLEN TREES

Mammals often use hollows in trees but rarely make them. Wood Mice and Yellow-necked Mice are known to inhabit old woodpecker holes and Northern Raccoons use hollow tree trunks for raising their young. Brown Bears and European Badgers break up and hollow out rotten tree trunks with their claws while searching for beetle larvae and other insects. The best-known hollows in trees are made by woodpeckers when carving out holes for nesting or searching for food.

Broken open areas on rotten tree trunks and stumps ▶ **Searching for food** ▶ **Brown Bears** (page 422) and **European Badgers** (page 444).

▲ *A European Badger has pulled apart this tree trunk while searching for larvae. The trunk has previously been used by a Black Woodpecker. Bieszczady, Poland.*

Other hollows in standing or fallen, living or dead trees
▶ **Searching for food** or **carving out hollows** ▶ **Woodpeckers** (page 712).

▼ *A Black Woodpecker was searching for food here.*
Lausitz, Germany.

▲ *Hollow made by a Great Spotted Woodpecker.*
Entrance holes are normally rounder than holes created
while searching for food. Drehna, Germany.

▲ *Black Woodpeckers can make impressively large*
hollows in trees whether searching for food or carving out
a nesting place. Bieszczady, Poland.

OTHER SIGNS

Animals leave other signs on trees including special feeding marks (such as ringing by woodpeckers), feeding patterns made by bark beetles or growth deformities caused by insects. Because their numbers are virtually unlimited, we have only given typical examples to make initial classification easier.

Roundish strike marks on the bark, often close together in a row ▶ **Woodpecker rings** ▶ **Woodpeckers** (page 715).

▲ *Ringing by a Great Spotted Woodpecker: The roundish strike marks are typical. Jämtland, Sweden. Laura Gärtner.*

Evenly stripped, horizontal browse line on the underside of trees in leaf
▶ **Browse line** ▶ **Deer** (page 508).

▲ *These neat and flat horizontal chewed edges on the underside of the crown foliage look as if they have been made by humans, but they are Common Fallow Deer feeding marks. Common Fallow Deer and Western Red Deer repeatedly visit trees and eat all the leaves and buds within reach. This kind of distinctive feeding pattern is created over time. West Sussex, England. John Rhyder.*

Large feeding marks on leaves of trees, torn or roughly cut
▶ **Even-toed ungulates** (page 492).

Small marks on the leaves of trees, cleanly cut, sawn or rasped ▶ **Insects**, such as caterpillars, beetles, leaf-mining moths, sawflies, crickets, grasshoppers, leafcutter bees and leafcutter ants (page 768).

▼ *Small feeding marks of various insects. Eucalyptus leaves are a popular food source. Coto de Doñana National Park, Spain.*

▲ *Even-toed ungulates leave roughly cut or torn feeding edges on leaves. This is the bite mark of a Western Roe Deer on a bramble. The feeding marks are identical to those on the leaves of trees. Vledder, Netherlands.*

Growth deformities on the leaf or bud of a tree ▶ **Galls** ▶ **Insects** (page 772).

▼ *Galls, like this beech gall, are found in different places and come in a variety of shapes and sizes. Spessart, Germany.*

Feeding tunnels under the bark, often with boreholes and bore dust ▶ **Insects**, such as bark beetles, longhorn beetles, metallic wood-boring beetles, cossid miller moths, clearwing moths, carpenter ants, carpenter bees and wood wasps (page 757).

▲ *The distinctive feeding pattern of a bark beetle. Heidelberg, Germany. Aaron Tiedemann.*

IDENTIFICATION GUIDE TO FEEDING MARKS ON FRUIT AND NUTS

Rich in valuable fats and carbohydrates, fruits and nuts are an important food source for a variety of animals. Because there are so many possible combinations, this is an introduction with a description of common or distinctive signs and some key questions that should help you interpret these feeding marks. It is usually easy to determine whether feeding marks on fruit and nuts have been made by a bird or mammal because the basic structure of a beak and a set of teeth are so different. It is often possible to identify the species by looking closely at the features of a mammal feeding mark. You should also think about where you found the feeding mark. Mice normally eat fruit and nuts in areas with plenty of ground cover, while squirrels often eat sitting up on tree trunks or stumps to give them a good view. Birds often wedge nuts and conifer cones into gaps in bark or wood so they can break them open with their beaks more easily.

PINE, FIR AND SPRUCE CONES

The female inflorescences of these coniferous trees lignify into solid cones with scales under which their seeds mature. Birds and mammals have different ways of opening these scales to get to the calorie-rich seeds. If birds have eaten pine, fir or spruce cones, the cones usually show clear traces of having been accessed by beaks. The scales have normally been picked apart or split but are usually still attached to the axis of the cone. By contrast, mammals almost always strip the scales with their teeth, so the axis and scales will be separate.

Evenly opened cones with scales split lengthwise ▶ **Crossbill** (page 710).

Frayed appearance, with scales sticking out in different directions. Also seen on green cones ▶ **Woodpecker** (page 709).

▼ *Scales that have been split lengthwise down the middle are a typical feature of crossbill feeding activity. This is the work of a Common Crossbill. Bavarian Forest National Park, Germany.*

▲ *Pine cones that have been hacked by a chisel-like beak. This is the work of a Great Spotted Woodpecker. Märkische Schweiz, Germany.*

Untidily frayed, unevenly stripped scales. Base with short or longer point
▶ **Squirrels** (page 236).

◀ This spruce cone has been eaten by a squirrel. Bieszczady, Poland.

Evenly stripped scales, rounded base ▶ **Mice and voles** (usually field mice and red-backed voles, page 327).

◀ Spruce cone that has been evenly eaten by a field mouse. Spessart, Germany.

HAZELNUTS, ACORNS, BEECHNUTS, WALNUTS AND CHERRY STONES

Different mammals use different strategies to open these fruits and nuts. Signs on the shells often enable us to identify the genus or species, especially if we take into consideration the place they were found.

Hazelnuts

Lots of toothmarks arranged in a circle and perpendicular to the edge of the hole. Irregular hole with lots of toothmarks around the edge
▶ **Field mice** (page 327).

▼ *The classic hazelnut feeding mark of a field mouse. Spessart, Germany.*

Few or no toothmarks, traces of gnawing running diagonally or almost parallel to the edge of the hole. Hole with an even, roundish, smooth edge
▶ **Dormice** (page 260).

▲ *Circular holes with diagonal gnaw marks are characteristic of dormice. Spessart, Germany.*

A few localised toothmarks at the opening. Hole with an even edge and lots of toothmarks
▶ **Red-backed voles** (page 304).

▲ *Note the clear contrast between the gnawed area and the otherwise undamaged edge. This is characteristic of Striped Field Mice and red-backed voles. Spessart, Germany.*

Shells split evenly into two halves
▶ **Squirrels** (page 244).

▼ *Two evenly split shell halves are characteristic of squirrels. An uneven split can be indicative of young animals. Spessart, Germany.*

Pecked open nuts and seeds, some with visible beak marks
▶ **Woodpecker** or **Nuthatch** (page 709)

▼ *Uneven edges and beak marks like these are characteristic of pecking by a beak. Spessart, Germany.*

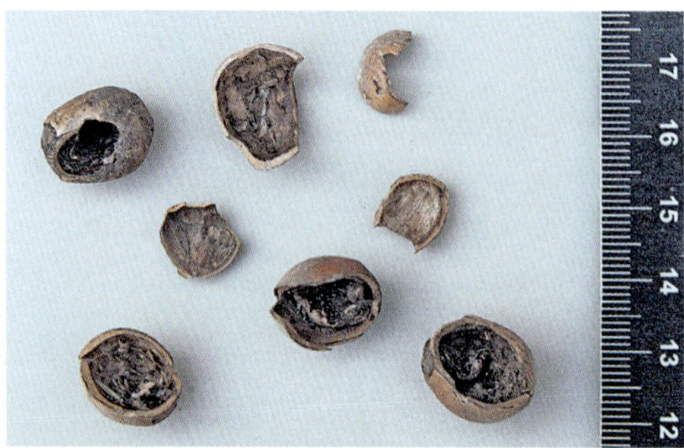

◀ *This hazelnut has been opened by a Great Spotted Woodpecker. Märkische Schweiz, Germany.*

◀ *Hazelnuts that have been opened by Nuthatches usually have smaller holes. Märkische Schweiz, Germany.*

Perfectly circular holes in hazelnuts, acorns and beechnuts
▶ **Nut weevil** (page 777).

◀ *The perfectly circular exit hole of a nut weevil. Spessart, Germany.*

Acorns, beechnuts, walnuts and cherry stones

Delicate toothmarks and shell damage on acorns, beechnuts, walnuts and cherry stones
▶ **Mice** (page 318).

▼ *Delicate gnaw marks on an acorn, made by a field mouse. Uhyst, Germany.*

▼ *A chewed edge is visible on the remains of these beechnut shells. If you look closely, you can also make out small scratch marks on the outer shell. These were made by a field mouse. Spessart, Germany.*

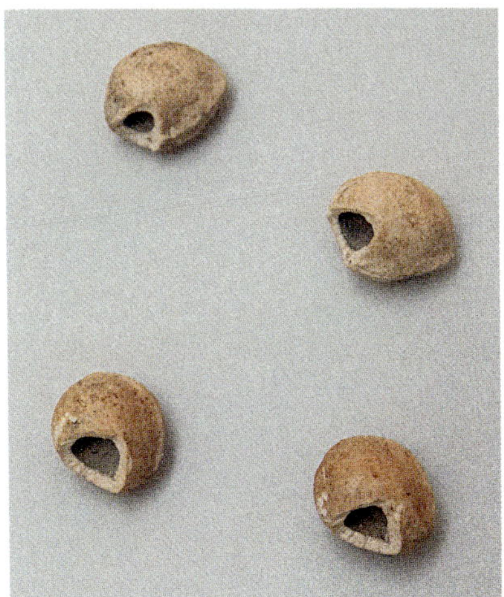

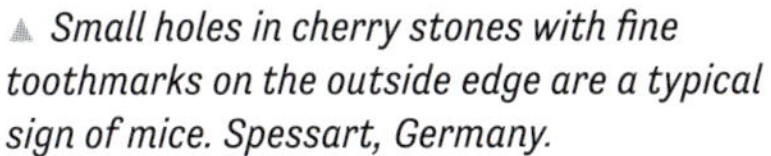

▲ *Small holes in cherry stones with fine toothmarks on the outside edge are a typical sign of mice. Spessart, Germany.*

▲ *Mice also leave a clear chewed edge with fine toothmarks on walnut shells. Spessart, Germany.*

Shells split cleanly into two halves ▶ **Squirrels** (page 244).

◀ *Walnuts split cleanly into two halves with fine toothmarks on the point are a sign of squirrels. Bielefeld, Germany. Ulrike Quartier.*

Coarse toothmarks on the flesh and wide, lengthways strips on the acorn shells ▶ **Squirrels** (page 244).

▲ *The toothmarks of a Grey Squirrel are clearly visible here. They are much coarser in comparison to those of mice. West Sussex, England.*

▲ *When squirrels eat acorns, they leave characteristic piles of lengthways strips of acorn shell. On closer inspection, you can see scratch marks made by claws. Midhurst, England.*

Green acorns that have been nibbled and then left
▶ **Squirrel** hunting for nut weevil larvae (page 244).

▶ *Squirrels can tolerate the high tannin content of green acorns. However,nut weevil larvae were the intended prey here. West Sussex, England.*

Pecked open with clear beak marks
▶ **Birds**, such as Eurasian Jay (page 708).

▶ *These acorns have been tackled by a Eurasian Jay. Kreba Teiche, Germany.*

IDENTIFICATION GUIDE TO FEEDING MARKS ON ANIMAL CARCASSES

Carnivores normally leave characteristic kill features on the carcass of their prey. Looking at these features can enable us to identify the predator and they provide fascinating information about different prey strategies. Determining cause of death or culprit is difficult without DNA analyses and misinterpretations are common. Avoid jumping to conclusions and give yourself time to interpret the findings. Imagine you work for a forensics unit and you have arrived at a crime scene where even the slightest carelessness could destroy important evidence.

Many trackers circle the surrounding area before they begin inspecting the carcass. Think about whether the animal died where the carcass is now located or whether it was moved there after death. Can you see marks where the body has been dragged? Are there signs of a pursuit or fight? Bears and lynx often carry their prey to a hiding place, while Wolverines, Grey Wolves and foxes hide individual parts of their prey or carry them away to eat. Lynx are known for covering their prey with surrounding material.

Carnivorous hunters are usually also carrion-eaters, so carcasses often also feature feeding marks made by secondary users. Some birds, including raptors (diurnal birds of prey), owls and corvids, also feed on carrion. Look out for any footprints, droppings or lying areas made by predators near the site before observing the carcass more closely.

The method used to kill and eat a prey animal can also give important indications about the predator. Has the carcass been fed on at several points or is there a localised opening? Are the stomach and intestines untouched or are the entrails scattered around? The positioning of scratch or bite marks on different parts of the body can also provide important information. In particular, look out for bite marks on the neck, throat or spinal column, as well as scratch marks on the back, flanks and rump. The distance between the canine teeth in a bite wound can often allow the number of possible predators to be narrowed down considerably. Species-specific carcass features can be found in the respective species descriptions.

If you come across bird remains, pay attention to the state of the bottom part of the quills. Bitten-off quills are a sure sign of carnivorous mammals. Birds of prey rip the feathers out and occasionally leave bite marks on the quills, but normally leave the quills intact.

Large carcass, taken apart and scattered
▶ **Canids** (page 384), ▶ **Brown Bears** (page 422).

▲ *Fresh wolf prey. Lausitz, Germany. LUPUS, Institute for Wolf Monitoring and Research in Germany.*

▲ *Grey Wolves have torn apart a Western Red Deer, old feeding marks. Bieszczady, Poland.*

European Rabbit to Western Roe Deer-sized carcass, in one piece and often covered with leaves, grass or snow. Normally heavy bleeding in the throat area, few injuries
▶ **Lynx** (page 364).

▶ *Lynx often bury their prey. There are normally few injuries and body parts aren't detached. Markus Schwaiger, Lynx Project Bavaria. Bavarian forest, Germany.*

Some carcass parts, in shallow indentations or hidden ▶ **Canids** (page 384) and **mustelids** (page 428)

Bitten off quills, feathers often stuck together with saliva
▶ **Carnivora** (page 360).

▼ *Here, a Great Spotted Woodpecker has been killed. Bitten off quills and feathers that are stuck together are signs of a carnivorous mammal, in this case probably a Red Fox. Märkische Schweiz, Germany.*

▲ *Small to medium-sized predators, in particular, often hide parts of their prey for later. There is a Western Roe Deer leg in this Red Fox food store. Vledder, Netherlands. René Nauta.*

Quills largely undamaged but ripped out and with some beak marks. Breast meat eaten
▶ **Plucking** ▶ **Raptors** (page 716).

◀ *A pile of feathers under the perch of a bird of prey. The quills are intact and the larger feathers are mainly separate instead of stuck together. Märkische Schweiz, Germany.*

Beheaded carcasses of small to rabbit-sized mammals or other animal remains, found on raised sites like tree stumps, fence posts or similar locations
▶ **Raptors** (page 716).

◀ *The raised location and the feeding pattern are typical of raptors. This Eurasian Blackbird was probably taken by a Eurasian Sparrowhawk.*
Vledder, Netherlands. René Nauta.

▲ *A Eurasian Goshawk has killed a pigeon.*
Vledder, Netherlands. René Nauta.

IDENTIFICATION GUIDE TO SIGNS ON THE GROUND

SLEEPING AND RESTING PLACES, DUST AND MUD BATHS, SCRATCH MARKS, DIGGING MARKS, MOUNDS OF EARTH, HOLES IN THE GROUND AND BURROWS

Animals lie on the ground to rest; they groom, make marks and build their shelters there, and, as they do so, they leave signs we can interpret. Below is an overview of some of the most important ones.

SLEEPING AND RESTING PLACES, DUST AND MUD BATHS

Animals lie down to rest and sleep and leave a print of the outline of their body on a flattened and often indented area. Resting places are normally only used once, for temporary breaks between periods of activity. Bedding sites are used for longer periods of sleep and some species return to the same one repeatedly. The size and shape of sleeping and resting places can be used for species identification.

Deer bedding sites

The bedding sites of deer (Cervidae) normally have a characteristic form. The animals rest on one side of their body so their back leaves a relatively long, curved print (a). By comparison, the opposite side of the print is shorter, noticeably straight and

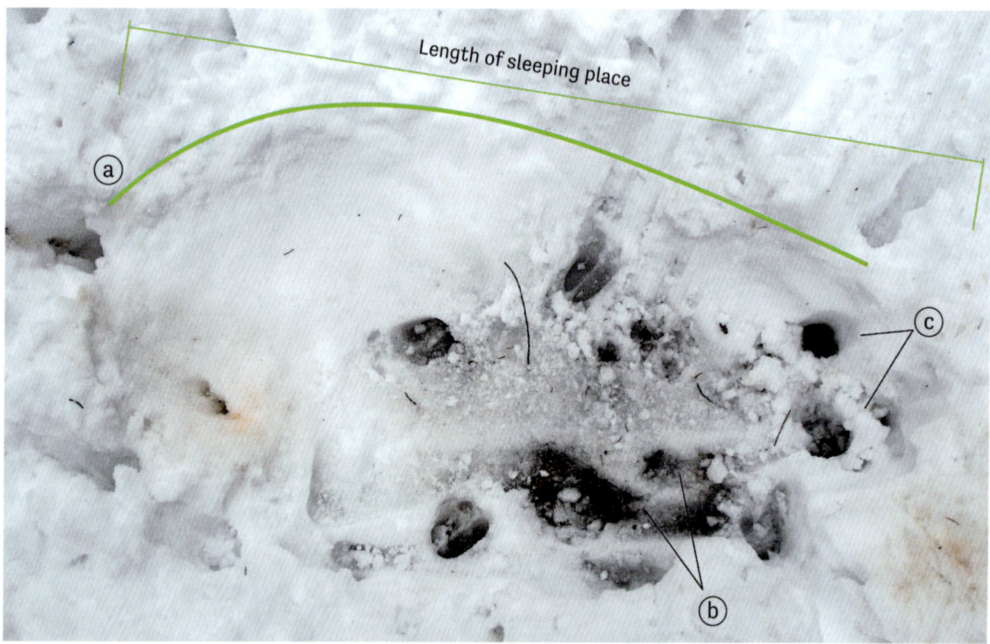

▲ *The bedding site of a Western Red Deer, which is typical of Cervidae. The wide, rounded end on the left is the rump and the narrow, tapered end on the right is the animal's front end. The straight, horizontal outside lines at the bottom of the image are made by the angled legs. Bavarian Forest, Germany.*

often features horizontal prints made by the legs that are folded up under the body (b). The outline of the bedding site is longer than it is wide and tapers to one end. At the tapered end, you can occasionally make out two distinctive, roundish prints, made by the carpal joints (c). In bedding sites with clear outlines, you might be able to make out the position of the animal's head, from the print of the muzzle. If you find the bedding sites of several animals, comparing different sizes and positions can provide information about the group structure. A single, relatively large bedding site further away than the others could be made by a male animal, while a clear size difference between several small and large prints could point to a herd of mothers and young animals. A series of bedding sites of similar size might come from a bachelor group. The bedding sites are often in well-covered border areas that offer protection and a good view.

Carnivore bedding sites

Canids and bears often curl up into a ball to sleep, resulting in almost perfectly circular bedding sites. Marks made by rolling and stretching are characteristic of cats in hot weather. They also make almost circular prints in low temperatures. Among carnivores, bears are most likely to use permanent bedding sites. The bedding sites of bears, lynx and wildcats are normally secluded and under cover. Cats often choose a slightly raised location on a slope or rocky outcrop with a good view. Wolves also prefer bedding sites that give a good view. Because they need to easily pick up scents, they often

▲ *The resting place of a Eurasian Lynx. The front legs are stretched out to the front and the hind legs are angled, which is typical of cats. Jämtland, Sweden. Laura Gärtner.*

▲ *Grey Wolves like to curl up in cold temperatures.*

choose slightly raised locations, such as large rocks or other elevated sites. Cover appears to be less important to them.

Dust and mud baths

Many animals take dust and mud baths to protect against insects and keep parasites at bay. Dust baths are used for coat care and often also for scent marking. They are typical of birds, rodents and lagomorphs but European Bison and Western Red Deer take dust baths too.

Bathing in mud, also known as wallowing, has similar benefits to dust bathing. It also helps to lower body temperature and the layer of mud that is applied protects against insect bites and stings. Animals often repeatedly visit popular wallowing sites and so the surrounding vegetation is usually heavily trampled, often with clear, wide paths leading to the site. Wallowing is typical of Wild Boar, Western Red and Sika Deer, as well as Reindeer, Elk and European Bison.

◄ *A Eurasian Otter dust bath. Elbauen, Germany.*

▼ *A Wild Boar mud bath. Spessart, Germany.*

SCRATCH MARKS, DIGGING MARKS AND MOUNDS OF EARTH

Scratch marks and digging marks in the ground range from shallow to deep, but are not connected to a tunnel or chamber. They can be difficult to identify because they are made by many different animals for many different reasons. Canids and many even-toed ungulates scrape and paw the ground to leave a scent trail and a visual sign as a marking. Many cats also leave scratch marks on the ground, both as a marking and to bury their droppings. Wild Boar, deer, Muskrats, Coypu, voles, Brown Bears and rabbits also dig for plant tubers and roots. Wild Boar can dig over considerable areas of ground with their snouts. Northern Raccoons, Brown Bears and European Badgers are known for digging out and raiding bee and wasp nests, as are Eurasian Honey-buzzards. Foxes and European Badgers also often dig for voles and other small mammals. Details that help to differentiate between the digging marks of these species can be found in the corresponding parts of the species descriptions. The following questions can be useful for determining and interpreting scratch and digging marks:

- Are there claw prints and, if so, have they been made by blunt or sharp claws?
- Why did the animal dig here? Were they leaving a marking or searching for food? If they were digging for food, what exactly were they searching for?
- What shape is the resulting hole in the ground?

Relatively large, flat area, scraped clean with clear scratch marks and without any mounds of earth ▶ **Scraping** ▶ **Western Roe Deer** (page 523).

◀ *Scraping is a form of territory marking. Waldsieversdorf, Germany.*

Small, flat, usually rectangular digging marks
▶ **Squirrels** (page 244), **mice** (page 326).

▶ *A field mouse has dug up an acorn. Vledder, Netherlands. Laura Gärtner*

Medium-sized digging marks ▶ **European Rabbits** (page 230), **Muskrat**, **Eurasian Beavers** and **Coypu** (pages 285, 268, 276), **canids** (page 384), **European Badgers** (page 450), **Beech Martens** (page 462) and **Northern Raccoons** (page 489).

◀ *A Red Fox has been digging for voles. European Badgers can leave similar digging marks (page 417). Lausitz, Germany.*

▲ *Coypu dig for roots near the riverbank when searching for food. Elbe, Germany.*

Medium to large digging marks ▶ **Brown Bears** (page 422)
and **European Badgers** (page 450).

◀ *European Badgers can dig very deep holes. A wasp nest was the goal here. Welzow, Germany.*

Large, rectangular sods of turf or areas of earth, dug out and turned over
▶ **Brown Bears** (page 426).

◀ *Brown Bears also turn over sods of turf while searching for food. Vogelgebirge, Slovakia. Immo Meyer.*

Lengthways, flat rooting marks, earth pushed forward and to the sides
▶ **Rooting** ▶ **Wild Boar** (page 506).

▲ *Wild Boar rooting.*
Märkische Schweiz,
Germany.

A mound is a pile of earth that initially appears to be unconnected with an excavation and doesn't have an obvious entrance. Molehills are probably among the better-known mounds, but a few other animals also leave small to large mounds of earth for a variety of reasons. Mounds made by moles and water voles are signs of their tunnelling lifestyle, while Eurasian Beavers, Muskrats and Eurasian Otters make mounds to mark territory. The distinctive communal mounds built by Steppe Mice are a familiar sign in drier regions of Eastern Europe.

Medium to large mounds of earth that have been scraped together with no entrance hole, often with secretions ▶ **Eurasian Beavers** (page 267), **cats** (page 363) and **Eurasian Otters** (page 435).

▲ *Mounds like this one made by a Eurasian Otter are a form of territory marking. Oderbruch, Germany.*

Normally blocked entrance hole in the middle of an evenly shaped mound of earth
▶ **Moles** (page 204).

◀ *The entrance hole at the top of a molehill is usually closed. Here, by way of an exception, is one with an open hole. Biebergemünd, Germany.*

Entrance hole on the side of an irregularly shaped mound of earth
▶ **Water voles** (page 290).

◀ *Side entrance hole made by a water vole. The irregularities in size and shape allow us to distinguish them from molehills. Hardegsen, Germany. Laura Gärtner.*

HOLES IN THE GROUND AND TUNNELS

This category describes openings in the ground that lead to a tunnel. Displaced earth can be found in front of them. Many animals use holes in the ground and subterranean tunnels and chambers for shelter. These can range from simple drop tubes with just one entrance to extensive complexes of chambers with several entrances and exits. European Rabbits dig warrens that are home to entire colonies of animals, connected by a complex tunnel system. Unused tunnels and chambers, such as an abandoned badger sett, are often used by other animals. The ejected material in front of the entrance hole is often the only sign of an underground dwelling. The size and shape of the entrance holes and the network of paths surrounding the structure, the presence of latrines or the positioning of territory markings give important indications of the species and the behaviour of the occupants. An occupied dwelling will show fresh signs of use, such as ejected material, tracks and droppings. Leaves will gather in an unoccupied dwelling and the entrance hole will soon be covered by cobwebs.

Ask yourself the following questions when identifying and interpreting holes in the ground:

- How large is the entrance hole and what shape is it?
- Is the entrance hole concealed or freely accessible?
- Is there ejected earth in front of it?
- Is there more than one entrance hole?
- Can you make out clearly trodden paths and are there other signs, such as latrines or dust baths, nearby?

Small holes in the ground, up to 8cm ▶ **Shrews** (page 199), **moles** (page 204), **squirrels** (page 242), **small Arvicolinae species** (page 278), **murids** (page 318), **amphibians** (page 741), **insects** (page 777) and **spiders** (page 778).

▲ *The often almost perfectly circular entrance hole of a wolf spider. A distinguishing feature is the material that has been gathered around the hole like a wall. Doñana National Park, Spain.*

▲ *The approximately 3cm-wide hole of a Wood Mouse. Wood Mice and Yellow-necked Mice can have a considerable pile of ejected material in front of the entrance hole to their dwelling. Märkische Schweiz, Germany.*

Medium-sized holes in the ground, up to 18cm ▶ **European Rabbits** (page 230), **Muskrats** (page 285).

◀ Entrance to a European Rabbit warren. The path to the entrance is trodden down and clearly visible. Hitzacker, Germany.

Large holes in the ground, wider than 18cm ▶ **Alpine Marmots** (page 247), **Coypu** (page 271), **canids** (page 384), **Wolverines** (page 442) and **European Badgers** (page 448).

◀ Entrance to a European Badger sett. Sett entrances come in different shapes and sizes. Vledder, Netherlands.

INSECTIVORES

In Europe, three families belong to the order of insectivores (Eulipotyphla): hedgehogs (Erinaceidae), moles and their close relatives (Talpidae) and shrews (Soricidae). The status of this order is the subject of ongoing debate, as, for example, there is dispute as to whether or not it should include hedgehogs.

There is no one common characteristic that clearly distinguishes insectivores from similar orders. Insectivores are relatively small mammals that mainly eat invertebrates. Their dentition is adapted to their predatory behaviour and consists of continuous rows of sharp, pointed teeth. Visual perception plays a secondary role in insectivores. Their highly developed sense of smell and touch are dominant. Many shrews, for example, use a form of echolocation to track down their prey.

TRACKS AND SIGNS OF INSECTIVORES

All insectivores are plantigrades with five toes and five claws on the front feet and hind feet. With the exception of moles, whose front feet are specialised for digging, insectivores have the most basal (closest to the mammalian common ancestor) foot structure of mammals living today.

Hedgehogs

Hedgehogs (Erinaceidae) are known for their spines and their ability to curl up into a spiky ball when threatened. They have a compact, roundish body, an elongated snout and a short tail. Two genera with three species are found in the region. The North African Hedgehog (*Atelerix algirus*), the Northern White-breasted Hedgehog (*Erinaceus roumanicus*) and the West European Hedgehog (*Erinaceus europaeus*) are the most common hedgehog species in Western and Central Europe. The North

African Hedgehog is only found in small numbers in southern France, the Balearics and the Canaries, and on the southeastern coast of Spain. West European Hedgehog tracks can also be found in these areas. An approximately 200km wide area where the range of the West European Hedgehog overlaps with that of the Northern White-breasted Hedgehog extends from the northern Baltic States, in western Poland, through the Czech Republic and Austria, to the Adriatic Coast in northern Italy (Neumeier 2006). The Northern White-breasted Hedgehog is mainly distributed in Eastern and Southeastern Europe. The difference between the tracks of different hedgehogs has not been sufficiently researched. The following looks more closely at the West European Hedgehog because it is the most widely distributed species in Western Europe.

TRACKS AND SIGNS OF HEDGEHOGS

Plantigrade. Asymmetrical. Normally five relatively short, wide, ribbed toe prints and a fused midfoot pad that resembles an L shape. Long, blunt claw prints are usually visible in the track. Hedgehog and rat tracks can be confused with each other.

- Track formula: 5f × 5H + C

WEST EUROPEAN HEDGEHOG

Erinaceus europaeus

HTL 18–28cm

TL 1–4cm

W 450–1,200g
Weight varies greatly according to season and availability of food. Over 2kg is exceptional. Average summer weight: 800–1,000g.

Hedgehogs are among the most well-loved wild animals in Europe. Their dominant sense is their sense of smell. They also have good hearing and are surprisingly adept at climbing and swimming, as well as being relatively fast runners. Their sight is comparatively weak and they are mainly nocturnal. During the day, hedgehogs normally sleep in the safety of their daytime hiding place, which can be piles of leaves and brushwood, hollow trees or cavities in stables and barns. Hedgehogs build a winter nest from leaves and brushwood before their five- to six-month hibernation begins in October/November. During hibernation, their body temperature falls from 35°C to ambient temperature. Their heart rate drops from an average of 147 beats per minute to 2–12 beats per minute. Before they hibernate, hedgehogs feed up to build fat reserves that they use up over winter. They lose around 25 per cent of their body weight during hibernation. Hedgehogs' natural predators include European Badgers, Red Foxes, Eurasian Eagle-owls, American Mink and European Polecats.

DISTINGUISHING FEATURES Characteristic spiky upperside made up of hairs that form spines. Round, thickset body with a pointed snout-like nose. The animal appears short-legged because its rather long legs are covered by fur at the sides. Short tail, strong claws.

DISTRIBUTION AND HABITAT Hedgehogs prefer deciduous and mixed forests with dense undergrowth. They are often found on the edges of forests with adjoining grassy areas, as well as in thickets, hedges, gardens and parks. Hedgehogs are found from sea level to over 2,000m. They tend to avoid damp areas and pine forests, which presumably lack sufficient nesting materials for hibernation.

DIET Almost exclusively ground-dwelling invertebrates. Their diet mostly consists of beetles, caterpillars and earthworms but they also eat other insects, snails, millipedes, centipedes and spiders. More rarely, hedgehogs will eat young mice, carcasses, frogs and birds' eggs. They also eat fungi, acorns and windfall fruit. Hedgehogs are often said to eat snakes but only a few observations of this have been documented. However, it has been possible to prove a certain degree of immunity to snake venom.

REPRODUCTION Mating season usually begins in March/April once the animals have woken up from hibernation. They normally only have one litter in a season. Hedgehogs usually give birth in June, when 2–7 (average 4–5) young are born after a gestation period of 31–36 days. The young are born altricial, becoming independent and beginning their life as solitary animals after 4–6 weeks. They reach sexual maturity after a year.

SIGNS

NESTS West European Hedgehogs make two types of nest: summer and winter. Summer nests are usually just simple hollows in the vegetation, while winter nests are made in large piles of leaves and brushwood.

DIGGING MARKS Digging marks can be found on grassy areas where the hedgehog has been searching for insect larvae with its snout-like nose.

EXCREMENT The cylindrical droppings are normally densely compressed, straight and blunt-ended. The remains of insects that the animal has eaten contain chitin, which give the excrement its usually shiny, black appearance. This is also why dried droppings can disintegrate and become crumbly. If an animal has eaten fur or feathers (mice or birds), droppings can be kinked, as is otherwise typical of mustelids, and matt grey in colour. In this case, hedgehog droppings can be confused with Stoat or European Polecat droppings, but are wider in comparison.

- L 1.5–5cm D 0.8–1.2cm

▼ *A summer hedgehog nest is often just a hollow that the hedgehog has made in the vegetation. Vledder, Netherlands. René Nauta.*

▲ *Winter hedgehog nests are usually larger piles of leaves and brushwood. Vledder, Netherlands. René Nauta.*

▼ *Hedgehog droppings normally contain the remains of many insects. Biebergemünd, Germany.*

TRACK

Front

L 1.6–4cm W 2.1–3.2cm

Small. Slightly asymmetrical. Five toes. Prints of toe 1 are unusual; it is often only possible to make out the corresponding claw. Toes 2–5 are short, powerful, ribbed and shorter than on the hind foot. They usually appear radial and connected with the midfoot pads. Some of the interdigital pads are fused and midfoot pads II-IV are noticeably large. Another thenar/hypothenar pad is very occasionally printed at the back of the track. The powerful, blunt claws are used for digging and are slightly shorter than on the hind foot. The tracks have a compact appearance with a small negative space. The front feet are slightly smaller than the hind feet and appear slightly wider in the print.

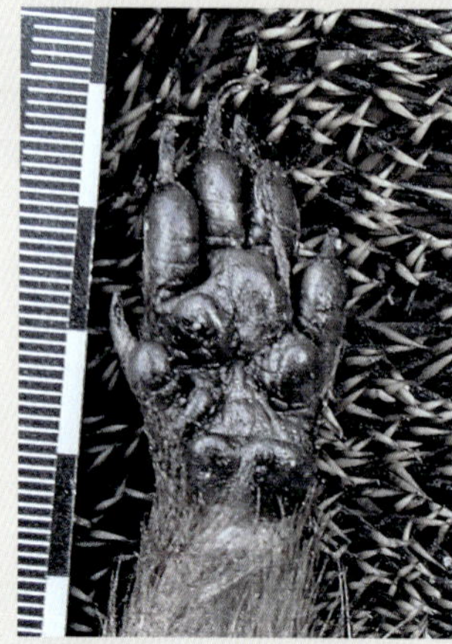

▲ *Left front.*
Märkische Schweiz, Germany.

▲ *Right hind.*
Märkische Schweiz, Germany.

▲ *Left front.*

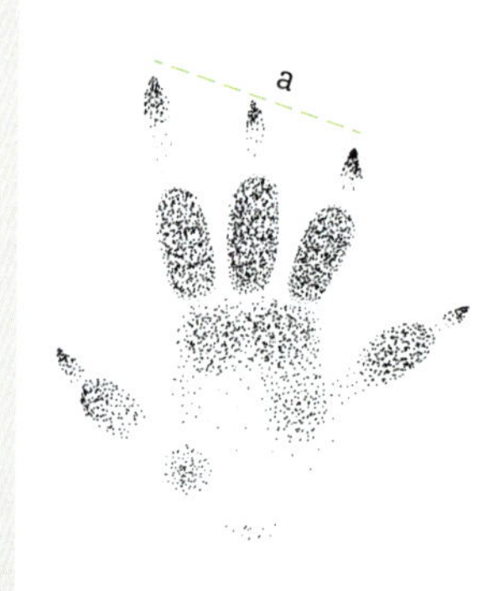

▲ *Right hind.*

▲ *Left front. Note the claw print of toe 1 on the left of the two cent coin.*
Märkische Schweiz, Germany.

▲ *Right hind.*
Märkische Schweiz, Germany.

Hind

L 2.5–4.5cm W 2.1–4.5cm

Small. Asymmetrical. Five toes. Toes 1 and 5 stick out to the side; toes 2, 3 and 4 are of a similar size and point forward. Toe 1 is much further back than on the front foot. The toes are short, powerful and ribbed but longer than on the front foot. Their prints usually merge with the midfoot pads. Some of the interdigital pads are fused and midfoot pads II-IV are noticeably large. The long, powerful claws often leave clear prints. The length of claw 2 is a characteristic distinguishing feature. It is the longest claw and slightly longer than claw 3. Claw 4 is slightly shorter than claw 3, creating an often prominent angle to the outside (a). Another thenar/hypothenar pad is very occasionally printed at the back of the track. The tracks have a compact appearance with a small negative space. The hind foot is slightly larger than the front foot and creates a longer, narrower print.

GAITS

Hedgehogs mainly walk. As they walk, the front feet are often turned inward and the hind feet are usually turned outward, resulting in a distinctive track pattern.

Walk
Stride length: 7–21cm
Trail width: 5.1–10cm

Trot
Stride length:
18.5–30cm
Trail width: 5–7.5cm

▲ *A typical hedgehog gait: overstep walk with front foot prints that turn inward. Vledder, Netherlands.*

▲ *Direct register walk.*

Similar tracks
Black and Brown Rat, Red and Grey Squirrel, Edible Dormouse.

SHREWS
Soricidae

HTL 3.5–9.7cm
TL 2.5–7.8cm
W 1.5–23g
Slight variation within individual species.

The shrew family includes very small to mouse-sized, diurnal and nocturnal insectivores. The majority are ground dwelling, but water shrews have adapted to swim and some species climb up smaller shrubs to find food. Four genera with a total of 19 species occur in Europe. Bicoloured Shrew, Greater White-toothed Shrew, Sicilian Shrew and Lesser White-toothed Shrew belong to the genus of white-toothed shrews (*Crocidura*). The Etruscan Shrew is a representative of the genus *Suncus*. The Iberian Water Shrew and the Eurasian Water Shrew belong to the genus of water shrews (*Neomys*). In Europe, the genus of long-tailed shrews (*Sorex*) includes the most species and is the most widespread in the region. Its members include Alpine Shrew, Valais Shrew, Common Shrew, Udine Shrew, Masked Shrew, Crowned Shrew, Taiga Shrew, Eurasian Least Shrew, Eurasian Pygmy Shrew and Tundra Shrew, as well as the Iberian Shrew and Appenine Shrew. Most species can only be reliably identified by detailed skull examination, which is not usually possible in the field. Tracks are rarely found.

TRACKS AND SIGNS OF SHREWS

It is difficult to tell the difference between the tracks of *Sorex* and *Crocidura* shrews. The distinguishing features given here are easier to make out in ink prints than in the field. Nevertheless, it is possible to distinguish between the two genera in the field if ground conditions are ideal and the prints are detailed. Clear tracks occasionally allow differentiation between water shrews and other shrews because their hind feet are noticeably larger than their front feet as they have adapted to be able to swim strongly. The Eurasian Water Shrew (*Neomys fodiens*) has stiff hairs on its hind feet to help it swim. These can be visible in very clear prints. The print of the keel of swimming hairs on the underside of the tail is another indication. The tracks of *Sorex* and *Crocidura* shrews are discussed on page 196.

- Track formula: 5f × 5H + C

Shrews have an excellent sense of smell and touch. They are incredibly fast and are capable of killing animals that could harm them, such as centipedes or wasps. Many shrews have venomous saliva that paralyses their prey. Some species are good at climbing. Shrews have rather weak eyesight; they can distinguish between light and dark dark but do not have detailed visual perception. They mainly use the tactile whiskers or hairs (vibrissae) on their snout and tail to navigate their way through their tunnels. An unusual behaviour of *Crocidura* shrews is that they form a chain or 'caravan' when in danger. Led by the mother, the young each bite onto the tail of the animal in front and they flee together. They are predated by owls and most smaller predators. Domestic Cats and some other mammals are known for killing, but not eating, shrews, so whole carcasses are often found.

DISTINGUISHING FEATURES   Very small to mouse-sized animals with elongated bodies, small eyes and small ears. They have a distinctive, almost trunk-like, flexible snout. Their tail is normally shorter than their body. The tips of the teeth of *Sorex* shrews are stained red by a ferrous pigment. The Etruscan Shrew (*Suncus etruscus*) is the smallest known mammal in the world by mass.

TRACK

Front

L 0.5–1cm   W 0.5–0.9cm

Tiny to very small. Plantigrade. Slightly asymmetrical. Five toes: toe 3 is longer than toes 2 and 4 so the tips form a slight arc. The midfoot pads consist of four individual interdigital pads, not all of which are always visible. Two other thenar and hypothenar pads are visible at the back of the track. Tracks with claw prints are slightly more common than tracks without. The front feet are slightly smaller than the hind feet.

Hind

L 0.7–1.4cm   W 0.5–1cm

Tiny to very small. Plantigrade. Symmetrical toes, asymmetrical midfoot pads. Five toes: toe 3 is approximately the same length as toes 2 and 4 so the tips of the toes are in a row. The midfoot pads consist of four individual interdigital pads, not all of which are always visible. Tracks with claw prints are slightly more common than tracks without. Two other thenar and hypothenar pads are visible at the back of the track. The hind foot is normally slightly larger than the front foot.

▲ *Common Shrew, left front. Märkische Schweiz, Germany.*

▲ *Common Shrew, left hind. Märkische Schweiz, Germany.*

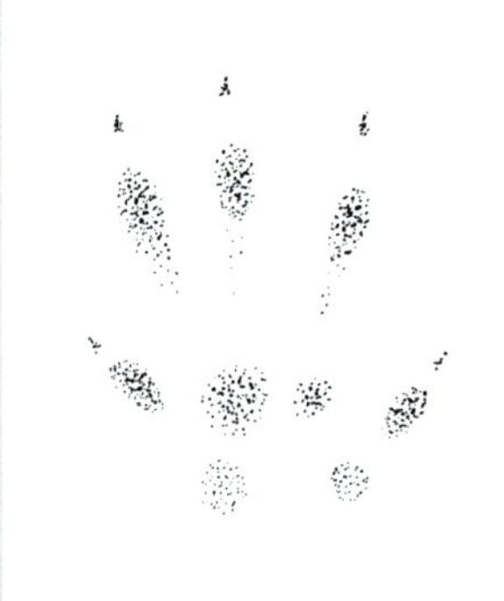

▲ *Common Shrew, left front.*

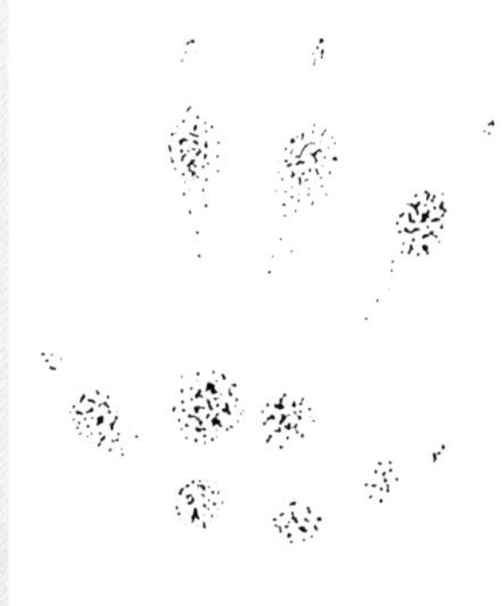

▲ *Common Shrew, left hind.*

▲ *Unknown shrew species, right front (left) next to right hind (right). Mühlheim, Germany. Simone Roters.*

GAITS

Shrews mainly walk or trot. They usually only bound or gallop over a short distance, especially when in danger or to cross an open space. Shrews have a higher variance of gaits compared to field mice. Voles exhibit similar behaviour in choice of gait, but shrew paths are much narrower (usually less than 2.5cm). In snowy conditions, shrews often move under the covering of snow. They normally stay close to the ground and leaf cover where they can hunt while staying safe from their own predators.

Walk and trot
Stride length: 3.5–10cm
Trail width: 1–4cm

Gallop
Group length: 1–4cm
Inter-group length: 4–14cm
Stride length: 6–16.5cm
Trail width: 2–4cm

▲ *Unknown shrew species in direct register walk. Midhurst, England.*

Similar tracks

The tracks of smaller European Common Vole species can be confused with those of shrews.

▲ Sorex *shrew species, direct register trot.*

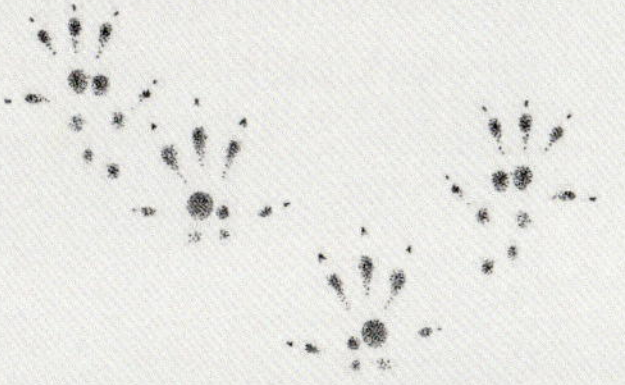

▲ Crocidura *shrew species, parallel bound.*

DIFFERENTIATING BETWEEN SHREWS (*SOREX* AND *CROCIDURA* SPP.)

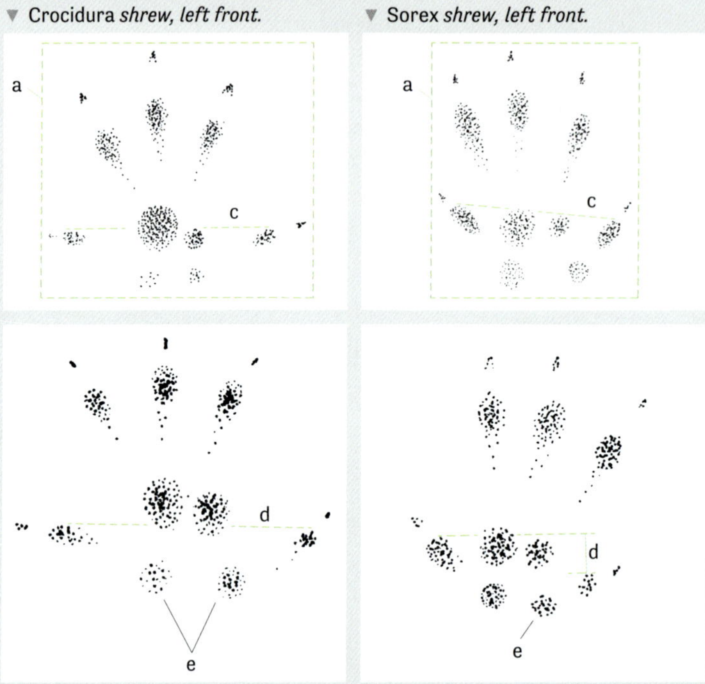

▼ Crocidura *shrew, left front.*

▼ Sorex *shrew, left front.*

▲ Crocidura *shrew, left hind.*

▲ Sorex *shrew, left hind.*

Crocidura shrews

a) Roundish-square track outline.
b) Toes more spread out, arranged in a star shape.
c) Front edge of toes 1 and 5 behind the front edge of the midfoot pad.
d) Toe 5 is relatively far back, almost in line with toe 1.
e) Metacarpal/metatarsal pads are relatively far apart, pads I and IV can appear out of line.

Sorex shrews

a) Rectangular track outline.
b) Toes tend to point forwards, toes 2–4 more parallel.
c) Front edge of toes 1 and 5 approximately level with the front edge of the midfoot pad.
d) Toe 1 is further back than toe 5.
e) Metacarpal/metatarsal pads are relatively close together, pad I can appear out of line.

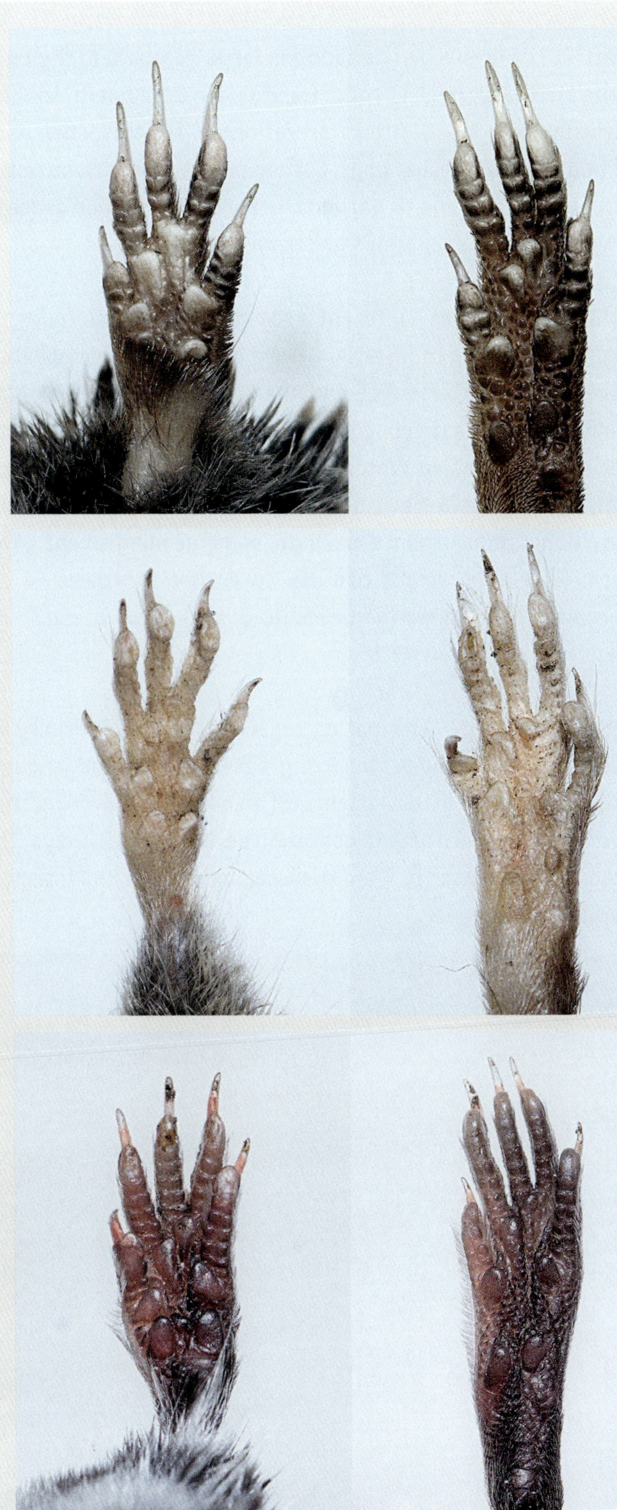

◀ *From top to bottom: left, the left front foot and right, the left hind foot of Common (Sorex araneaus), Greater White-toothed (Crocidura russula) and Iberian Water (Neomys anomalus) Shrews. Ennstal, Austria. Stefan Resch.*

HABITAT Shrews can be found in large numbers in the most diverse habitats throughout Europe. Habitat can be a useful tool for species identification. Most shrews prefer damp habitats, such as the banks of bodies of water and bogs. Some species prefer to be near buildings or in cultivated land or gardens, while others favour dense forests or dry areas, such as steppes and semi-deserts.

DIET Insectivores. Different species eat different foods, but a diet mainly consisting of insects, worms and other invertebrates, such as spiders, beetles, woodlice and snails, is common to all of them. They sometimes also eat young mice and carrion. Some species occasionally eat plants. Eurasian Water Shrews catch small fish and frogs in the water. All shrews have an exceptionally fast metabolism so their nutritional requirements are high, some needing to eat several times their own body weight per day so as not to starve to death. The Eurasian Least Shrew (*Sorex minutissimus*) needs to eat 2–5 times its own body weight every day.

REPRODUCTION The mating season lasts from February to October with the main activity occurring between May and August. Shrews usually have 1–4 litters, each with an average of 3–5 (up to 11) young. The young are born after a gestation period of 20–30 days. The animals reach sexual maturity from their second year at the latest, but often during the first year.

▼ *Etruscan Shrew (Suncus etruscus) in an underground passage, followed by its young. Southern Europe.*

SIGNS

ENTRANCE HOLES   The entrances to shrews' tunnel systems are very small, normally no wider than the diameter of an index finger. Diameter approx. 2cm.

NESTS   Shrews build simple nests under stones and roots, usually by piling up nesting material, such as grass and leaves. The entrances are usually not clearly visible.

EXCREMENT   Shrew droppings vary greatly according to diet. If the animal's diet mainly consists of earthworms or snails, the droppings are usually just a small, mushy patch. If the animal has eaten more solid food, it will produce very small, cylindrical droppings that are either rounded or taper to a point at one end. As in the case of other insectivores, the excrement is relatively brittle and often has an irregular and grainy surface. It is brownish to blackish in colour. Shrew droppings are usually found along their paths and in larger piles near the entrance holes to their burrows. Water shrew droppings can be distinguished from that of other shrews. They are mainly found on the banks of bodies of water and in wetland areas and can be curved in a V-shape. They have a characteristic pungent odour reminiscent of grilled crab. Measuring 6–12mm in length and with a diameter of 3-4mm, they tend to be larger than droppings from other shrews, which are usually less than 4mm long and 3mm in diameter. The droppings of the genera *Suncus*, *Crocidura* and *Sorex* are usually relatively straight and have a slightly sweet odour.

- L 3–12mm   D 1.5–4mm

◄ *Shrew dropping.*
Ennstal, Austria.
Stefan Resch.

EURASIAN MOLES
Talpa

HTL 9–16cm
TL 1.4–4.5cm
W 30–130g
Males larger than
females.

In the region covered by this book, the family Talpidae (moles, shrew-moles and desmans) is represented by two genera. The Pyrenean Desman, which is highly adapted to an aquatic lifestyle, is the only representative of the genus *Galemys* and will not be discussed further due to a lack of data. All the moles described in greater detail belong to the genus of Eurasian moles (*Talpa*), of which the Blind Mole (*Talpa caeca*), European Mole (*T. europaea*), Levantine Mole (*T. levantis*), Iberian Mole (*T. occidentalis*), Roman Mole (*T. romana*) and Balkan Mole (*T. stankovici*) are found in our region. Eurasian moles all make almost identical tracks and signs. They are described here using the European Mole as an example.

- **Track formula: 5F × 5h + C**

Eurasian moles are active both during the day and at night and tend to be solitary. They spend most of their life underground and are superbly adapted to their subterranean way of life. They use their shovel-like front feet to dig extensive tunnel systems, as well as nesting and storage chambers. These digging activities result in the molehills that we are all familiar with. Moles find their way underground using their excellent sense of smell and touch. The many tactile hairs (vibrissae) on their snout give them a kind of 'tactile image' of their surroundings. Their vision is poor and most moles' eyes are covered with skin. The European Mole is the only mole species with open eyes that enable it to distinguish between light and dark. Moles do not have external ears but they do have an external fold of skin that allows them to close their ears to keep out soil when digging. They can hear low-pitched sounds very well. Their predators are owls, such as Barn, Tawny and Eurasian Eagle-owl, as well as Domestic Cats, Red Foxes, Wild Boar, mustelids, raptors and White Storks.

DISTINGUISHING FEATURES Slightly larger than mice. Cylindrical body with short legs and highly developed feet. The front feet turn outwards and have developed into 'shovels'. Moles have a short black coat with a velvety shine. Their snout, tail and neck are short.

DISTRIBUTION AND HABITAT The Blind Mole is found in Switzerland, Italy and the Balkans. It prefers higher altitudes and loamy or sandy soils. The European Mole is dominant over the Blind Mole, so the two species do not occur in the same area. The European Mole is found throughout Europe, where it lives in meadows, fields, gardens and deciduous forests. Levantine Moles only occur on the European continent in a small population west of the Bosporus and prefer moist ground at the edge of water bodies. The Iberian Mole is only found on the Iberian Peninsula. It likes moist ground in meadows, fields and forests. The Roman Mole lives in Italy, in similar habitats to the Iberian Mole. The Balkan Mole lives in the southern part of the Balkan Peninsula, in northern Greece, Macedonia, Montenegro, Serbia and Albania. It prefers meadows and forests.

DIET Insectivores. Moles find most of their prey on regular searches of their hunting grounds. The earthworms and insect larvae that make up most of their prey enter the moles' tunnels where they are easy to capture. If there is a surplus of food, moles store live earthworms in underground storage chambers, having immobilised the worms with a bite to the head. Their diet also includes snails, young mice and carrion.

TRACK

Front

L 1–2.5cm   W 1–1.9cm

Very small. Plantigrade. Very asymmetrical. Five flat, short toes with broad, powerful claws and a large, hairless midfoot pad. The front foot is wider than it is long. Prints are normally only left by the large, powerful claws of toes 3–5. In rare cases, the prints of the claws of toes 1 and 2 are also visible. In very rare cases, the print of a kind of 'sixth toe' can be recognised. It is a small, sickle-shaped bone on the inside of the body in front of toe 1 that adds extra width to the foot, and which has evolved into a digging shovel. To propel themselves along, moles move their front feet right in to the middle of their body so they are almost parallel to the direction of travel. In this case, toe 1 will be at the back and toe 5 at the front of the track. The track shows an elongated set of claw prints arching to the outside. They can resemble the hind foot print of a toad (page 727). The front feet are much larger and wider than the hind feet.

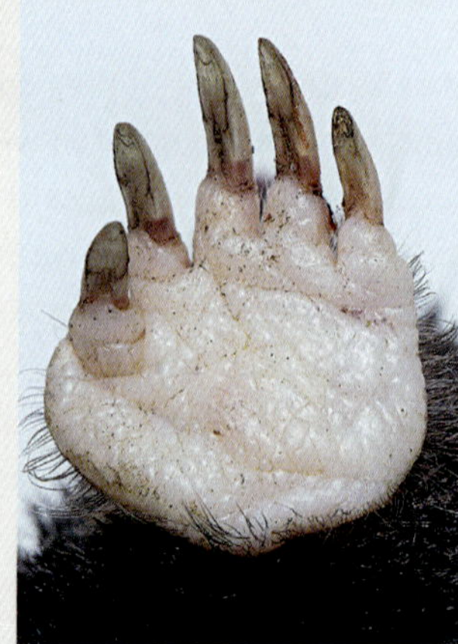

▲ *European Mole, left front. Ennstal, Austria. Stefan Resch.*

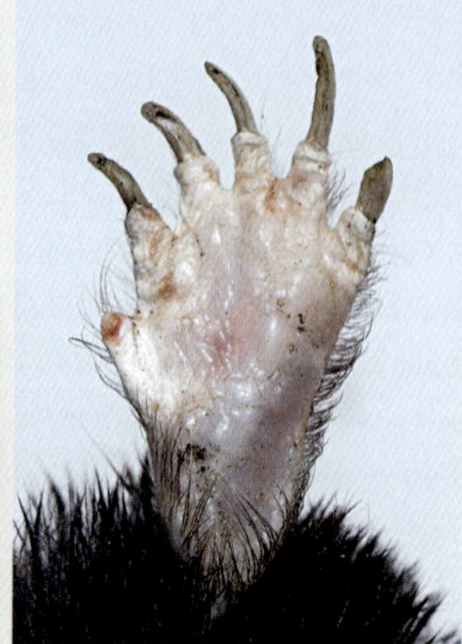

▲ *European Mole, left hind. Ennstal, Austria. Stefan Resch.*

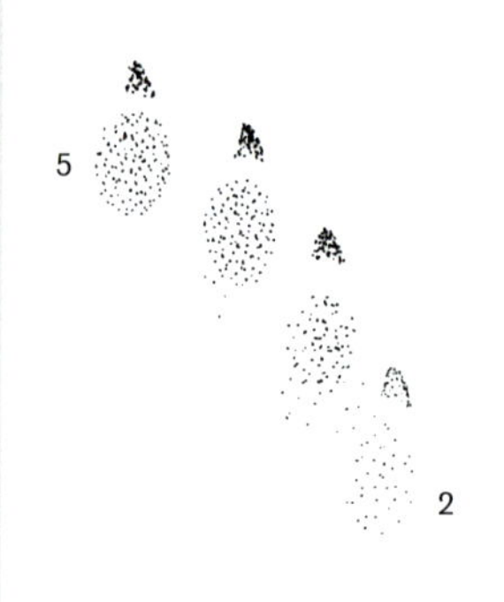

▲ *Left front.*

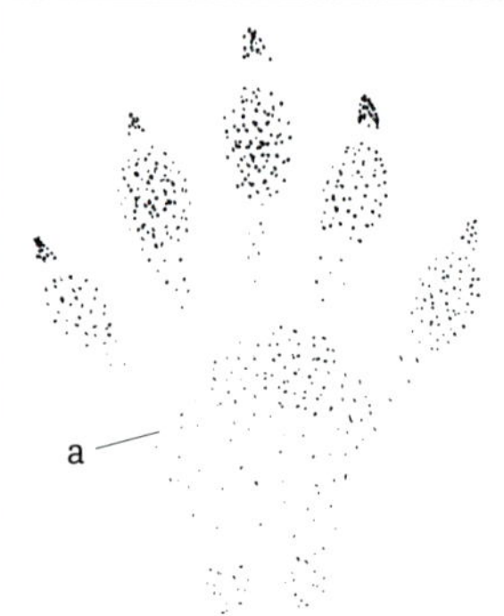

▲ *Left hind.*

▲ *From left to right and from top to bottom: LF, RH, LH, RF. Göttingen, Germany. Heide Ulrich.*

Hind

L 0.5–2.2cm W 1.1–1.6cm

Very small. Plantigrade. Symmetrical. Five toes with claws pointing forwards. The claws are often the only part of the hind feet that leave visible tracks. They are long, but shorter and finer than on the front feet and reliably leave prints. The midfoot pad almost never leaves a print (a) and the digital pads are very rarely recognisable. In rare cases, the print of a 'sixth toe' can be seen, made by a small bone on the inside of the body below toe 1. The hind feet are longer than they are wide. If the rearmost thenar/hypothenar pad leaves a print, the track will clearly taper towards the back. The hind feet are much smaller and narrower than the front feet.

GAITS

Moles rarely come above ground, so their tracks are equally rare. The track pattern corresponds most closely to an understep walk. The animals can pull and push themselves along the ground surprisingly quickly. Prints made by the animal's fur and body can be found in soft ground. Moles come to the surface to gather nesting materials, to hunt in the leaf layer or to make latrines. Their paths start and finish at an entrance hole.

Walk
Stride length:
7–12cm
Trail width:
3.5–6cm

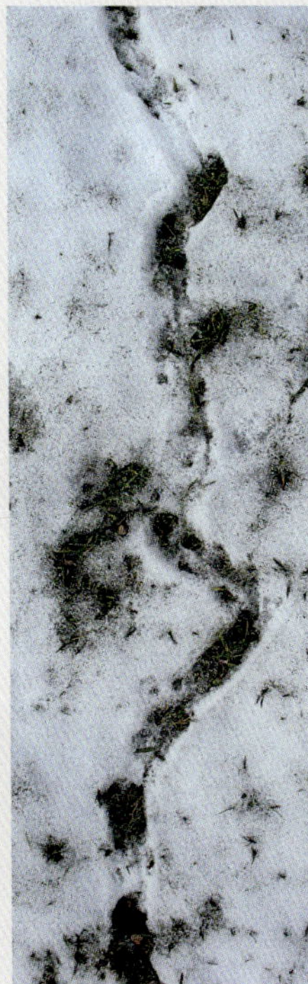

▶ *Moles can cover ground surprisingly fast. Their paths can make it seem as if they are clumsily 'searching'. Bielefeld, Germany. Ulrike Quartier.*

▲ *Finding a perfect mole track is exceptionally rare. Note the print of the 'sixth toe' on the hind foot prints. Göttingen, Germany. Heide Ulrich.*

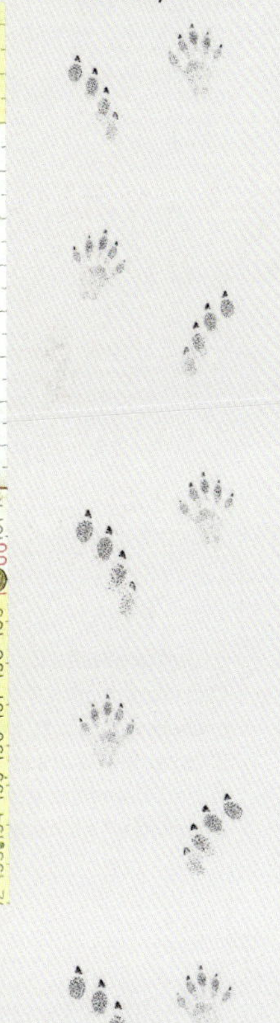

▼ *Understep walk.*

Similar tracks

Mole tracks can be mistaken for toad tracks.

REPRODUCTION Mating season is normally between March and May, resulting in a litter of 3–4 (2–8) young in May/June. The animals reach sexual maturity after about ten months.

SIGNS

MOLEHILLS These very obvious mounds of earth are very familiar, although few people have ever seen the animals themselves. The mounds, which normally occur in large numbers, are usually about 25cm high, roundish and almost symmetrical. They are formed when moles dig nesting chambers and tunnels and transport the loosened soil to the surface with their front feet. They push the fine, crumbly soil out of a vertical tunnel in the middle of the hill. A central entrance hole is characteristic of mole burrows, but it is rarely visible as the animals usually close it up again with soil. Water vole hills look similar and can be mistaken for molehills, but are longer and narrower in shape and the exit hole is on the side. The ejected soil is also coarser and interspersed with roots and other plant remains. In rare cases, the entrance hole of a molehill can also be on the side. The entrance holes have a diameter of 4–6cm.

SURFACE TUNNELS Relatively shallow tunnels that are normally either used for hunting or created when expanding the tunnel system. Collapsed tunnels are also a distinctive sign and can mainly be found on forest paths and in fields and meadows. They often stand out because of the clear colour contrast with the ground around them.

EXCREMENT Mole droppings vary greatly according to diet. If the animal's diet mainly consists of earthworms or snails, the droppings

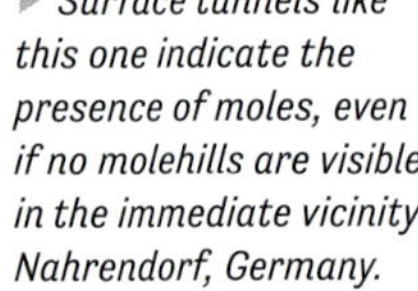

▶ *Surface tunnels like this one indicate the presence of moles, even if no molehills are visible in the immediate vicinity. Nahrendorf, Germany.*

◀ *Molehills usually occur in large numbers and in different sizes. West Sussex, England.*

▲ *If the animal has eaten more solid food, mole droppings will be sausage-shaped with an irregular surface. Vledder, Netherlands. René Nauta.*

▲ *A characteristic mole latrine. Midhurst, England. Nate Harvey.*

are usually a small, mushy patch. If the food eaten was firmer, the animal produces long, sausage-shaped droppings that are either rounded at both ends or taper to a point at one end. As in the case of other insectivores, the droppings are very brittle and often have an irregular and grainy surface. They are brownish to blackish in colour. Some moles make above-ground latrines. These are usually found in the shelter of fallen branches, pieces of wood that are lying around or above-ground tree roots.

• L 1–2.5cm D 0.2–0.5cm

BATS

After rodents, bats (Chiroptera) make up the world's most species-rich mammal order. Around 40 species are found in Europe and most of them belong to the family of vesper bats (Vespertilionidae).

ultrasonic echolocation, which works like radar. As bats spend most of their life in the air and very rarely move on the ground, there is very little data on their tracks. It is often only possible to tell the difference between individual species using molecular biology. For this reason, the entire order is treated here in a single portrait.

DISTINGUISHING FEATURES Brown to grey. Large, stretchy patagium (wing membrane) on the forearm. Large or very large ears.

DIET Insectivores. Most species almost exclusively hunt flying insects like moths, but may also pick beetles and spiders from vegetation. Bats can eat up to 50 per cent of their body weight in insects in one night.

DISTRIBUTION AND HABITAT
Varies according to species. Some species prefer to live close to settlements, while others favour continuous woodland areas. Bats roost in tree holes, caves, attics, crevices in houses and under bridges.

Bats in Europe are nocturnal and almost all species hibernate, normally in more sheltered winter quarters. To overwinter, they need stable temperatures above freezing, high humidity and quiet. Bats are the only mammals capable of true flight and they can often be seen hunting insects on the wing on warm summer nights. They find insects using

HTL 3–9cm
TL 2–7cm
W 4–45g

REPRODUCTION   Almost all species gather in autumn in preparation for hibernation. During this time, the males mate with the females so that they can all hibernate together. However, the females store the sperm, with fertilisation taking place the following spring. When ready to give birth to their pups in May–June, the females gather in so-called maternity roosts, holding sometimes more than 1,000 females. They produce an average of 1–2 young. The pups are born hairless and blind and become independent after 4–6 weeks.

TRACKS AND SIGNS OF BATS

Bat tracks are rarely found. It is currently only possible to tell different species apart by process of elimination according to size. Large hind foot prints and large droppings tend to be produced by the larger species, such as the Common Noctule (*Nyctalus noctula*) or the Greater Mouse-eared Bat (*Myotis myotis*). Measurements in the lower ranges tend to come from smaller species, such as the Soprano Pipistrelle or the Common Pipistrelle. However, there is an extensive area of overlap between large and small species where identification based on tracks and signs is not yet possible. In very rare cases, their tracks can be found in dusty attics or on the ground near muddy forest paths and puddles.

- **Track formula: 1f × 5 H + C**

FRONT FOOT   The highly specialised front foot has developed into a wing. Toes 2–5 have become part of the wing and help to stretch the wing membrane on the arm. Toe 1 is unattached, very mobile and has a strong claw.

HIND FOOT   Five toes with powerful claws, can be locked for hanging upside down to rest.

Bats can carry rabies, but this poses little danger to humans unless bats are handled. All native bat species are protected.

GAIT   Bats usually move on the ground in a walking gait. The claw of toe 1 of the front foot, the five clawed toes of the hind foot and the usually visible tail print leave a distinctive track pattern.

EXCREMENT   Tiny to small, dark brown or mostly black, depending on the species. The elongated, cylindrical faecal pellets resemble mouse droppings with slightly pointed ends. Unlike mouse droppings, however, the faecal pellets are dry, porous, crumbly and often in several parts. They are made up of tiny insect remains (chitin), which give the droppings their characteristic features. Bats keep returning to successful resting and roosting sites, where very large piles of pellets can be found.

- L 0.2–1.5cm D 0.2–0.4cm

◀ *Large piles of pellets can be found under popular resting and roosting sites. West Sussex, England.*

▶ *Close-up of bat droppings. Extertal, Germany.*

LAGOMORPHS

Lagomorphs (Lagomorpha) were formerly categorised as rodents, but are now regarded as a separate order, consisting of the pika family (Ochotonidae) and the family of rabbits and hares (Leporidae).

Seven species are found in Europe, all of which belong to the family Leporidae (rabbits and hares). We focus on the three most common species. Their skull structure and their inability to hold food with their front feet are the main features that distinguish them from rodents. Their teeth are also very different. Like rodents, lagomorphs do not have any canine teeth and their incisors grow continuously. However, they also have a pair of small, peg-like incisors in the upper jaw, behind the incisors. Other distinguishing features of lagomorphs are very long hind feet, long ears, a short, hairy tail and a split upper lip. Lagomorphs are pure herbivores.

FRONT FOOT Clear asymmetry, toe 1 much further back and often only recognisable as a claw print.

HIND FOOT Slightly asymmetrical, toes 2 and 5 and toes 3 and 4 are in more of a horizontal line than in the more asymmetrical front foot. Nevertheless, the track is more asymmetrical than the hind foot print of a canid.

GAITS Usually a half/angled bound, where the front feet touch the ground one behind the other. The hopping that is typical of rabbits and hares is common when the animal is calm and relaxed.

FEEDING MARKS Sharp incisors usually leave clean cuts on plants. Typical feeding marks are left when the animals eat cambium, for example on shrubs or on branches with thin bark that have been blown down. In contrast to cambium feeding marks made by rodents and cervids, the damage caused by hares and rabbits extends deeper into the wood. In areas with high hare or rabbit population densities, conspicuous, even, horizontal browse lines can be found on trees and shrubs. This distinctive sign occurs when the animals have eaten all the leaves, shoots, buds and twigs within their reach.

TRACKS AND SIGNS OF RABBITS AND HARES

The soles of the feet of all rabbits and hares are completely covered with fur, so the midfoot pads are covered and the tracks often appear indistinct. The toes of the front feet and hind feet normally leave deeper prints than the midfoot pads and, on harder ground, often only the claw prints can be seen.

- Track formula: 5f × 4H + C

▲ *The appearance of hare tracks can vary greatly depending on the ground and the animal's speed and gait. Compare these left hind foot prints (including page opposite).*

EXCREMENT Hares and rabbits are coprophagous. Their caecum produces soft, vitamin-packed caecal pellets that the animals usually ingest again immediately after passing them. This enables better digestion and nutrient absorption. Finding caecal pellets is rare. The second type of droppings are oval to spherical, usually slightly flattened, dry faecal pellets of various sizes. Unlike even-toed ungulates, rabbits and hares excrete their faecal pellets individually. Larger piles of pellets mean that an animal has been in one place for a longer time or has returned there repeatedly, or that the site has been visited by several individuals (group feeding sites).

RESTING PLACES Brown and Mountain Hares rest in shallow depressions in the ground, under grass or bushes, called 'forms'. European Rabbits tend to retreat into their burrows to rest.

BROWN HARE
Lepus europaeus

HTL 49–75cm
TL 7–13cm
W 2.2–6.5kg (up
to 8kg)
Females slightly
larger than males.

The Brown Hare is widespread throughout Europe and one of the region's best-known wild animals. Its hearing, sense of smell, taste and touch are especially well developed. Hares' vision is based on motion perception and they have difficulty recognising motionless bodies. Brown Hares are active both during the day and at night. They are less gregarious than European Rabbits but cannot be described as solitary animals as they gather in groups to feed. Brown Hares do not dig burrows, but scrape a shallow depression in the ground, known as a 'form'. Hares will sometimes remain quietly sitting in the form until a threat is almost upon them. Only then do they take flight, reaching speeds of up to 80km/h (50mph) and performing their classic swerving and zigzag manoeuvres. Their main predators include foxes, lynx, martens, bears, hawks, crows, Eurasian Eagle-owls and eagles. Predators rarely catch healthy adults. By contrast, mortality rates among young animals are very high.

DISTINGUISHING FEATURES Significantly larger than European Rabbits. Elongated body with very long hind legs. Ears longer than those of Mountain Hares and much longer than those of European Rabbits. The ears have large, black tips.

DISTRIBUTION AND HABITAT Brown Hare are found throughout almost the whole of Europe and are only absent from the southern Iberian Peninsula, Ireland and north-west Scandinavia. As a primary steppe dweller and species associated with human activities, the Brown Hare has successfully adapted to monoculture farming. Hares prefer to inhabit open areas with low vegetation and sufficient cover, as well as a temperate, dry climate. Meadows, fallow and arable land, pastures and sparse forests are suitable habitats. Large, unbroken areas of woodland and extensive areas of fields are unsuitable. In the Alps, Brown Hares can be found of altitudes of up to around 1,600m. At these altitudes, their habitat overlaps with that of Mountain Hares.

DIET Pure herbivore that depends on a high level of diversity. In summer, 40–60 per cent of their diet consists of foliage. They also eat various arable crops, roots, fungi and berries. Grasses can account for up to 90 per cent of their diet in winter. Brown Hares also eat bark and buds.

REPRODUCTION During the breeding season from January to October, group courtship displays sometimes see more than 50 animals come together. Brown Hares have a total of 3–4 litters per year. 1–4 (up to 6) young are born after a gestation period of 38–44 days. Unlike European Rabbits, the young are precocial and born densely furred, mobile and able to see. On average, a female has 4–12 young per year but only around 2–3 offspring per doe survive to adulthood. Superfetation (fertilisation of new egg cells in an already pregnant female) is possible.

SIGNS

FORMS Brown Hares use their front feet to scrape approximately 10cm deep forms under tufts of grass, the edges of areas of cover or behind stones. Forms can also be found in open fields. They often contain hairs. When the hare sits, its body fits into the oval shape of the form.

▼ *A Brown Hare form in a deciduous forest. Maintal, Germany. Simone Roters.*

▲ *The form of a Brown Hare. Märkische Schweiz, Germany.*

TRACK

Front

L 4–7.5cm (up to 9cm) W 2.5–7cm

Medium-sized. Digitigrade. Very asymmetrical. Classic Leporidae shape. Five toes: toe 1 is very short, the corresponding claw sometimes leaves a print ⓐ. The track has a pointed outline, with toe 3 clearly at the tip. Toes 2–5 form a kind of '1' shape. This is a characteristic distinguishing feature. The '1' of the left footprint is a mirror image. In clear prints, it is easy to make out the thick hair on the soles of the feet. Distinct claws that are often or sometimes the only part of the foot to leave a print. The front feet are smaller than the hind feet.

Hind

L 4.5–9cm (up to 15cm)

W 3.2–9cm

Medium to large. Digitigrade. Slightly asymmetrical. Four large toes: toe 1 is absent. Track appearance varies greatly. Depending on ground conditions, there will often be a complete print of a hairy thenar/hypothenar pad. If there is, the hind foot track will be much larger than the front foot track. If there is no thenar/hypothenar pad print, the front foot and hind foot prints will be a similar size. In this case, the hind foot print can be mistaken for that of a canid. Occasionally, only claw prints are visible.

▲ *Right front.*
Vledder, Netherlands. René Nauta.

▲ *Right hind.*
Vledder, Netherlands. René Nauta.

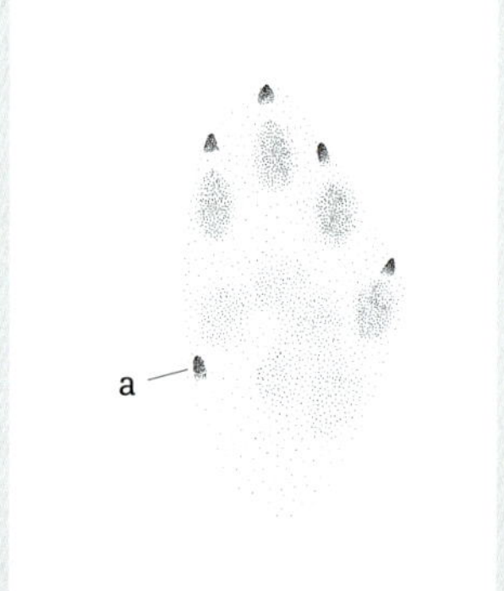

▲ *Right front.*

▲ *Right hind.*

▲ *A perfect print of the right front foot of a Brown Hare. The thick hair on the foot and the '1' shape are clear. Lausitz, Germany.*

▲ *Right hind, the very hairy sole is typical of hares. Lausitz, Germany.*

GAITS

The preferred gait of the Brown Hare is the parallel bound, but they usually move around their resting area by hopping. Unlike European Rabbits, Brown Hares are long-distance runners. When in flight, hares can abruptly change direction and achieve considerable stride lengths of over 6m.

Parallel bound and gallop
Group length: 19–231cm
Inter-group length: 36–115cm
Stride length: 42–350cm (up to over 6m)
Trail width: 8.5–25cm

Similar tracks
Mountain Hare and European Rabbit. Depending on the surface and the quality of the track, hind foot prints of Brown Hares can also be mistaken for canid footprints.

◀ *Brown Hare in a half/angled bound. Lausitz, Germany.*

▶ *Left, the track of a Brown Hare in a parallel bound; right, the track of a crane.*

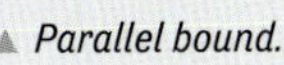

▲ *Parallel bound.*

FEEDING MARKS Frequent cambium feeding and feeding on buds and shoots. Cambium feeding by hares is characterised by relatively deep damage to the wood. Unlike cervids, Brown Hares' upper and lower incisors leave a clean cut edge at a 45° angle to the branch or stem. Where population densities are high, a clear horizontal browse line can occur on shrubs and hedges, similar to that caused by even-toed ungulates. This browse line is usually found at a height of around 75–95cm.

▼ *Snow can allow hares to reach previously inaccessible food sources. Here we see the feeding marks of a Brown Hare. Bavarian forest, Germany.*

▶ *The height of the feeding mark is an important indicator for identifying the animal that caused it. Vledder, Netherlands.*

▼ Unlike deer droppings, hare droppings are often almost spherical. The surface is coarser and more fibrous than that of ruminant droppings. Lausitz, Germany.

▲ Adult Brown Hare faecal pellets (left) and the pellet of a young animal (right). Droppings of young animals can be mistaken for rabbit droppings. Lausitz, Germany.

DUST BATHS Brown Hares take dust baths to rid themselves of parasites. They roll on the ground, sometimes leaving large, flattened areas (up to 120×80cm). Look out for any sand thrown up on the surrounding vegetation and for the outline of the hare's body.

EXCREMENT Roundish or round, usually light brown to brown and, in rare cases, dark green to black faecal pellets. Very rough fibrous surface. Normally much larger than rabbit droppings. However, there is some overlap so size alone is not always a reliable distinguishing feature. As a rule, rabbit droppings are darker in colour and their surface is smoother. Unlike rabbits, hares do not create latrines. Their droppings can be found sporadically along their paths or in piles at their feeding places.

- D 1.2–2cm

MOUNTAIN HARE
Lepus timidus

HTL 40–68cm

TL 4–9cm

W 1.6–5.8kg

Females slightly larger than males.

Mountain Hares have excellent hearing and a well-developed sense of taste, smell and touch. Like Brown Hares, they have excellent motion vision but can barely recognise motionless creatures. They can run fast for long distances and may be active both during the day and at night, depending on season and geographical location. They are more gregarious than Brown Hares and temporary meetings of up to 200 animals can occur in the north of their range. Like Brown Hares, Mountain Hares rest during the day in a shallow depression known as a 'form' that they scrape out themselves. The animals will sometimes even remain in their forms until the snow covers them over. Unlike Brown Hares, Mountain Hares dig straight, up to 2m long burrows in the earth and snow, which they use as hiding places and for protection. Occasionally, they also inhabit burrows abandoned by other species. Their main predators include foxes, lynx, wildcats, Wolverines, wolves, martens, eagles, hawks, Common Ravens and owls.

DISTINGUISHING FEATURES Slightly smaller than Brown Hares, although size varies greatly according to region. In many areas, the otherwise brownish fur turns completely white in winter, with only the tips of the ears remaining black. The ears are shorter than those of Brown Hares but longer than those of European Rabbits.

DISTRIBUTION AND HABITAT In Europe, Mountain Hares are found in Scandinavia, Scotland, Ireland, the Baltic states, Eastern Europe and in the Alps. They live in snowy areas and prefer sparse mixed forests. They also inhabit riverbanks with dense undergrowth, isolated patches of woodland in the steppe and reed beds on lakes. These adaptable animals can also be found on moors and heaths, in the stony *Krummholz* belt of the Alps up to the snow line and in the tundra up to the seashore.

DIET Similar to that of Brown Hares, although Mountain Hares generally inhabit more barren areas and therefore have access to a poorer range of foodstuffs. They are pure herbivores that feed on herbs, leaves and grasses. Heather, berries, twigs, tree bark, needles and buds are also eaten, while lichen and moss play an important role in the diet of animals living in barren areas.

REPRODUCTION Group courtship takes place during the mating season from February to June. It can involve even more animals than in the case of Brown Hares. Mountain Hares normally have 2–3 litters per year; 2–5 (up to 6) precocial young are born after a gestation period of 45–55 days. Superfetation is possible.

Mountain Hares can breed with Brown Hares, but it is not yet clear whether the hybrids are capable of reproducing.

SIGNS

FORMS Mountain Hares use their front feet to dig approximately 10cm deep forms in the snow or in the ground between stones and bushes. Forms can also be found in open fields. They often contain hairs. When the hare sits, its body fits into the oval shape of the form.

DUST BATHS Mountain Hares will make and use dust baths for grooming (see Brown Hares, page 219).

FEEDING MARKS Bark gnawing and bud browsing as with Brown Hares (page 218).

▼ *A Mountain Hare dust bath. Jämtland, Sweden. Heide Ulrich.*

◄ *Cambium feeding on a birch tree. Deep snow enables Mountain Hares to access food that was previously out of their reach. Jämtland, Sweden. Heide Ulrich.*

TRACK

Front

L 5.5–9cm   W 4–7.5cm

Medium-sized. Digitigrade. Very asymmetrical. Classic Leporidae shape. Five toes: toe 1 is very short, the corresponding claw can occasionally be seen (a). Pointed track outline with toe 3 clearly at the tip. Toes 2–5 form a kind of '1' shape. This is a characteristic distinguishing feature. The '1' of the left footprint is a mirror image. Thick hair on the sole of the foot is visible in clear prints. Distinct claws that are usually visible and sometimes the only part of the foot to leave a print. The front feet are much smaller than the hind feet and can spread out slightly more than in European Hares.

Hind

L 5.5–9.5cm (up to 18cm)
W 4–10cm

Medium to large. Digitigrade. Slightly asymmetrical. Four large toes. Toe 1 is absent. Tracks vary greatly in appearance. The entire hairy rear part of the track often leaves a print, depending on the ground conditions. Without a heel print, hind foot tracks are wider and rounder than front foot tracks. Occasionally, only claw prints are visible.

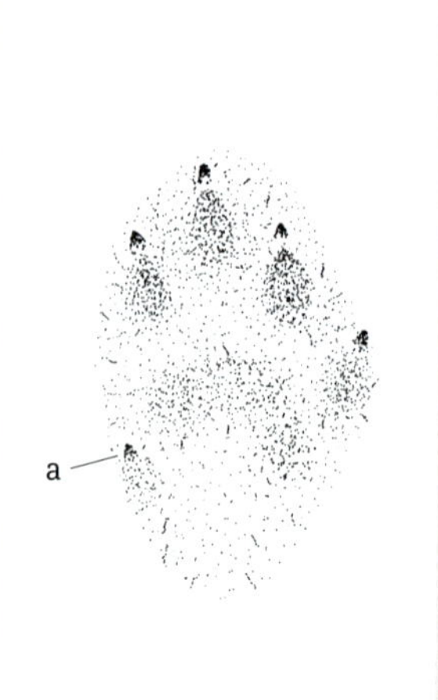

▲ *Right front.*

▲ *Right hind.*

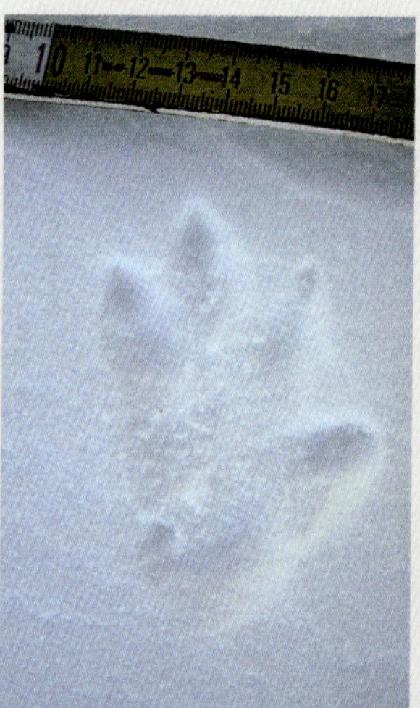

▲ *Right front.*
Jämtland, Sweden. Heide Ulrich.

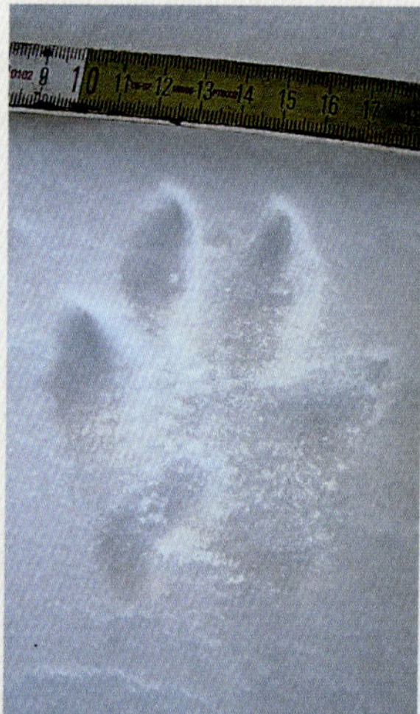

▲ *Right hind. Hind foot prints of Mountain Hares are usually much wider than their front foot prints and than the hind foot prints of Brown Hares. Jämtland, Sweden. Heide Ulrich.*

Similar tracks

Brown Hare, European Rabbit. Depending on the surface and the quality of the track, the hind foot prints of Mountain Hares can also be mistaken for canid prints. The hind foot prints of Mountain Hares tend to be larger and wider than those of Brown Hares. When spread out, the considerable width can make a strong print and provides a helpful distinguishing feature from the Brown Hare.

GAITS

The preferred gait of the Mountain Hare is the parallel bound. They are usually seen hopping around in their resting area. As a rule, the group length is shorter and the straddle/trail width wider than that of the Brown Hare. Mountain Hare paws are also hairier and able to spread out more. They act as a kind of 'snowshoe' in deep snow, preventing the lighter Mountain Hare from sinking as deeply as the Brown Hare.

Parallel bound and gallop
Group length: 20–166cm
Inter-group length: 22–98cm
Stride length: 42–262cm (up to 6m possible)
Trail width: 10–25cm

▼ *This Mountain Hare was sitting to defecate, so the complete hind prints, with the front prints in front of them, are visible. Jämtland, Sweden. Laura Gärtner.*

▷ *Parallel bound with increasing speed. Jämtland, Sweden. Laura Gärtner.*

▲ *A Mountain Hare searching for food. From the bottom to the top of the right-hand side of the image are also the tracks of a Red Fox moving in side trot. Jämtland, Sweden. Heide Ulrich.*

▲ *Parallel bound.*

▲ *Hop.*

DIFFERENTIATING BETWEEN BROWN HARES AND MOUNTAIN HARES

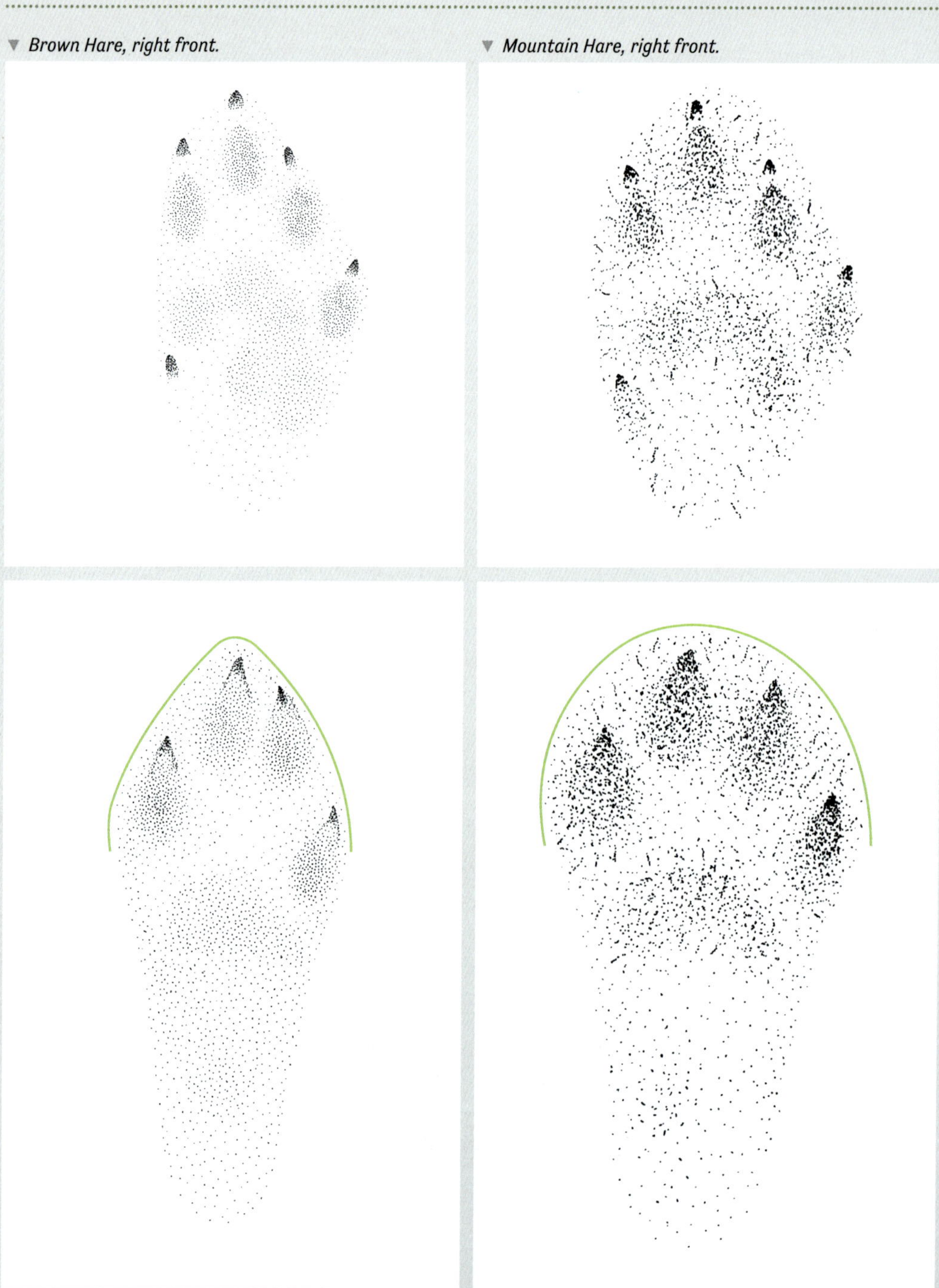

▼ *Brown Hare, right front.*

▼ *Mountain Hare, right front.*

▲ *Brown Hare, right hind.*

▲ *Mountain Hare, right hind.*

Brown Hare

- Clear footprints with hair on the soles of the feet.
- Hind foot track outline rather elongated.
- Size difference between front foot and hind foot slightly less pronounced.
- Hind foot toes less spread out.

Mountain Hare

- Blurred-looking footprints with very hairy soles.
- Hind foot track outline roundish and wide.
- Size difference between front foot and hind foot more pronounced.
- Hind foot toes can spread out widely, snowshoe effect.

◄ *Mountain Hare (left), distinguishable from Brown Hare tracks (right) by the broader hind foot track and the more noticeable size difference between the front foot and hind foot. Other distinguishing features are the smaller group length and the larger trail width.*

EXCREMENT

Like other members of Leporidae, Mountain Hares are coprophagous (they eat their excrement). Mountain Hare droppings are almost identical to those of Brown Hare (see page 219).

- D 1.2–2cm

▶ *If Mountain Hares stay at a feeding site for a long period of time, they usually leave the almost-spherical faecal pellets that are typical of Leporidae. Jämtland, Sweden. Heide Ulrich.*

EUROPEAN RABBIT
Oryctolagus cuniculus

HTL 35–45cm
TL 4–7.5cm
W 1.2–2kg
Exceptional
specimens can
weigh over 3kg.

European Rabbits are often thought of as being a kind of 'small hare', although their way of life is in some ways very different. Unlike Brown Hares, European Rabbits are avid diggers. They are social animals that live colonially in burrow systems that they dig together. Their vision is attuned to movement but they also have excellent hearing and a keen sense of smell. They are generally crepuscular and nocturnal, but are also active during the day when undisturbed. Rabbits used to be valued as a source of food and fur, but they now tend to be regarded as pests because they can cause considerable economic damage, especially in agriculture. Drastic drops in population due to diseases such as myxomatosis and RHD (Rabbit Haemorrhagic Disease) have demonstrated the ecological importance of rabbits. Declines in the population density of European Wildcats, Stoats and Red Kites were observed following slumps in rabbit population. In England, the breeding success of the Common Buzzard depends strongly on the population density of wild rabbits.

DISTINGUISHING FEATURES Body much smaller and stockier than that of Mountain or Brown Hares. Ears and hind legs much shorter in proportion. The outermost edges of the ears are black, rather than just the tips. The top of the tail is black, with a white underside.

DISTRIBUTION AND HABITAT Originally native to the Iberian Peninsula and southern France, European Rabbits have been able to spread throughout Europe, thanks to humans; they have been introduced to Britain and many other islands. As steppe dwellers, they prefer a dry climate and dry, diggable soil for their burrows. Their ideal habitat is rich in cover, flat to slightly hilly with sandy soil and offers a plentiful supply of food. As an adaptable species associated with human activities, European Rabbits can also be found in parks, green spaces, railway embankments, dykes and cemeteries, even in large cities. They also inhabit hedgerows, thickets, pine plantations and dunes on the North Sea islands. European Rabbits tend to avoid areas of dense forest, large wetland areas, mountains and habitats above the tree line.

DIET A varied selection of stems and leaves, especially grasses. European Rabbits prefer nutrient-rich young leaves, shoots and buds. Crops such as maize and other cereals are also on the menu. European Rabbits peel bark to get to the cambium, especially in spring and winter. They meet their daily water requirements with juicy food, so are rarely seen drinking.

REPRODUCTION Reproduction parameters vary strongly according to population density and environmental factors such as soil type. The breeding season in Europe is mainly from January/February–July/August. European Rabbits have 2–5 (up to 7) litters, with 2–13 young in each (average 5–6). Each female raises an average of 8–12 young per year. European Rabbits can respond to low population numbers by increasing the birth rate by more than 30 offspring per female per year. The gestation period lasts 28–31 days. Unlike hares, young European Rabbits are altricial. They reach sexual maturity after 4–5 months.

TRACK

Front

L 2–4.8cm W 2.1–3.2cm

Small. Digitigrade. Very asymmetrical. Classic Leporidae shape. Five toes: toe 1 is short and the corresponding claw is often visible. Compared to the Brown Hare, toe 1 is further forwards ⓐ. The overall shape of the track is pointed, with toe 3 clearly at the tip. Toes 2–5 form a kind of '1' shape. This is a characteristic distinguishing feature. The '1' of the left footprint is a mirror image. Thick hair on the sole of the foot is visible in clear prints. Clear, relatively large claws that are usually visible and sometimes the only print. The front feet are much smaller than the hind feet.

Hind

L 2.5–6cm (up to 9.5cm)
W 2.1–3.5cm

Small to medium sized. Digitigrade. Slightly asymmetrical. Classic Leporidae shape. Four toes: toe 1 is absent. Depending on the ground conditions or when the animal is sitting, the hairy heel can leave a complete print. If it does, the print will be much larger than the front foot track. Occasionally, only claw prints are visible. The claws are relatively large.

▲ *Left front.*

▲ *Left hind.*

▲ *Left front. The print of toe 1 is missing. Depending on ground conditions, prints can show the smallest details; here the prints of individual hairs. Nahrendorf, Germany.*

▲ *Left hind. The prints of hares and rabbits can vary considerably. As here, sometimes only three toes and claws are visible. Midhurst, England.*

Similar tracks

Brown Hare and Mountain Hare, although European Rabbit footprints are much smaller and can really only be confused with the footprints of young hares. Hares' hind feet are much larger than their front feet. This size difference is less pronounced in European Rabbits. The more powerful front feet of European Rabbits can be attributed to their digging lifestyle.

GAITS

The preferred gait is the parallel bound. European Rabbits can usually be seen hopping when they are near their burrows. Their track pattern is very similar to that of the Brown Hare, but it is smaller overall and consists of proportionately more group lengths and inter-group lengths per track. European Rabbits are very territorial. Unlike Brown Hares, they are not long-distance runners, so their paths are correspondingly shorter. They generally do not tend to move more than 500m away from their burrow, which is their main destination when fleeing.

Parallel bound and gallop
Group length: 10–72cm
Inter-group length: 36–78cm
Stride length: 26–150cm
Trail width: 6.3–14cm

▼ *Parallel bound track of a European Rabbit in the snow. Nahrendorf, Germany.*

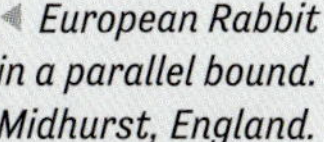

◄ *European Rabbit in a parallel bound. Midhurst, England.*

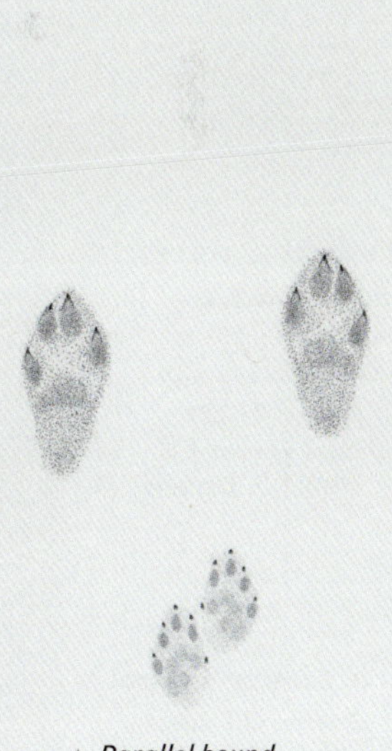

▲ *Parallel bound.*

SIGNS

BURROWING   A rabbit colony significantly changes its environment. The size of a warren depends on the soil type and the age of the colony. Burrows can be up to 3m deep and tunnels over 45m long. The entrance holes have a diameter of 10–50cm (usually 15cm).

▲ *European Rabbits colonies can build extensive warrens with many entrance holes in different shapes and sizes. Midhurst, England.*

▶ *Entrance to a rabbit burrow. Latrines are usually located in front of or next to the entrance hole. Nahrendorf, Germany.*

▲ *These digging marks are used for territory marking. They can often be found along the edges of fields. Nahrendorf, Germany.*

DIGGING MARKS To mark their territory, European Rabbits make two parallel scrapes with their front feet, especially at the boundaries to their territory and near burrows.

PATHS Look out for worn paths that link up different entrance holes. Depending on the size and activity of a colony, this 'road network' can be very distinctive.

DUST BATHS Like Brown Hares, European Rabbits use dust baths to clean themselves. They roll on the ground and use their hind feet to kick up sand, which is then often found on the surrounding vegetation. The area flattened by the rabbit's body is significantly smaller than the flattened area made by a hare.

▲ *European Rabbit paths are much smaller than those made by hares. Nahrendorf, Germany.*

▲ *European Rabbit feeding marks on cambium. Fresh feeding marks like these have a shiny appearance and are often noticeable from a distance. Nahrendorf, Germany.*

▶ *A fine cut at a 45° angle is sign of European Rabbits.*

FEEDING MARKS Buds, shoots and thin branches are snipped off at the 45° angle typical of Leporidae. Rabbits gnaw bark to get to the cambium of trees. Storm damage or snow, for example, can allow them to reach tree bark and buds that would otherwise be inaccessible. You can often find traces of feeding in these cases. Where population densities are high, a clear horizontal browse line can sometimes be found on shrubs and hedges, like that made by even-toed ungulates. This browse line is usually found in the undergrowth at a height of around 50–60cm, near the burrow.

EXCREMENT Droppings are not always distinguishable from those of Brown Hares, but they are usually much smaller and darker in colour, with a finer texture. They may be dark green, brown or black in colour. Unlike hares, rabbits make latrines. These are usually in an elevated location next to burrow entrances and at territory boundaries. There are some other signs that will help you distinguish European Rabbits droppings from the droppings of young Brown

Hares: a European Rabbit colony shapes the landscape with distinctive territory markings, numerous burrows, digging marks and a network of paths connecting the various entrance holes. Latrines are situated along paths, on elevations or in prominent places. Urine produced at different points in the reproduction cycle of female European Rabbits can cause discoloration. The colour is related to diet and varies from orange to red to blue. Damp blue patches in the sand may be caused by European Rabbits.

- D 0.7–1cm

▼ *European Rabbit droppings are much smaller than those of hares. However, the droppings of young hares can be hard to distinguish from the droppings of adult rabbits without additional signs. Coto de Doñana National Park, Spain.*

▼ *European Rabbit droppings, close-up. Nahrendorf, Germany.*

◄ *Unlike hares, European Rabbits scrape together small piles of earth where they then defecate. These latrines also serve the purpose of territory marking. Nahrendorf, Germany.*

▲ *A European Rabbit's diet can influence the colour of its urine. Nahrendorf, Germany.*

RODENTS

With around 2,300 described species at present, rodents (Rodentia) are by far the most species-rich order of mammals. In the region covered by this book, there are nine families, which together hold more than 60 species.

One of the most striking characteristics of this group is their teeth: rodents have a single pair of constantly growing, rootless incisors in the upper and lower jaw that are usually yellowish-reddish in colour. They do not have canine teeth. There is a large gap (diastema) that extends all the way from the incisors to the premolars and molars. Rodents have evolved to live in a wide variety of habitats. They can live above ground,

underground or even in trees, and different groups have adapted accordingly. There are also some semi-aquatic species that have specialised in living in and around water. Most rodents are herbivores, although there are some omnivorous exceptions. The majority of rodents have a high reproduction rate and are the main food source for many raptors, owls and other carnivores.

TRACKS AND SIGNS OF RODENTS

Plantigrade. Their adaptation to a wide variety of habitats is reflected in how rodents' feet have developed differently across species. The feet of climbing rodents are different to those of rodents that spend a lot of time in water, so this order is full of exceptions to the generally applicable characteristics. For this reason, the species covered in this book are summarised in groups with similar foot structure and similar tracks and signs. These groups generally correspond to taxonomic classification, but in some exceptions, members of one family have been examined together with members of another family to make identifying tracks in the field easier and to allow a more direct comparison of similar signs. In the following, squirrels, dormice, larger semi-aquatic rodents, Arvicolinae species (voles, muskrats and lemmings) and murids (mice and rats), as well as some unusual species such as blind mole rats and porcupines, are considered as distinct groups.

- Track formula: 5f × 5H + C

FRONT FOOT Asymmetrical, but often with the appearance of symmetry, as toe 1 is very short and often not visible or only in rudimentary form. Toes 2 and 5 point to the sides, toes 3 and 4 point forwards. The midfoot pad consists of three interdigital pads that can be separate or partially or fully fused. There are usually two more thenar/hypothenar pads, but they are often not visible.

HIND FOOT Slightly asymmetrical, toes 1 and 5 point to the sides, toes 2–4 are in a row of three that is characteristic of rodents. The shape of the midfoot pad varies greatly, but usually consists of four recognisable interdigital pads. One or two other thenar/hypothenar pads can also leave prints.

GAITS Mostly a walk, trot or parallel bound. The typical gaits and track patterns are shown for each rodent group described.

FEEDING MARKS All rodents use their incisors to gather food or nesting material, make paths and mark their territory. If they eat grasses and shrubs, they leave a characteristic clean cut and a sharp angle. They eat the bark of trees and shrubs to get to the nutritious cambium, known as bark gnawing. Rodents' specialised form of gnawing on roots is called root gnawing. Root gnawing often only becomes apparent when subsequent signs appear, such as weakened leaf shoots or loose soil. Rodents also eat the roots of herbaceous plants, tubers, nuts and fruit, usually leaving conspicuous furrows caused by their incisors. Old bones or antlers found in the field may also have toothmarks, as many rodents use them as a source of calcium. The size and position of the toothmarks and the spacing of the furrows can provide information about which species made the marks.

EXCREMENT Almost all species produce multiple elongated faecal pellets each time they defecate, but their shape and size can vary greatly. The size of the individual pellets corresponds to the size of the animal. Use of latrines is characteristic of some species. Latrines are often found along their paths, at feeding places or at resting places.

Squirrels

In Europe, the squirrel family (Sciuridae) comprises six genera: red and grey squirrels; flying squirrels; oriental tree squirrels; marmots; sousliks; and chipmunks. There is a lack of data on the tracks of flying squirrels so they are not discussed here. Likewise, the isolated populations of oriental tree squirrels and chipmunks are also not considered. We cover a total of nine different species.

RED AND GREY SQUIRRELS
Sciurus

The Red Squirrel (*Sciurus vulgaris*) and the Grey Squirrel (*Sciurus carolinensis*) both occur in the region. These excellent climbers spend most of their life in trees. They are active during the day and have good hearing, a good sense of touch and an excellent sense of smell. Their wide field of vision enables them to see extremely well above them, presumably to spot airborne predators. In very hot weather, you will occasionally see Red Squirrels lying on a branch to cool off. Their main predators are hawks and European Pine Martens. Squirrels can jump from treetops to the ground to escape predators. They can easily cope with the impact of jumping from a height of 20m.

DISTINGUISHING FEATURES Slender, supple body with powerful hindlegs and a bushy, hairy tail for steering and balance. Coat colour is not a clear distinguishing feature because Red Squirrels can be grey to black in colour and Grey Squirrels can be reddish-brown or black. Red Squirrels have visible ear tufts from late summer to spring, whereas Grey Squirrels have no or minimal ear tufts. Grey Squirrels are also usually recognisable by the white tips of their tail hairs and their more robust build.

DISTRIBUTION AND HABITAT Red Squirrels are found throughout almost all of Europe, where they are limited to coniferous and mixed forests that provide sufficient food. Grey Squirrels are native to the eastern US but have been introduced in England, Ireland and Italy. They live in deciduous and mixed forests and are highly adaptable. Both species prefer forests with different tree species, as different mast intervals ensure a more reliable food supply. Grey and Red Squirrels also live in urban parks, gardens and cemeteries. In areas where both species occur, Grey Squirrels displace Red Squirrels.

Red Squirrel
HTL 18–26cm
TL 14–21cm
W 180–480g

Grey Squirrel
HTL 23–30cm
TL 19–25cm
W 350–800g

TRACK

Front

L 3–4.8cm W 1.5–3.5cm

Small. Plantigrade. Asymmetrical. Five toes in a classic rodent structure: toe 1 is very short and rarely visible in the track ⓐ. Toes 2 and 5 point to the sides, toes 3 and 4 point forwards. A special characteristic of squirrels is that toe 4 is longer than toe 3 ⓑ. Toes 2–5 are long, thin and have a slightly rounded thickening at their tip (digital pad). The interdigital pads have fused to form a midfoot pad, but three distinct metacarpal pads are usually visible in the track; there will occasionally also be prints of two thenar/hypothenar pads. Both are chunky, appear disproportionately large and are positioned at an open angle to the centre of the path ⓒ. The strongly developed thenar/hypothenar pads are an adaptation to squirrels' arboreal way of life. Fine, sharp claw prints are sometimes visible.

Hind

L 3–6cm W 2.3–4.5cm

Small. Plantigrade. Slightly asymmetrical. Five toes in a classic rodent structure: toes 1 and 5 point to the sides, with toe 1 slightly further back than toe 5. Toes 2–4 are in a line, almost parallel to each other and usually point straight forwards. Toe 4 is longer than toe 3. The interdigital pads have fused to form a C-shaped midfoot

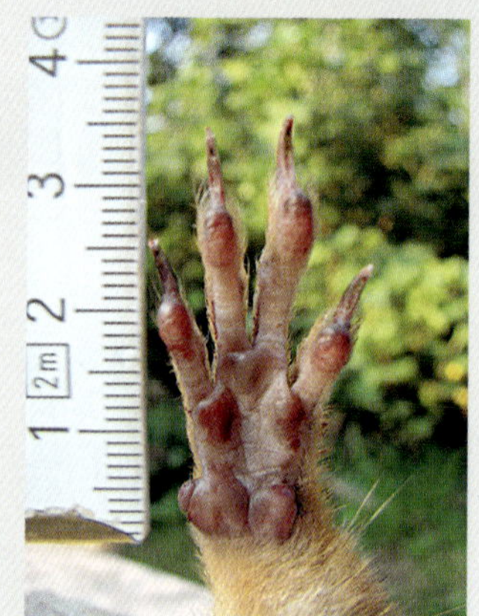

▲ *Left front.*
Vledder, Netherlands. René Nauta.

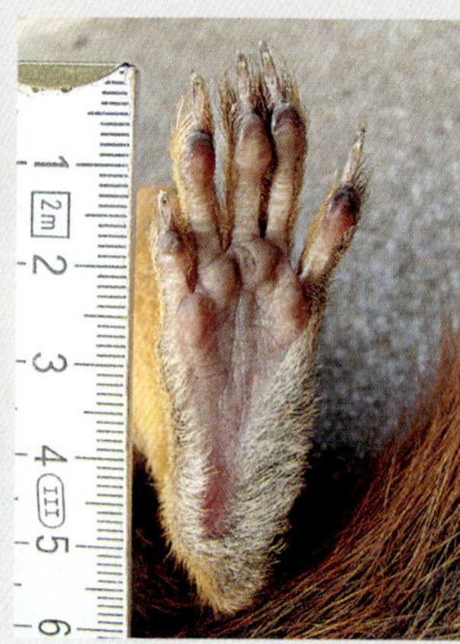

▲ *Left hind.*
Vledder, Netherlands. René Nauta.

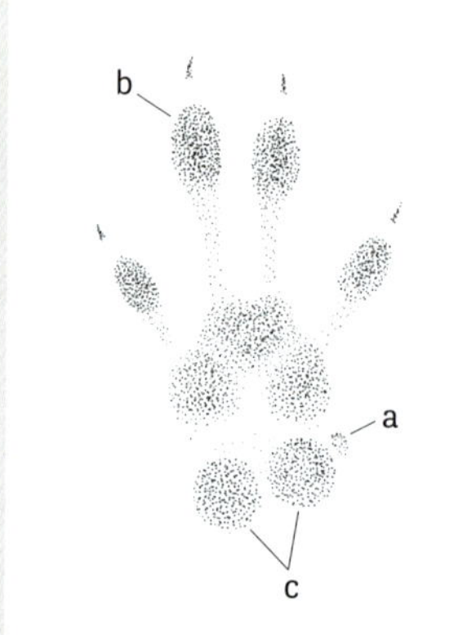

▲ *Left front.*

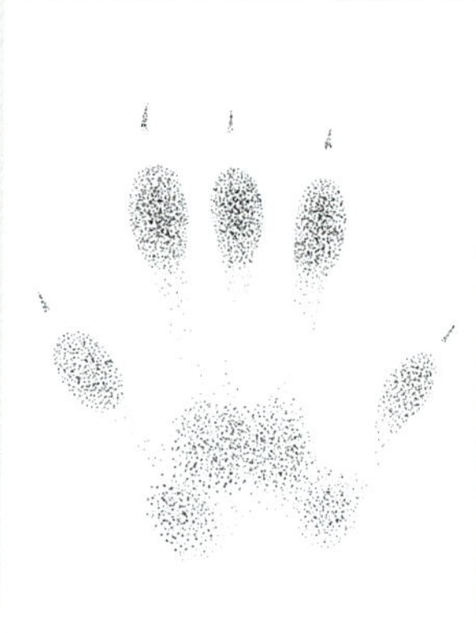

▲ *Left hind.*

▲ *Left front.*
Jetzendorf, Germany.

▲ *Left hind.*
Lausitz, Germany.

pad, but four distinct metatarsal pad prints are usually visible in the track. Sharp claw prints are sometimes visible, but they are much shorter than on the front foot. If the rear part of the foot makes a print, the hind foot print will be longer, but otherwise the front foot and hind foot prints appear similar in size.

GAITS

Grey and Red Squirrels generally move along the ground in a parallel bound, but they can also walk. They often move from tree to tree, but Grey Squirrels also cover longer distances on the ground. In winter, squirrels' storage holes, which have been scraped clear, are the main sign of their presence. In deep snow, the group length is shorter and the front feet land directly between or closer to the prints of the hind feet. This can give the impression that there are two rather than four tracks. In addition, the feet often leave drag marks when they step in and out of the snow, so a group of tracks can have the V-shape typical of rodents: the top points of the V are made by the adjacent prints of the hind feet and the bottom point is made by the front foot prints, which are closer together.

Walk
Stride length:
16–27cm
Trail width: 8–10cm

Parallel bound
Group length: 4–20cm
Inter-group length:
12.5–105.5cm
Stride length: 15–120cm
Trail width: 8.4–17.5cm

▶ *This squirrel crossed the forestry track in a full/paired bound. Bavarian Forest, Germany.*

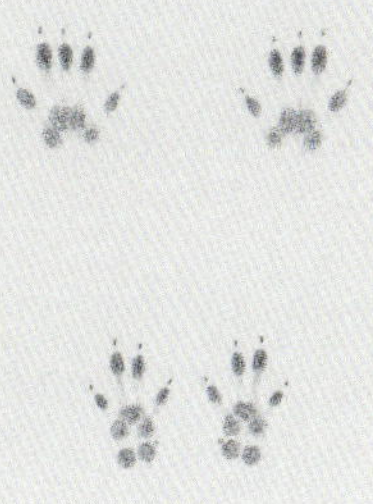

▲ *Squirrel, full/ paired bound.*

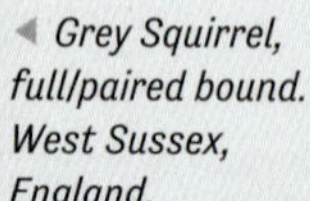

◄ *Grey Squirrel, full/paired bound. West Sussex, England.*

► *The full/paired bound is a characteristic gait of mammals with an arboreal (tree-dwelling) lifestyle. In contrast to hares and rabbits, the front foot and hind foot prints are often next to each other. Bavarian forest, Germany.*

DIFFERENTIATING BETWEEN RED AND GREY SQUIRRELS

Red Squirrels have longer, narrower toes, whereas the toe prints of the heavier Grey Squirrels are shorter and deeper. The interdigital pads and thenar/hypothenar pads of Grey Squirrels are more robust and often leave clearer prints. Being larger, Grey Squirrels generally also leave larger tracks. They spend more time on the ground and in walk, so their tracks can be found more often and over longer distances. Deciduous and mixed forests with a low proportion of conifers virtually rule out the presence of Red Squirrels.

THE EUROPEAN ANIMAL TRACKS HANDBOOK

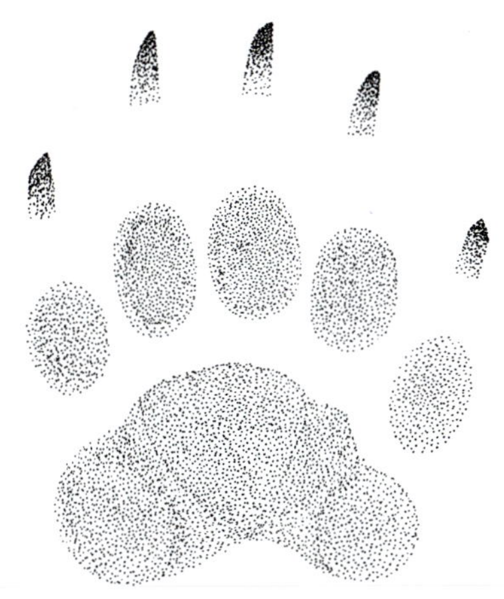

Joscha Grolms

THE EUROPEAN ANIMAL TRACKS HANDBOOK

BLOOMSBURY WILDLIFE

LONDON · OXFORD · NEW YORK · NEW DELHI · SYDNEY

BLOOMSBURY WILDLIFE
Bloomsbury Publishing Plc
50 Bedford Square, London, WC1B 3DP, UK
Bloomsbury Publishing Ireland Limited,
29 Earlsfort Terrace, Dublin 2, D02 AY28, Ireland

BLOOMSBURY, BLOOMSBURY WILDLIFE and the Diana logo are trademarks of
Bloomsbury Publishing Plc

First published in 2021 in Germany as *Tierspuren Europas* by Eugen Ulmer
First published in the United Kingdom 2025

A catalogue record for this book is available from the British Library

ISBN: HB: 978-1-3994-1577-4;
ePub: 978-1-3994-1576-7;
ePDF: 978-1-3994-1579-8

2 4 6 8 10 9 7 5 3 1

Typeset in the UK by Susan McIntyre
Printed and bound in Dubai by Oriental Press

To find out more about our authors and books visit www.bloomsbury.com
and sign up for our newsletters
For product safety related questions contact productsafety@bloomsbury.com

DIET Conifers, especially spruce, fir and pine, play an important role for Red Squirrels. Grey Squirrels particularly like to eat the buds of broad-leaved trees such as beech and birch. Both species mainly eat seeds, buds, fruits and fungi, but are also opportunistic in their diet and can switch to a wide range of foods in the absence of seeds and buds. Seasonal food includes insects and other invertebrates, young birds, birds' eggs, shoots, flowers, bark, berries and other parts of plants in summer and cambium in winter. Both species store food. They gather tree seeds, nuts or acorns and bury them or hide them in hollow trees. They hang fungi from branches high up in the trees. Squirrels use their keen sense of smell to locate their stores.

REPRODUCTION Males can be seen chasing females during the mating season from December/January–August/September. The gestation period is 36–44 days. Usually 1–2 litters per year. A winter mating results in a spring litter; a spring mating results in a summer litter. There can be 1–8 young per litter (average of 2–3). The young are altricial and reach sexual maturity after 8–12 months.

SIGNS

NEST Both squirrel species build nests, known as dreys. These are usually found high up in trees, near the trunk or between the forks of the main branches. Dreys are spherical, made of twigs and leaves and padded inside with soft material like moss and dry grass. Red Squirrel dreys have a diameter of 30–40cm. Grey Squirrel dreys tend to be slightly larger, with a diameter of 40–60cm.

▶ *A Grey Squirrel drey in the classic position close to the trunk in the fork of a tree.*
West Sussex, England.

▼ An old Grey Squirrel nest that has fallen out of a tree. West Sussex, England.

▶ A rare find: a Grey Squirrel has dug up some fungi. West Sussex, England.

▲ Squirrel marking. Note the well-positioned area and the fine toothmarks made by the incisors. West Sussex, England.

▲ Grey Squirrel territory marking. The appearance of these markings can vary depending on age, stage and tree species. West Sussex, England.

MARKINGS   Squirrels chew tree bark along frequently used routes, at distinctive places and on preferred food sources, such as walnut trees. They usually chew a small, well-positioned patch, leaving a smooth transition to the bark and clearly recognisable toothmarks. This is territory marking and different from ringing.

DIGGING MARKS   Fresh digging marks near a tree trunk can indicate a squirrel food store, especially in winter. Fresh soil on the snow is a distinctive sign of this behaviour, especially if there are nut shell remains next to the digging marks. Small, roundish holes in the ground that squirrels leave when they dig up fungi are a rarer find.

CONE FEEDING To get at the calorie-rich seeds of conifer cones, squirrels must first remove the scales that protect the seeds underneath. After biting a cone off a branch or selecting one that has fallen, the animal holds it firmly in its front feet and begins to remove the scales by biting and tearing with its teeth. Squirrels normally remove the looser scales at the basal end of the cone first. This often results in a bare area at the base of the cone where the scales have been removed. The squirrel works its way to the tip of the cone, scale by scale, often leaving frayed, uneven remains of scales on the cone axis, which give it a shaggy appearance. As it works its way along the cone, the squirrel turns it around, often securing it against a surface. Usually all that is left of the cone is a short tip with a few undamaged scales. The identification guide to feeding marks on fruit and nuts (page 164) compares cone feeding by squirrels and mice.

PILES OF CONES Remains of conifer cones piled up at the feeding site are a distinctive sign. If a squirrel was sitting in a tree to eat, we would find a pile of gnawed cone axes and scales at its base. These feeding marks can often be found on a tree stump, because squirrels prefer to eat their cones in a slightly elevated position, to be on the lookout for danger. Large piles of hundreds of cones can sometimes be found at preferred feeding sites.

▼ *Squirrels repeatedly return to popular feeding places. Large piles of characteristic leftovers can accumulate over time. Bieszczady, Poland.*

▲ *This distinctive collection of spruce shoots indicates feeding by a Red Squirrel. Laura Gärtner, Sweden.*

▶ *This is where a Grey Squirrel has been collecting nesting material. They prefer tree bark with soft, insulating properties. West Sussex, England.*

BUD BROWSING Collections of small green twigs can occasionally be found, especially under firs and spruces. Squirrels bite them off to access the male flower buds. The twigs are between 5–10cm long and bitten off in even lengths. This sign can easily be mistaken for storm damage, but the twigs would have been scattered more randomly and broken off more irregularly in that scenario.

BARK GNAWING Both squirrel species gnaw off bark to access cambium or for nesting material. They pull off the bark in spiral loops around the trunk (ringing). In contrast to territory marking, the areas where the bark and cambium have been removed are larger. The edges are frayed and pieces of bark hang down from the tree. Rejected bark piles up on the ground.

OTHER FEEDING MARKS Grey and Red Squirrels leave a variety of other feeding marks, for example on fungi, walnuts, hazelnuts and acorns. Squirrels open walnuts and hazelnuts in a characteristic way using their incisors. They gnaw at the tip of the nut until a split opens in the shell. They then split the nut cleanly into two halves by inserting their lower incisors into the opened slit and rotating the nut with their hand-like front feet. Irregular splitting may be a sign of young, less skilled animals. In the identification guide to feeding marks on fruit

▲ *Squirrels leave two neatly split hazelnut halves. Spessart, Germany.*

▲ *Squirrels use the slit at the point to split open a hazelnut. Bielefeld, Germany. Ulrike Quartier.*

▲ *Squirrels also open walnuts by inserting their incisors into the slit at the point and then splitting them into two halves. Bielefeld, Germany. Ulrike Quartier.*

and nuts, hazelnuts opened by squirrels are compared with those opened by field mice, red-backed voles and dormice (see page 165).

EXCREMENT The excrement of Grey and Red Squirrels can take different forms depending on their diet and the time of year. Droppings normally take the form of elongated to almost round faecal pellets, flattened at one end. The elongated variant is sometimes reminiscent of much larger mouse droppings, while the flat-sided, roundish variant can resemble rabbit or deer droppings. Colour ranges from grey and dark grey to brown and black.

- L 0.5–1.7cm D 0.4–0.7cm

▼ *Squirrel droppings can vary greatly in appearance and size. North Rhine-Westphalia, Germany. Ulrike Quartier.*

▲ *Squirrel droppings. North Rhine-Westphalia, Germany. Ulrike Quartier.*

ALPINE MARMOT
Marmota marmota

HTL 42–60cm
TL 13–20cm
W 3.2–6kg
(up to 9kg)

Alpine Marmots are Europe's largest squirrels. They live in family groups of up to 20 animals and form colonies with other families. There can be more than 200 animals in a large colony. Alpine Marmots are active during the day. They have very good eyesight and their ability to perceive movement is excellent. Their hearing and sense of touch are well developed. They can also climb and are very agile. Alpine Marmots hibernate in their winter burrows. Depending on altitude and climate, they do so for 5–6 months from October to April. Vital functions, such as body temperature, respiration and heartrate slow down during hibernation. The animals live on their fat reserves and lose 30–50 per cent of their body weight. In summer, marmots can often be seen playing or feeding around their burrow. They emit various sequences of shrill whistles to warn others about the presence of predators in the air or on the ground. They are mainly predated by Golden Eagles and more rarely foxes, Eurasian Eagle-owls, hawks, Common Ravens, bears, lynx, wolves and martens, but these predators mainly prey upon inexperienced young animals.

DISTINGUISHING FEATURES About the size of a hare with a stocky, strong body and a broad stance. Short, bushy tail with a black tip and small ears that are almost hidden in the fur.

DISTRIBUTION AND HABITAT Widespread in the Alps and the High Tatras, isolated populations in the Pyrenees, the Vosges, the Black Forest, the Bavarian Forest, the Massif Central in southern France and the Jura Mountains. Marmots prefer meadows and rocky regions above the tree line in the mountains. They require diggable soil to be able to excavate their burrows, which they usually create in south-facing areas, as areas with permanent snow cover are unsuitable.

DIET Pure herbivores that mainly feed on low-cellulose, energy-rich food such as shoots, flowers and roots. They mainly eat roots and bulbs in spring, followed by grasses, sedges and herbs, such as dandelion, clover, plantain and yarrow, later in the year. Marmots seem to prefer plants that are rich in unsaturated fatty acids, along with berries and seeds.

REPRODUCTION Mating season begins after hibernation, usually in mid-April. Female marmots do not reproduce every year, but have a break between pregnancies, sometimes for up to four years. They are only receptive for one day, when several males mate with each female. After a gestation period of 31–34 days, 2–4 (up to 7) altricial young are born. Young animals reach sexual maturity in their second year at the earliest but normally only after their third winter. In some areas and under harsh environmental conditions, animals sometimes only become capable of reproduction after their fourth hibernation. Offspring do not leave the family group to start their own family until their third year at the earliest.

In many places, people used to eat Alpine Marmots and they were coveted for their fat and fur. Today, their fat is still sold in rheumatism ointments but their meat is only consumed in a few regions.

SIGNS

BURROWS There are three different types of marmot burrow. The most obvious is the winter burrow, which can extend down to 5–7 metres below the surface in a hillside location. It consists of an insulated hibernation chamber, many entrances and exits, dead-end passages for defecation (latrines) and an extensive system of tunnels. Large quantities of earth that have been thrown out can be found in front of the entrance holes. The animals use stones and earth to close up the entrance to the hibernation chamber, within which they spend the winter curled up together. The winter burrow is also a permanent burrow that can be used all year round.

TRACK

Front

L 4.8–7.8cm W 3.5–5.1cm

Medium-sized. Plantigrade. Asymmetrical. Five toes in a classic rodent structure: toe 1 is very short and rarely visible. Toes 2–5 may appear slightly curved inwards and end in strong, rounded digital pads. Toes 2 and 5 point sideways, toes 3 and 4 point forwards. Interdigital pads II and III are fused so the print of the midfoot pad consists of three distinct metacarpal pad prints, but they can sometimes appear to be four individual pad prints. Two chunky thenar/hypothenar pads are often visible at the rear of the track (a). Both are at an open angle to the inside of the path. All the pads are hairless and robust. Strong, blunt claw prints are often visible. The front feet and hind feet are a similar size.

Hind

L 5.5–9cm W 4–5.5cm

Medium-sized. Plantigrade. Slightly asymmetrical. Five toes in a classic rodent structure: toes 1 and 5 point to the sides, toes 2–4 usually point forwards. They form an almost parallel line. The interdigital pads have fused to form a C-shaped midfoot pad, but four distinct metatarsal pad prints are usually still visible in the track. Occasionally, two other thenar/hypothenar pads can be seen, the outer print of which is made by the much larger H pad. If these pads

▲ *Right front. Vledder, Netherlands. René Nauta.*

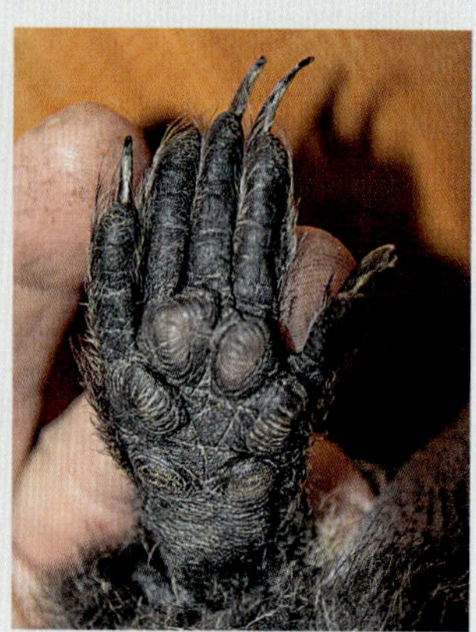

▲ *Right hind. Großglockner High Alpine Road, Austria.*

▲ *Right front.*

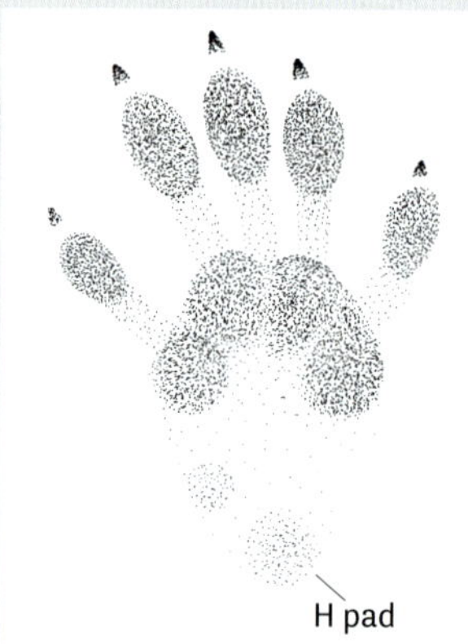

▲ *Right hind.*

▲ *Right front. The midfoot pad consists of three distinct interdigital pads. Großglockner High Alpine Road, Austria.*

▲ *Right hind. Note the row of three toes typical of rodents. Großglockner High Alpine Road, Austria.*

leave a print, the hind foot track will be longer than the front foot track, otherwise the front foot and hind foot print will appear similar in size. Claw prints are often visible and shorter than on the front foot.

GAITS

Alpine Marmots' preferred gaits are various types of walk, with the hind feet typically pointing inward. They often cross open ground in bounds or at a trot. When threatened, the animals flee to the nearest burrow at a gallop.

Walk
Stride length: 23–54cm
Trail width: 12–18cm

3 × 4 bound (rotary) and 4 × 4 bound (transverse)
Group length: 14–36cm
Inter-group length: 12–55cm
Stride length: 26–91cm
Trail width: 14–20cm

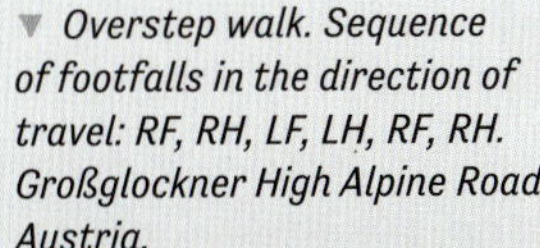

▼ Alpine Marmot in a 4 × 4 bound (transverse). Großglockner High Alpine Road, Austria.

▼ Overstep walk. Sequence of footfalls in the direction of travel: RF, RH, LF, LH, RF, RH. Großglockner High Alpine Road, Austria.

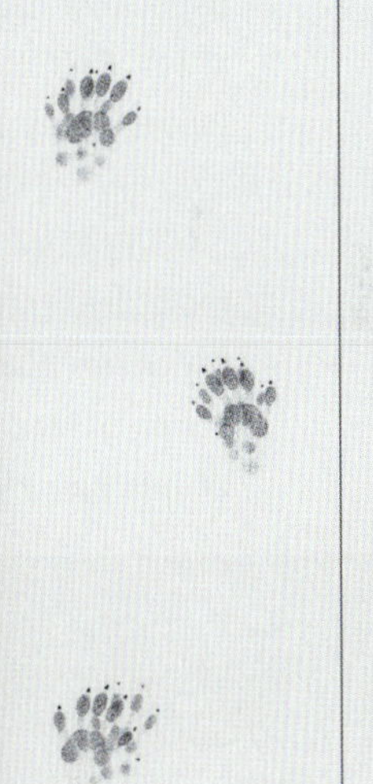

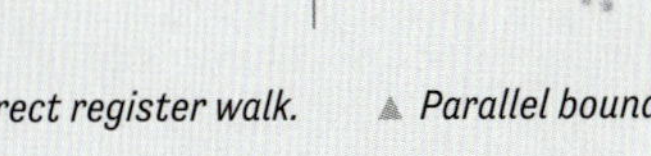

▲ Direct register walk.   ▲ Parallel bound.

Similar tracks
Squirrel prints can appear similar, but they are less robust and smaller overall.

▲ Entrance hole to an Alpine Marmot burrow. Großglockner High Alpine Road, Austria.

Heavily trampled paths sometimes radiate from a winter burrow like the spokes of a wheel or link up different entrance holes. The shallower summer burrows have shorter paths and smaller nest chambers. The main purpose of a summer burrow is to avoid heat stress. Alpine Marmots are sensitive to temperature and can fall into a kind of heat torpor if temperatures become too high. This is partly why they spend a lot of time in their burrows in summer. In total, marmots spend 90 per cent of their lives underground. The third type of burrow is the simple escape tunnel. These usually only reach one metre in length and are used as a means of escape in the event of danger. The entrance holes have a diameter of 20–40cm (usually 25cm).

MARKINGS   Alpine Marmots leave different types of markings. They leave their odour on small bushes and stones, especially near their burrows, by rubbing their mouth and cheeks on them. Rolling spots similar to those of otters can also be found in loose, sandy soil (see page 177).

EXCREMENT   Shape and colour are strongly influenced by the moisture content of the food. Droppings are usually greenish to brownish-blackish and turn grey with age. A common form is a long string of faecal pellets that can be mistaken for raccoon, cat or badger droppings. However, marmot droppings consist almost exclusively of plant remains. Fresh droppings have an unpleasant smell, recalling fermenting vegetables. Very moist faeces may look like a dark green blob. Alpine Marmot latrines are mainly used as territory markings at boundaries and near burrows.

• **L 2–6cm   D 0.8–2cm**

▼ Latrines like these are mainly found at territory boundaries and near burrows. Großglockner High Alpine Road, Austria.

SOUSLIKS
Spermophilus

These ground-dwelling squirrels, similar in size to *Sciurus* species, are active during the day. They live gregariously in communities of many animals. They are known for standing up on their hindlegs. This book covers the European Souslik (*Spermophilus citellus*) and the Spotted Souslik (*Spermophilus suslicus*). At least one adult normally looks out for predators while others forage for food, groom, play or sunbathe outside their dwellings. The lookout makes loud whistling sounds to warn the other animals, who respond by diving into their escape tunnels. Male sousliks are territorial and will chase away other males from their dwelling. Females live in territories established by the males but do not defend these territories themselves. Unlike other souslik species and Common Hamsters, European Sousliks rarely store food. Instead, they begin feeding more intensively in late summer, to build up their body's fat reserves. Sousliks hibernate for 6–7 months, beginning in September/October. Before hibernation, they block the entrance holes of their dwelling with soil to increase protection against the cold and predators. They are predated by many species of raptors, as well as mustelids, foxes and other canids.

European Souslik

HTL 18–23cm

TL 3.1–9cm

W 150–380g

Spotted Souslik

HTL 18–26cm

TL 2.9–5.6cm

W 200–360g

Males are slightly larger and heavier than females.

TRACK

Front

L 1.8–3.3cm W 1.2–2.2cm

Small. Plantigrade. Asymmetrical. Five toes in a classic rodent structure: toe 1 is very short and rarely visible. Toes 2–5 may appear to curve inwards slightly. Toes 2 and 5 are to the side. Toes 3 and 4 point forwards. Metacarpal pads II and III are fused, so the midfoot pad is made up of three distinct metacarpal pad prints arranged in a triangle. Occasionally, there can appear to be four metacarpal pads. Two further thenar/hypothenar pads are often visible at the back of the track. Both are at an open angle to the inside of the path and are approximately the same size as the interdigital pads. The overall outline of the track can look long and pointed, like hare tracks. The strong, relatively long claws are usually visible in the track. The front feet and hind feet are a similar size, although the front foot print often appears longer because it is more common for the thenar/hypothenar pads to be visible.

Hind

L 2-3.2cm W 1.3-2.6cm

Small. Plantigrade. Slightly asymmetrical. Five toes in a classic rodent structure: toes 1 and 5 point to the sides. Toes 2–4 usually point forwards and are aligned and almost parallel. The interdigital pads have fused to form a C-shaped midfoot pad, although four individual metatarsal pad

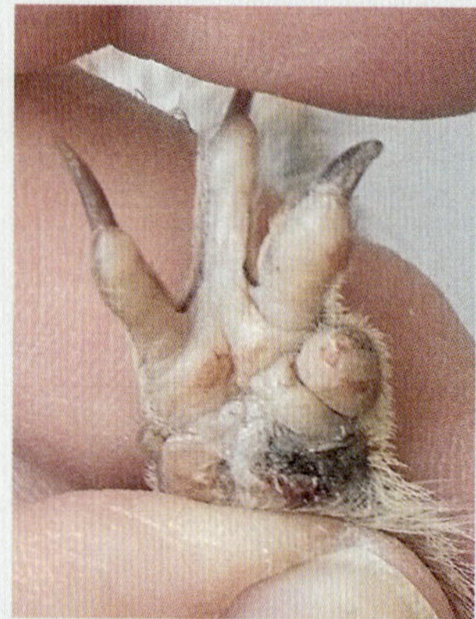

▲ *European Souslik, left front. Austria, Andreas Wenger.*

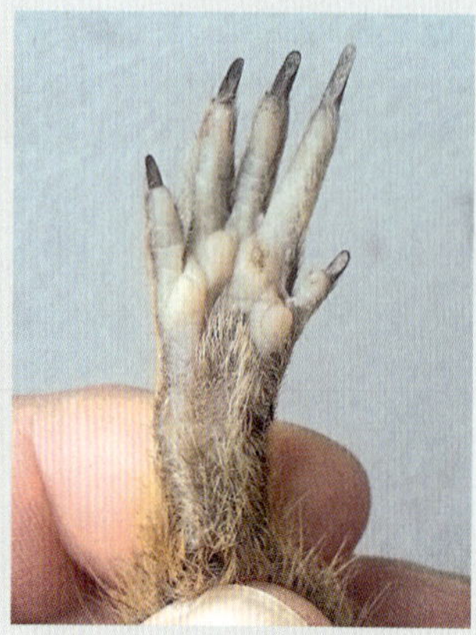

▲ *European Souslik, right hind. Austria, Andreas Wenger.*

▲ *Left front.*

▲ *Right hind.*

▲ *Left front. As with Leporidae, the track can resemble a '1' (here mirror image). Austria, Andreas Wenger.*

▲ *Right hind. Austria, Andreas Wenger.*

prints are occasionally visible in the track. However, often only the two foremost and much larger metatarsal pads leave a print. In rare cases, two further thenar/hypothenar pads are visible at the back of the track. If both leave a print, the hind foot print will be longer than the front foot print, otherwise the front foot and hind foot tracks appear similar in size. The claws are large and usually visible, but slightly shorter and finer than on the front foot. The claw on toe 2 is longer than the one on toe 3 (a), giving the track the impression of sloping from toe 2 to toe 5 (see hedgehogs, page 190).

GAITS

Sousliks tend to cover the ground in a 4 × 4 bound (transverse) or a parallel bound. If they need to cross open spaces quickly, they usually gallop or choose a 2 × 2 bound. They will often slow down to a walk when foraging for food. Sousliks only move evenly in one gait over short distances. Their paths begin and end at entrance holes to their dwelling or escape tunnels. Footprints often turn outwards slightly.

Walk
Stride length: 7–10cm
Trail width: 7–8.6cm

2 × 2 bound
Group length: 5–10cm
Inter-group length:
20–45cm
Stride length: 30–55cm
Trail width: 8–9cm

**4 × 4 bound (transverse)
and gallop**
Group length: 5.5–13cm
Inter-group length:
9.3–53cm
Stride length: 15–63cm
Trail width: 5–8cm

▲ *European Souslik in a 4 × 4 bound (transverse). Direction of travel from bottom to top. Austria, Andreas Wenger.*

Limited data available, primarily from the European Souslik.

▲ *4 × 4 bound (transverse).*

▲ *Parallel bound.*

Similar tracks
Marmot tracks are similar but much larger. The tracks of the Spotted Souslik tend to be slightly smaller than those of the European Souslik.

▶ *Sousliks typically dig the entrances to their dwelling on a slope. Austria, Andreas Wenger.*

▲ *Entrance hole to a European Souslik dwelling. Austria, Andreas Wenger.*

The European Souslik is a strictly protected species in Europe. The Spotted Souslik is also declining throughout its entire range.

DISTINGUISHING FEATURES Tend to be slightly smaller than tree squirrels, with a slim build, short ears and large eyes surrounded by a light-coloured ring. The Spotted Souslik can be distinguished from the European Souslik by the pale spots on its fur and shorter tail.

DISTRIBUTION AND HABITAT Two populations of European Souslik occur, separated by the Carpathian Mountains: a western population in south-west Poland, the Czech Republic, Slovakia, Austria, Hungary and Romania and a south-east European population in Bulgaria, Montenegro, Greece, Macedonia and European Turkey. The Spotted Souslik is found in eastern and south-eastern Europe, in Poland, Moldova and Ukraine. Both souslik species prefer open landscapes with varied, low vegetation. These include steppe-like terrain, uncultivated land and dry grasslands. They avoid wetlands and forests.

DIET Sousliks prefer green parts of plants, seeds and roots of alfalfa, grasses, clover and dandelion. They also eat crops such as beet, potatoes and cereals, as well as insects such as beetles, caterpillars, ants and grasshoppers. The genus name *Spermophilus* comes from the Greek *spermatos* (seed) and *phileo* (to love), referring to their diet.

REPRODUCTION Sousliks usually mate from March to May, after hibernation. The single annual litter of 4–9 (up to 13) hairless young is born after a gestation period of 22–27 days. The animals reach sexual maturity after about a year. Sexually mature males are driven away, while female offspring take over their mother's dwelling.

SIGNS

DWELLINGS Sousliks live in burrows that they dig themselves. The burrows may be up to 2m deep and can have several entrances. At the end of the main tunnel, which can be up to 4m long, is a living chamber lined with dry grass. This is also where the young are born. A separate chamber is used as a latrine. In addition to these maternity and winter chambers, there are also several escape tunnels. The animals use them as temporary refuges and places to rest. The entrance holes measure 5–7cm in diameter.

EXCREMENT Sousliks usually defecate in underground latrines, but droppings can also be found scattered around the dwelling's vicinity. Droppings vary in shape and colour depending on the food consumed. Elongated faecal pellets with a conical end are typical. Several pellets often clump together when moisture content is higher. The colour is usually dark brown to black. Souslik droppings can resemble Western Roe Deer droppings, but their surface is usually rougher and more fibrous as sousliks do not digest their plant food as thoroughly.

- **L 1–1.5cm D 0.4–0.5cm**

▼ *A selection of different shapes of souslik droppings. Austria, Andreas Wenger.*

▲ *Souslik droppings in the field. Austria, Andreas Wenger.*

DORMICE
Gliridae

There are five species of dormice in Europe: Hazel Dormouse (*Muscardinus avellanarius*), Roach's Mouse-tailed Dormouse (*M. roachi*), Forest Dormouse (*Dryomys nitedula*), European Garden Dormouse (*Eliomys quercinus*) and Edible or Fat Dormouse (*Glis glis*). These mouse- to squirrel-sized nocturnal rodents mainly live in trees and bushes and rarely spend any time on the ground. They are characterised by large black eyes, small ears and a long bushy tail, from which they can shed the skin if a predator seizes it. Dormice in cold and temperate climatic zones hibernate for 6–7 months. Dormouse tracks are very rarely found.

All five species are very good to excellent at climbing and jumping. They also have an excellent sense of hearing. Edible Dormice and European Garden Dormice are highly vocal and make a rich repertoire of hissing, whistling, squeaking and cheeping sounds. During the mating season, they make various angry sounds, as well as murmurs and grunts. Forest Dormice hiss and whistle when excited, but otherwise make much quieter, drawn-out sounds. Dormice usually hibernate in earth burrows or rock crevices around 0.5–1m deep. They dig their own burrows. During hibernation, the animals curl up into a ball and their body temperature can drop to 1°C. On rare occasions, dormice also use Black Woodpecker holes for hibernation. European Garden Dormice in Spain do not hibernate when temperatures are mild. The main predators of dormice are owls, such as Tawny and Tengmalm's Owls, as well as martens, foxes, Least Weasels and cats.

DISTINGUISHING FEATURES In terms of appearance, can only be confused with other dormice or squirrels. The Hazel Dormouse is the smallest species, being smaller than a Wood Mouse. The Roach's Mouse-tailed Dormouse is slightly larger and the only European dormouse with an almost hairless tail. The Forest Dormouse is about

Hazel Dormouse
HTL 6.5–9cm
TL 5.3–8.4cm
W 15–40g

Roach's Mouse-tailed Dormouse
HTL 8.5–13.5cm
TL 6.5–9.5cm
W 21–70g

Forest Dormouse
HTL 8–13cm
TL 6–11cm
W 17–95g

European Garden Dormouse
HTL 10–18cm
TL 8–18cm
W 50–180g

Edible Dormouse
HTL 12–21cm
TL 10–19cm
W 58–200g

the same size as a Wood Mouse and the Edible Dormouse is the largest species, approaching the size of a Red Squirrel. European Garden Dormice are little larger than a field mouse.

DISTRIBUTION AND HABITAT Hazel Dormice are found in many parts of Europe. They prefer mixed deciduous forests with plenty of fruit-bearing shrubs, such as brambles and hazel trees. Roach's Mouse-tailed Dormice occur in Bulgaria and Turkey, where they are limited to western coastal areas. They favour riverbanks and open and semi-open landscapes with mulberry bushes and fig trees. Patchy populations of Forest Dormice occur in central, eastern and south-eastern Europe. They prefer mixed deciduous forests, but also inhabit spruce forests. European Garden Dormice live in south-west Europe from Portugal and Italy to southern Germany and in north-west Europe as far as the Ural Mountains. Endemic to Europe, they prefer rocky areas (karst landscapes), as well as coniferous and mixed forests with sparse ground vegetation and rocky subsoil. Unlike other dormice, the European Garden Dormouse spends a lot of time on the ground. Edible Dormice are found throughout most of Europe and were introduced to southern England in 1902. They prefer deciduous and mixed forests with dense undergrowth. All five species are associated with human activities and can be found in orchards, vineyards and buildings.

DIET Seeds such as nuts and beechnuts, leaves, buds, bark and fruit. They also eat snails, insects and their larvae, lizards and occasionally mice, birds' eggs and young birds. European Garden Dormice seem to predominantly eat animals, while Edible Dormice feed more on plants.

REPRODUCTION Depending on temperature, dormice normally hibernate from October–April (September–May). The mating season usually begins afterwards. There is normally one litter per year but a second litter is possible with sufficient food and warm weather. This is common, especially in more southerly regions. Edible Dormice will not reproduce if there is a lack of food. The first young are usually born in June after a gestation period of 21–25 days. The gestation period of Roach's Mouse-tailed Dormice and Edible Dormice lasts around 30 days. Litters are normally of 3–6 (up to 10) young, which reach sexual maturity after their first hibernation.

Edible Dormice were traditionally considered a delicacy in some areas. They were reared in captivity and fattened, hence the common name.

TRACK

Dormice provide a striking example of how animals' feet adapt to a certain way of life. The arrangement and size of the metacarpal/metatarsal pads, as well as the development of the claws and even the type and extent of any hair on the soles of the feet, provide valuable information about the animals' lifestyle. Dormouse metacarpal/metatarsal pads are very pronounced, with a robust, strong appearance. This distinctive adaptation to a tree-dwelling lifestyle characterises animals that spend a lot of time climbing and jumping. The strong metacarpal/metatarsal pads give a firm grip and good support.

Front

Hazel Dormouse
L 0.7–1.3cm   W 0.7–1cm

Forest Dormouse
L 1–1.5cm   W 0.8–1.2cm

European Garden Dormouse
L 1.5–2 cm   W 1–2.2cm

Edible Dormouse
L 1.5–2 cm   W 1–2.2cm

Very small to small. Plantigrade. Asymmetrical. Four toes in a classic rodent structure, with toe 1 completely absent. Toes 2 and 5 are slightly forwards and point clearly to the side. The print of digital pads 2 and 5 are well separated from the three fused interdigital pads. Toes 3 and 4 point forwards. Toes 2–5 are very short and the negative space between the metacarpal pads and

▲ *Edible Dormouse, Right front. Spessart, Germany.*

▲ *Edible Dormouse, right hind. Spessart, Germany.*

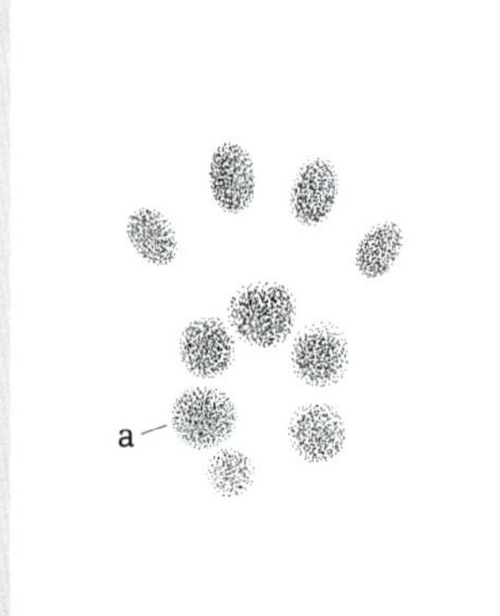

▲ *Right front.*

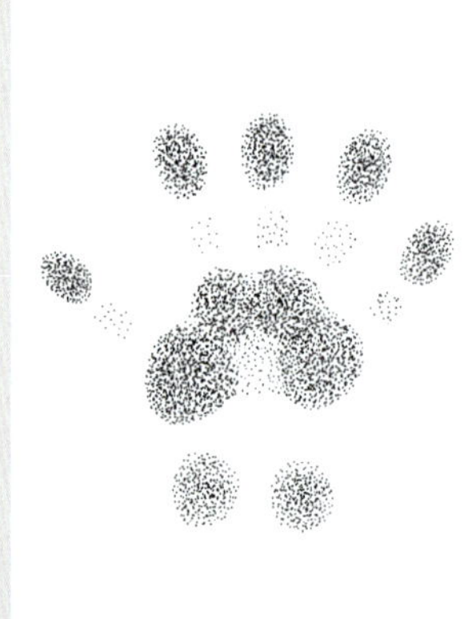

▲ *Right hind.*

▲ *Edible Dormouse, Right front. Spessart, Germany.*

▲ *Edible Dormouse, right hind. Spessart, Germany.*

the digital pads is correspondingly small. The six metacarpal pads that are arranged in a rosette shape, with four in front and two behind, are a characteristic feature. The strongly developed metacarpal pad I is clearly set further to the back (a).The two thenar/hypothenar pads are only visible occasionally. The claws are small and rarely leave prints. The front feet are slightly smaller than the hind feet.

Hind

Hazel Dormouse   L 1–1.6cm   W 1–1.2cm

Forest Dormouse   L 1.2–2cm   W 1.2–1.4cm

European Garden Dormouse   L 2–3cm   W 2–2.5cm

Edible Dormouse   L 2.5–3.8cm   W 2–2.5cm

Small. Plantigrade. Slightly asymmetrical. Five toes in a classic rodent structure: toes 1 and 5 point diagonally to the sides. Toes 2–4 are usually in a row and almost parallel to each other. The metatarsal pads have fused to form a midfoot pad but four strong interdigital pads arranged in a circle are still visible in the track. Two further adjacent thenar/hypothenar pads may also be visible. The hind feet are slightly larger than the front feet. Claw prints are small and not usually visible.

GAITS

With the exception of the European Garden Dormouse, the dormice found in Europe are rarely active on the ground. They only travel short distances, usually from tree to tree. Preferred gaits are the parallel bound for covering longer distances quickly, and a walk for covering the ground more slowly. The tracks are often slightly angled to the inside or outside, which gives the track pattern its special character.

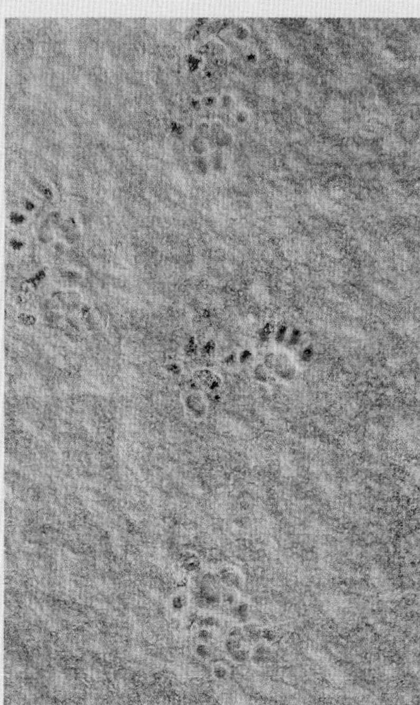

▶ *Edible Dormouse, at a walk. Spessart, Germany.*

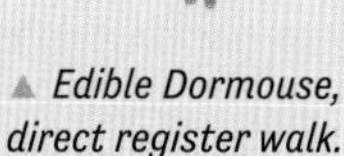

▲ *Edible Dormouse, direct register walk.*

▲ *Edible Dormouse, parallel bound.*

SIGNS

NEST   The often spherical or oval nests are usually made of leaves and thin branches and normally padded with moss and dry grass. The nest of the Hazel Dormouse is about the size of a fist and padded with more finely chewed up plant material than that of the larger dormice. Forest Dormice build their nests high up in the trees, while European Garden and Edible Dormice build nests in tree hollows, crevices and nesting boxes, rarely in the open. The entrance holes are usually on the side.

HAZELNUT FEEDING   The dormice usually prop up the hazelnut against something and hold it with their front feet. They make the first opening in the shell of the nut by rapidly gnawing it with their lower incisors. They then use their lower incisors to gnaw parallel or slightly diagonal to the edge of the hole, leaving corresponding toothmarks. The edge of the hole appears even, rounded and smooth and is mainly free of toothmarks (see page 165). Hazel Dormouse toothmarks are much finer than those of European Garden and Edible Dormice.

▼ *The summer nest of a Hazel Dormouse. Rimberg, Germany. Simone Roters.*

► *The spherical nest of an Edible Dormouse. Midhurst, England.*

▲ *Gnaw marks running parallel to the edge of the hole are a sign of dormice. Here, an Edible Dormouse has opened a hazelnut. Midhurst, England.*

EXCREMENT Elongated faecal pellets, often coiled up like a rope, with an irregular surface and diverse contents. This coiled form is characteristic of dormice. One end is usually pointed, the other blunt or broken off. The colour varies according to diet, but is normally yellowish-brown, brownish to black. If the animal has been feeding on fruit, droppings appear shiny and occasionally olive green. They are found in trees in forests and orchards, on stone walls along forest edges, on woodpiles and in huts and house attics. Fresh Edible Dormouse and European Garden Dormouse droppings can be mistaken for rat droppings by appearance. However, the odour is not as unpleasant as that of rats and the droppings are found in different places, which can also be a distinguishing feature. Rat droppings are rarely kinked and are less coiled than dormouse droppings, although dormouse droppings are not necessarily always coiled and kinked. Old droppings appear dull grey to grey-black and look like bat droppings, but less porous. Dormouse droppings shrink considerably when dry, becoming barely larger than field mouse droppings.

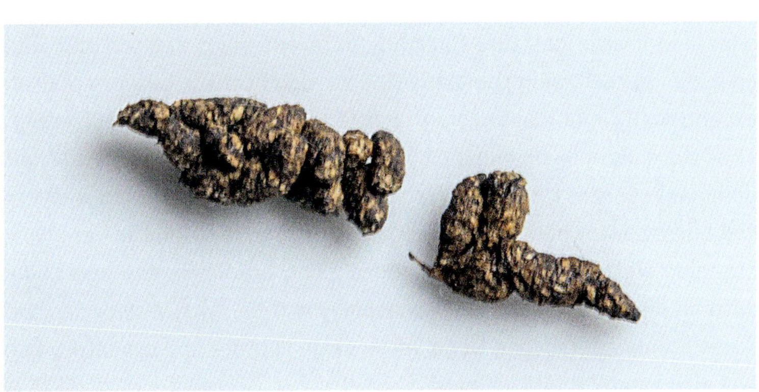

◄ *The coiled appearance of the elongated faecal pellets is characteristic of dormice. These are the droppings of an Edible Dormouse. Spessart, Germany.*

EURASIAN BEAVER
Castor fiber

HTL 80–105cm

TL 29–38cm

W 15–30kg (up to 38kg)

Size and weight vary greatly from region to region.

The Eurasian Beaver is one of two representatives of the genus *Castor* within the beaver family (Castoridae), along with the non-native North American Beaver (*Castor canadensis*). It is the largest rodent in Europe and the second largest rodent in the world. Beavers are best known for their impressive landscaping. They build dams to regulate the water level and keep the entrances to their dwellings under water as protection against predators, as well as to make other food sources accessible, and to increase the occurrence of their preferred forage plants. Their activities can create new wetlands or enlarge existing ones and create habitats for many different plant species and wild animals. By reshaping the landscape so significantly, beavers play an exceptionally important ecological role and contribute to biodiversity. They are slow-moving on land but exceptionally good swimmers. Their dominant sense is touch. If riverbanks are suitable for digging, beavers will burrow into them. If this is not possible, as is often the case with standing water, beavers build lodges using branches, and seal them with mud and clay. They are predated by wolves, lynx, Wolverines and foxes. Common Raccoon Dogs, mink, large raptors and predatory fish will also attack young animals.

DISTINGUISHING FEATURES Small, roundish head that merges into the strong body with almost no neck. Beavers are characterised by their broad, horizontally flattened tail, which is hairless and covered with leathery skin. Large, orange-coloured incisors.

DISTRIBUTION AND HABITAT Eurasian Beavers are restricted to standing water or flowing water that is not prone to flooding. It is crucial that standing water does not freeze over completely in winter or dry out completely in summer. Eurasian Beavers are very adaptable and can also colonise banks with few larger trees if, for example, willows, reeds or bulrushes are present.

DIET Pure herbivores that feed mainly on the cambium of trees, branches and buds. Their preferred tree species are softwoods such as willow, hazel and poplar. Herbaceous and aquatic plants, such as ground elder, stinging nettles, bulrushes and reeds also play an important role. Eurasian Beavers will also eat crops such as beet, maize or cereals if they are close and safely accessible.

REPRODUCTION Eurasian Beavers are monogamous and pair for life. The mating season lasts from the end of December to April. After a gestation period of 103–108 days, 2–4 young are born, usually in May/June. The precocial young are born with hair and able to see.

Eurasian Beavers are strictly protected and due to a population rebound are now listed as being of 'least concern' on the IUCN Red List of Endangered Species. They were formerly widely hunted for their fur and the castoreum they secrete.

SIGNS

Although beavers spend most of their time in water, they leave many obvious signs above the water. Look out for dams and lodges, paths between bodies of water, feeding marks along riverbanks and marking mounds within the territory.

▶ *A classic beaver lodge. Spessart, Germany. Udo Klein.*

TRACK

Front

L 5.4–8cm W 4.5–7.5cm

Medium-sized. Plantigrade. Asymmetrical. Five toes: toe 1 slightly reduced and only leaves a print occasionally. Toes 2 and 5 are to the sides. Toes 3 and 4 point forwards and are curved towards the inside of the path. All toes continue to the midfoot pad. The midfoot pad is made up of some of the fused interdigital pads. Two further thenar/hypothenar pads are occasionally visible and can give the back of the track a boxy appearance (a). Long, powerful, blunt claws that are used for digging leave clearly visible marks. The front foot prints are often destroyed by the hind feet or the tail.

Hind

L 12–18cm W 7.5–11.5cm

Large to very large. Plantigrade. Asymmetrical. Five long toes: toe 1 is significantly further back than the other toes and only leaves a print occasionally. Toe 2 is split to make a grooming claw but the split is rarely visible in the track. Toes 3–5 usually leave clearer and deeper prints than toes 1 and 2. Often only toes 3–5 are visible. Toes 3 and 4 often appear to be the same length. The distal webbing between all toes is often visible. The rounded area at the back of the foot usually leaves a print and is an important identification feature. Powerful,

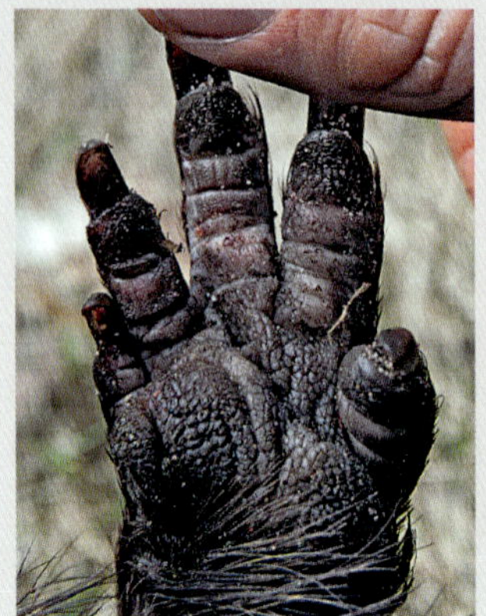

▲ *Left front.*
Märkische Schweiz, Germany.

▲ *Left hind.*
Märkische Schweiz, Germany.

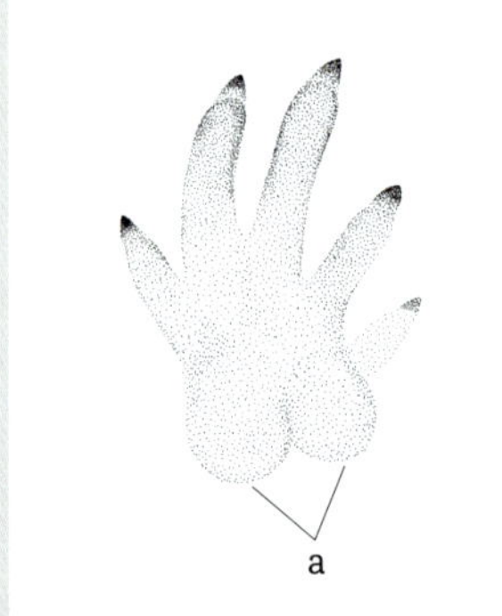

▲ *Left front.*

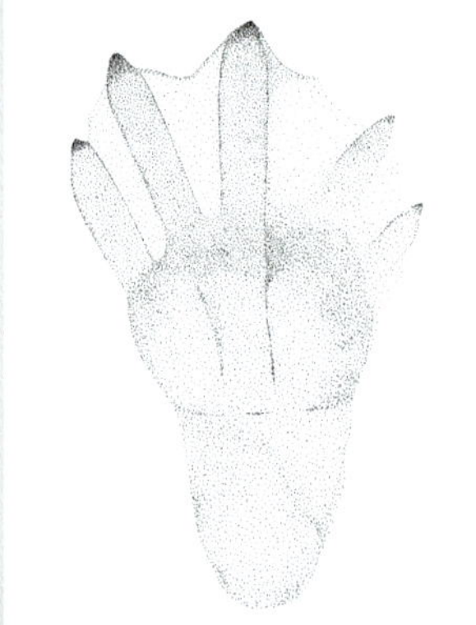

▲ *Left hind.*

▲ *Left front. The two thenar/ hypothenar pads and the reduced toe 1 are faintly visible. Elbtalaue National Park, Germany.*

▲ *Left hind. Toes 3–5 are clearly visible. Elbtalaue National Park, Germany.*

short, blunt claw prints are usually visible. The hind feet are very large in relation to the size of the animal and more than twice as big as the front feet.

GAITS

The preferred gait is a walk. The hind feet normally land just behind, in front of or on top of the prints of the front feet. The hind feet often turn inwards. Paths with drag marks from branches are common. If threatened, beavers bound or gallop to safety. The quality of beaver tracks largely depends on how much the footprints are destroyed by the drag marks made by the tail, which sometimes completely cover the tracks.

Walk
Stride length: 24–55cm
Trail width: 16–26cm

Parallel bound
Group length: 15–35cm
Inter-group length: 22–78cm
Stride length: 37–102cm
Trail width: 16–31.5cm

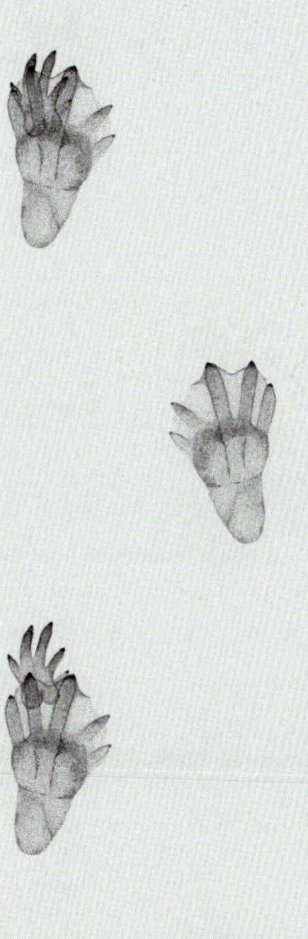

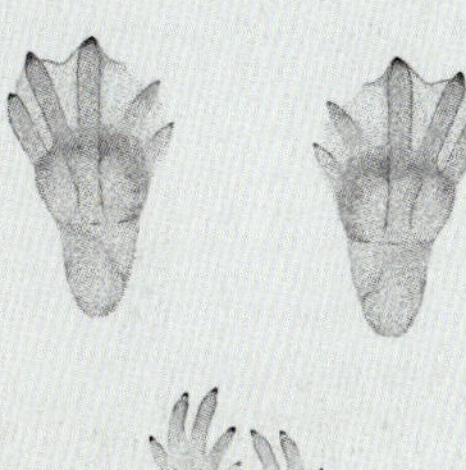

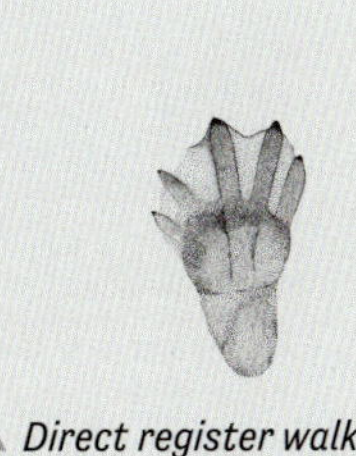

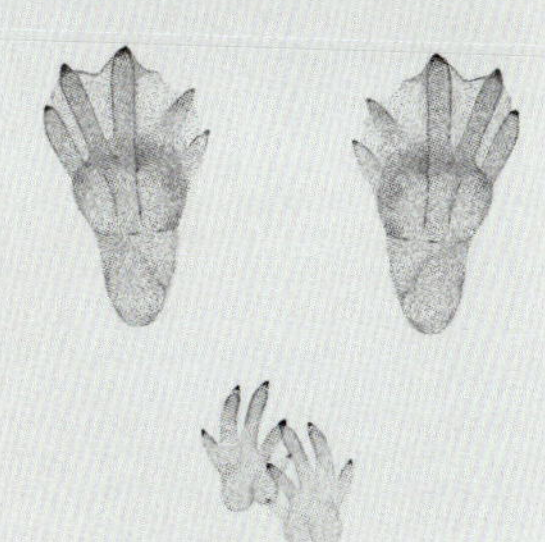

▲ *Direct register walk.*　　▲ *Full/paired bound.*

◄ *Here, the front foot prints can be seen in front of the hind foot prints. Elbtalaue National Park, Germany.*

Similar tracks
Muskrat, Coypu.

▶ *Eurasian Beaver in direct register walk. The large hind feet completely cover the front feet. Elbtalaue National Park, Germany.*

▼ *A Eurasian Beaver dragging a small tree behind it. The animal's tracks are on the left and the drag mark of the tree is next to them on the right. Elbtalaue National Park, Germany.*

▲ *Riverbanks are a preferred habitat, for example along the Oder. Frankfurt (Oder), Germany.*

◄ Beaver dams can reach a considerable size...

▼... and dam entire rivers. Offenbach, Germany. Simone Roters.

▲ A marking mound on the banks of the Elbe. Elbtalaue National Park, Germany.

DAMS AND LODGES Eurasian Beavers build dams and lodges from tree trunks, branches, stones and mud. Eurasian Beaver lodges can measure between 2m and 12m in diameter and up to 2m in height. They may be multi-storey and used by several generations. Eurasian Beaver dwellings usually tend to go unnoticed because the entrances are normally under water. However, the entrance holes, which are usually 30–60cm in size, can become visible when water levels are low. Large holes can form when dwellings collapse. The largest known beaver dam in Europe is in Belgium and is 180m long. The largest beaver dams in the world are around 850m long.

MARKING MOUNDS These are created within a territory as markers. Eurasian Beavers build the mounds by piling up mud and then excrete a tallow-like substance (castoreum) and/or a strong-smelling substance from their anal glands on top of them.

DIGGING MARKS Eurasian Beavers leave digging marks when searching for roots of bulrushes and other aquatic plants. However, they can be difficult to distinguish from the digging marks of a Muskrat if there are no associated tracks and signs.

FEEDING MARKS Among the most obvious signs of beavers are their feeding marks. The animals bite off young branches with their incisors to get to the cambium. The 45° angle is typical of rodents. The toothmarks

◀ *This large oak has been gnawed by a Eurasian Beaver and almost felled.*
Märkische Schweiz Nature Park, Germany.

▲ *The 45° cutting angle typical of rodents and the 'toothmark staircase' characteristic of Eurasian Beavers are clearly visible. Märkische Schweiz, Germany.*

◁ *The wide furrows made by the incisors are a characteristic sign. Frankfurt (Oder), Germany.*

▽ *Debarked branches often collect on the banks of rivers and lakes where Eurasian Beavers are active. Märkische Schweiz, Germany.*

make a kind of 'staircase' along the edge. Large piles of debarked branches are often found in the water and in the surrounding area where beavers are active.

Beavers have the largest incisors of all rodents in Europe. They make correspondingly large toothmarks that measure 6–12mm wide. The large wood shavings that can be found at a beaver feeding site are another distinctive sign. The large trees on which beavers leave their toothmarks are generally not used as a food source, but as material for building dams and lodges.

EXCREMENT Light to dark brown, oval with a rough fibrous surface. Mainly consists of small pieces of wood that look like sawdust. Beavers usually defecate in water so their droppings are rarely found.

- L 2–4cm D 1.5–2cm

▲ *Eurasian Beaver droppings are rarely found. The egg shape and coarse contents are characteristic. Vledder, Netherlands. René Nauta.*

COYPU
Myocastor coypus

HTL 45–65cm

TL 30–45cm

W 4–9kg (up to 12kg)

Males larger than females.

The Coypu is the only species in the family Myocastoridae. Although its closest relative in Europe is the Crested Porcupine, it is considered here so that its signs can be compared alongside those of other larger, semi-aquatic rodents. Coypus are native to South America but were brought to Europe in the 1930s to be farmed for their fur. This non-native species has successfully adapted to climatic conditions in Europe and stable populations have become established. Coypus are strong swimmers with good hearing but rather poor eyesight. They can stay underwater for up to five minutes. They are often active during the day and live gregariously in large family groups. Coypus dig dwellings with complex tunnel systems in riverbanks, where they sleep, seek refuge and give birth to their young. A dwelling measuring 6m deep was recorded in South America. Coypus' natural predators are foxes, wolves, mink, Stoats and dogs. Their young may also be taken by owls, raptors and cats.

CHARACTERISTICS Long, rounded rather than flattened, hairless tail. Large, orange-coloured incisors.

DISTRIBUTION AND HABITAT Coypus were able to spread throughout Europe when they escaped from fur farms. In many European countries, there is an isolated spread near bodies of water. However, isolated populations are often unable to become permanently established. Coypus need to live near water and favour bodies of water with vegetation, such as reeds, bulrushes and other aquatic plants, close to the bank. They colonise both standing and flowing water.

DIET Purely vegetarian except for occasional consumption of fresh-water mussels. They eat both the green parts and the roots of aquatic plants, such as reeds, waterlilies and bulrushes. Meadow grasses and foliage make up a large part of the rest of their diet. Coypus will also eat crops such as beet, maize or cereals if they are close and safely accessible. In winter, Coypus dig more specifically for roots.

REPRODUCTION Coypus can mate all year round. After a gestation period of 128–135 days, 4–7 young are born. The young are born with teeth and fur and can eat vegetation and see. They can also swim and walk immediately after birth. These highly developed precocial young can reproduce after 3–10 months. Females can resorb individual embryos to reduce litter size. This is presumably done in response to food shortages. Some complete losses of a litter appear to be a mechanism for manipulating the sex ratio of the offspring.

Coypus use the river current to flee if they are in danger. Coypus are also known as 'nutria', which is Spanish for 'otter'. This can cause some confusion.

SIGNS

DWELLINGS AND PLATFORMS A Coypu dwelling can have several above-water entrances with a diameter of 20–25cm. The entrance holes are often found near water under a bush or hidden by dense vegetation. During the day, the animals rest, groom and feed on platforms in the water made of reeds and other plants. The platforms look like small Mute Swan nests (see page 693)

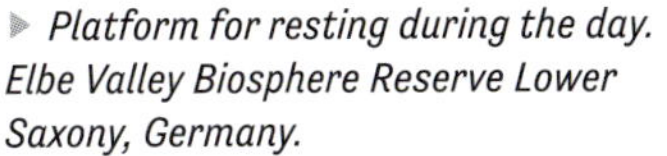
▶ *Platform for resting during the day. Elbe Valley Biosphere Reserve Lower Saxony, Germany.*

TRACK

Front

L 4–7.2cm W 3.5–6cm

Medium-sized. Plantigrade. Asymmetrical. Five toes in a classic rodent structure: toe 1 is short, but the corresponding claw is often visible (a). Toes 2 and 5 point to the sides. Toes 3 and 4 point forwards. Toes 2–5 are long, slim and usually appear connected to the midfoot pad. The midfoot pad is formed of three distinct, partially fused interdigital pads. Two further thenar/hypothenar pads are often visible at the back of the track, relatively close to the interdigital pads. Long, powerful, sharp claws that usually leave clearly visible prints. The front feet are much smaller than the hind feet, although the size difference is less pronounced than in Eurasian Beavers. A Coypu front foot looks like a very large Muskrat front foot, but less curved.

Hind

L 5.5–13cm W 4.3–10cm

Medium to large. Plantigrade. Asymmetrical. Five long, slim toes: toes 1 and 5 are not always clearly visible and occasionally don't leave prints. Toes 1–4 are connected by distal webbing that is often visible. There is no webbing between toes 4 and 5 (b). The midfoot pad can have clearly visible interdigital pad prints that are either incompletely fused or very indistinct. The strong claw prints are usually clearly visible and much sharper than

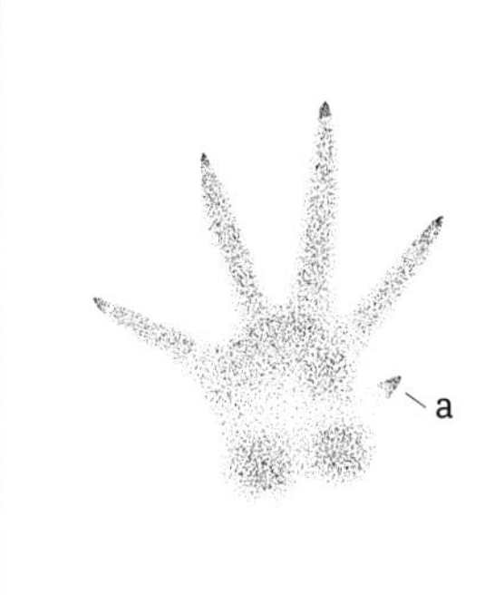

▲ *Left front.*

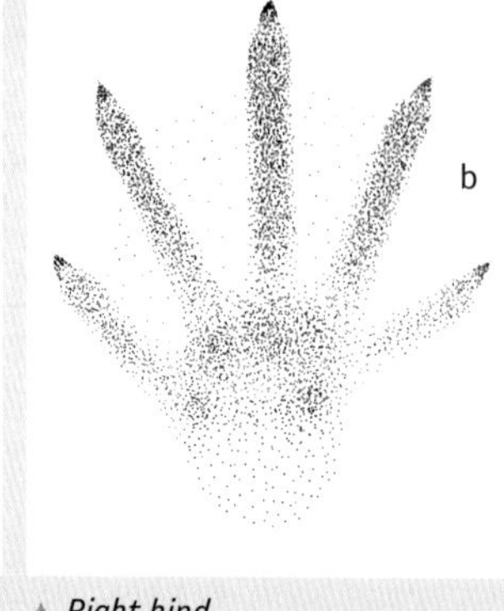

▲ *Right hind.*

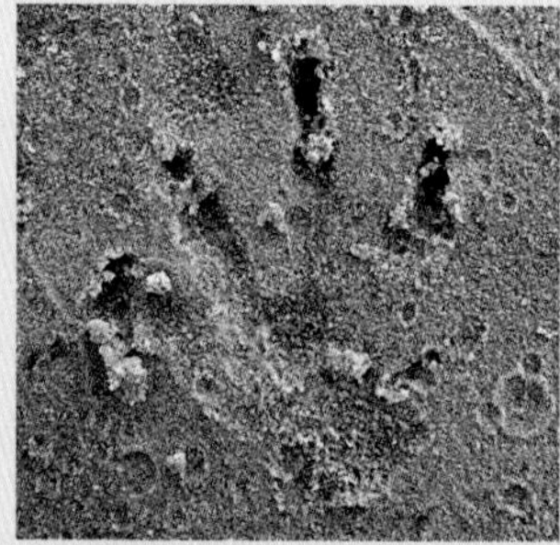

▲ *Left front.*
Elbtalaue, Germany.

▲ *Right hind. The back is missing; the sharp claws are clearly visible. Elbtalaue, Germany.*

▲ *Left front.*
Elbtalaue, Germany.

▲ *Right hind. The back rarely leaves such a complete print. Bielefeld, Germany. Ulrike Quartier.*

those of Eurasian Beavers. The back of the track very rarely leaves a visible print when the Coypu is walking.

GAITS

Coypus usually walk. The hind feet generally land slightly behind or on top of the prints of the front feet. Paths with shallow, approximately 2cm wide drag marks made by the tail are common. Eurasian Beavers leave much wider tail drag marks. If threatened, Coypus bound or gallop to safety. They usually prefer a parallel bound.

Walk
Stride length: 22.5–55cm
Trail width: 9.8–21.5cm

Parallel bound
Group length: 15–35cm
Inter-group length: 22–72cm
Stride length: 37–98cm
Trail width: 16–25cm

As Coypus mate all year round, juvenile tracks can often be found among the tracks of adult animals. In Europe, tracks of larger groups of 3–6 Coypus can occasionally be found. Muskrats are more likely to be found in small families of 2–3 animals.

▲ *The characteristic track patterns of a Coypu (left) and a Eurasian Beaver (right) side by side. Elbe, Germany. Frank Jermis.*

▲ *Direct register walk.*

Similar tracks
The footprints of juvenile Coypus can be mistaken for those of adult Muskrats (for differentiation, see page 284). The tracks of adult Coypus can be mistaken for those of Eurasian Beavers.

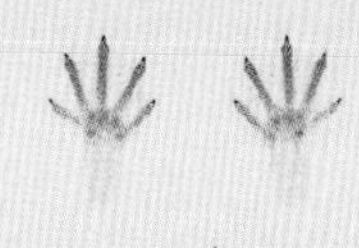

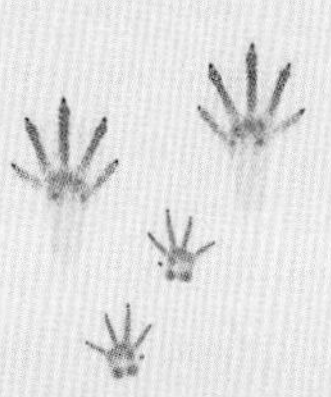

▲ *Parallel bound.*

DIFFERENTIATING BETWEEN COYPUS AND EURASIAN BEAVERS

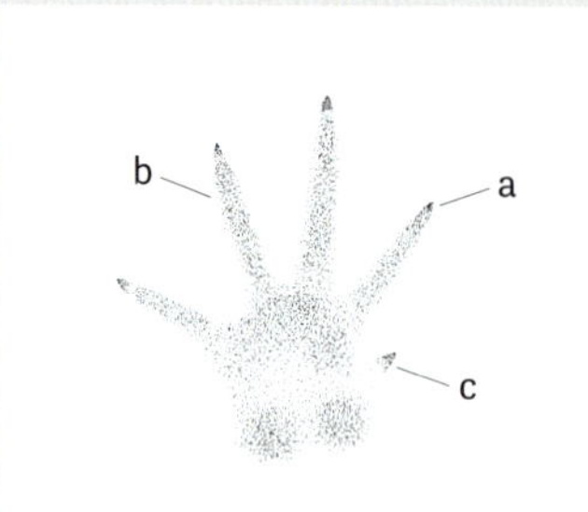

▶ *Coypu, left front.*

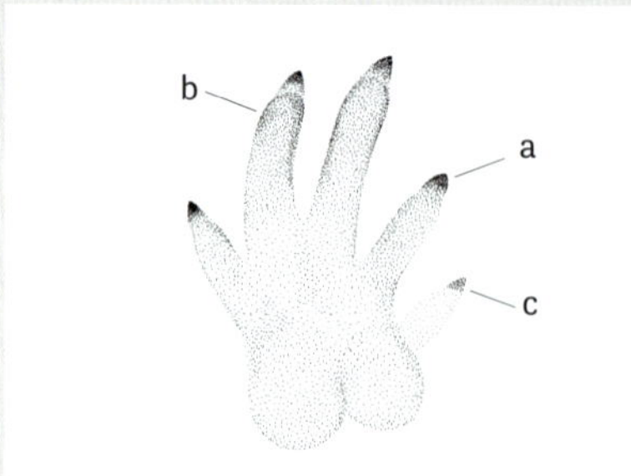

◀ *Eurasian Beaver, left front.*

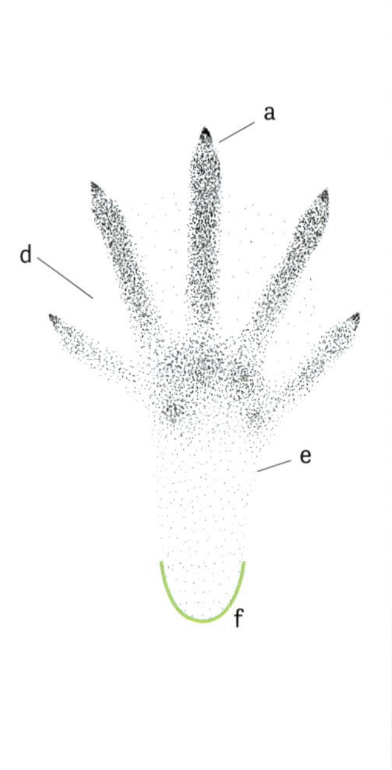

▶ *Coypu, left hind.*

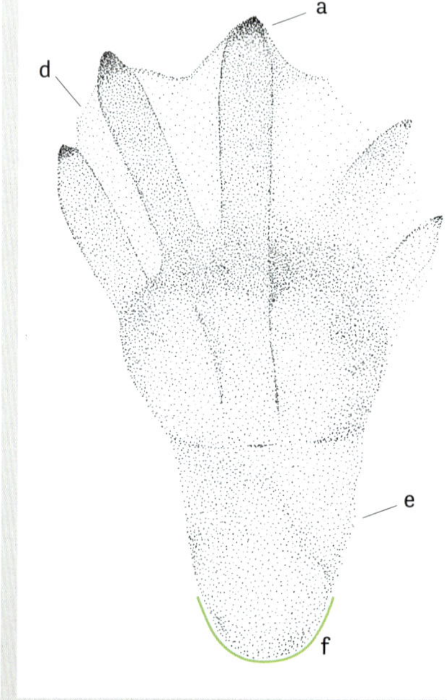

◀ *Eurasian Beaver, left hind.*

Coypu

a) Powerful, long, sharp claws.
b) Toes 2–5 are relatively long, narrow and straight.
c) The claw of toe 1 is frequently visible.
d) No webbing between toes 4 and 5.
e) The back of the track rarely leaves a print in walk.
f) Long, narrow curve.
g) The difference in size between the front feet and hind feet is less pronounced.

Eurasian Beaver

a) Powerful, short, blunt claws.
b) Toes 2–5 are relatively short, broad and curved inwards.
c) The claw of toe 1 is rarely visible.
d) Webbing between toes 4 and 5.
e) The back of the track usually leaves a print.
f) Short, wide curve.
g) The difference in size between the front feet and hind feet is more pronounced.

▲ Entrance hole
to a Coypu burrow.
Bielefeld, Germany.
Ulrike Quartier.

▶ Grazed areas like
this are a classic sign
of Coypu presence. Elbe
Valley Biosphere Reserve
Lower Saxony, Germany.

FEEDING MARKS Coypus leave conspicuous, cropped patches of vegetation when feeding intensively on grasses close to water. By regularly eating off the tips, the animals keep the grass short and the nutritious regrowing parts of the plant near the banks remain available for longer.

DIGGING MARKS   Coypus leave distinctive digging marks in meadows near the shore as they try to reach the roots of various dock species (*Rumex* spp.), especially in autumn/winter. You will usually also find trampled vegetation and tracks in these areas.

PATHS   The approximately 15cm wide paths between the dwelling and preferred feeding sites are usually very pronounced and easily recognisable.

▲ *Droppings are often found along the conspicuous paths.*
Elbe Valley Biosphere Reserve Lower Saxony, Germany.

▲ The parallel furrows make
Coypu droppings unmistakable.
Nahrendorf, Germany.

▲ Latrines may be
created in prominent
places such as raised
areas created by tufts
of grass. Elbe Valley
Biosphere Reserve Lower
Saxony, Germany.

EXCREMENT Like European Rabbits and Eurasian Beavers, Coypus are coprophagous. They eat their droppings to better absorb nutrients from the plants. Their droppings are long, cylindrical, usually slightly curved and very symmetrical. The clearly visible, fine, parallel furrows give the droppings their distinctive appearance. The colour varies from dark green to brown to black, depending on diet.

- L 3–7cm D 1–1.5cm

Voles, lemmings and muskrats

The subfamily of voles, lemmings and muskrats (Arvicolinae) belongs to the species-rich family Cricetidae that also includes hamsters and comprises a total of 130 genera and 681 species.

Two subfamilies – Arvicolinae, with 26 species in eight genera, and hamsters (Cricetinae), with three species – occur in the area this book covers. The vole genus *Microtus* contains the most species and is the most widespread. These voles are mainly small species with a compact body and a barely visible neck. The tail is shorter than the body. The animals have small, dark eyes and short ears, which are partially hidden in the fur. Arvicolinae species in Europe do not hibernate. They are mainly pure herbivores and almost of them dig their own dwellings. With the exception of Muskrats and water voles, all species are ground-dwelling with a burrowing lifestyle. Unlike many murids (mice and rats of the family Muridae), voles rarely climb and generally not very high when they do. People are most familiar with Arvicolinae species as pests in farmland and gardens.

TRACKS AND SIGNS OF ARVICOLINAE SPECIES

Plantigrade. Five toes on the front foot and five on the hind foot. The first toe of the front foot rarely leaves a print. The toes are slim and may appear ribbed. Compared to murids of a similar size, the digital pads are delicate and the claws quite long. An exception are the feet of *Clethrionomys* voles, which represent early intermediate stages of the transition from a terrestrial to an arboreal lifestyle. Correspondingly, some adaptations to climbing are visible in their tracks (see page 306).

- Track formula: 4f × 5H + C

FRONT FOOT Asymmetrical. Five toes in a classic rodent structure. Toe 1 is very short and rarely visible. Toes 2 and 5 protrude slightly and usually point clearly to the sides. The print of digital pads 2 and 5 are normally in front of the three fused interdigital pads that are arranged in a triangle. Toes 3 and 4 are angled forwards and fused at the foremost metacarpal pad, so they often have a characteristic V-shape.

HIND FOOT Slightly asymmetrical. Five toes in a classic rodent structure. Toes 1 and 5 point straight to the sides. Toes 2–4 are slim, often ribbed and usually point forwards. They are aligned and almost parallel. They normally appear to be fully fused with the midfoot pad.

GAIT Arvicolinae species usually move in walk or trot. In rare cases, they will bound over long distances. If they bound or gallop, the two hind feet almost never overstep the two front feet. This can help to distinguish them from murids.

PATHS Clear, usually pronounced paths under turf, in meadows or fields or on the edge of bodies of water are typical of Arvicolinae. These often well-maintained paths connect entrance holes with feeding sites and latrines.

FEEDING MARKS Arvicolinae species create characteristic feeding sites to where they carry food before eating it. Look out for neat piles of shredded vegetation along the usually well-established paths.

EXCREMENT Uniform faecal pellets with blunt, rounded ends consisting almost exclusively of fine plant fibres. The colour varies from light brown to dark green to black. Depending on the size, the shape of the excrement can resemble a grain of rice or a large Tic Tac sweet. Unlike mice and rats, Arvicolinae create latrines.

DIFFERENTIATING BETWEEN ARVICOLINAE SPECIES USING THE SIZE OF THEIR FAECAL PELLETS

The size of the faecal pellets is a useful criterion for differentiating between Arvicolinae species.

- The excrement of **Short-tailed Field Voles** (*Microtus agrestis*) and **Common Voles** (*Microtus arvalis*) is a maximum of 0.8cm long and 0.3cm thick. It resembles a grain of rice in terms of shape and size.
- **Water vole** droppings resemble a Tic Tac sweet in shape and size. Their droppings are at least 0.7cm long and 0.2cm thick.
- **Muskrat** droppings are at least 1.2cm long and 0.4cm wide – about twice the size of water vole droppings.

MUSKRAT
Ondatra zibethicus

HTL 22–36cm

TL 20–28cm

W 0.6–2kg, average 0.8–1.1kg (females), 1–1.3kg (males)

The Muskrat is the largest living Arvicolinae species and the only species within the genus *Ondatra*. The name 'muskrat' is misleading as the species is not a true rat nor closely related to them. Muskrats are excellent swimmers that can stay under water for up to 20 minutes, and are well adapted to life in the water and on land. They have hairs on their hind feet that enlarge the rudder area for swimming. The highly developed claws on all their feet make them excellent burrowers who make their dwellings in diggable riverbanks. The entrance to the dwelling is below the surface of the water, with a diagonally upward-sloping tunnel leading to one or more living chambers lined with vegetation. Muskrats that live in biotopes without diggable banks build lodges. When water levels are low, Muskrats create channels to provide water access to their favourite food sources.

Muskrats are mainly crepuscular and nocturnal, but they may also be spotted during the day. Their territorial behaviour is particularly pronounced during the breeding season. Other members of the same species tend to be tolerated in winter, when they share burrows to benefit from each other's body heat.

They are mainly predated by mink because these semi-aquatic mustelids can easily reach the underwater entrances to Muskrat dwellings. Mink have been known to inhabit abandoned Muskrat burrows. Other predators include foxes, otters and other mustelids, as well as Wild Boar, Common Raccoons, lynx and wildcats, and also raptors and owls, storks, herons and snakes. Northern Raccoons deliberately open Muskrat lodges to prey on young animals.

DISTINGUISHING FEATURES About the size of a European Rabbit, brown with a paler underside. Blunt-nosed and stocky. Wedge-shaped tail with compressed sides, which acts as a rudder. Solid claws.

DISTRIBUTION AND HABITAT Native to North America, Muskrats were introduced near Prague in 1905. They escaped from fur farms and are now found in large parts of Europe and throughout Germany. They prefer to live in large ponds and lakes with heavy aquatic vegetation but can colonise almost any body of water.

DIET Almost exclusively herbivorous. The diet is mainly made up of the shoots, leaves and bark of aquatic plants, as well as meadow grasses, foliage and fruits. They prefer reeds, bulrushes, sedges and horsetails. In winter, Muskrats feed on roots, which can sometimes cause serious damage to vegetation around buildings and on riverbanks. Muskrats' main animal food sources are freshwater mussels and crustaceans. They eat more animal-based food during spring and winter, when edible vegetation is scarcer.

REPRODUCTION Reproduction varies greatly according to geographical latitude. In Germany the first litter of an average of 5–7 young (1–16) is born between the end of March and the beginning of April, following a gestation period of 28–30 days. Muskrats have a total of 3–4 litters per year. In optimal habitats, Muskrats can breed almost continuously all year round. The altricial young are fully grown after the first winter.

Muskrats clear large areas of vegetation, which they use as a source of food or building materials. This lets more sunlight into the marsh, allowing animal and plant species to become established, so Muskrats contribute to the regeneration and biodiversity of marshland.

▼ *Muskrats build their characteristic lodges on floodplains. Vledder, Netherlands. René Nauta.*

TRACK

Front

2.1–4.1cm W 2.2–4.3cm

Small. Asymmetrical. Five toes in a classic rodent structure: toe 1 is very short, but the claw print is often visible. Toes 2–5 are long, slim, ribbed and usually continue all the way to the midfoot pad. The track may appear to curve inwards and the digital pads can have a robust appearance. The midfoot pad is made up of the three distinct and partially fused interdigital pads. Two further thenar/hypothenar pads are often visible at the back of the track. Long, strong, blunt claws that are used for digging and usually leave visible prints. The front feet are much smaller than the hind feet.

Hind

L 3.2–6.2cm W 3.7–5.9cm

Small to medium sized. Asymmetrical. Five toes: toes 1 and 5 point to the side. Toes 2–4 point forwards and are angled to the inside, because toe 4 is longer than toe 3 ⓐ. The toes are long, broad, ribbed and usually continue all the way to the midfoot pad. Although the interdigital pads are not fused completely, individual metatarsal-pad prints are rarely visible. The midfoot pad is often faintly visible and can appear blurry. The toes and edges of the sole have thick hairs to aid swimming that give the track a kind of border ⓑ. There is no webbing between the toes. In rare cases, a further, relatively large

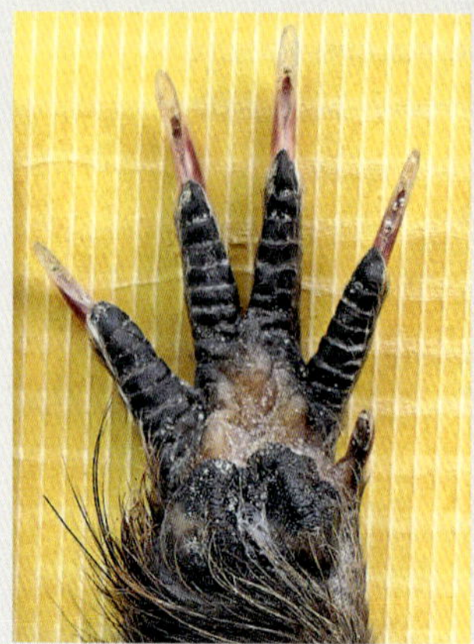

▲ *Right front.*
Vledder, Netherlands. René Nauta.

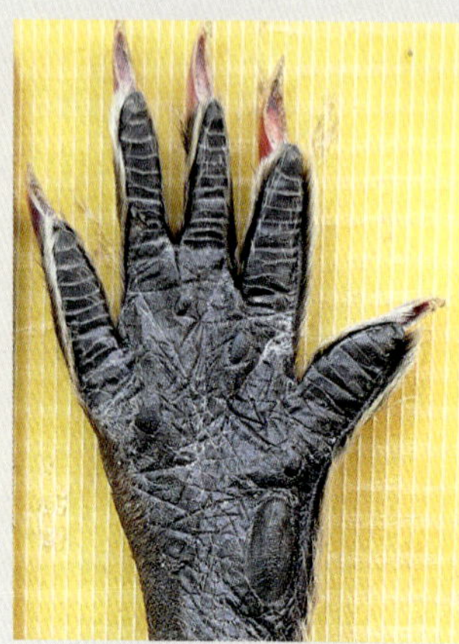

▲ *Right hind.*
Vledder, Netherlands. René Nauta.

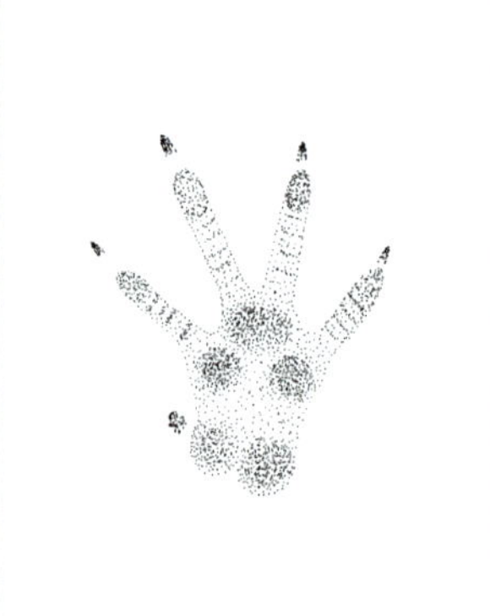

▲ *Right front.*

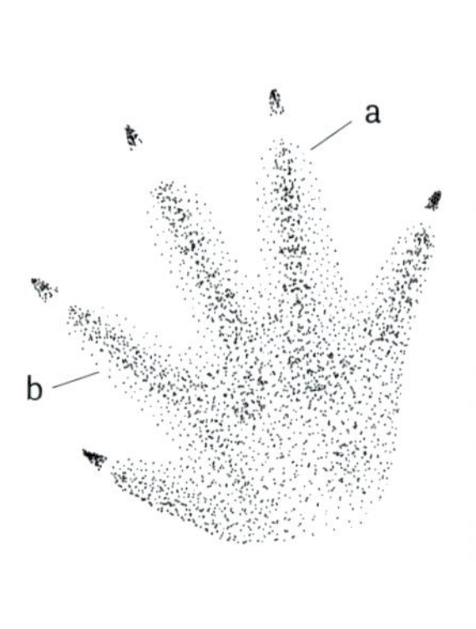

▲ *Right hind.*

▲ *Right front.*
Bielefeld, Germany. Ulrike Quartier.

▲ *Right hind.*
Jetzendorf, Germany.

thenar/hypothenar pad leaves a print at the back of the track. The strong, long claws are usually visible. The hind feet often turn in. The hind feet are much larger than the front feet.

GAITS

On land, Muskrats usually move at a slow walk or in some form of bound. The narrow drag mark of the tail is often visible when the animal walks.

Walk
Stride length: 16–29.3cm
Trail width: 6.8–13.1cm

Bound
Group length: 18–26cm
Inter-group length: 5–15cm
Stride length: 35–40.5cm
Trail width: 12–14.5cm

▼ A Muskrat kit at walk, showing the typical tail drag mark. Vledder, Netherlands. René Nauta.

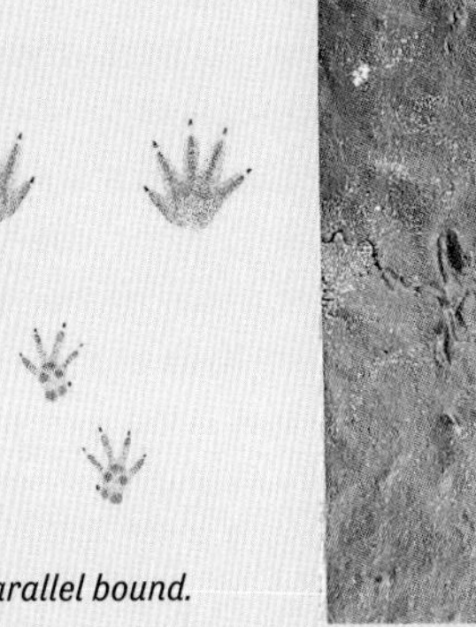

▲ A classic Muskrat direct register walk, here without a tail drag mark. Midhurst, England. Nate Harvey.

▲ *Direct register walk.* | ▲ *Parallel bound.*

Similar tracks
The tracks of a young Coypu can easily be mistaken for those of a Muskrat.

DIFFERENTIATING BETWEEN COYPUS AND MUSKRATS

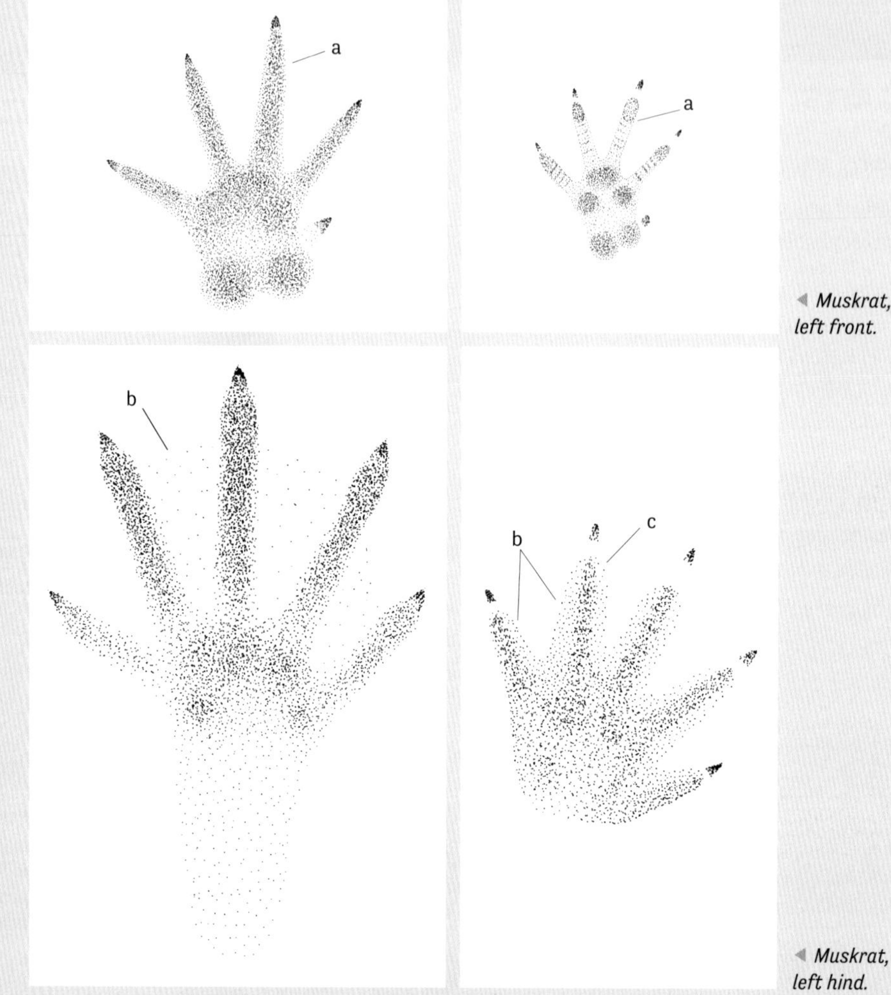

▶ *Coypu, left front.*

◀ *Muskrat, left front.*

▶ *Coypu, left hind.*

◀ *Muskrat, left hind.*

Coypu

a) Toes 2–5 are straight.
b) Webbing between toes 1–4.

Muskrat

a) Toes 2–5 curve inwards.
b) No webbing.
c) Swimming hairs on the edges of the sole can give the impression of a border.
d) A further thenar/hypothenar pad can be visible at the back of the track.
e) Toe 4 is longer than toe 3, angled to the inside.

SIGNS

MUSKRAT DWELLINGS Muskrat lodges resemble beaver lodges (see page 267), but are made using softer marsh plants like reeds, bulrushes and sedges. Because their jaws are weaker, Muskrats avoid woody branches. The heavy use of the surrounding vegetation and the channels they build characterise the landscape surrounding a lodge and are conspicuous signs of Muskrat presence. They may also dig tunnels into the riverbank. The entrance holes, which are usually hidden about 15cm below the surface of the water, can become visible when the water level is low and have a diameter of 9–15cm.

FEEDING MARKS Muskrats occasionally leave conspicuous piles of plants and mussels when they eat. They make other characteristic feeding marks when digging for roots. Remains of aquatic plants floating in the water are a classic sign of Muskrats. They resemble the piles of driftwood found in a beaver habitat, but consist almost exclusively of softer, non-woody plants. Cut stems, for example from bulrushes or reeds, can be found along the water's edge.

EXCREMENT Muskrats mark their territory with piles of droppings, for example on frequently used entrances and exits or trunks of fallen trees protruding out of the water, during the breeding season. At other times, they defecate in the water. The cylindrical faecal pellets vary in colour between dark green and light brown depending on diet.
* L 1–2.2cm D 0.4–0.7cm

▼ *Typical feeding mark. Vledder, Netherlands. René Nauta.*

▲ *Plant remains floating in the water are a typical sign of Muskrat activity. Vledder, Netherlands. René Nauta.*

▶ *Cylindrical droppings of this size found near water come from Muskrats. Offenbach, Germany.*

WATER VOLES
Arvicola

Montane Water Vole
HTL 12–17cm
TL 5.5–10cm
W 65–130g

European Water Vole
HTL 12–23cm
TL 6.5–14.5cm
W 70–320g

**South-western
Water Vole**
HTL 16–22cm
TL 9.5–15cm
W 150–300g

After the Muskrat, water voles are the largest Arvicolinae species in Europe. The European Water Vole (*Arvicola amphibius*), the South-western Water Vole (*A. sapidus*) and the Montane Water Vole (*A. monticola*) occur in Europe. It is only possible to reliably distinguish between the three species using molecular biology or by close examination of skull features. The number of species is open to debate, as the European and Montane Water Voles could be different ecotypes of the same species; they can interbreed and produce fertile offspring.

Water voles tend to live a solitary life and are active both during the day and at night. These excellent swimmers and divers dig tunnel systems into the riverbank. All three species are ecologically highly adaptable and can live both in water (aquatic) and on land (terrestrial). On land, their presence is usually evident by the piles and walls of earth created by their shallow tunnels.

Their main predators are mustelids such as Stoats, Least Weasels, American Mink, European Polecats and Eurasian Otters, as well as White Storks, Grey Herons, various owls and pike.

DISTINGUISHING FEATURES   Large Arvicolinae with very short ears, short legs and a tail about half the length of the body. Coat colour can vary greatly.

DISTRIBUTION AND HABITAT European Water Voles are found throughout Europe, except for the Iberian Peninsula. The South-western Water Vole is found on the Iberian Peninsula and in large parts of France, where its habitat can overlap with that of the European Water Vole. The Montane Water Vole usually lives at higher altitudes and inhabits mountains in Spain, France, Switzerland, Germany, Slovakia and Romania. Water voles inhabit slow-flowing or standing waters, preferably with dense vegetation on the banks. Burrowing (fossorial) populations live in fields, meadows and gardens. European Water Voles in Britain seem to be exclusively aquatic.

DIET In aquatic animals, mainly aquatic plants. Pulpy stems like reeds and bulrushes are preferred. They also eat foliage, crops, tree bark and grasses. They eat more roots and tubers in autumn. Animal prey is rarely eaten, but can include insects and snails, as well as small fish, frogs or crabs. Water voles store food for winter.

REPRODUCTION During the mating season from March to October, there may be 2–5 litters of 2–8 (up to 14) young each. The young are born hairless and blind after a gestation period of 20–22 days. They reach sexual maturity at the age of two months.

Where European Water Voles and South-western Water Voles occur together, the latter is usually larger, with a longer tail. The Montane Water Vole is smaller than the European Water Vole, but the dimensions can overlap.

SIGNS

As the tracks of water voles are often not reliably distinguishable from those of Brown Rats, signs such as entrance holes, paths and droppings are important and often reliable distinguishing features. There are several signs you should consider. Entrance holes can be helpful for identification if you look at them together with the paths that lead from them and their position in relation to the water.

▶ *Water vole toothmarks are about twice as wide as those of the Short-tailed Field Vole. Vledder, Netherlands. René Nauta.*

TRACK

Front

L 1.3–2cm W 1.5–2.3cm

Very small to small. Plantigrade. Asymmetrical. Five toes in a classic rodent structure: toe 1 is very short and very rarely visible in the track. Toes 2 and 5 are slightly to the side, but usually less spread out than in Brown Rats. Toes 3 and 4 point forwards. The print of the midfoot pad consists of three easily recognisable interdigital pads, arranged in a triangle. Two further thenar/hypothenar pads can be visible at the back of the track. These are proportionately much smaller than those of squirrels or Common Hamsters. The claws are fine, pointed and long, but not always reliably visible. The front feet are smaller than the hind feet.

Hind

L 1.7–3.3cm W 1.5–3cm

Small. Plantigrade. Slightly asymmetrical. Five long toes in a classic rodent structure: toes 1 and 5 point straight to the sides. Toes 2–4 are usually in a row and almost parallel to each other. They appear elongated and slim. The interdigital pads have fused into a midfoot pad but four metatarsal pad prints can still be visible in the track. However, often only the two foremost metatarsal pads are visible. The whole foot rarely touches the ground, but if it does, a further thenar/hypothenar pad can be visible at the back of the print. A perfect print has a total of five metatarsal pads. This can help us to tell them apart from rats, which have six metatarsal pads. This distinguishing feature is particularly useful for species identification using ink prints. It can rarely be applied in the field, as the thenar/hypothenar pads of these species

generally do not leave a print. In addition, a second thenar/hypothenar pad can occasionally occur in water voles, but it is only rudimentary and therefore significantly smaller than in rats. The claws are fine, pointed and long, but not always reliably recognisable. The hind foot is larger than the front foot.

◄ *Water vole, left hind (bottom) and left front (top). Vledder, Netherlands. René Nauta.*

Similar tracks

Brown Rat and Yellow-necked Mouse, but Yellow-necked Mouse tracks are smaller. Distinguishing them from Brown Rats can be difficult or even impossible, as their foot morphology is very similar and there is also a clear overlap in terms of size. The prints of adult Brown Rats tend to be larger than those of adult water voles. However, this only applies to the maximum size values for Brown Rats, where water voles can be reliably excluded. There is a risk of confusion where dimensions are smaller. The frequent occurrence of large numbers of juveniles makes identification even more difficult.

DIFFERENTIATING BETWEEN WATER VOLES AND BROWN RATS

Water voles belong to the subfamily Arvicolinae and Brown Rats to the family Muridae. Some of the basic features for distinguishing between Arvicolinae and Muridae species can therefore be used to differentiate between the water voles and Brown Rats (see page 330).

Water vole

a) Long, fine claws.
b) Front foot more symmetrical, toe 3 slightly ahead.
c) Toe 5 almost level with toe 2.
d) Toes slightly longer and narrower.
e) Toe 1 very rarely visible.
f) Toe 1 much further back than toe 5.
g) Toes 2–4 slightly shorter and narrower.
h) Five metatarsal pads, H pad usually absent.

Brown Rat

a) Short, relatively blunt claws.
b) Front foot more asymmetrical, toe 3 clearly ahead.
c) Toe 5 much further behind toe 2.
d) Toes slightly shorter and wider.
e) Toe 1 visible relatively often.
f) Toe 1 further back than toe 5.
g) Toes 2–4 slightly longer and wider.
h) Six metatarsal pads, H-pad usually present.
i) The foremost two metatarsal pads are enlarged and can create a long, narrow print.

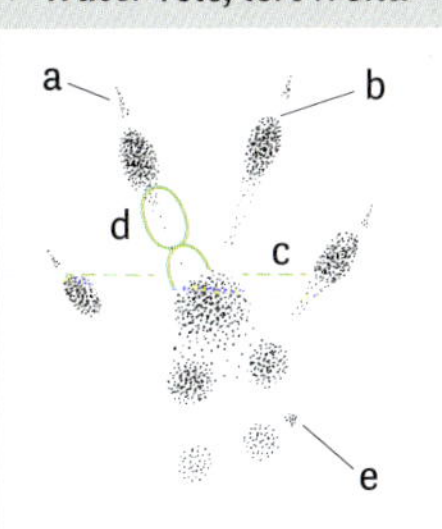

▼ Water vole, left front.

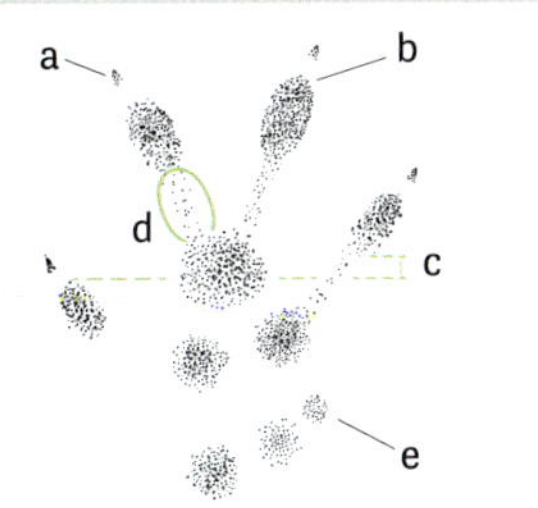

▼ Brown Rat, left front.

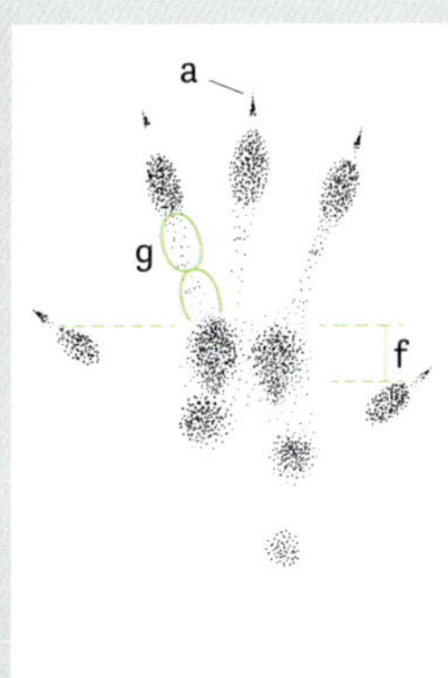

▲ Water vole, left hind.

▲ Brown Rat, left hind.

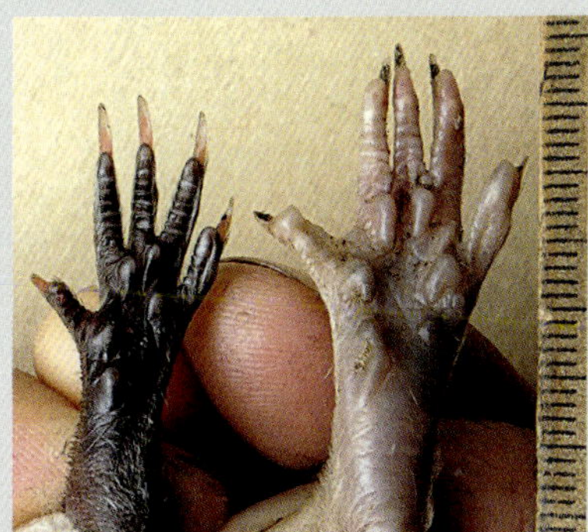

▲ Water vole, left hind foot (left) and, for comparison, the left hind foot of a Brown Rat (right). Vledder, Netherlands. René Nauta.

Almost perfect track conditions with detailed prints are required to differentiate between these features. You should also consider signs such as latrines, paths and feeding sites.

GAITS

Water voles prefer a form of walk. They only travel short distances and do so under cover and often in trot. They often switch to a 2 × 2 bound in snow and on ground where they could sink. To escape danger or to cross open spaces quickly, they use a kind of gallop.

Walk
Stride length: 5–16cm
Trail width: 5–9cm

Trot
Stride length: 14–25cm
Trail width: 4–8cm

3 × 4 bound (rotary) and
4 × 4 bound (transverse)
Group length: 5.5–10cm
Inter-group length:
5–26cm
Stride length: 10.5–36cm
Trail width: 5–8cm

▲ *The understep walk of a water vole. Vledder, Netherlands. René Nauta.*

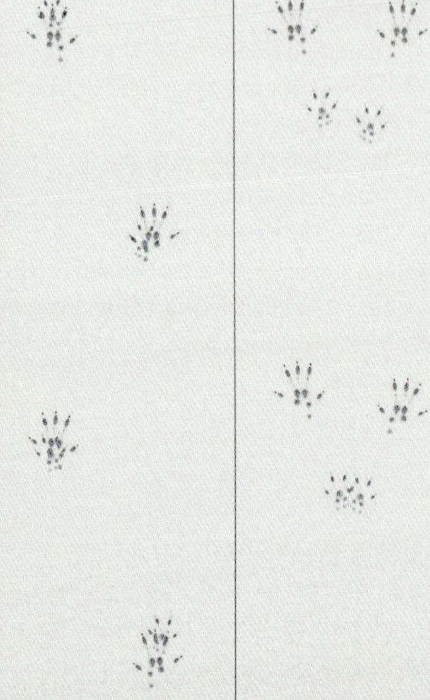

▲ *Direct register walk.* | ▲ *Full/paired bound.*

DWELLING AND NEST The spherical nest, which measures around 20cm in diameter and is made of dry plants, is usually found underground, but rarely also in dense patches of reeds and bushes. Dwellings close to the water are often built into embankments. Entrances to the dwellings can be above or below the surface of the water. The dwellings of aquatic and terrestrial animals can sometimes consist of tunnel systems measuring 100m long, one or two nest chambers and several storage chambers. The tunnels are 5–60cm deep and the nest chambers usually 40–60cm deep. When they dig these burrows, water voles make piles of earth that can be mistaken for molehills. If this is the case, it is important to pay attention to the position of the entrance holes. The entrance holes of molehills are usually in the middle and lead down vertically. The entrance holes to water vole dwellings are on the side and lead down diagonally. Almost complete removal of the plants and roots in the entrance area is typical of Arvicolinae species. The plants around the entrance hole to a water vole dwelling are usually shredded and eaten away. This is not the case with entrance holes to rat burrows. Another difference is the fan-shaped ejected material, often recognisable in front of rat holes, that is not found in front of water vole dwellings. Water vole tunnels have a round to oval cross section and the entrance holes have a diameter of 4–8cm and are usually wider than they are high.

PATHS   Water voles make well-established paths and feeding sites in dense vegetation. Unlike rat paths, water vole paths are usually hidden under vegetation and are often only visible when you specifically push it aside. Underneath, they are usually wide, well trodden and clearly visible. Paths are 4–9cm wide. They can even be visible in the water if there is a lot of vegetation on the surface. Characteristic feeding sites along the path are conspicuous.

FEEDING MARKS   Water voles create distinctive feeding sites because they carry food to their favourite sites before eating it. Look out for flattened areas in and near the water with neat piles of shredded vegetation. These plant remains can be up to 1cm in diameter and are usually between 8 and 10cm long. Comparable feeding marks of smaller Arvicolinae species are usually thinner and shorter and consist of less robust plants such as foliage and grasses. The 45° cutting angle is also typical.

CAMBIUM FEEDING   Water voles often gnaw on the bark of roots to get at the nutritious cambium. They prefer deciduous trees but also gnaw at the roots of conifers. More rarely, you can also find evidence of cambium feeding above ground on tree trunks, branches and twigs. To my knowledge, reliable evidence has only been found on ash trees (*Fraxinus*). Unlike red-backed voles, water voles do not climb. They usually debark areas in the lower parts of trees at a height of up to 25cm. They can gnaw at trees with a diameter of over 30cm. The furrows left by the gnawing teeth of the upper jaw measure 3.5–5mm wide in total.

Gardeners complain about these species tunnelling in flower beds and vegetable patches. South-western Water Voles can reproduce in vast numbers and cause significant damage, e.g., to fruit trees. The animals eat the roots and can destroy entire orchards.

▲ *Piles of approx. 10cm long, neatly cut plant stalks are a sign of water voles. Hoher Fläming Nature Park, Germany. Ingolf König-Jablonski.*

EXCREMENT Typical elongated, cylindrical Arvicolinae excrement. Faecal pellets with blunt, rounded ends that resemble a large Tic Tac sweet in terms of size and shape. Generally, very similar to Muskrat droppings, but only about half the size. The colour varies from dark green to black. In summer, droppings tend to consist of green plant parts, so they are often green. In winter, the animals feed more on bark, roots and similar, so their droppings are often brownish. Can be mistaken for rat droppings, but the latter have a more unpleasant odour. Water vole droppings are often more homogeneous and generally contain more plant fibres. They are usually deposited in latrines next to entrances or at feeding sites. From February to November, the latrines are also used for territory marking.

The animals' hind feet bear scent glands, so they often trample on the odourless droppings to give them a distinctive odour. This characteristic sign is left by water voles during the mating season to indicate their presence. Except for terrestrial water voles, most water voles deposit their droppings near water. Nick Baker mentions that light green concentric circles appear when water vole excrement is broken up.

- L 0.7–1.2cm D 0.2–0.5cm

▼ *Characteristic cambium feeding by a water vole on an ash tree. Vledder, Netherlands. René Nauta.*

▼ *Water vole droppings can be found on branches or roots at favourite feeding sites. Vledder, Netherlands. René Nauta.*

TRUE LEMMINGS AND WOOD LEMMINGS

Lemmus, Myopus

The tribe (genus group) Lemini belongs to the subfamily Arvicolinae and comprises three genera. The Norway Lemming (*Lemmus lemmus*) is the most common representative of the true lemmings and the Wood Lemming (*Myopus schisticolor*) is the only species of wood lemming. These species will be discussed here as examples of the genera. Lemmings are best known for their mass reproductions that occur every 3–5 years and the mass migrations that occasionally follow, although Wood Lemmings tend to migrate much less than Norway Lemmings. An abrupt increase in population can result in a shortage of food and, consequently, to mass migration. This sometimes results in enormous processions of lemmings that swim across rivers and lakes to settle in new areas. Although this isn't a case of 'mass suicide' which is a widely popular misconception, animals sometimes drown when crossing large lakes. Norway Lemmings often behave aggressively towards each

Norway Lemming

HTL 7–15cm

TL 1–2cm

W 45–130g

Wood Lemming

HTL 8–12.5cm

TL 1–2cm

W 18–45g

other during migration. They are known to hiss and squeak and even attack larger opponents. Wood Lemmings are less aggressive, which is probably related to their less pronounced mass reproduction. Norway Lemmings are active during the day and night, while Wood Lemmings are only active at night. Both are good swimmers and are very fast. They are predated by Stoats, wolves, bears, Wolverines, lynx, foxes, ravens, skuas, raptors and owls.

DISTINGUISHING FEATURES   Small to medium-sized Arvicolinae with short, partially fur-covered ears, small eyes and a very short, stubby tail. Norway Lemmings are brightly coloured, while Wood Lemmings are ash grey with a reddish to rust-brown colouring on the uppperside. Wood Lemmings are much smaller than Norway Lemmings. There is no size difference between the sexes.

DISTRIBUTION AND HABITAT   Northern Europe, Scandinavia and northern Finland. Norway Lemmings mainly live in the lichen zone above the tree line and spend most of the year under a covering of snow. For the very short summer period after the snow melts, they inhabit marshes, dwarf shrub heaths and, in the case of mass reproduction, birch and coniferous forests. Wood Lemmings prefer damp, mossy spruce forests. They also migrate to marshes with dwarf shrubs in spring.

DIET   Herbivores that mainly eat grasses and moss, rushes, sedges and tree bark, as well as berries, flowers and fungi. In autumn, Wood Lemmings may gather considerable stocks of moss for the winter. Norway Lemmings are not known to stockpile.

REPRODUCTION   The main mating season is in summer from May to August, but young are also born in winter. There are usually 3–5 litters in a year with 3–5 (up to 10) young in each, following a gestation period of 20–21 days. Females reach sexual maturity after three weeks, males after five weeks. Cyclical fluctuations in population density and mass reproduction are possible, but rarely occur in Wood Lemmings and, if so, only to a small extent. Large populations of Norway Lemmings usually collapse. The occurrence of females that can only give birth to females because of a special set of sex chromosomes is unique to Wood Lemmings.

TRACK

Front

L 0.8–1.5cm W 0.7–1.4cm

Very small. Plantigrade. Asymmetrical. Five toes in a classic rodent structure: toe 1 is very short and rarely visible. Toes 2 and 5 are slightly forwards and point clearly to the sides. Toes 3 and 4 point forwards. Two further thenar/hypothenar pads can be visible at the back of the track. The three interdigital pads are fused and arranged in a triangle. In Norway Lemmings, the metacarpal pads are covered with hair, whereas the soles of the feet of Wood Lemmings tend to be less hairy. The claws are strong, pointed and only visible occasionally. Martin Görner states that the claws of the front feet are longer than those of the hind feet in Norway Lemmings, but the other way around in Wood Lemmings (Görner & Hackethal 1987, page 199–200). The front feet are slightly smaller than the hind feet.

Hind

L 1.2–1.8cm W 1–1.5cm

Small. Plantigrade. Slightly asymmetrical. Five toes in a classic rodent structure: toes 1 and 5 point straight to the sides. Toes 2-4 usually point forwards and are aligned almost parallel. The interdigital pads have fused to form a midfoot pad. Three adjacent, similarly sized metatarsal pads and a fourth, smaller, slightly offset metatarsal pad are usually visible in the track. In rare cases, two further thenar/hypothenar pads are visible at the back of the track, with the T pad further back than the outer H pad. The hind feet are slightly larger than the front feet. The claws are strong, pointed and only visible occasionally.

Based on limited number of specimens.

GAITS

Lemmings mainly move at a walk or trot, especially over longer distances. This is characterised by relatively pronounced 'winding'. Bounding and galloping are relatively rare, except when fleeing danger. As lemmings spend most of the year under snow, their tracks are rarely found. The tail does not usually leave a print. Lemming paths mainly link up the entrance holes of their tunnel systems.

Similar tracks

Lemming tracks are similar to those of *Microtus* voles, but with much stronger claw prints.

SIGNS

ENTRANCE HOLES AND TUNNEL SYSTEMS   Norway Lemmings dig a network of open tunnels on the surface of the ground or under the snow. After the snow melts, these tunnel systems usually remain clearly visible. Wood Lemmings create their tunnel system in the thick moss of the forests. Entrance holes are usually cleared of vegetation and have a diameter of 3–6cm.

NESTS   In summer, lemmings build their nests from grass and moss in short, underground passages. The nests measure around 15–18cm in diameter. In winter, they make their nests on the surface of the ground, under the snow or between stones or in hollow tree stumps. The winter nests are slightly larger and have a diameter of 20–30cm. Old nests can be a conspicuous sign after the snow has melted.

FEEDING MARKS   Wood Lemmings are known for eating away the top layers of moss, causing light patches. They often carry food away and eat it in safe places, such as under tree roots or stones. Characteristic piles of uneaten stalks, similar to those left by water voles, and droppings are found at popular feeding sites.

CAMBIUM FEEDING   Like *Microtus* voles, lemmings leave feeding marks on bark (see page 303).

EXCREMENT   The droppings are like those of *Microtus* voles and water voles (see page 303). Typically for Arvicolinae, communal latrines, as well as larger piles of droppings, can be found in tunnels and at distinctive places, such as feeding sites.
- L 6–9mm D 3mm

MICROTUS VOLES
Microtus

The genus *Microtus* contains the most species within the subfamily of Arvicolinae and is the most widespread with the highest population density. There are 16 species of *Microtus* vole in the area covered by this book. The Short-tailed Field Vole (*Microtus agrestis*) and the Common Vole (*Microtus arvalis*) are given here as examples. The other representatives of *Microtus* are similar to the species discussed here in terms of their characteristics and way of life and their measurements usually fall within comparable ranges. Species identification is very difficult or even impossible, even when an individual animal is spotted, and it is rarely possible to distinguish between different *Microtus* vole species by looking at their tracks. It can occasionally be possible to differentiate between Short-tailed Field Voles and Common Voles based on their different habitats. *Microtus* voles are often active all day with their activity interrupted several times by irregular sleep phases (polyphasic activity). The animals are often active during the day in winter. They can swim well and have a particularly good sense of touch. Most species live gregariously in colonies. They predominantly live below ground in a complex system of burrows. The females can form nesting communities. Predators are cats, mustelids, foxes, Wild Boar, owls, raptors, storks, herons and snakes.

HTL 8–14cm
TL 2–5.5cm
(about one third
of HTL)
W 12–75g

TRACK

Front

L 0.7–1.4cm W 0.7–1.4cm

Very small. Plantigrade. Asymmetrical. Five toes in a classic rodent structure: toe 1 is very short and rarely visible. Toes 2 and 5 are slightly forwards and usually point to the sides. They can create the appearance of a straight horizontal line through the track. The print of digital pads 2 and 5 is normally in front of the three fused interdigital pads that are arranged in a triangle. Toes 3 and 4 are angled forwards and join up at the foremost metacarpal pad to create a V shape. Toes 2–5 are slender and can appear ribbed. In rare cases, two further thenar/hypothenar pads are visible at the back of the track. They are much further behind the interdigital pads than they are in field mice (*Apodemus* spp.). The claws are long, fine and pointed, but do not regularly leave prints. If visible, their prints are much further forward than those of field mice claws. The front feet are only marginally smaller than the hind feet and, in some species, they are almost the same size.

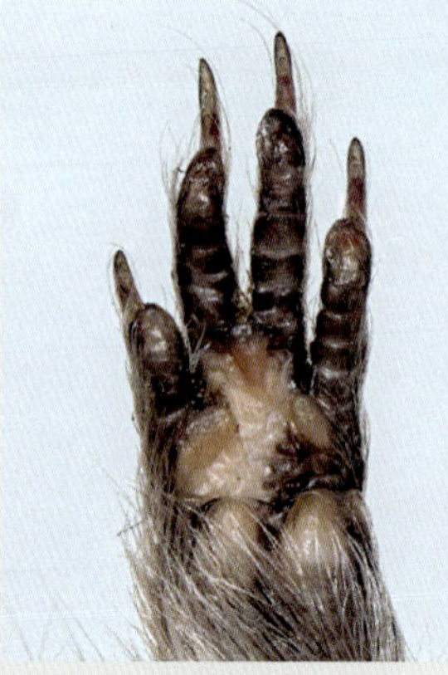

▲ *Common Vole, right front.*
Ennstal, Austria. Stefan Resch.

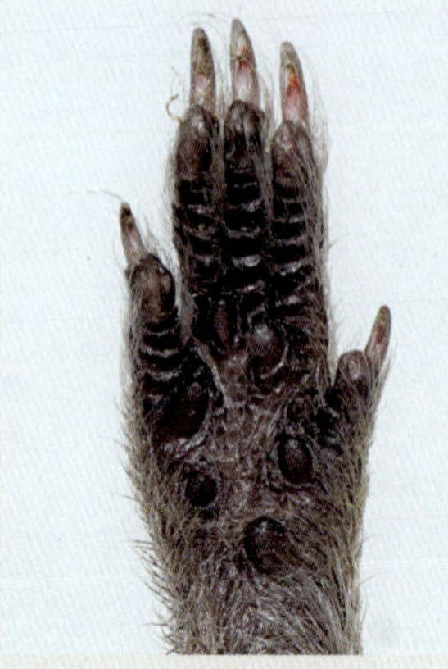

▲ *Common Vole, right hind.*
Ennstal, Austria. Stefan Resch.

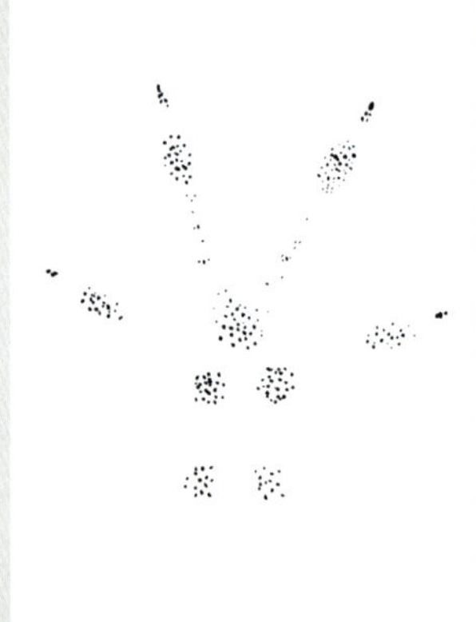

▲ *Common Vole, right front.*

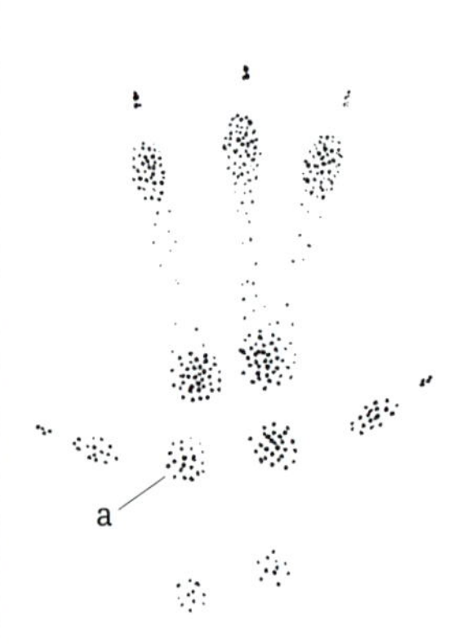

▲ *Common Vole, right hind.*

▲ Microtus *vole, right front (left) and right hind (right).*
Elbtalaue National Park, Germany.

Hind

L 0.8–1.4cm (up to 1.7cm) W 0.7–1.5cm
Very small to small. Plantigrade. Slightly asymmetrical. Five toes in a classic rodent structure: toes 1 and 5 point straight to the sides. Toes 2–4 are slim, often ribbed and usually point forwards. They are in a row and almost parallel to each other. They normally appear to be fully merged with the midfoot pad. The interdigital pads have fused to form a midfoot pad. However, three adjacent metatarsal pads of similar sizes and a fourth, smaller and slightly offset metatarsal pad are usually still visible in the track (a).

In rare cases, two further thenar/hypothenar pads are visible at the back of the track. The inner T pad is further back than the outer H pad. The hind feet are only marginally larger than the front feet. The fine, pointed claws are slightly shorter than on the front feet and are only visible occasionally.

GAITS

Unlike many field mice and red-backed voles, *Microtus* voles hardly ever climb and do not climb high when they do. They mainly travel in a walk or trot, often including over longer distances. Rarely, they will bound over longer distances. *Microtus* voles often move between vegetation and snow cover, which creates the characteristic tunnels.

Walk and trot
Stride length: 5.2–14cm
Trail width: 2.2–4.8cm

Parallel bound
Group length: 3.5–5.5cm
Inter-group length: 8–18.5cm
Stride length: 11–24cm
Trail width: 3–5cm

Similar tracks

At first glance, the tracks can be mistaken for those of any other small rodents. Closer examination and clear prints occasionally make it possible to distinguish between the genera of *Microtus* voles, typical mice (*Mus* species), field mice and red-backed voles (see page 330). The prints of smaller *Microtus* vole species can also be mistaken for those of shrews.

▼ *Direct register trot of a* Microtus *vole. Lausitz, Germany.*

▲ *Direct register trot.*

▲ *Full/paired bound.*

DISTINGUISHING FEATURES Small to medium-sized Arvicolinae species with short, partially hair-covered ears and a short, hairy tail. The Root Vole (*Microtus oeconomus*) is the largest species of *Microtus* vole in Europe. It can reach a weight of up to 90g and a head-trunk length of 16cm. The Common Vole is slightly smaller than the Short-tailed Field Vole.

DISTRIBUTION AND HABITAT The Short-tailed Field Vole and Common Vole live in northern and central Europe, but the Common Vole is absent from Britain and Scandinavia. The Short-tailed Field Vole is very adaptable in its choice of habitat, favouring damp forests and meadows, bogs, swamps, pastures, sedge and rush stands. As the name suggests, the Common Vole is Europe's most common Arvicolinae species. It usually avoids wetlands and denser forests. It prefers gardens, fields, roadsides and open grassland.

DIET Almost exclusively herbivorous. Mainly seeds, foliage, rushes, green parts of plants, grasses and roots. They eat more tree bark in winter, which can cause damage to the forest. *Microtus* voles also eat crops, such as potatoes and cereals. They are known to stockpile food. Insects can make up a significant part of their diet at times.

▼ *The freshly dug entrance hole of a* Microtus *vole. Lausitz, Germany.*

REPRODUCTION The mating season usually lasts from February until November. They usually produce 2–6 litters with 3–8 (up to 13) young in each. Cyclical fluctuations in population density and mass reproduction are possible. In some cases, these occur every 3–4 years, almost as a cycle. The young are born naked and blind following a gestation period of 19–21 days, and reach sexual maturity after 2–7 weeks.

SIGNS

ENTRANCE HOLES AND TUNNEL SYSTEMS Entrance holes usually have a diameter of 2.5–3.5cm. In contrast to entrance holes made by Wood Mice and Yellow-necked Mice, only small amounts of ejected material are scattered around the hole. Typically for Arvicolinae, *Microtus* vole species dig with their front feet and push the earth under their belly towards their hind legs, which they then use to move the material. The animals keep moving forwards as they dig, which further packs the loose earth (Dieterlen 2005). Entrance holes are often connected by clear paths that are identifiable because the vegetation has been eaten away (Jenrich *et al.* 2010). Distinctive tunnel systems under turf and in meadows and fields are characteristic and the most common signs of this genus. The paths are normally hidden under vegetation. They are carefully maintained and link up entrance holes and feeding sites. Piles of snipped stems and latrines can often be found along these paths. After the snow melts, these tunnel systems usually remain clearly visible and can offer a fascinating insight into the microcosm of *Microtus* voles.

▼ *We can discover feeding sites, latrines and lookout points by following their paths. Spessart, Germany.*

▲ Microtus *vole tunnels are often hidden under the vegetation in meadows. Spessart, Germany.*

DWELLINGS AND NESTS *Microtus* voles usually dig their own shallow burrows at a depth of 20–40cm. These have several exits and often form an extensive network of tunnels. The tunnel systems are usually close to the surface. Voles use the tunnels for their nests and sometimes to store food.

The spherical nests are made of dry, chewed up grass and can measure up to 20cm in diameter. If the ground is completely covered in snow, *Microtus* voles also nest on the ground underneath it.

FEEDING MARKS The feeding marks of *Microtus* voles can resemble those of field mice. Seeds appear to have been gnawed neatly rather than coarsely shredded as they would be by squirrels or rats. *Microtus* voles often carry food away and eat it in safe places, such as under tree roots or stones. Grass, sedge and rush stalks are usually carefully snipped into pieces measuring 2.5–5.5cm long. Characteristic piles of uneaten stalks, similar to those left by water voles, and droppings are found at popular feeding sites.

CAMBIUM FEEDING Like red-backed voles, *Microtus* voles frequently gnaw on the bark of roots, branches and twigs to get at the nutritious cambium, especially in winter. A thick layer of the outer bark that the animals cannot use can be found near the trunk of trees where cambium has been eaten. Unlike red-backed voles, *Microtus* voles rarely climb and never very high, so areas that have been debarked by *Microtus* voles are usually found in the lower part of the tree at up to 10–15cm high. They prefer deciduous trees, but also eat the cambium of coniferous trees with soft bark. The toothmarks left by the incisors of the upper jaw are 1.4–2.3mm wide in total.

EXCREMENT Faecal pellets with blunt, rounded ends, the size and shape of a grain of rice. Overall, very similar to water vole droppings, but much smaller. Short-tailed Field Vole excrement appears thick (D usually > 1.7mm). Thinner faecal pellets are also regularly found in piles of droppings, but these do not allow any differentiation. The colour varies from dark green to blackish. Droppings in summer tend to consist of green plant parts, so they are often green in colour. Winter droppings are often brownish. *Microtus* voles usually deposit their droppings in latrines at feeding sites or in specific chambers within their tunnel system. *Microtus* vole tunnels are often clearly visible in meadows after the snow melts. Chambers containing droppings along the path are a clear sign of *Microtus* vole activity.

- **Common Vole** L 2.2–6.2mm D 1–1.8mm
- **Short-tailed Field Vole** L 5–7.7mm D 1.5–2.4mm
- **Bank Vole (for comparison)** L 2.5–7mm D 0.8–1.8mm

▲ *Cambium feeding by a Short-tailed Field Vole. The toothmarks are about half as wide as those of a water vole. Jetzendorf, Germany.*

▼ *These carefully chewed off seeds are the work of a* Microtus *vole. Elbtalaue National Park, Germany.*

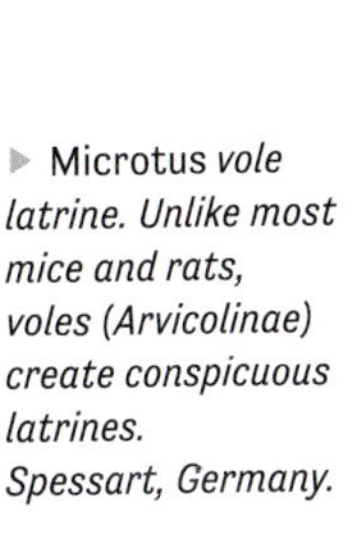

▶ Microtus *vole latrine. Unlike most mice and rats, voles (Arvicolinae) create conspicuous latrines. Spessart, Germany.*

▲ *Droppings of a Root Vole (*Microtus oeconomus*). Elongated faecal pellets with rounded ends, a uniform shape and a rough fibrous surface are typical of Arvicolinae species. Årjäng, Sweden.*

RED-BACKED VOLES
Clethrionomys, Craseomys

HTL 5.6–13cm
TL 3–7cm
(about 50 per cent
of HTL)
W 9.5–40g

There are three species of red-backed vole (genera *Clethrionomys* and *Craseomys,* formerly *Myodes)* which occur in Europe: the Bank Vole (*Clethrionomys glareolus*), the Northern Red-backed Vole (*Clethrionomys rutilus*) and the Grey Red-backed Vole (*Craseomys rufocanus*). Here, we will be looking at the Bank Vole as a representative of this genus. Red-backed voles are medium-sized Arvicolinae species with clearly visible ears and relatively large eyes. The fur on their back is usually reddish. The animals are active all year round, both during the day and at night. They are good climbers and live almost exclusively in forests. Unlike *Microtus* voles, they do not usually construct distinctive tunnel systems and instead prefer to move around under the protection of leaves and fallen branches. Their ability to find their way around is excellent. They often cover distances of 600m and experiments have shown that they can find their way back to their nest site from over 700m away. These orientation skills improve with age and experience (Gipps 1985). Predators as for *Microtus* voles.

DISTINGUISHING FEATURES Medium-sized Arvicolinae with a slightly longer tail than the *Microtus* vole species and usually with reddish-coloured fur on their back. Like Short-tailed Field Voles, but with a clearer colour contrast to the belly and larger ears. The Grey Red-backed Vole is slightly larger and the Northern Red-backed Vole slightly smaller than the Bank Vole. Populations from different regions can vary in size and weight by up to 300 per cent.

DISTRIBUTION AND HABITAT Apart from the Mediterranean islands, red-backed voles are found throughout almost the whole of Europe and are among the continent's most common mammals. Grey Red-backed Voles and Northern Red-backed Voles both occur in Scandinavia, but the distribution of Northern Red-backed Voles begins further north. Red-backed voles prefer damp deciduous and mixed forests, as well as alder swamps, but they also live in gardens, parks and hedges. Grey Red-backed Voles inhabit rockier areas of the taiga forest zone and subpolar birch forests. Northern Red-backed Voles live in subarctic birch forests and in coniferous forests. Both are absent from pure lichen tundra. Compared with other Arvicolinae species, Red-backed voles tend not to tunnel into the ground and need a layer of grass, leaves or moss for their tunnels. They favour moist biotopes over dry ones.

DIET Underground parts of fungi, truffles and roots. They also eat grasses, moss, lichen, foliage, leaves, fruits and seeds, such as nuts and beechnuts, depending on availability, as well as insects, snails and worms. In autumn, Bank Voles stockpile small amounts of nuts, acorns, beechnuts and similar in holes in the ground and hollow trees. Grey Red-backed Voles and Northern Red-backed Voles possibly exhibit similar behaviour. They eat more tree bark, for example from elder, in winter.

REPRODUCTION Mating season varies according to region but usually occurs in the summer. Females exhibit territorial behaviour during the breeding season. Between March and September, but mostly between May and August, 3–7 litters with 4–6 (up to 11) young each are born. Young are rarely born in winter. They are born hairless and blind following a gestation period of 17–23 days. The animals reach sexual maturity after 4–9 weeks. Females from the first litters usually reproduce the same year. Cyclical mass reproduction occurs in Grey Red-backed Voles and Northern Red-backed Voles every 4–6 years.

Northern Red-backed Voles often climb high up in the trees, which is why they are known as 'squirrel mice' in Lapland.

TRACK

Front

L 0.8–1.2 cm   W 0.7–1.4cm

Very small. Plantigrade. Slightly asymmetrical. Five toes in a classic rodent structure: toe 1 is very short and rarely visible. If toe 1 does not leave a print, the track may appear noticeably symmetrical. Toes 2 and 5 are slightly forwards and usually point clearly to the sides. The print of digital pads 2 and 5 is normally in front of the three fused interdigital pads that are arranged in a triangle. Toes 3 and 4 are angled forwards and fused at the foremost metacarpal pad, occasionally forming a V shape. Toes 2–5 are rather short, broad and have strong digital pads compared to *Microtus* voles, which are an adaptation for climbing. In rare cases, two further thenar/hypothenar pads can be visible at the back of the track, almost in a horizontal line next to each other. If the thenar/hypothenar pads don't make a print, then the track will have a noticeably roundish appearance. The claws are fine, pointed, rather short and only leave prints occasionally. The front feet are only slightly smaller than, and in some species almost the same size as, the hind feet.

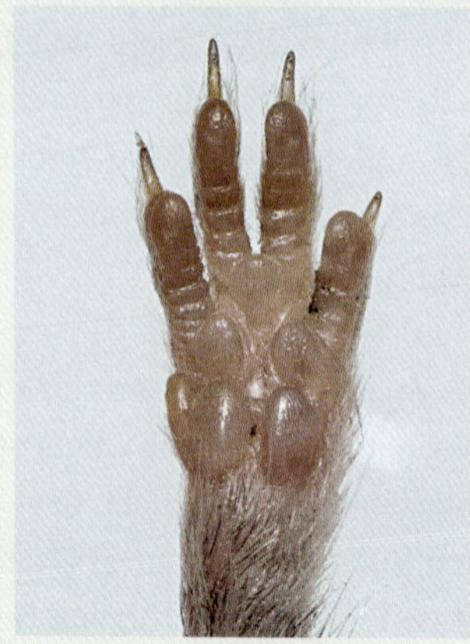

▲ *Bank Vole, left front. Ennstal, Austria. Stefan Resch.*

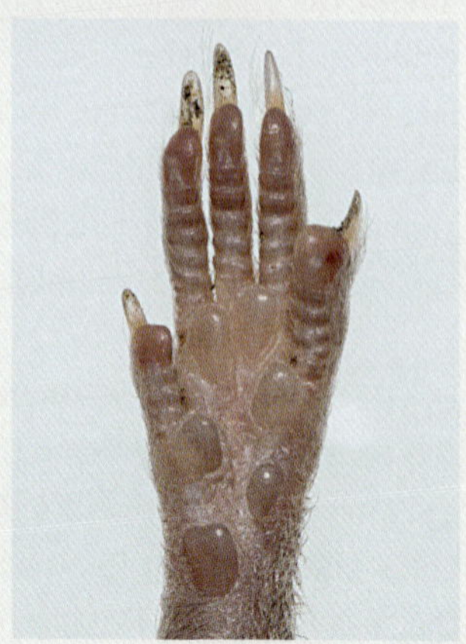

▲ *Bank Vole, left hind. Ennstal, Austria. Stefan Resch.*

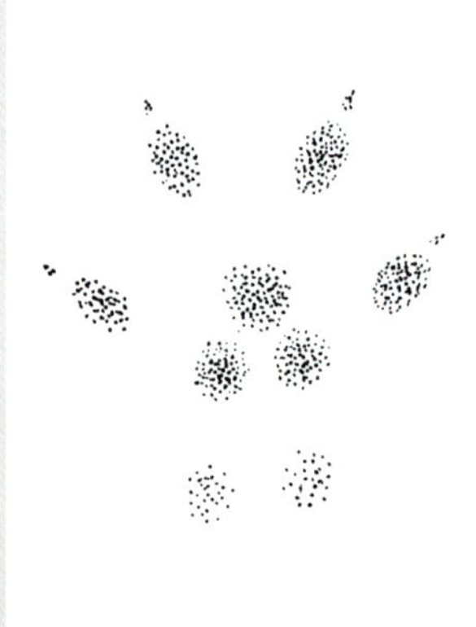

▲ *Left front.*

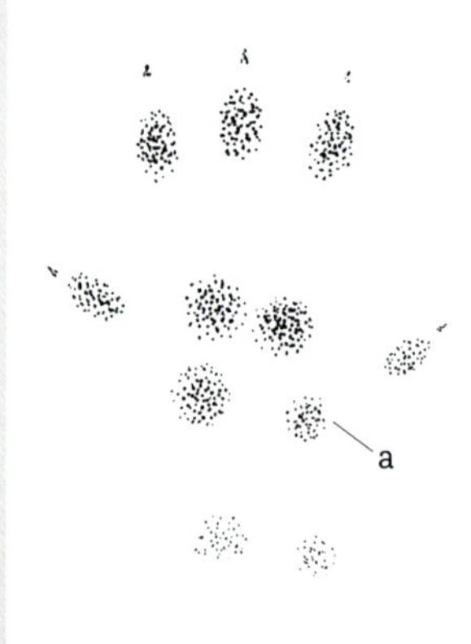

▲ *Left hind.*

▲ *Left front. Bielefeld, Germany. Ulrike Quartier.*

▲ *Left hind. Bielefeld, Germany. Ulrike Quartier.*

Hind

L 1–1.4cm   W 0.7–1.5cm

Very small to small. Plantigrade. Slightly asymmetrical. Five toes in a classic rodent structure: toes 1 and 5 point straight to the sides. Toes 2–4 point forwards and are aligned and almost parallel. They are also rather short, relatively broad and, compared to *Microtus* voles, have strong digital pads as an adaptation for climbing. The digital pads usually appear separate from the midfoot pad. The interdigital pads have fused to form a midfoot pad. However, three adjacent metatarsal pads of similar sizes and a fourth, smaller and slightly offset, metatarsal pad are usually still visible in the track (a). In rare cases, two further thenar/hypothenar pads can be seen at the back of the track. The inner T pad is further back than the outer H pad. The hind feet are only marginally larger than the front feet. The claws are fine, pointed, relatively short and only leave prints occasionally.

GAITS

Unlike field mice (*Apodemus* spp.), red-backed voles often also cover longer distances in a walk or trot. They often cross layers of leaves, moss or snow in a 2 × 4 bound or a 3 × 4 bound (rotary). They make tunnels in the moss and leaves or shallow tunnels in the ground. If there is snow on the ground, these voles often move underneath it, creating their typical tunnels. The dimensions are identical to those of *Microtus* voles (see page 299).

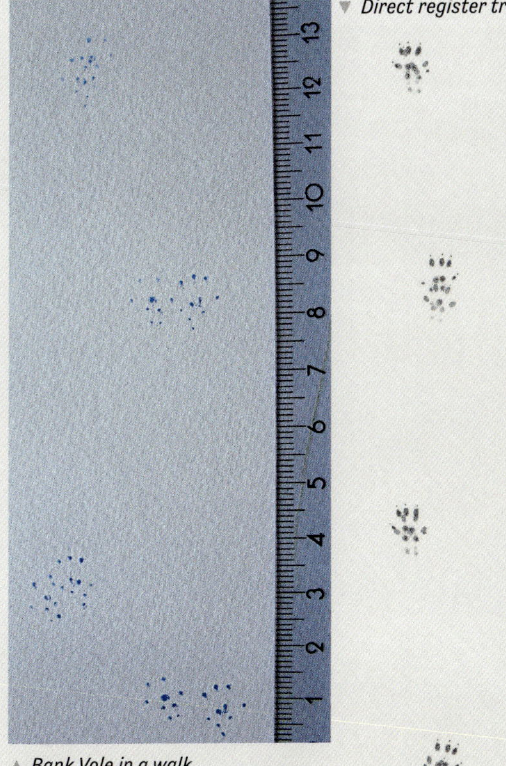

▲ *Bank Vole in a walk.*
Bielefeld, Germany. Ulrike Quartier.

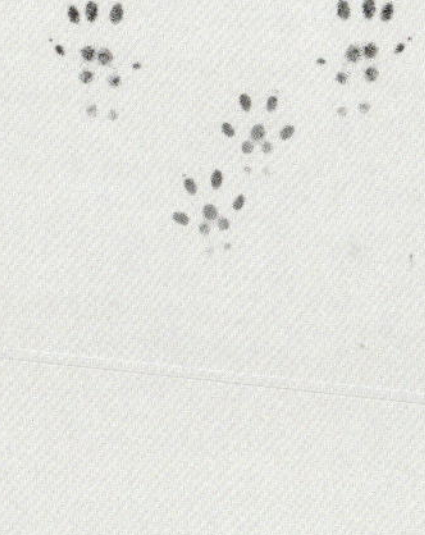

▼ *Direct register trot.*

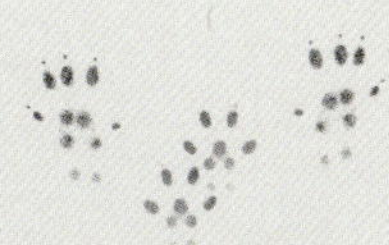

▼ *Parallel bound.*

Similar tracks

At first glance, the tracks can be mistaken for those of any other small mice. Closer examination and clear prints occasionally make it possible to distinguish between the genera of *Microtus* voles, typical mice (*Mus* genus), field mice and red-backed voles (see page 330).

SIGNS

NESTS   Spherical nests made of leaves, moss, bark fibres and grasses measuring about 8–15cm in diameter. Nests are usually built on the ground in hollows, under stones and in piles of brushwood or tree stumps. They are often hidden underground at depths of up to 45cm. In rare cases, they are found above the ground in trees.

FEEDING MARKS   The feeding marks of red-backed voles are very similar to those of field mice and *Microtus* voles. Seeds also appear to have been gnawed neatly rather than coarsely shredded as they would be by squirrels or rats. The cone feeding behaviour of these voles is similar to that of field mice (see page 164). Red-backed voles feed more on animals compared with *Microtus* vole species. They usually open snail shells carefully from the top and eat their way down, following the thread of the shell. The lower mouth of the shell is usually preserved, presumably because the voles pull out the rest of the snail (Brown 1985).

▼ *Feeding site of a Bank Vole. Snails have been eaten here. Lausitz, Germany.*

▲ *The snail shells, which have been carefully opened from the top, are typical of red-backed voles. Jetzendorf, Germany. Laura Gärtner.*

▲ *Characteristic behaviour: This black tree fungus has been scraped off by red-backed voles. West Sussex, England.*

This is a clear difference to the snail shells left by field mice (see page 321), which are usually opened at the mouth of the shell, and the smashed snail shells found at a Song Thrush anvil (see page 712). Species identification is often only possible in conjunction with other signs, such as markings, droppings or tracks.

CAMBIUM FEEDING Like *Microtus* voles, red-backed voles often gnaw on the bark of branches and twigs, mainly in winter, to get to the nutritious cambium. In other seasons, they gnaw bark when food supply is limited, for example in areas that are purely used for forestry. A thick layer of the outer bark that the animals cannot use can be found on the ground near the trunk of trees where cambium has been eaten. Red-backed voles usually start gnawing in a sitting position at the base of branches and, from there, feed on the cambium of the trunk and the surrounding, tapering branches. Areas debarked following this pattern are usually in the higher parts of the tree and have conspicuous, light-coloured patches of wood. They prefer elder and other deciduous trees, but also eat the cambium of firs, spruces and other conifers. They seem to favour trees with soft bark. Although they do gnaw trees with a trunk diameter of 20–30cm, smaller diameters of 2–15cm are typical. The toothmarks left by the incisors of the upper jaw are 1.5–2mm wide in total.

◀ *Bark gnawing by a Bank Vole. West Sussex, England.*

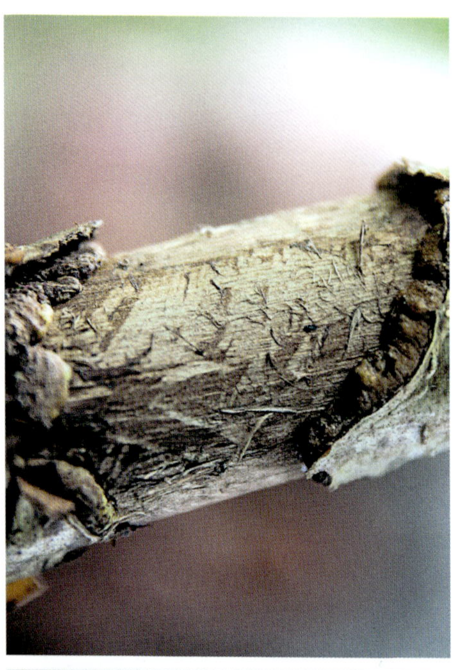

HAZELNUT FEEDING The vole usually props up the hazelnut on the ground and holds it with its front feet. It quickly gnaws the first opening in the shell with its lower incisors. Then, the vole inserts its upper jaw into the inside of the nut and begins to gnaw the side of the hazelnut closest to its body from the outside in, using its lower incisors. Meanwhile, it turns the nut with its front feet, gradually increasing the size of the hole. In contrast to field mice, the marks made by the upper incisors are located on the inside of the shell and are rarely visible. Apart from a few localised furrows that originate from the first opening of the nut by the lower incisors, there

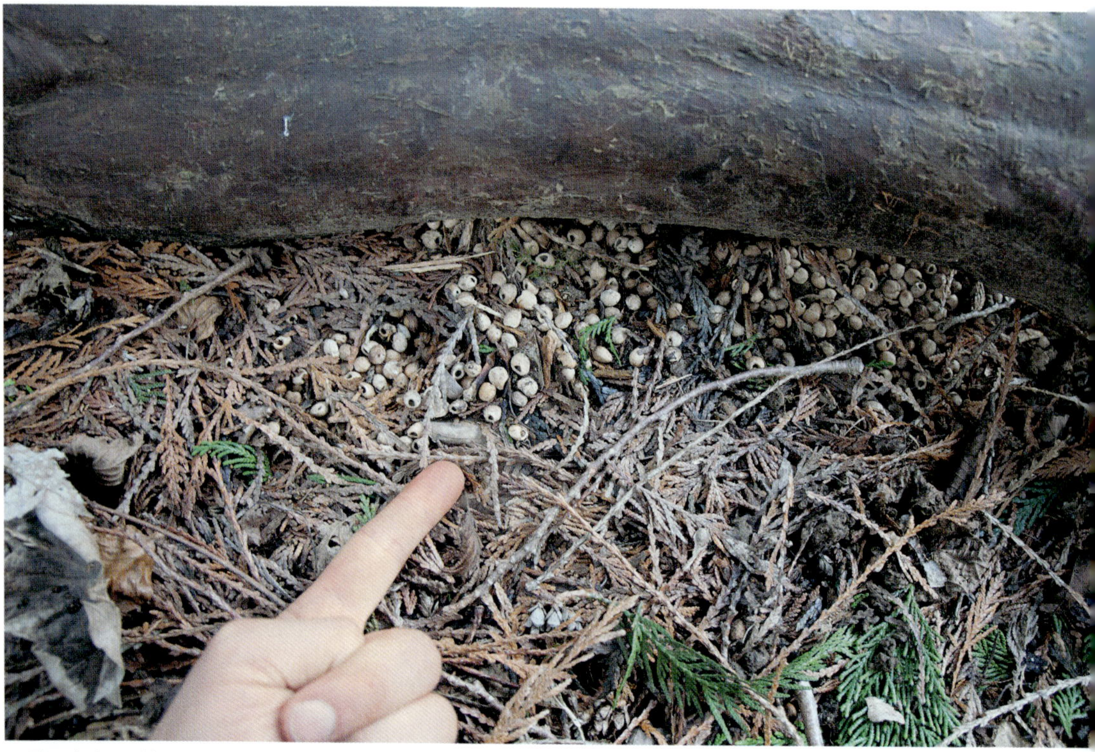

▲ *The sheltered feeding site of a Bank Vole, under a fallen tree. West Sussex, England.*

are hardly any toothmarks on the outside of the shell. The perforated edge has an even appearance with fine furrows (see page 165).

FOOD STORES Red-backed voles stockpile food for winter. They put acorns, beechnuts, nuts and seeds in small holes in the ground and hollow trees and sometimes cover them with leaves. Digging up the stores creates distinctive feeding and digging marks. Digging them up leaves small, shallow, usually rectangular holes in the ground, surrounded by the remains of shells. If a food store is not emptied, the seeds in it can germinate. Uneaten stores of beechnuts can produce dense stands of beech trees, which are called 'mouse beeches' in some countries.

EXCREMENT See *Microtus* vole excrement (page 303).
- L 2.5–7mm D 0.8–1.8mm

▲ *The typically even cylinder shape tells us that these are Arvicolinae droppings. The position on a branch at some height is indicative of a red-backed vole. West Sussex, England.*

Caution!

Red-backed voles are carriers of the Puumala virus (hantavirus). This virus is mainly transmitted via animal droppings. It can live in droppings for several weeks and is often inhaled with dust when cleaning out barns and during forestry work. Although the hantavirus found in Europe is less dangerous than the one found in the US, it should not be underestimated. After an incubation period of 2–4 weeks, one third of infected people will develop flu-like symptoms with a high fever. Renal dysfunction develops in around half of these patients. These complications can be serious and lead to death in 0.2 per cent of cases. The disease does not generally cause any permanent damage and is easily curable according to the Robert Koch Institute. If contact with vole droppings is unavoidable, please take the necessary precautions.

COMMON HAMSTER
Cricetus cricetus

HTL 17–30cm
TL 3–7cm
W 150–650g
(up to 1kg)

The Common Hamster is the only species of the genus *Cricetus*. This solitary, territorial rodent spends most of its life underground. Each animal has its own burrow, which it digs itself and fiercely defends including against other members of the same species. Common Hamsters spend six months of the year sleeping in their burrow. They wake up from hibernation at irregular intervals every 1–2 weeks to defecate and to eat from their food supplies. Common Hamsters stockpile food in late summer and autumn. They fill their cheek pouches with large quantities of food and then carry them to their dwelling. Most people will be familiar with the image of a hamster using its cheeks to store food. An individual hamster generally stores 2–3kg of food but, in rare cases, they can store 15kg or more. They use up these stocks over winter. Common Hamsters lose 20–30 per cent of their body weight during hibernation. They are mainly crepuscular and nocturnal but can also be seen during the day. Their sense of touch is particularly well developed. Common Hamsters are agile, fast, and adept jumpers and climbers. They are very defensive and will attack larger animals and even humans if threatened. In fights, they stand on their hind legs, raise their front legs, grind their teeth and hiss. This shows off the contrast between the black belly and the powerful white front feet and is presumably intended to intimidate. Common Hamsters also inflate their cheek pouches when they fight. They are predated by almost all small to medium-sized, land-based predators, such as mustelids, foxes and cats, as well as birds such as Eurasian Eagle-owls, crows, herons, buzzards, kites and storks.

Common Hamsters were formerly hunted for their fur and were traditionally considered to be agricultural pests. Today, they are classified as Critically Endangered and are strictly protected.

DISTINGUISHING FEATURES Approximately rat-sized, stocky body and colourful fur with a striking black belly and a very short tail. The males are slightly larger and heavier than the females.

DISTRIBUTION AND HABITAT From Belgium to central and eastern Europe, where they inhabit open landscapes with loamy or clay-based, diggable soils. Rarely found at altitudes above 600m. The Common Hamster was originally an eastern European steppe dweller but has become a species associated with human activities in central Europe, and has moved into agricultural areas. In western Europe, increasing industrialisation of agriculture and the resulting soil compaction have led to a drastic decline in Common Hamster populations. Intensive hunting and fragmentation of habitat by human infrastructure have also contributed to this sharp decline.

DIET Common Hamsters mainly eat plants, especially seeds and fruit, but also the green parts. They eat almost all crops, including cereals, potatoes, maize and beet. They also eat invertebrates, such as beetles, snails, earthworms, and even small vertebrates, such as *Microtus* voles. However, these only make up a small part of their diet. Cannibalism has been observed in high population densities.

TRACK

Front

L 1.4–2.2cm W 1–1.7cm

Small. Plantigrade. Asymmetrical. Five toes in a classic rodent structure: toe 1 is very short and rarely visible in the print. Toes 2–5 are relatively short and broad. Toes 2 and 5 point to the side and toes 3 and 4 point forwards. The interdigital pads have fused to form a midfoot pad, but this usually leaves a print consisting of three clearly visible, strong metacarpal pad prints. Two strong thenar/hypothenar pad prints are often visible at the back of the track. The claws are long, strong and usually leave clear prints.

Hind

L 1.8–3.5cm W 1–2cm

Small. Plantigrade. Slightly asymmetrical. Five toes in a classic rodent structure: toes 1 and 5 point to the sides. Toes 2-4 are relatively short and broad. They usually point forwards and are aligned and almost parallel. The interdigital pads have fused to form a midfoot pad, but you can usually make out four distinct metatarsal pad prints in the track. Two thenar/hypothenar pads are occasionally visible at the back of the track, making the hind foot print noticeably longer than the front foot. The claws are usually visible, but weaker compared with those on the front foot.

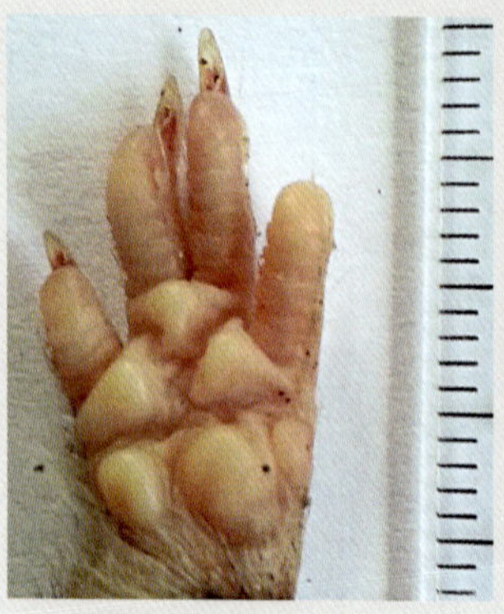

▲ *Right front. Strong metacarpal pads, short, wide toes and long claws are characteristic. Austria, Andreas Wenger.*

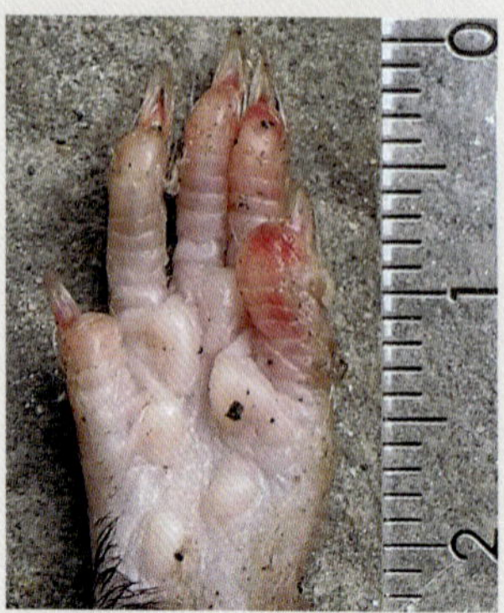

▲ *Left hind. Austria, Andreas Wenger.*

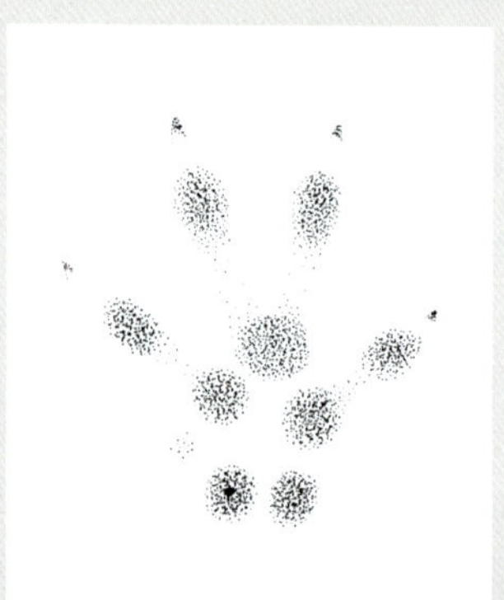

▲ *Right front.*

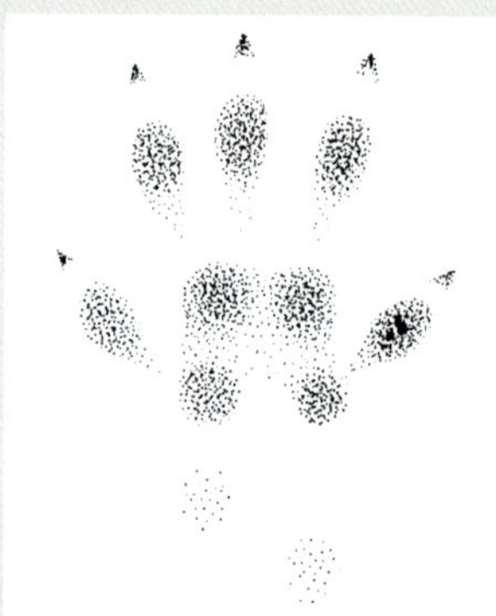

▲ *Left hind.*

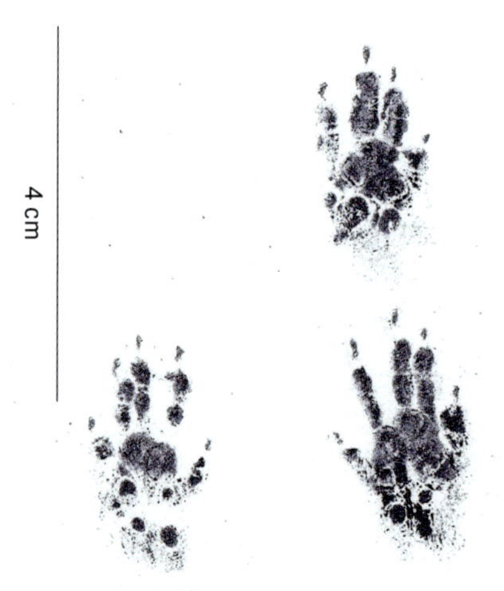

◀ *Ink prints of a Common Hamster. Bottom, both hind foot prints; top, the print of the right front foot. Austria, Andreas Wenger.*

GAITS

Common Hamsters usually bound over longer
distances. They usually walk when foraging for
food. Hind foot prints appearing on the outside of
the front foot prints are a characteristic sign.

Walk
Stride length: 12–20cm
Trail width: 6–9cm

Bounds
Group length: 7–12cm
Inter-group length: 7–25cm
Stride length: 14–32cm
Trail width: 5–9.5cm

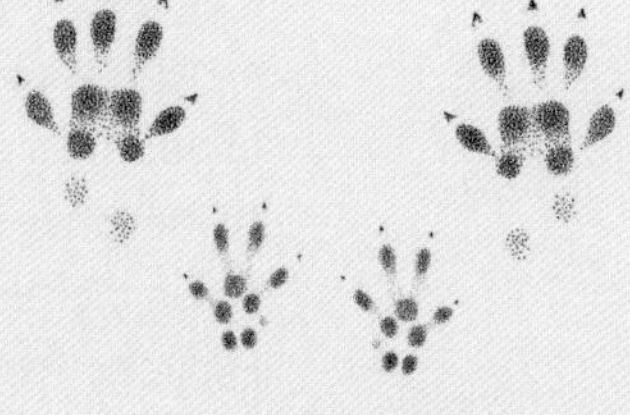

Similar tracks
Rats and other hamsters. There are currently
no findings that enable us to differentiate
between different hamster species.

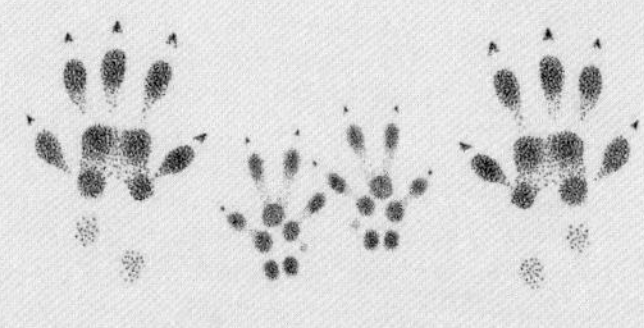

▲ *Full/paired bound.*

DIFFERENTIATING BETWEEN HAMSTERS AND RATS

Hamster tracks resemble those of rats in terms of shape and size. Their toes are like those of
Black Rat, but shorter and wider. Brown Rat toes, on the other hand, appear long and slender.
Hamster front feet have very well-developed claws and pads. However, it can be difficult to tell
the difference when tracks are unclear. Habitat and behaviour provide further important clues.
Hamsters live in open landscapes with diggable loamy soil and hibernate for long periods. They
therefore rarely leave fresh tracks in the winter months. In addition, hamsters are only active in
a very small radius around their dwelling, whereas rats sometimes travel long distances.

REPRODUCTION The mating season begins immediately after hibernation and lasts from around April to the beginning of September. Mating is the only time in the annual cycle when a hamster can enter another's territory. The males visit the females in their burrow, where mating takes place. The females chase the males away again just a few days after mating. The litter is born following a gestation period of 18–20 days. Each litter (up to three a year) has 4–10 (up to 18, but usually 6–8) young that are born hairless and blind. After around a month, the young leave the mother's dwelling to become independent. The animals reach sexual maturity after three months at the earliest but usually after their first hibernation. Females from the first litter can sometimes even mate during the same reproductive period.

▼ *A Common Hamster escape tunnel. Maintal, Germany. Simone Roters.*

SIGNS

DWELLINGS Common Hamsters live in complex, self-dug burrows. These can be up to 2m deep in dry, hard soil. In addition to the nest chamber, which is lined with dry grass, there are usually one or more storage chambers and several latrines. A dwelling like this would have around 4–8 upward sloping entrance tunnels and at least one vertical escape tunnel. Common Hamsters seal up the entrances with soil shortly before hibernation. Entrance holes are between 6–10cm in diameter.

FEEDING CIRCLE Common Hamsters create a distinctive area of clipped vegetation, caused by feeding, within a 5–8m radius around their burrow entrance.

EXCREMENT Typical cylindrical, elongated Arvicolinae faecal pellets, usually with blunt ends. The colour varies from yellowish-brown to dark green to black. Common Hamster droppings are generally similar to water vole droppings but tend to be slightly larger. They usually have a rather rough surface texture. Rat droppings are more irregular in size and often taper to a point at one end. As hamsters live in sandy loam or loess soil, their droppings are often surrounded by a thin layer of sand. This can also help to distinguish them from rats and water voles. Common Hamster droppings are found on the edges of fields, paths and pastures, while water vole droppings are usually found near water. However, droppings of terrestrial water voles may be found in hamster habitat.
* L 0.5–1.2cm D 0.2–0.4cm

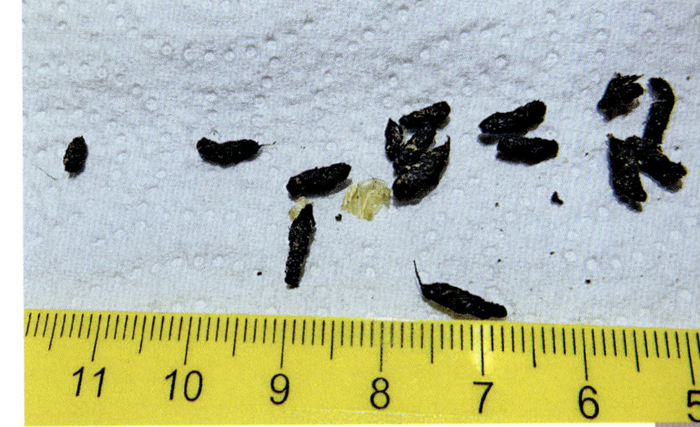

▶ *Common Hamster droppings are a rare find. It is difficult to distinguish them from Brown Rat droppings. Maintal, Germany. Simone Roters.*

Mice and rats (murids)

Muridae

A species-rich family with 150 genera and at least 730 species. A total of four genera with 14 species occur in the area we are looking at.

All representatives are between the sizes of the familiar House Mouse and Brown Rat, with relatively large eyes and ears. They usually have a long tail that sometimes even exceeds the animal's head-trunk length. Unlike Arvicolinae species, many murids climb frequently and while Arvicolinae species are mainly herbivorous, some murids are omnivores.

TRACKS AND SIGNS
OF MURIDS

Plantigrade. Five toes on the front foot and five on the hind foot. Compared to Arvicolinae species, the digital pads are strong and usually create a print that is out of line with the midfoot pad. The fine, pointed claws are rather short.

- **Track formula: 5f × 5H + C**

FRONT FOOT Asymmetrical. Five toes in a classic rodent structure: toe 1 is very short but often still visible in many species. Toes 2 and 5 point slightly forwards and slightly to the sides. The print of digital pads 2 and 5 is usually in line with the three fused interdigital pads, which are arranged in a triangle. Toes 3 and 4 point forwards, with toe 3 usually clearly ahead.

HIND FOOT Slightly asymmetrical. Five toes in a classic rodent structure: toes 1 and 5 point straight to the sides and can appear slightly offset to the rear. Toes 2–4 usually point forwards and are almost parallel.

GAITS The most common gait is the parallel bound. Unlike Arvicolinae species, murids rarely travel in walk and trot and, if they do, it is only for short distances.

PATHS Unlike Arvicolinae, murids do not make permanent paths.

EXCREMENT Irregular faecal pellets, often with pointed ends and a rough surface. They are brownish to greyish-black in colour. While Arvicolinae droppings are found in latrines or along their usually clearly visible paths, murid droppings tend to be scattered around.

FIELD MICE
Apodemus

HTL 6.5–12cm
TL 6.5–14cm
W 12–50g

Six species of field mouse occur in the area we are looking at: Striped Field Mouse (*Apodemus agrarius*), Alpine Field Mouse (*A. alpicola*), Yellow-necked Mouse (*A. flavicollis*), Wood Mouse (*A. sylvaticus*), Ural Field Mouse (*A. uralensis*) and Western Broad-toothed Field Mouse (*A. epimelas*).They are mainly crepuscular and nocturnal, but the Striped Field Mouse is also active during the day. They are good at climbing, jumping and swimming. Most species dig their own shallow dwellings with tunnels where they make nests and store food. Their nests are made of dry plant material, such as leaves and grasses. The Western Broad-toothed Field Mouse lives in gaps between stones and in crevices, while the Yellow-necked Mouse can also nest in hollow trees. They are predated by cats, mustelids, foxes, wild boar, owls and raptors.

DISTINGUISHING FEATURES Medium-sized mice with large, dark eyes, large ears and a long, hairless tail. The tail is usually as long as the body but can be shorter or longer. The length of the tail in relation to the head-trunk length varies depending on the species and can help with identification. The Striped Field Mouse and the Ural Field Mouse are the smallest field mouse species, while the Yellow-necked Mouse and the Western Broad-toothed Field mouse are among the slightly larger representatives.

DISTRIBUTION AND HABITAT   The Striped Field Mouse lives in central and eastern Europe. The Alpine Field Mouse is found in the east and north-west of the Alps. The Yellow-necked Mouse is found throughout almost all of Europe, except in the north of Britain, Scandinavia and Spain. Field mice are widespread in western and southern Europe, although in Scandinavia they are only found in southern Sweden and Norway. The Ural Field Mouse is limited to central and eastern Europe and the Western Broad-toothed Field Mouse inhabits the Balkan Peninsula. Despite the group's name, field mice do not live predominantly or exclusively in fields. Their habitats can be deciduous and mixed forests, hedges, fields, meadows and wetlands. These adaptable animals often move into buildings during winter.

DIET   Seeds, fruits, fungi, roots, buds and other plant parts. Field mice mainly store hazelnuts, beechnuts and acorns. They also eat insects, spiders, worms, snails and carrion.

REPRODUCTION   The mating season varies according to geographical location but is normally between March and October. The Alpine Field Mouse has a slightly shorter breeding season (April–August). There are usually 2–3 litters with 2–6 (up to 11) young each. There can be a sharp increase in population after mast years, for example in beech forests. The young are born hairless and blind after a gestation period of 20–23 days. They reach sexual maturity after 2–3 months.

◀ *These snail shells have been opened at the mouth, which is typical of field mice. Also note the fine scratch marks. Biebergemünd, Germany.*

TRACK

The dimensions for Wood Mice and Yellow-necked Mice are given here, because they are similar to the other representatives of this group.

Front

Wood Mouse
L 0.8–1.3cm
W 0.8–1.3cm

Yellow-necked Mouse
L 0.9–1.3cm
W 0.9–1.3cm

Very small. Plantigrade. Asymmetrical. Five toes in a classic rodent structure: toe 1 is very short but still frequently visible (a). Toes 2 and 5 point slightly forwards and slightly to the sides. The print of digital pads 2 and 5 is usually in line with the three fused interdigital pads, which are arranged in a triangle. Toes 3 and 4 point forwards, with toe 3 usually clearly ahead. Digital pads 2–5 are strongly developed compared to those of *Microtus* voles and usually make a print that appears out of line with the midfoot pad. Two further thenar/hypothenar pads can be visible at the back of the track. They are relatively close to the interdigital pads. The fine, pointed claws do not regularly leave prints and are rather short compared to those of *Microtus* voles. The front feet are smaller than the hind feet.

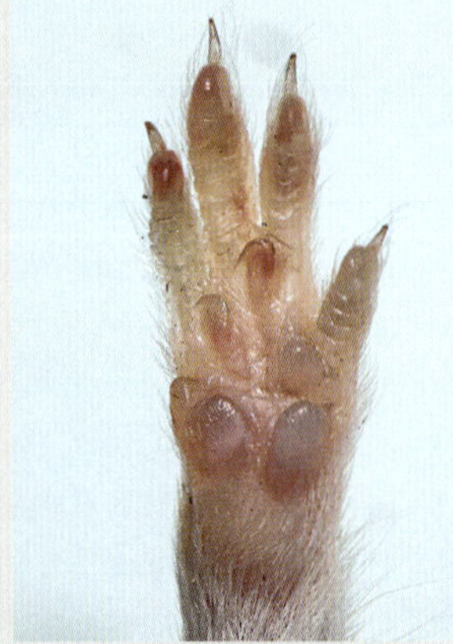

▲ *Wood Mouse, left front. Ennstal, Austria. Stefan Resch.*

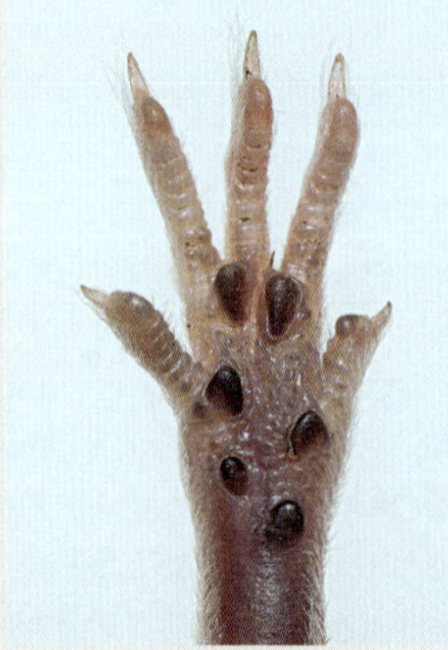

▲ *Wood Mouse, right hind. Ennstal, Austria. Stefan Resch.*

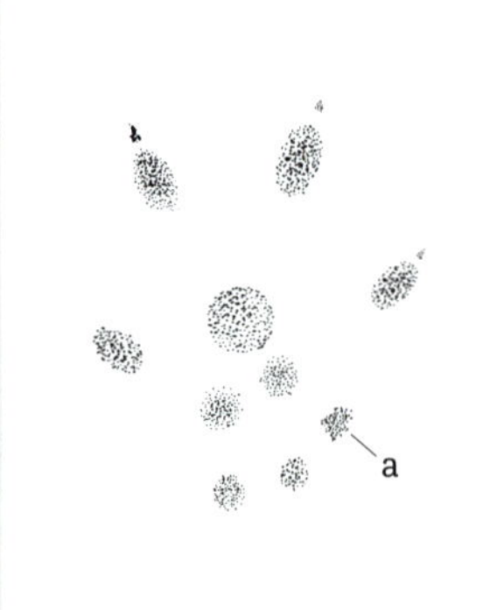

▲ *Left front.*

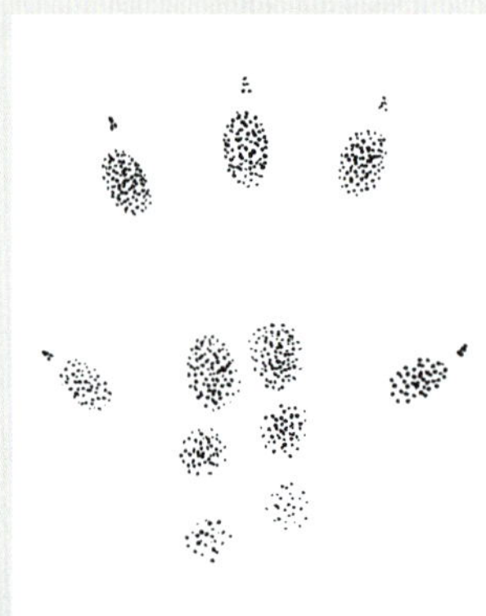

▲ *Right hind.*

▲ *Field mouse, left front. Oderbruch, Germany.*

▲ *Field mouse, right hind. Oderbruch, Germany.*

Hind

Wood Mouse

L 0.9–1.5cm (with heel up to 2cm)

W 1–1.4cm

Yellow-necked Mouse

L 1.1–1.7cm (with heel up to 2.3cm)

W 1–1.7cm

Very small to small. Plantigrade. Slightly asymmetrical. Five toes in a classic rodent structure: toes 1 and 5 point straight to the sides. They are relatively short and appear slightly offset to the rear. Toes 2–4 appear long compared with toes 1 and 5, usually point forwards and are almost parallel to each other. Compared with *Microtus* voles, the digital pads are strongly developed and usually leave a print that is out of line with the midfoot pad. The interdigital pads have fused to form a midfoot pad. The track usually has three adjacent, similarly sized pads and a fourth, smaller, slightly offset interdigital pad. Two further thenar/hypothenar pads can be visible at the back of the track. The inner T pad is further back than the outer H pad. The hind feet are slightly larger than the front feet. The fine, pointed claws do not always leave prints.

Similar tracks

The following applies to field mice: front foot L ≥ 0.8cm and W ≥ 0.8cm. Typical mice (*Mus* genus) usually have smaller front feet (front foot L ≤ 1.1cm and W ≤ 1cm). Field mouse tracks are often mistaken for rat tracks, although rats leave much larger footprints (see photo below). At first glance, field mouse tracks can be mistaken for those of many other small rodents. Identifying small rodents using their tracks is difficult and often impossible. It is only possible to distinguish between the genera of typical mice (*Mus*), *Microtus* voles, field mice and red-backed voles under almost perfect track conditions (see page 330). The number, arrangement and size of the metacarpal/metatarsal pads, as well as the shape of the toes, the length of the claws and hair on the soles, can make it possible to identify these genera. It is only possible to differentiate between field mouse species in exceptional cases, mainly based on their size.

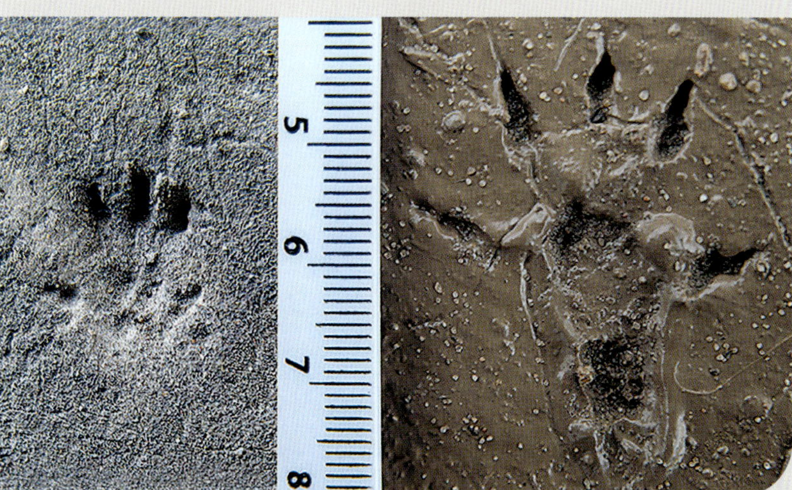

▶ *Left hind foot print of a field mouse (left) and a Brown Rat (right) for size comparison. Aaron Tiedemann. Germany.*

DIFFERENTIATING BETWEEN YELLOW-NECKED MICE AND WOOD MICE

Track size is a distinguishing feature: hind foot track with L > 1.5cm and W > 1.5cm can be assigned to a Yellow-necked Mouse. In the overlap area below, it is not possible to differentiate reliably based on size alone. However, identification is occasionally possible based on foot morphology:

Yellow-necked Mouse

a) Toe 3 is clearly ahead.
b) More asymmetrical.
c) Toes 2–4 slightly longer.
d) Toes 1 and 5 clearly offset towards the back and often splayed out.
e) Clear difference in size between front foot and hind foot, the hind foot appears more elongated.

Wood Mouse

a) Toe 3 is ahead.
b) Less asymmetrical.
c) Toes 2–4 slightly shorter.
d) Toes 1 and 5 slightly offset towards the back and less splayed.
e) Less pronounced difference in size between front foot and hind foot, the hind foot appears squarer.

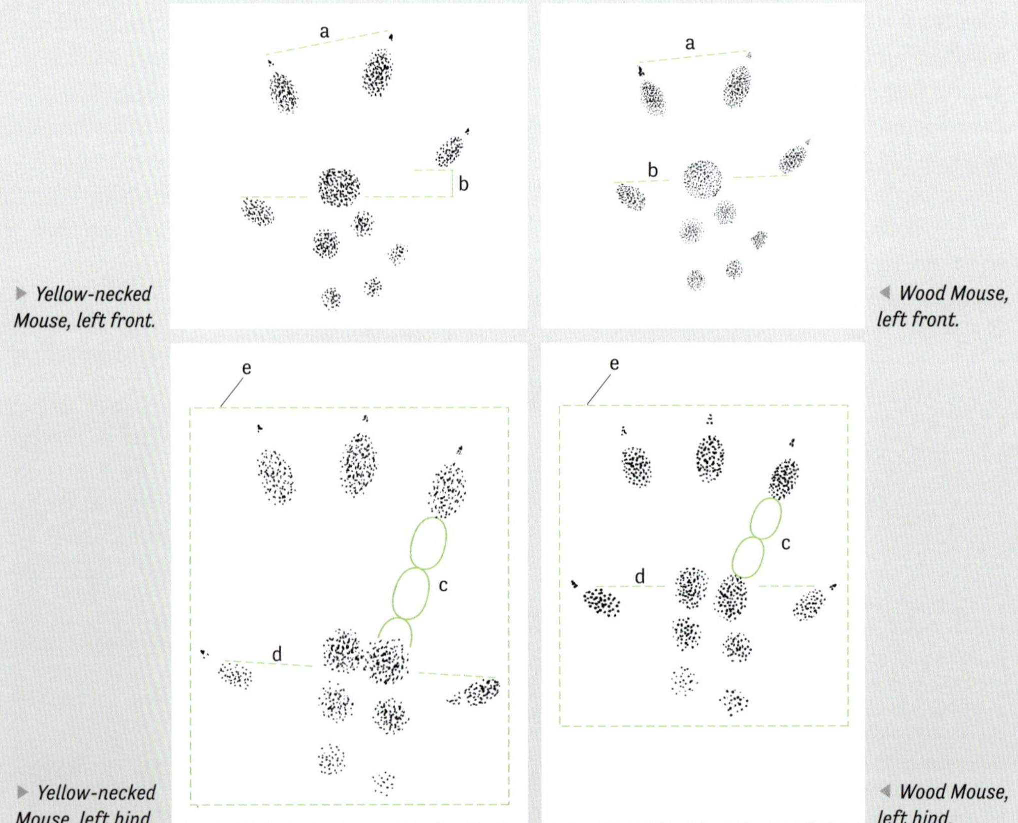

▶ *Yellow-necked Mouse, left front.*

◀ *Wood Mouse, left front.*

▶ *Yellow-necked Mouse, left hind.*

◀ *Wood Mouse, left hind.*

GAITS

The most common gait is the parallel bound. Unlike *Microtus* voles and typical mice (*Mus* genus), field mice very rarely walk or trot and, if they do, it is usually only for a short distance. Their paths usually lead to cover such as branches lying on the ground, tree stumps and other elements that protect them from being seen. However, they can also cross larger open spaces. When there is snow on the ground, field mice move over the top of it, often leaving a visible tail print.

▼ *Yellow-necked Mouse in parallel bound. The outer toes of the hind feet are further back than those of Wood Mouse. Steyerberg, Germany.*

Parallel bound
Group length: 3–6.5cm
Inter-group length: 4.5–40cm
Stride length: 7.5–45cm,
Up to 80cm in exceptional cases
Trail width: 3.5–4.5cm

▲ *Parallel bound track pattern of the same Yellow-necked Mouse. Steyerberg, Germany.*

▲ *Wood Mouse in a parallel bound. Footfall from bottom to top: LF, RF, RH, LH. Lausitz, Germany.*

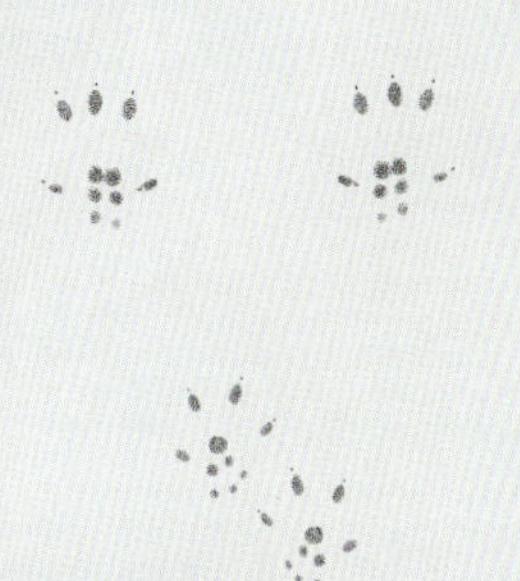

▲ *Parallel bound.*

SIGNS

FEEDING MARKS The feeding marks of field mice differ greatly according to their varied diet. Seeds appear to have been gnawed neatly rather than coarsely shredded as they would be by squirrels or rats. Field mice often carry their food to safe places, such as under tree roots or stones, and eat it there, so small, distinctive feeding marks are left behind along with the fine remains of mast fruit, such as beechnuts or acorns. This is another clear difference to squirrels, whose feeding places are often elevated on a tree stump, giving them a good view. In contrast to *Microtus* vole species, field mice feed more on invertebrates. They usually open snail shells at the mouth of the shell and eat from the bottom up (see photo on page 321). This is a clear difference to snail shells eaten by red-backed voles, which are carefully opened from the top to the bottom (see page 309) or the smashed snail shells found at a Song Thrush anvil (Brown 1985).

If bones and antlers have been eaten as a source of calcium, they usually show clear furrows made by the incisors. The distance between the toothmarks measures a total of 1–3mm and can be an important indicator for species identification but is generally not enough to distinguish between different mice. You should also consider the habitat and associated signs, such as droppings.

▼ *Old food store of a field mouse. Some of the acorns have started to germinate. Midhurst, England.*

HAZELNUT FEEDING   In the identification guide to feeding marks on fruit and nuts (page 165), hazelnuts opened by squirrels are compared with those opened by field mice, red-backed voles and dormice.

CONE FEEDING   In order to get at the calorie-rich seeds of various conifer cones, field mice first have to remove the scales that protect the seeds underneath. Unlike squirrels, field mice rarely bite cones from the trees; instead they mainly eat from fallen ones. The animals remove the scales with their teeth, usually beginning with the looser scales at the basal end of the cone. They gnaw evenly so the base of the cone is evenly rounded, rather than leaving a torn out point like squirrels do. Field mice gnaw off each scale individually, leaving a relatively even gnawed area with no scale remains sticking off the axis of the cone. The cone axis appears neat and smooth. Like squirrels, field mice often leave a short tip with few undamaged scales. However, completely eaten cones can also be found. In the identification guide to feeding marks on fruit and nuts (page 164), cone feeding by mice and squirrels is compared.

DIGGING MARKS Field mice leave distinctive feeding and digging marks when they dig up their own food stores, e.g., acorns buried in the ground. This leaves small, shallow, usually rectangular holes in the ground, surrounded by shell debris (see photo on page 327).

DWELLINGS The entrance holes of field mouse dwellings measure about 3–5cm. They mainly dig their burrows using their lower incisors, and almost exclusively use their front legs to get rid of rejected material. They push it under their belly and then kick it away with their hind feet (Turni 2005). The size and depth of the dwellings varies, for example according to soil conditions and water balance. Large quantities of loose soil can pile up in front of the entrance, usually in distinctive conical heaps. This mainly occurs in sandy soils. This can be a clear difference to the entrance holes of *Microtus* voles.

▼ *The entrance hole to the dwelling of a Yellow-necked Mouse. A considerable amount of soil has piled up in front of it. Märkische Schweiz, Germany.*

EXCREMENT Small droppings in the elongated cylindrical shape typical of rodents, often with pointed ends. They are brownish to greyish-black in colour. Compared to typical *Mus* species droppings, field mouse droppings tend to be larger, denser and more regular in shape. However, the faecal pellets of Arvicolinae vole species are even more regular. There is also a second, oval dropping shape with rounded ends, which resembles the regular shape of Arvicolinae vole droppings but has a more irregular surface. The presence of both types is a typical characteristic of field mouse excrement. In addition, field mouse droppings tend to be scattered, whereas Arvicolinae vole droppings are found in latrines or along their usually clearly visible paths. Field mouse droppings are typically found in forests, gardens, parks, attics and similar.

- L 4–10mm D 1.7–3mm, usually D > 2mm

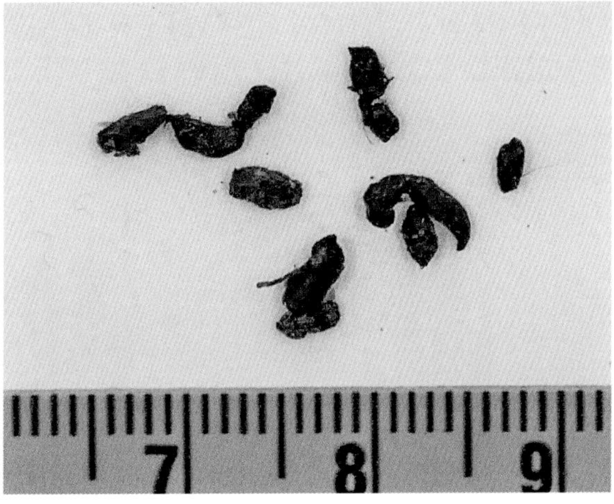

▲ *Wood Mouse droppings. The irregular shape and rough surface enable us to distinguish them from the more uniform and smooth droppings of voles (Arvicolinae). Midhurst, England. John Rhyder.*

DIFFERENTIATING BETWEEN *MICROTUS* VOLES, RED-BACKED VOLES AND FIELD MICE BY THEIR TRACKS

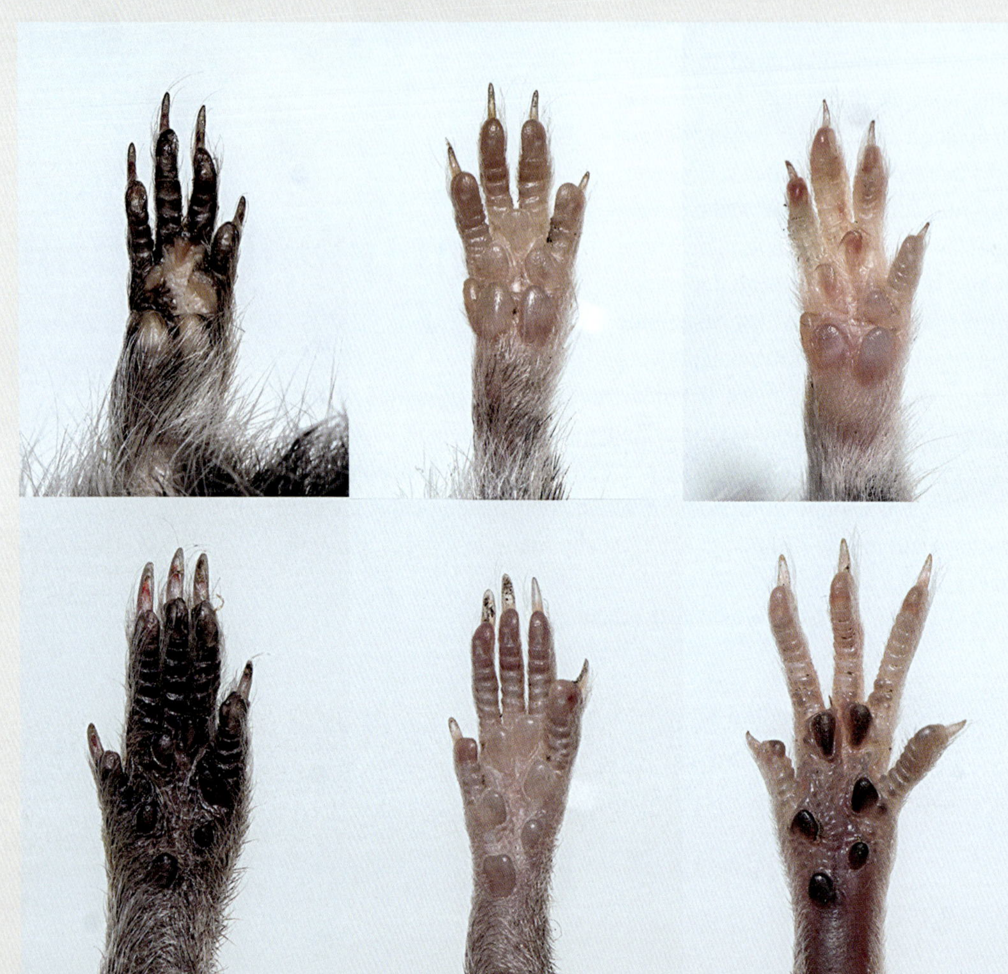

▲ *From left to right: top, the left front feet of* Microtus *vole, red-backed vole and field mouse; bottom, the corresponding left hind feet. Offenburg, Germany. Stefan Resch & Aaron Tidemann.*

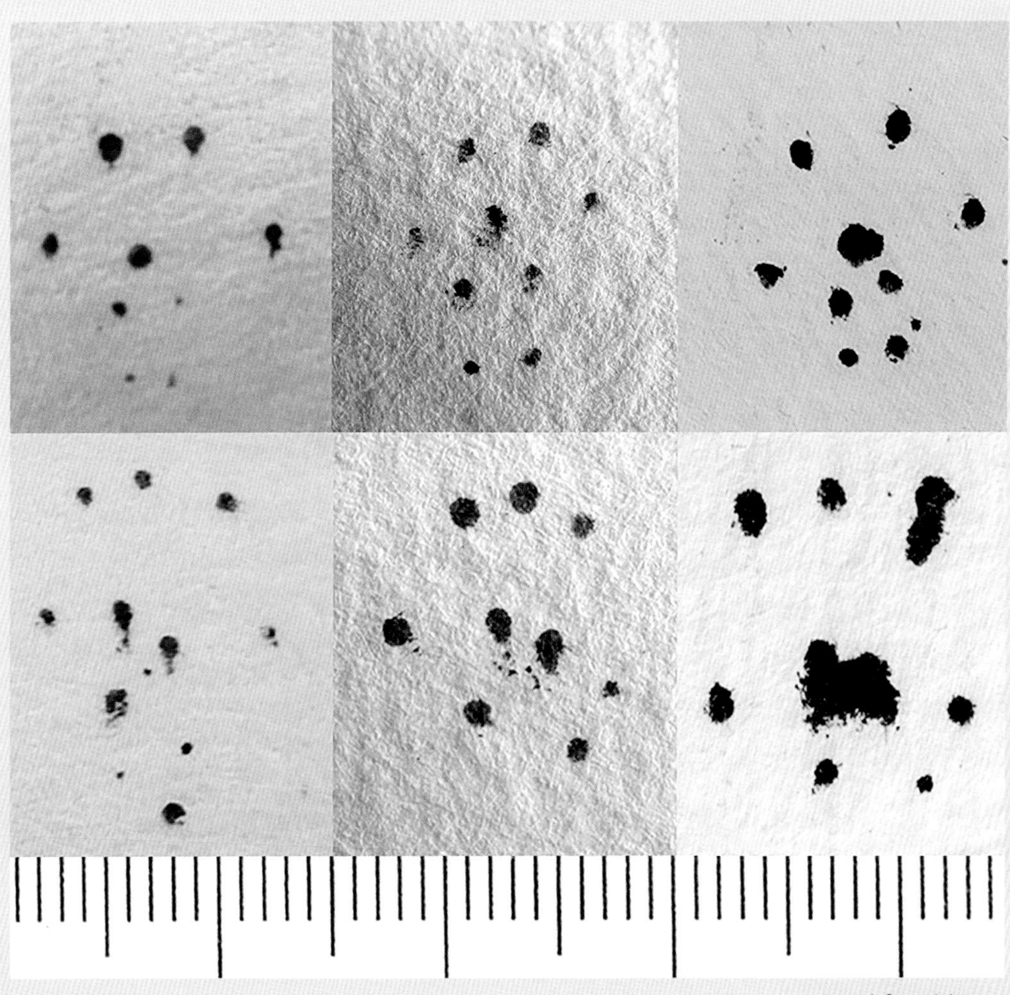

▲ *From left to right: top, ink prints of the left front feet of* Microtus *vole, red-backed vole and field mouse; bottom, the corresponding left hind foot prints.*

The feet of *Microtus* vole species are adapted to a burrowing lifestyle, so they are different from the feet of many field mice, which are more adapted to climbing. Long claws help with digging and are a distinguishing feature of *Microtus* voles. Long, narrow toes, more bent claws, strong digital and metacarpal/metatarsal pads, as well as spreadable outer toes that are set further back, are typical adaptations for climbing and characteristic of climbing field mouse species. Some of these characteristics enable us to distinguish between Arvicolinae species and murids and can also be used to differentiate between water voles and Brown Rats (see page 289). The behaviour of red-backed voles (*Clethrionomys* spp.) illustrates the early intermediate stages of the transition from a terrestrial to an arboreal lifestyle. Their tracks show some adaptations to climbing which makes it possible to distinguish them from field mouse and *Microtus* vole tracks. The following characteristics relate to *Microtus* voles, red-backed voles and field mice.

Microtus voles	Red-backed voles	Field mice
a) More symmetrical, finer track, toes 3 and 4 are often arranged next to each other in a V-shape.	a) Tracks often very symmetrical in comparison and noticeably roundish if the thenar/hypothenar pads do not leave a print. Toes 3 and 4 are often arranged next to each other in a V-shape.	a) More asymmetrical, strong track with toe 3 in front.
b) Long, narrow toes with fine pads that often appear to be fused with the interdigital pads.	b) Toes noticeably short with strong pads that occasionally appear to be fused with the interdigital pads.	b) Long, narrow toes with strong pads that rarely appear to be fused with the interdigital pads.
c) Fine interdigital pads. Thenar/hypothenar pads tend to be further back.	c) Strong interdigital pads. Thenar/hypothenar pads tend to be further back.	c) The interdigital pads are strong. The thenar/hypothenar pads are further forward.
d) Toe 1 very rarely leaves a print.	d) Toe 1 rarely leaves a print.	d) Toe 1 often leaves a print.
e) Front foot and hind foot very similar in size. The hind foot appears long and narrow, even without the thenar/hypothenar pads.	e) The front foot is slightly smaller than the hind foot. The hind foot looks squarish without the thenar/hypothenar pads.	e) The front foot is slightly smaller than the hind foot. The hind foot looks squarish without the thenar/hypothenar pads.

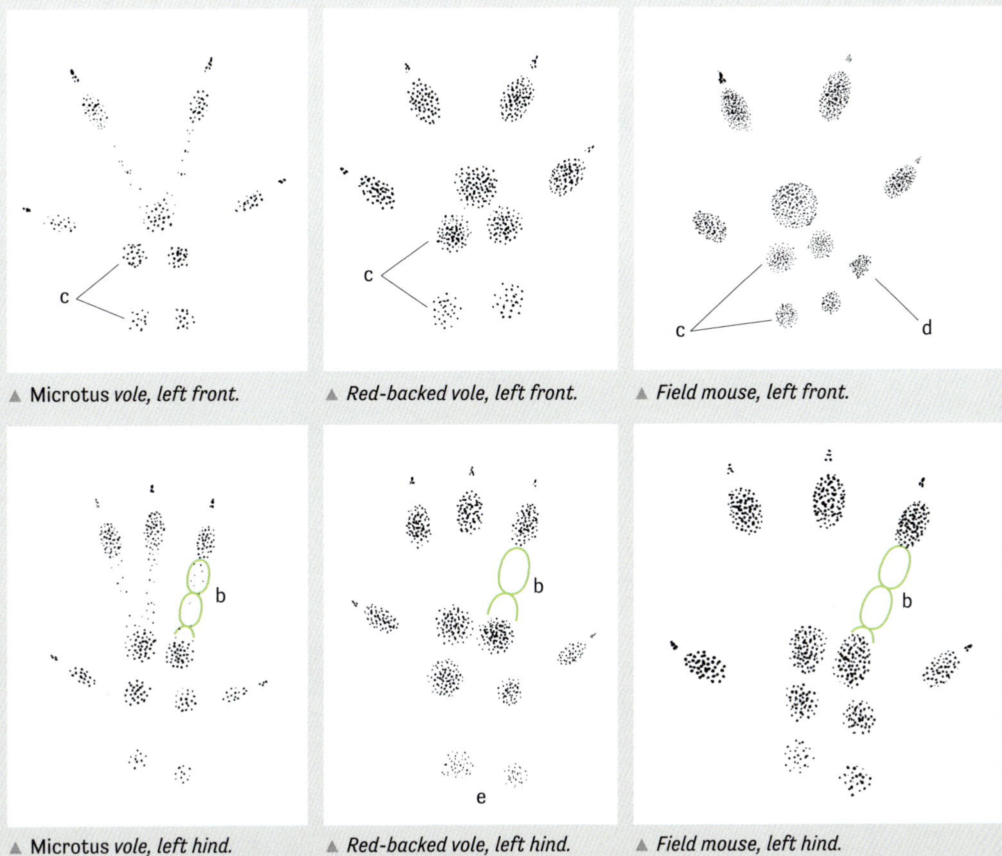

▲ Microtus *vole, left front.*

▲ *Red-backed vole, left front.*

▲ *Field mouse, left front.*

▲ Microtus *vole, left hind.*

▲ *Red-backed vole, left hind.*

▲ *Field mouse, left hind.*

EURASIAN HARVEST MOUSE
Micromys minutus

HTL 5–8cm
TL 4.5–7.5cm
W 4–15g (usually 7–9g)

The non-gregarious Eurasian Harvest Mouse is active both during the day and at night, with its main activity occurring at dusk. This highly specialised climber scales stalks and branches and spends most of its life above ground. It uses its long prehensile tail to help it climb by wrapping it around stalks for support. Its paws are adapted for climbing and have large sole pads that give it additional support. Like our thumbs, toe 1 is opposable in relation to the other toes. This gives the Eurasian Harvest Mouse an exceptionally good grip. It is predated by mustelids, foxes, owls, raptors and corvids.

DISTINGUISHING FEATURES A very small and dainty mouse with a pointed nose, short, furry ears and a long, hairless, prehensile tail.

DISTRIBUTION AND HABITAT The Eurasian Harvest Mouse is found throughout almost the whole of Europe. It is absent from northern Britain and the Iberian Peninsula, apart from northern Spain. In Norway and Sweden, it is only found in the south. It prefers wetlands with dense vegetation, such as patches of sedge, reed canary grass, reeds and bulrush belts beside water. It is rarely found higher than 800m above sea level and avoids mountains. The edges of forests and fields with hedges or ditches with tall vegetation can also provide habitat. Buildings such as barns and stables provide shelter in winter.

DIET Mainly grass seeds, cereals and berries, but also insects and beetles.

REPRODUCTION During the mating season between April and September, each female can give birth to 2–6 litters of 3–9 (up to 12) young. The young are born hairless and blind after a gestation period of around 21 days. The animals reach sexual maturity after 1–2 months.

SIGNS

NEST Eurasian Harvest Mice build summer and winter nests. Their summer nests are usually located in vegetation 30–60cm above the ground. They are spherical, 6–10cm in diameter and woven into the grasses and stems of the surrounding vegetation. The inner nest chamber is lined with fine material, such as plant seeds and wool. The outer shell consists of blades of grass that are split lengthways and woven into the nest. The mice do not snip the blades of grass off their stalks, so the outer shell of the nest remains green throughout summer and only turns a brownish colour when the vegetation wilts. The nests are well camouflaged in both summer and autumn. They are usually only discovered when abandoned nests fall to the ground or are blown away. They look like small balls of hay. In winter, Eurasian Harvest Mice build sheltered ground nests (for example, under roots) or dig short tunnels in the ground that they make their winter dwelling.

TRACK

Front
L 0.8–1cm W 0.5–0.9cm

Tiny. Plantigrade. Asymmetrical. Five toes in a classic rodent structure: toe 1 is very short and rarely visible in the print. Toes 2 and 5 are slightly ahead and point clearly to the sides. The print of toe pads 2 and 5 is in front of the three front fused metacarpal pads, which are arranged in a triangle and form the midfoot pad. Toes 3 and 4 point forwards at an angle. Toes 2–5 are shorter and stronger than the comparatively long toes of field mice and the negative space between the midfoot pad and digital pads is correspondingly small. Two further thenar/hypothenar pads can be visible at the back of the track. The fine, pointed claws do not always leave prints. The front feet are slightly smaller than the hind feet.

Hind
L 1–1.3cm W 0.9–1cm

Tiny to very small. Plantigrade. Slightly asymmetrical. Five toes in a classic rodent structure: toes 1 and 5 point straight to the sides. Toes 2–4 usually point forwards. They are in a line and almost parallel to each other. They are shorter and stronger than the long, narrow toes of field mice and the negative space between the midfoot pad and digital pads is correspondingly small. The interdigital pads have fused to form a midfoot pad but four metatarsal pad prints can still be visible in the track. In rare cases, two further thenar/hypothenar pads can make a print at the back of the track. The hind foot is slightly longer than the front foot. The claws are fine, pointed and only visible occasionally.

GAITS

Eurasian Harvest Mice rarely move on the ground. They only travel short distances, usually moving from stalk to stalk. Their preferred gait is the parallel bound. If they want to move more slowly, they usually walk.

Walk
Stride length: 3.2–5.2cm
Trail width: 1.8–2.2cm

Limited data available.

Similar tracks
At first glance, the footprints of Eurasian Harvest Mice can be mistaken for those of any other small mice. However, clear prints normally enable unambiguous identification, because they are small with relatively strong metacarpal/metatarsal pads and a small negative space. Hazel Dormice leave prints with the metacarpal/metatarsal pads in the rosette-shaped arrangement that is typical of rodents, as well as a track pattern featuring inward and outward tracks.

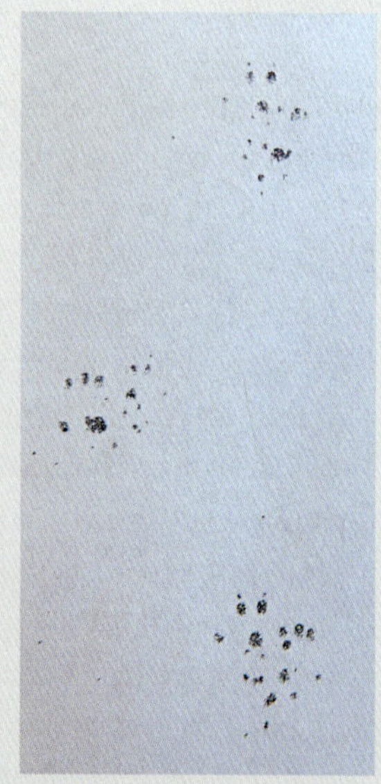

▲ *Eurasian Harvest Mouse walk. Biebergemünd, Germany.*

◀ *The characteristic summer nest of a Eurasian Harvest Mouse.*
Vledder, Netherlands.
René Nauta.

EXCREMENT Eurasian Harvest Mouse droppings are so tiny they are rarely found. You might find them in abandoned nests or reed beds. They are brownish to black in colour, with an uneven, slightly curved shape and a grainy surface.

- L 2.5–5.5mm D 0.9–1.7mm

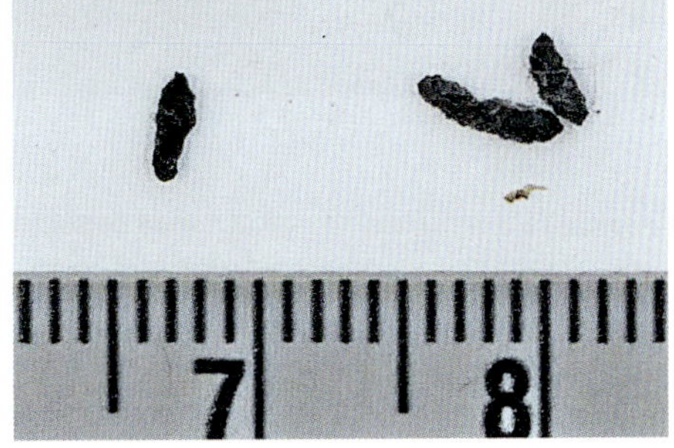

▲ *The tiny droppings are a useful criterion for detecting Eurasian Harvest Mice. Droppings are usually less than 5mm long and less than 1.5mm thick.*
Midhurst, England.
John Rhyder.

TYPICAL MICE
Mus

HTL 6–11cm
TL 5–10cm
W 12–32g

When we talk about mice, we usually mean a species from the genus *Mus*. *Mus* species correspond most closely to the classic mouse depiction in numerous illustrations, drawings and photographs. Four different species of the *Mus* genus occur in the region covered by this book. All very similar in appearance and behaviour, they are the House Mouse (*Mus musculus*), the Western Mediterranean Mouse (*M. spretus*), the Steppe Mouse (*M. spicilegus*) and the Macedonian Mouse (*M. macedonicus*). Each species is active all year round and is mainly nocturnal. They are extremely adaptable. They are good at climbing, swimming and, except for the Steppe Mouse, can dig shallow burrows. They are threatened by humans and their pet cats, as well as mustelids, foxes, owls and raptors.

DISTINGUISHING FEATURES Medium-sized mice with relatively large eyes and ears and a long, hairless tail. The House Mouse is the only species of this genus to emit the typical unpleasant, musty odour.

DISTRIBUTION AND HABITAT The House Mouse is the most widespread species of the genus and its habitat overlaps with that of some other species. The Western House Mouse subspecies (*Mus musculus domesticus*) is more common in western and southern Europe and the Eastern House Mouse subspecies (*Mus musculus musculus*) is more common in central Europe and Scandinavia. Hybrid populations occur in areas where both subspecies live. This highly adaptable species, associated with human activities, usually lives close to settlements, in buildings, gardens and livestock facilities. It can even be found in cold stores, as well as in outdoor habitats such as fields, hedgerows or in the mountains up to an altitude of 2,000m. The Steppe Mouse, Macedonian Mouse and Western Mediterranean Mouse avoid buildings, human settlements and dense forests. The Western Mediterranean Mouse lives in southern France and on the Iberian Peninsula. The Steppe Mouse is found in Austria, Hungary, Slovakia and the Balkans. Both favour dry habitats such as steppes, hedges, roadsides and field margins, as well as maquis. The Macedonian Mouse is widespread in the Balkans and also inhabits karst areas, olive groves and more humid habitats, such as stream banks and meadows.

DIET Omnivorous, eat seeds, fruits, fungi, roots and other parts of plants. They also eat food intended for humans, as well as insects such as beetles.

REPRODUCTION The mating season for Steppe Mice and Macedonian Mice is from March to October. The other *Mus* species can reproduce all year round. The gestation period is 18–21 days. There are normally 3–5 litters per year, with 4–6 young per litter. An increased birth rate of 4–12 litters with 5–14 per litter young can occur. Such exceptionally high reproduction rates are mainly related to food supply. The proximity to human settlements, buildings and associated abundant supplies of food could explain the high reproduction rates. The animals reach sexual maturity after 6–8 weeks. Steppe Mice develop slightly slower and take 2–3 months to reach sexual maturity.

TRACK

Front

L 0.5–1.1cm W 0.6–1cm

Tiny. Plantigrade. Asymmetrical. Five toes in a classic rodent structure: toe 1 is very short and occasionally visible in perfect prints. Toes 3 and 4 point forwards. Toes 2 and 5 point slightly to the sides. Toe pads 2–5 are narrow in comparison to those of field mice and, in contrast to *Microtus* voles, are offset in relation to the midfoot pad. The print of digital pads 2 and 5 is normally in front of the three fused interdigital pads, which are arranged in a triangle and are less well-developed than in field mice. Two further thenar/hypothenar pads are often visible at the back of the track. The claws are short, fine, pointed and do not regularly leave prints. The front foot is much smaller than the hind foot.

Hind

L 1–1.3cm (up to 1.5cm)

W 0.8–1.2cm

Tiny to very small. Plantigrade. Slightly asymmetrical. Five toes in a classic rodent structure: toes 1 and 5 point to the sides. In contrast to *Microtus* voles, toe 1 is less offset to the rear and more in line with toe 5 (a). Toes 2–4 usually point forwards and are aligned and almost parallel. The interdigital pads have fused to form a midfoot pad. Three adjacent, similarly sized metatarsal pads and a further, smaller, clearly offset metatarsal pad are usually visible in the track. Two further thenar/hypothenar pads can be visible at the back of the track. The inner T pad is further back in the track than the outer H pad. The hind foot is much larger than the front foot. The claws are short, fine, pointed and not regularly visible.

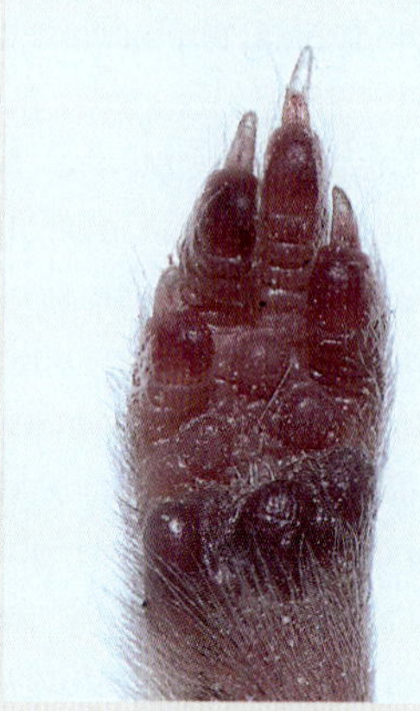

▲ *House Mouse, left front. Ennstal, Austria. Stefan Resch.*

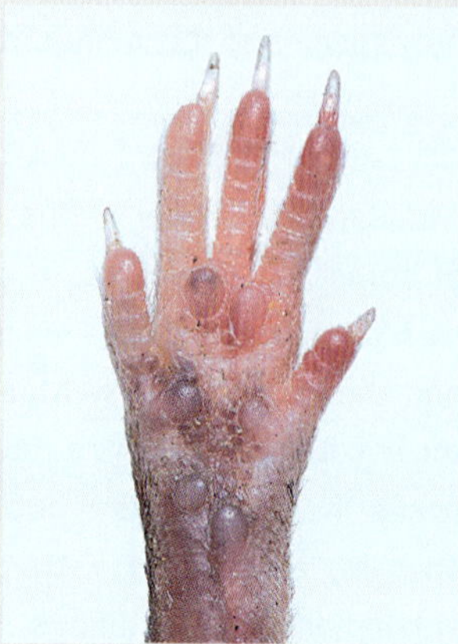

▲ *House Mouse, left hind. Ennstal, Austria. Stefan Resch.*

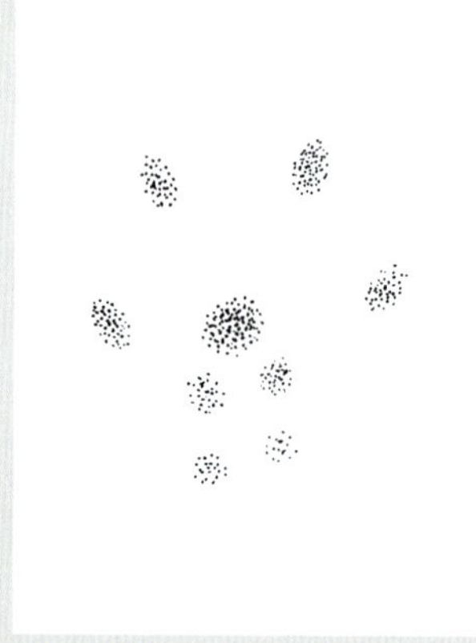

▲ *Left front.*

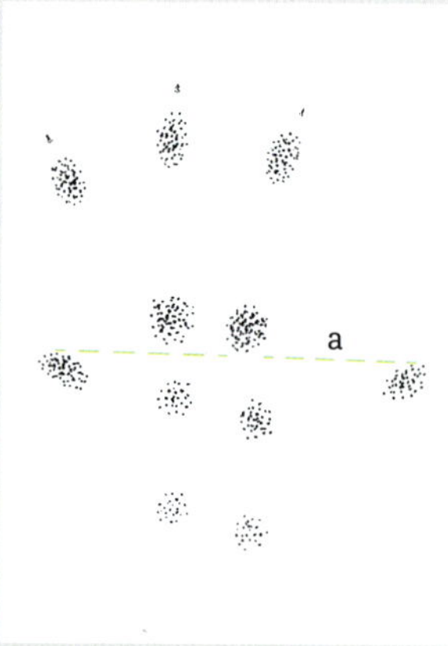

▲ *Left hind.*

GAITS

Their preferred gait is the parallel bound. When exploring or gathering food, they tend to walk more often than field mice. Tracks are usually found close to settlements and livestock facilities. They are rarely found out in the open far from human settlements.

Walk
Stride length: 4.5–8.5cm
Trail width: 2–3cm

Parallel bound
Group length: 2.5–5.5cm
Inter-group length: 3.5–20cm
Stride length: 6–25cm
Trail width: 3–4cm

Limited data available.

Similar tracks

At first glance, the tracks can be mistaken for those of any other small mice. However, clear prints are normally easily identifiable. The digital pads appear quite slim compared to those of field mice and are usually offset in relation to the midfoot pad compared with those of *Microtus* voles.

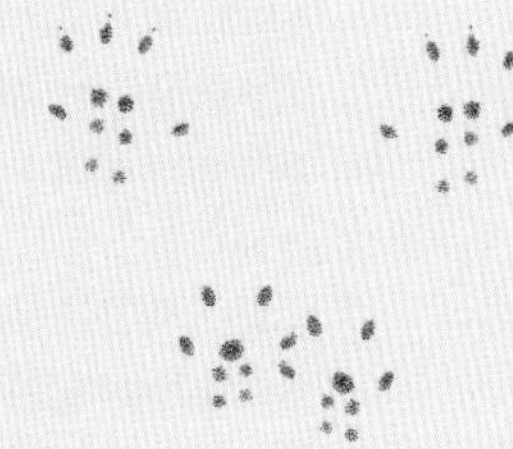

▶ *Parallel bound*

DIFFERENTIATING BETWEEN TYPICAL MICE (*MUS*) AND FIELD MICE (*APODEMUS*)

We can differentiate between typical mice and field mice using the front foot. An imaginary line connecting the front edges of digital pads 2 and 5 of a typical mouse usually runs in front of the foremost edge of the midfoot pad. In field mice, the outer digital pads are often in line with or behind the foremost metacarpal pads. However, size is the most obvious feature. The front foot length of a typical mouse can only exceed 0.9cm if the track includes both thenar/hypothenar pads and claws. The maximum length of 1.1cm is rare even with a complete print. Front feet generally measure less than 1cm. If length is greater than 1.1cm and width is greater than 1cm, then *Mus* species can usually be ruled out.

SIGNS

NESTS AND DWELLINGS   Nests are usually built in well-protected hiding places above ground, such as niches and cavities between walls, ceilings and upholstery. Any suitable material, such as fabric scraps, paper, leaves or straw, is used to build the nest, provided it is gnawable and offers a certain degree of protection. Outdoors, nests can also be found in hedges close to the ground. All *Mus* species can dig their own shallow burrows, for example under roots, logs or stones. Their tunnel systems are around 25cm deep and have a nest chamber. Steppe Mice are an exception because they build communal mounds.

COMMUNAL MOUNDS   Steppe Mice are the only mammals (apart from humans) that build above-ground storage chambers for winter. These communal mounds, which can measure up to 50cm high and 60–180cm in diameter, are built from August and can contain up to 50kg of seeds, grain and similar. The stocks are covered with dry material such as leaves and grass and then with a 5-10cm thick layer of soil. The nesting chamber is located underneath. This sign allows us to reliably distinguish between Steppe Mice and other *Mus* species.

▼ *Large communal mounds are a clear sign of Steppe Mice. Lake Neusiedl, Austria. Apodemus – Private Institute for Wildlife Biology, Stefan Resch.*

EXCREMENT Very small faecal pellets. Long, thin cylindrical shape, often with pointed ends. Usually black or blackish. Droppings can be confused with those of field mice, but field mouse droppings tend to be brownish to greyish-black, larger, more compact and more regular. Field mouse droppings also often occur in a second, oval shape with rounded tips, which is uncommon in *Mus* species droppings. As *Mus* species more commonly live near human settlements, their droppings are also mainly found in inhabited buildings, stables or breeding enclosures. They can also be found on open ground near human settlements, in copses, gardens and grain fields. They are rarely found in dense forests. This can also help us to distinguish them from field mouse droppings.

- L 3.6–7.5mm D 1.2–2.5mm, usually D < 2mm

▶ *The long, thin faecal pellets of a House Mouse. Midhurst, England. John Rhyder.*

RATS
Rattus

Black Rat

HTL 16–24cm

TL 18–26cm (≥ HTL)

W 100–280g

Brown Rat

HTL 18–28cm

TL 13–23cm (≤ HTL)

W 150–580g

Significant sexual dimorphism in Brown Rats, with males on average 20 per cent heavier and with an HTL about 8–10 per cent longer.

Black Rats (*Rattus rattus*) and Brown Rats (*R. norvegicus*) are species associated with human activity and both are widespread throughout the world. They tend to be crepuscular and nocturnal in the vicinity of humans but can also be encountered during the day if undisturbed. The animals live in packs in buildings. These packs usually originate from one family and can be made up of 60 or more animals. Pack size depends on various factors, for example on food supply.

A pack will defend its territory against other rats. Brown Rat packs have a clear social structure. The animals play together and groom each other. The hierarchical pecking order is maintained by fighting. To date, no evidence of such a hierarchy has been found in Black Rats. Brown Rats can dig burrows outdoors, but this is not known in Black Rats. Both species build nests in buildings between beams and in cavities, under bulky goods or in other safe hiding places. They use any suitable material for nesting, such as fabric scraps, paper, leaves, straw or thin branches. They occasionally build large communal nests. In the warmer Mediterranean region of Europe, Black Rats also nest in hollow trees, bushes and rock crevices. Sense of smell and touch are important to both rat species. They are intelligent and agile, quick to learn and good at swimming, diving and jumping. Both rat species can climb, but Black Rats more adept. In addition to humans, predators include mustelids, cats and dogs, foxes and owls. Unlike Brown Rats, Black Rats tend to flee upwards when threatened. Brown Rats are very defensive and can even attack humans if they feel cornered.

DISTINGUISHING FEATURES Both species are much larger than mice. They have a hairless-looking, relatively long, round tail, large ears and large eyes. Brown Rats are larger and more robust than Black Rats. Their muzzle is blunter and their eyes and ears smaller in comparison. If the ears of the Brown Rat were folded forwards, they would not reach the eyes, whereas the ears of a Black Rat would cover the eyes. Black Rats have a slimmer shape that is emphasised by the pointed snout.

DISTRIBUTION AND HABITAT Brown Rats are native to north-east Asia and were introduced to Europe in the eighteenth century. Today they are found all over the world and are one of the most common mammal species on earth. They occur more or less everywhere in Europe. The Black Rat is also a non-indigenous species and came to Europe from India, probably by ship, around 2,000 years ago. Isolated populations of Black Rats are found all over the world, but in Europe they are largely absent in Britain and Scandinavia. The number of Black Rats in Europe is in decline. This is partly due to intensive pest control and the intensification of agriculture. In some areas, Black Rats are also being displaced by the more dominant Brown Rat.

Brown Rats prefer human settlements, where they live near sewers, in warehouses, harbours, stables or on rubbish dumps. They are also found out in the open, where they live by rivers, the coast and other bodies of water. In its native Asia, the Black Rat usually lives in forests, where it builds nests in trees. In Europe, Black Rats prefer warm, dry habitats. However, only a few populations live out in the open, mainly in the Mediterranean region. In the rest of Europe, Black Rats are restricted to human habitats and live in the upper floors and attics of buildings. Very large populations can accumulate in grain stores, feed stores, stables or pig fattening facilities. Unlike Brown Rats, Black Rats avoid basements.

TRACK

Black Rats and Brown Rats leave similar tracks. Close examination of the foot morphology of clear prints can allow us to tell the difference. Otherwise, identification can be difficult. Habitat is an important additional indicator for differentiation. Except for the Mediterranean region of southern Europe, Black Rats in Europe are found almost exclusively near humans and buildings. Brown Rats are often also found close to humans but can also be found more frequently in the wild near bodies of water.

Front

Black Rat
L 1–1.8cm   W 1.3–1.8cm

Brown Rat
L 1.3–2.4cm   W 1.3–2.1cm

Very small to small. Plantigrade. Asymmetrical. Five toes in a classic rodent structure: toe 1 is very short and rarely visible in the print. Toes 2 and 5 are often positioned clearly to the sides and can create almost horizontal prints. Toes 3 and 4 point forwards at an angle. When spread out, the toes appear arranged in a star shape. Some metacarpal pads have fused. The print of the midfoot pad is made up of three clearly recognisable metacarpal pad prints arranged in a triangle. In rare cases, two further thenar/hypothenar pads can be visible at the back of the track. These

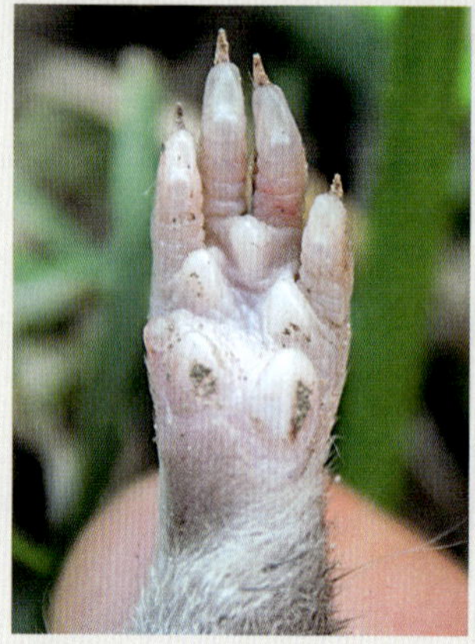

▲ *Black Rat, left front. Märkische Schweiz, Germany.*

▲ *Black Rat, left hind. Märkische Schweiz, Germany.*

▲ *Black Rat, left front (bottom) and left hind (top). Lausitz, Germany.*

▲ *Black Rat, left hind. The strong digital pads are an important feature that distinguishes them from the Brown Rat. Lausitz, Germany.*

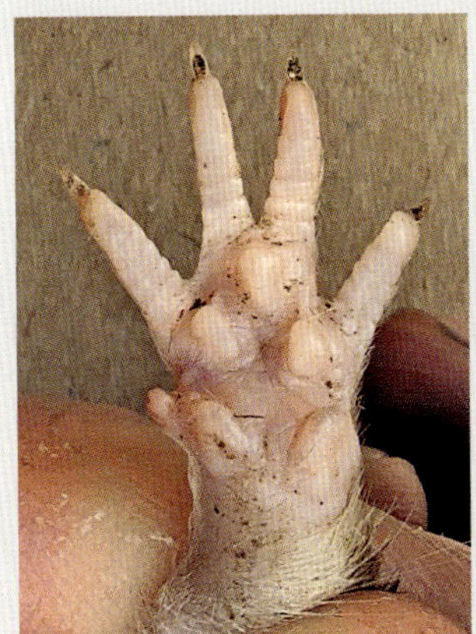

▲ *Brown Rat, left front. Vledder, Netherlands. René Nauta.*

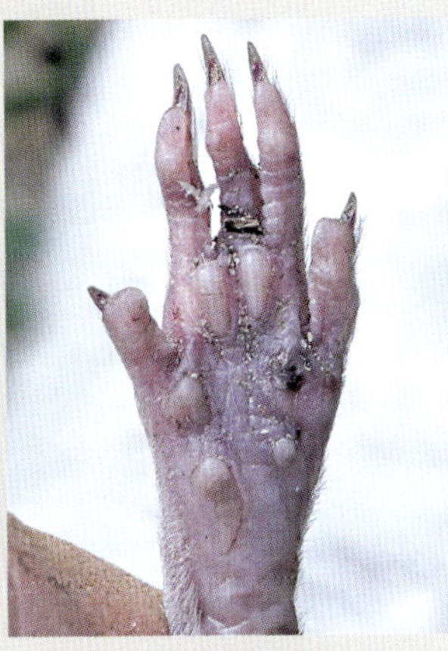

▲ *Brown Rat, left hind. Vledder, Netherlands. René Nauta.*

▲ *Brown Rat, left front. Vledder, Netherlands.*

▲ *Brown Rat, left hind. The back of the track has left a complete print and the long toes 2–4 are clearly visible. Spessart, Germany.*

are proportionately smaller than those of squirrels or Common Hamsters. The claws rarely leave prints. They are fine, pointed and shorter than those of water voles. The front feet are smaller than the hind feet.

Hind

Black Rat
L 1.4–2.7cm   W 1.6–2.6cm

Brown Rat
L 2–3.7cm   W 1.8–3cm

Small. Plantigrade. Slightly asymmetrical. Five long toes in a classic rodent structure: toes 1 and 5 point straight to the sides. Toes 2–4 usually point forwards and are aligned and almost parallel. The interdigital pads have fused to form a midfoot pad, although four individual metatarsal pads are occasionally visible in the track. However, often only the foremost two metatarsal pads leave a print. It is rare for the whole foot to touch the ground. If it does, two further thenar/hypothenar pads can be visible at the back of the track. The inner T pad is further back in the track than the outer H pad. The hind foot is longer than the front foot. The claws are fine, pointed and rarely visible.

DIFFERENTIATING BETWEEN BLACK AND BROWN RATS

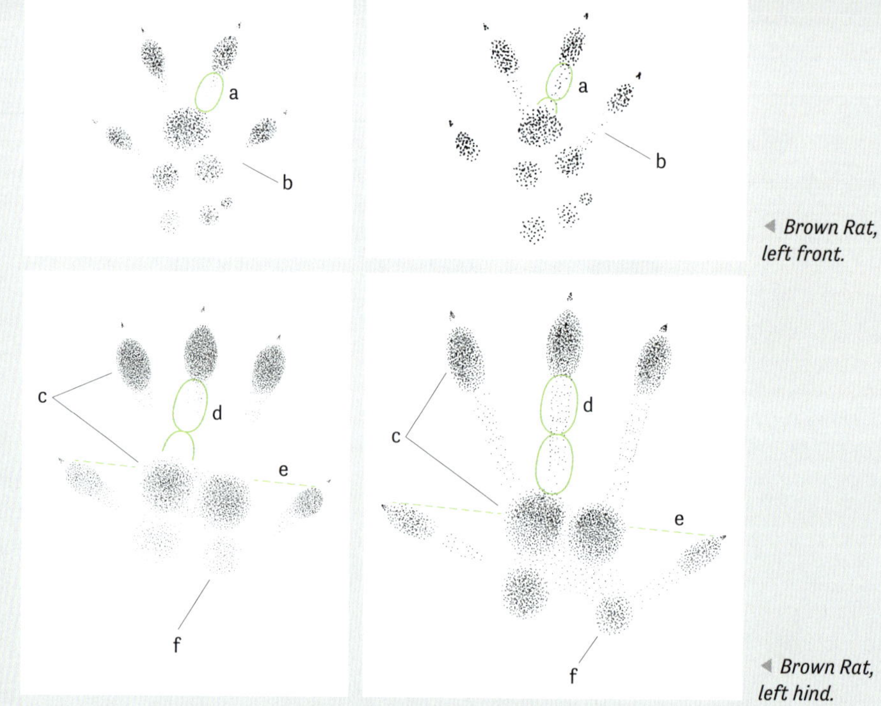

▶ *Black Rat, left front.*

◀ *Brown Rat, left front.*

▶ *Black Rat, left hind.*

◀ *Brown Rat, left hind.*

Black Rat

a) Toes slightly shorter and wider.

b) Toes rarely leave a full print all the way to the midfoot pad.

c) Digital and metatarsal pads are stronger and take up proportionally more space in the track.

d) Toes proportionately shorter and wider, clearly visible in toes 2–4. Negative space between digital pads 2–4 and the front two metatarsal pads correspondingly smaller.

e) Outer toes 1 and 5 only slightly offset to the rear.

f) Metatarsal pad I close to II and IV.

g) The size difference between front foot and hind foot is less pronounced.

Brown Rat

a) Toes slightly longer and narrower.

b) Toes more frequently leave a full print all the way to the midfoot pad.

c) Digital and metatarsal pads weaker and take up proportionally less space in the track.

d) Toes proportionally longer and narrower, clearly visible in toes 2–4. Negative space between digital pads 2–4 and the front two metatarsal pads is correspondingly larger.

e) Outer toes 1 and 5 clearly behind the front edge of the midfoot pad.

f) Metatarsal pad I out of line with II and IV.

g) The size difference between front foot and hind foot is more pronounced.

GAITS

Rats usually move at a walk when exploring or searching for food. They travel over longer distances under cover and usually in trot. The animals often switch to a 2 × 2 bound on snow and ground where they could sink. They usually use a kind of parallel bound to escape danger or to cross open spaces quickly.

Walk
Stride length: 11–21.5cm
Trail width: 4.5–8.5cm

Trot
Stride length: 17–28cm
Trail width: 6.1–7.4cm

3 × 4 bound (rotary) and 4 × 4 bound (transverse)
Group length: 6.5–11cm
Inter-group length: 6.5–28cm
Stride length: 13–36cm
Trail width: 4–8.5cm

Parallel bound
Group length: 4.2–14cm
Inter-group length: 15–45cm
Stride length: 20–59cm
Trail width: 6–9cm

▼ *Direct register walk.*

▼ *Parallel bound.*

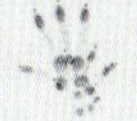
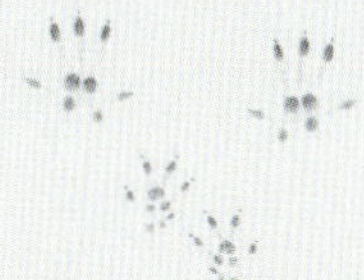
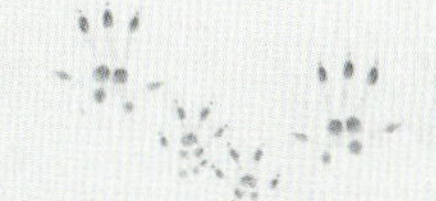

▲ *Brown Rat in a parallel bound. Footfall from bottom to top: RF, LF, LH, RH. Spessart, Germany.*

◄ *Typical track pattern of a Black Rat. Footfall from bottom to top: LF, LH, RH, RF. Lausitz, Germany.*

Similar tracks
Sousliks, Common Hamsters, squirrels, water voles.

DIET   Rats are omnivores with very variable eating habits. If sufficient plant food is available, they eat a predominantly vegetarian diet. Their diet includes seeds, fruits, roots, grains and grasses, but also maize, sunflower seeds, potatoes, human food and food waste. They eat a relatively low proportion of vertebrates and invertebrates. Depending on food supply, rats can also switch to a predominantly animal-based diet if little or no plant-based food is available. This can include snails, eggs, mice and fish. Brown Rats can be predatory and prey on birds, catch fish and gnaw on helpless mammals, such as small piglets, lambs or mice. There have even been individual cases where defenceless human adults and babies have been eaten.

REPRODUCTION   Both rat species can reproduce all year round. Their mating season peaks in March–April and September–October. Three to nine (up to 16) hairless, blind young are born after a gestation period of 20–24 days. There are usually 4–6 litters per year, although more are possible. The animals reach sexual maturity after 2–4 months. Two or more females may rear their young together. Orphaned young are suckled by other females.

REMARKS   The Black Rat is best known as the host of the plague flea. It probably had a decisive influence on the spread of the bubonic plague in the Middle Ages. The Brown Rat is the wild ancestor of the Fancy Rat (*Rattus norvegicus domestica*), which is still bred today as a pet and for use in medical and biological research in laboratories. Many people see Brown Rats as a pest because they cause considerable damage to food and can transmit various diseases. They can also gnaw through plastic, aluminium, copper and lead, causing technical issues. Despite intensive pest control by humans, eradication does not seem possible. In evolutionary terms, the Brown Rat is an outstanding model for success. It is extremely adaptable. Some populations have even become resistant to certain poisons, through genetic mutations. Their unspecialised way of life enables them to survive almost anywhere and to make use of almost any food.

SIGNS

NESTS AND DWELLINGS Rat nests usually measure 12–15cm in diameter. Brown Rat dwellings are normally 40–50cm underground. They consist of a lined living chamber measuring around 20–30cm in diameter, one or more storage chambers and a tunnel system with several entrances. Food is kept in the storage chambers, whose entrances are often concealed by boards or other obstacles. Above ground, well worn, clear paths usually connect several entrance holes measuring 6–9cm in diameter. In contrast to the rather neat-looking entrance holes of water vole burrows, fan-shaped piles of earth are often found in front of the entrance holes of rat dwellings. Water voles also gnaw off plant stems and roots in the entrance area, while rats just squeeze past them. Plants around the entrance hole to a water vole burrow are usually shredded and eaten away. This is not the case with rat entrance holes.

▼ *CyberTracker evaluator George Leoniak points to the entrance hole of Brown Rat burrow.*
West Sussex, England.

PATHS   Rats use established paths between their burrow system and their food sources. These paths are about 5–10cm wide and usually clearly visible. They are characterised by urine and scent markings. Black Rat paths can even appear black in colour as a result. Brown Rat paths in buildings are usually close to walls and obstacles, so the animals can feel their surroundings with their whiskers. Unlike Brown Rats, Black Rats usually raise their tail so they rarely leave a visible drag mark. Brown Rat tails leave drag marks more often.

FEEDING MARKS   Rat feeding marks are difficult to distinguish from those of other rodents and it is often impossible to make a definite identification. Bang and Dahlström mention a difference in feeding marks on oat kernels. Mice (*Mus* genus) nibble the grain from the side while holding it by the ends with their front feet. They leave behind elongated shavings reminiscent of crushed grain. Rats, on the other hand, gnaw the grains from the end. They tend to leave behind the end pieces, which can be large or small.

EXCREMENT   Rat droppings have a strong smell that humans usually find unpleasantly pungent and rancid. Texture and content vary according to the animal's diet. Typical colours are brown, grey-brown, black-grey and black. With experience, Black and Brown Rat droppings can be distinguished relatively reliably. Brown Rat droppings are large, cylindrical and usually have one blunt and one pointed end. The faecal pellets are granular and often have an irregular surface. Superficially, they resemble the faeces of large bats. Despite their firmness, they appear brittle, but are less porous than bat droppings. Unlike water vole and Muskrat droppings, they are not very homogeneous, contain fewer plant fibres and are not necessarily found near water.

◄ *An irregular surface as well as a blunt and a pointed end are typical characteristics of Brown Rat droppings. West Sussex, England.*

While Muskrat and water vole droppings are usually found near water, Brown Rats often deposit their faecal pellets in a kind of latrine, which are mainly found on paths, in prominent places such as trees and stones, at feeding sites and near nests. However, droppings can also be scattered individually. Black Rat droppings are thinner and shorter than those of Brown Rats. They are more irregularly shaped, often slightly curved and rather blunt at both ends.

If both animals occur in the same habitats, such as attics, warehouses, stables or barns, Black Rat droppings are more scattered and deposited in more varied places compared with those of Brown Rats. In this case, Brown Rat droppings are more likely to be found along walls and in corners, whereas the Black Rat with its climbing ability tends to deposit its droppings on sloping beams and along high roof battens. Unlike Brown Rats, the droppings of Black Rats are not found in open fields. They are almost always deposited in or near buildings or ships. A diameter of more than 4mm is indicative of Brown Rats.

- **Black Rat** L 0.6–1.5cm D 0.2–0.5cm
- **Brown Rat** L 1–2.4cm D 0.4–0.7cm

Caution!

Rats can transmit the infectious disease leptospirosis through contact with their faeces, urine and blood. The disease is rare in Europe. Nevertheless, it poses a serious risk and, in exceptional cases, can cause damage to the liver and kidneys and lead to death.

BLIND MOLE-RATS
Spalax, Nannospalax

HTL 15–25cm
W 140-365g

To date, little research has been done into the family of blind mole-rats (Spalacidae), which includes at least 36 species, all of which burrow and mainly live underground. The Bukovina Blind Mole-rat (*Spalax graecus*) and the Lesser Blind Mole-rat (*Nannospalax leucodon*) are found in Europe. Like moles, blind mole-rats are highly specialised to live almost exclusively underground, with short-legged, cylindrical bodies that are ideally suited to moving through underground tunnels. The animals have no visible tail and their outer ears are reduced to a fold of skin hidden in their fur. Their eyes are also covered with fur and are probably non-functional evolutionary remnants. Their large incisors protrude visibly from the mouth. Blind mole-rats dig extensive underground burrows with nesting and storage chambers, as well as tunnels and feeding tunnels. They have an excellent sense of touch that enables them to feel the finest earth tremors and they can probably orientate themselves using the earth's magnetic field. They also have a well-developed sense of smell that they use to follow scent trails in their tunnels. The animals are active during the day and at night, all year round. Remains of blind mole-rats have been found in owl pellets. Otherwise, little is known about the animals' predators. As blind mole-rats rarely come to the surface, I have never seen any tracks.

DISTINGUISHING FEATURES Can only be confused with other blind mole-rats, otherwise unique. Larger than a mole. The Bukovina Blind Mole-rat is very similar to the Lesser Blind Mole-rat in appearance, size, weight and lifestyle.

DISTRIBUTION AND HABITAT The Bukovina and the Lesser Blind Mole-rats are found in Romania. The Lesser Blind Mole-rat is also found on the Balkan Peninsula and in Hungary. Both blind mole-rats are originally steppe dwellers, but as species associated with human activity, they are now also found in meadows, pastures and fields.

DIET Roots, tubers and bulbs of plants and trees. Crops such as potatoes, carrots and peas, as well as fruits, leaves and plant stems. Blind mole-rats can store up to 20kg of food in large underground chambers.

REPRODUCTION The mating season lasts from January to April. After a gestation period of 29–32 days, 2–4 (up to 6) hairless and blind young are born, mainly in March/April. The offspring can reproduce after about a year.

▼ A line of mounds of earth is a distinctive sign of the presence of blind mole-rats.

SIGNS

DWELLINGS The complex underground burrows feature tunnel systems measuring up to 200m long, several storage and nest chambers, latrines, as well as feeding tunnels and passageways. They may extend as far down as 4m below ground. The shallower feeding tunnels are only 5–30cm deep. The tunnels have a diameter of 6–12cm.

MOUNDS Like moles, blind mole-rats carry the soil from their tunnels up to the surface and create mounds. These piles of earth, which are often aligned in a row, are a distinctive sign of the presence of these otherwise secretive animals.

CRESTED PORCUPINE
Hystrix cristata

HTL 57–68cm
(up to 85cm)
TL 5–15cm
W 10–15kg
(up to 18kg)

Porcupines (Hystricidae) are large rodents that are unmistakable due to their long quills and size. The species discussed here is the only one in Europe. These almost exclusively nocturnal animals live in small family groups, usually consisting of the parents, and young from the current year and previous year. They live in rock crevices, hollows under tree roots or in underground burrows that they dig themselves or take over from other animals. While their sense of hearing and smell are well developed, their eyesight seems to be of little importance. In contrast to New World porcupines (Erethizontidae), Crested Porcupines rarely climb but are good at swimming. They are noisy animals that often growl, snort and grunt. At night, you will often hear them before you see them. Crested Porcupines have hollow spines known as 'rattle quills' on the end of their tail that they rattle as a warning signal when excited. When threatened, porcupines growl, rattle, stamp their hind feet and raise their quills to make themselves appear much larger. If the threat continues, the animals move sideways or backwards towards their attacker with their spines raised. The sharp spines can easily penetrate the skin of an attacker and cause injuries that often become infected and can even lead to the predator's death. Natural predators are wolves and other large canids.

DISTINGUISHING FEATURES About the size of a European Badger with a distinctive spiny coat. The black-and-white banded quills are formed from the same substance as hairs. They are used for self-defence, are about 2–5mm thick and can grow up to 40cm long.

DISTRIBUTION AND HABITAT Central and southern Italy, Sicily, Albania and northern Greece. These adaptable animals prefer dry and rocky areas but can also be found in forests or near agricultural land.

DIET Predominantly herbivorous, feeding mainly on roots, tubers, fruits and leaves. They also feed on cambium and crops such as potatoes and maize. However, frogs, insects and other small animals, as well as carrion, make up part of their diet. Occasionally, their toothmarks can be found on bones, which they use as a source of calcium and to sharpen their teeth (Grzimek 1990).

REPRODUCTION Crested Porcupines usually live in strong pair bonds. After a gestation period of around 93–105 days, 1–2 (up to 4) precocial young are born. They can see and walk almost immediately after birth. Their quills are soft at first but harden within a few days. The young reach sexual maturity in their second year.

SIGNS

DWELLING A Crested Porcupine family's underground burrow can be reused by several generations and can become very large. In Africa, they frequently use Aardvark burrows. The entrance holes have a diameter of around 25–35cm.

CAMBIUM FEEDING Like beavers, porcupines gnaw on the bark of branches and trunks to get to the nutritious cambium. Areas debarked by porcupines are usually up to 80cm high. They prefer deciduous trees. The gnawing teeth of the upper jaw leave grooves of about 6–10mm wide.

▲ *A Crested Porcupine by its burrow.*

EXCREMENT The colour and shape of droppings depend strongly on the animal's diet. They are usually brown to black in colour. Piles of elongated faecal pellets, which can be mistaken for Domestic Cat or European Badger droppings, are typical. However, porcupine droppings consist almost entirely of plant remains. Distinctive lengthways grooves are occasionally recognisable.
- L 1.9–3.5cm D 0.7–1.4cm

TRACK

Front

L 7.2–8.5cm W 5.2–6cm

Medium-sized. Plantigrade. Asymmetrical. Five toes in a classic rodent structure. Toe 1 is very short and rarely visible. Toes 2–5 are short and strong and may appear slightly curved inwards. Toes 2 and 5 are slightly to the side. Toes 3 and 4 point forwards at an angle. Interdigital pads II and III have fused, so the print of the midfoot pad consists of three metacarpal pad prints, which can sometimes look like a single pad print. The negative space between the digital pads and metacarpal pads is short and wide. Two strong thenar/hypothenar pads can be seen at the back of the track. Both are at an open angle to the inside of the path. All the pads are hairless and robust. Long, strong, blunt claws that often leave a print. The front foot leaves a slightly larger and wider print than the hind foot.

Hind

L 6.7–7.8cm W 5–5.6cm

Medium-sized. Plantigrade. Slightly asymmetrical. Five toes, usually pointing forwards and almost parallel to each other. Toes 2–5 can point slightly outwards, which makes the track appear more asymmetrical overall than that of the front foot. However, the front foot track is actually more asymmetrical if toe 1 is taken into account.

▲ *Right front.*

▲ *Right hind.*

▲ *Right front.*
Italy, Toni Romani.

▲ *Right hind (top) and right front (bottom).*
Italy, Toni Romani.

Interdigital pads II and III have fused, so the print of the midfoot pad consists of three metatarsal pad prints that can sometimes look like a single pad print. Compared to the front foot print, the negative space between the digital and metatarsal pads is elongated and narrow. Two strong thenar/hypothenar pads can be seen at the back of the track. Both are at an open angle to the inside of the path. All the pads are hairless and robust. Strong claw marks are often recognisable and shorter than on the front foot. The hind foot print is slightly smaller and narrower than the front foot print.

GAITS

The preferred gait is a slightly overstep walk.

Walk
Stride length: 39–65cm
Trail width: 10–15cm

Based on limited data.

▲ *Overstep walk. The cycle of footfalls from bottom to top is: RF, RH, LF, LH. Italy, Toni Romani.*

▼ *Overstep walk.*

Similar tracks
The tracks can be mistaken for European Badger tracks, but the front foot claws are much shorter.

CARNIVORES

In the the region covered by this book, there are nine families in the order of carnivores (Carnivora): cats, viverrids, mongooses, dogs or canids, bears, mustelids, procyonids, Walrus and earless seals.

All members of this order feed at least partly on meat. However, some species are omnivores rather than pure carnivores and also eat plant food. The common feature of carnivores is their four long canine teeth which have developed into fangs, and the sharp premolars and molars, which have developed into carnassial teeth. We will look at each family individually, except for Walrus and earless seals, which are grouped together because of their similar tracks and signs.

Cats

Cats (Felidae) are charismatic carnivores with a lithe body and a small, roundish head with a short snout. Cats have excellent eyesight and are primarily visual hunters that sneak up on and ambush their prey. They have sharp, retractable claws, which they mainly use for pinning down prey and climbing. There are two genera in the region, with three species. All the species discussed here are pure carnivores that, unlike many canids, never eat plant matter. Cats use their urine, droppings and other glandular secretions to leave scent marks. They do this to communicate, for example to signal their presence and willingness to mate. Adult males are larger and stronger than adult females (sexual dimorphism).

TRACKS OF CATS

Digitigrade. Asymmetrical. Normally four relatively small toes and a large, fused midfoot pad. Cats that kill their prey without a prolonged pursuit can retract their claws. As a result, claw marks are not usually visible in the track. Exceptions are common on slippery ground and at high speed. Tracks of cats and canids can easily be confused. The presence or absence of claw marks alone is not enough to tell them apart.

- **Track formula: 4F × 4h**

FRONT FOOT Asymmetrical and rounded. Four asymmetrically arranged digital pads that reliably leave a print. Toe 1 is very short and rarely visible. The midfoot pad often tends to be slightly angled towards the outside of the foot. A single thenar/hypothenar pad may be visible in deep prints or at high speeds.

DIFFERENTIATING BETWEEN CANID AND CAT TRACKS

Canids

a) Rather symmetrical arrangement of the digital pads.

b) Relatively large digital pads and smaller midfoot pad. Canids are 'pursuit hunters' that chase their prey over longer distances. Strong digital pads give them the traction they need.

c) Midfoot pad single-lobed on the front edge and double-lobed on the back edge. Rather triangular.

d) Negative space between the midfoot pad and the digital pads is much larger and appears H-shaped.

e) Claws usually leave deep prints, in line with the prints of the digital pads.

Cats

a) Rather asymmetrical arrangement of the digital pads.

b) Relatively small digital pads and large midfoot pads. Cats are generally 'ambush hunters' that sneak up on their prey and rarely chase it over longer distances. The large midfoot pad, which accounts for up to 50 per cent or more of the total surface area of the track, helps them to tread quietly.

c) Midfoot pad double-lobed on the front edge and triple-lobed on the back edge. Rather square.

d) Negative space between the midfoot pad and the digital pads is narrower and C-shaped.

e) Claw marks are not usually visible. Tend to leave more superficial prints than the digital pads.

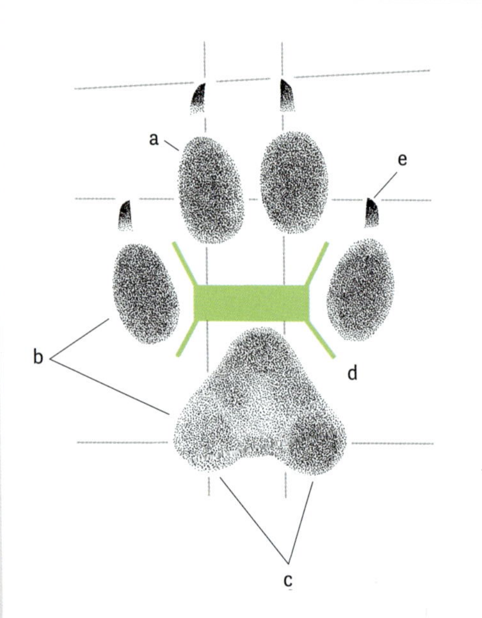

▲ *Track of a canid, left front.*

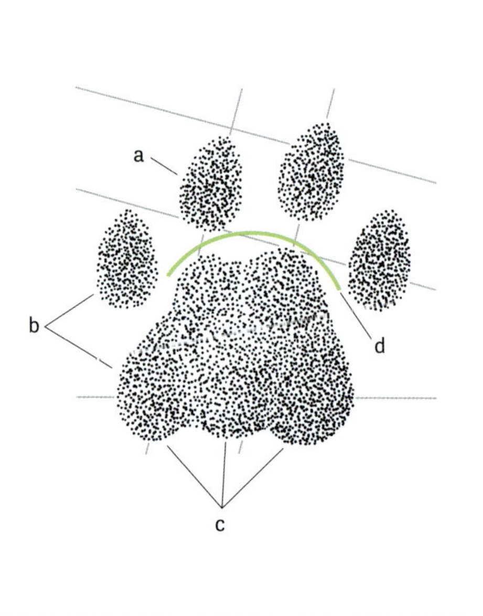

▲ *Track of a cat, left front.*

▲ *Domestic Dog, left front. Symmetrically arranged digital pads, strong claw marks and the H-shaped negative space typical of canids are clear characteristics. Lausitz, Germany.*

▲ *Domestic Cat, left front. The asymmetry and the three-lobed shape of the midfoot pad are clearly recognisable. Lausitz, Germany.*

HIND FOOT Slightly asymmetrical and more like an upright oval than round. The four toes that reliably leave prints are arranged more symmetrically than on the front foot. The individual digital pads can be longer and narrower than the rather rounded, droplet-shaped digital pads of the front foot. The midfoot pad is slightly smaller and weaker than on the front foot.

GAITS Walk is usually preferred: understep, direct register or overstep. Cats also trot, bound and gallop.

SIGNS OF CATS

FEEDING MARKS Cats occasionally create food stores and sometimes very carefully bury their prey. Cleanly removed tufts of hair near the carcass can be a typical sign of predation by a cat. This can mainly be observed in the thicker winter fur of predated even-toed ungulates.

EXCREMENT Sausage-shaped, usually clearly segmented and with blunt ends. Has a relatively smooth surface and is not usually kinked. The contents are often very compressed. Cats often bury their droppings and adjacent scratch marks can also be seen. The strong odour of the excrement and urine of European Wildcats is comparable with the odour of Domestic Cats.

SCENT MARKING Males and females leave scent marks by scratching the ground or scratching trees. Territorial males seem to mark more frequently.

LYNX

Lynx

Eurasian Lynx
HTL 80–110cm (up to
140cm)
TL 11–25cm
W 15–26kg (up to
38kg)

Iberian Lynx
HTL 84–100cm (up to
110cm)
TL 8–14cm
W 8–15kg (up to
18kg)

The secretive lynx species are Europe's largest wild cats, with males larger and much heavier than females. The Eurasian Lynx (*Lynx lynx*) and the Iberian Lynx (*L. pardinus*) occur in the region covered by this book. As predominantly crepuscular and nocturnal solitary animals, lynx have excellent eyesight and very good hearing. They can hear the rustling of a mouse over 50m away. Lynx hunt by ambushing or sneaking up on their prey. If a prospective kill is in sight, a lynx will launch a surprise attack, usually consisting of a few bounds. They normally reach prey within 20m. Lynx only rarely pursue prey over longer distances. They kill larger animals with a bite to the neck or throat. Lynx prefer to rest in elevated and sheltered sites, such as rocky outcrops or under the roots of fallen trees. Kittens may be preyed on by Grey Wolves and Wolverines. Humans are still the main threat to lynx.

DISTINGUISHING FEATURES Typical short cat face with whiskers and black ear tufts. Long legs and short stumpy tail with a blunt black tip. The Iberian Lynx is smaller, slimmer and leggier than the Eurasian.

DISTRIBUTION AND HABITAT The Eurasian Lynx was formerly considered to be extinct in large parts of Europe. Reintroductions and migrations have since occurred in various regions. Eurasian Lynx are once again widespread in Europe, from Scandinavia and the Baltic states to the Balkans. In central Europe, mainly isolated populations occur in the Alps, the foothills of the Alps and other mountain regions such as the Harz Mountains, the Bohemian Forest and the Vosges Mountains. The Eurasian Lynx is a shy animal that avoids human activity and prefers unbroken, extensive and relatively undisturbed forest and rocky landscapes. It can also be found in moorland and heathland, as well as in non-forested areas, provided they offer sufficient cover and food. The Iberian Lynx is one of the most endangered cat species in the world, with a total population of around 400 animals in 2016. It is found in south-western Spain and in Portugal and its range is fragmented into many small, isolated populations. It favours rocky scrubland and grassland, heathland and Mediterranean dry woodlands as its habitat.

DIET Pure carnivores that mainly eats mammals. Lynx prefer medium-sized mammals, such as Mountain and Brown Hares, European Rabbits and ungulates up to Western Roe Deer size. Ground birds like grouse and small mammals such as voles and mice, squirrels or marmots can make up a large part of their diet, depending on the season and region. Lynx also eat insects like grasshoppers, crickets and beetles.

REPRODUCTION Female lynx are in heat for up to 14 days during the mating season from February to April. A second heat may occur if they are not impregnated. After a gestation period of 67–75 days, an average of 2–3 (up to 6) furred but blind and helpless young are born in the shelter of a cave, tree root hollow or abandoned badger sett or fox den. They begin hunting with their mother from the third month. The young move away to find their own territory before the next mating season. Females reach sexual maturity in their second year and males in their third year.

Eurasian Lynx have excellent eyesight and can spot a flying bird of prey 3km away.

TRACK

Typical cat structure. Roundish overall outline. The teardrop-shaped digital pads are arranged asymmetrically and take up relatively little space in the entire track. The large midfoot pad can constitute up to 50 per cent of the track. In relative terms, the area of all four digital pads almost fits into the area of the midfoot pad. The midfoot pad is double-lobed on the front edge and triple-lobed on the back edge. Claw marks are not usually visible. Exceptions may occur on slippery ground and at high speed.

Front

Eurasian Lynx
L 6.5–9.5cm W 5.5–9.9cm

Iberian Lynx
L 5–6.5cm W 5–5.9cm

Medium to large. Digitigrade. Asymmetrical. Five toes. Toe 1 is very short and sits high up on the inside of the foot. It is rarely recognisable. Toes 2–5 almost always leave a print. The interdigital pads have fused to form a large midfoot pad. A single thenar/hypothenar pad can occasionally be visible at high speed or in deep ground. The claws are retractable and rarely leave a print. They appear fine and sharp. The front foot is larger and rounder than the hind foot, although the toes of the hind foot are longer. The negative space between the midfoot pad and digital pads is short, wide and C-shaped.

▲ *Eurasian Lynx, left front.*

▲ *Eurasian Lynx, left hind.*

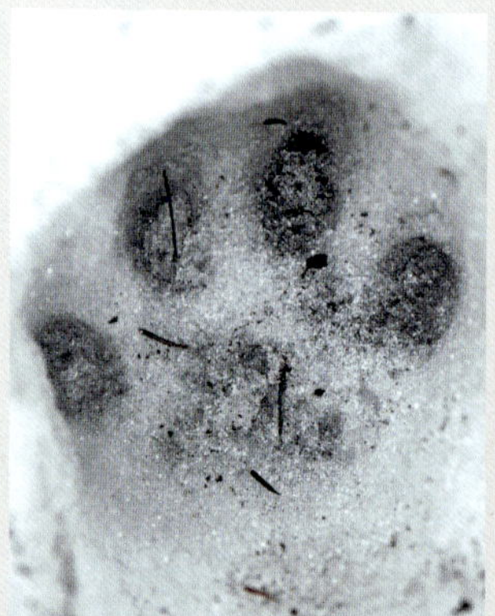

▲ *Eurasian Lynx, left front. Bieszczady, Poland.*

▲ *Iberian Lynx, left hind. Jaén, Spain. Paloma Troya (SERAFO).*

▲ *Eurasian Lynx, left front (bottom)
and left hind (top).
Jämtland, Sweden. Laura Gärtner.*

▲ *Iberian Lynx, left front (top)
and left hind (bottom).
Spain, Paloma Troya (SERAFO).*

Hind

Eurasian Lynx
L 6.4–8.5cm   W 5.5–8.5cm

Iberian Lynx
L 5–6cm   W 4.3–5.5cm

Medium to large. Digitigrade. Slightly asymmetrical. Four toe prints that often appear more elongated than on the front foot. The interdigital pads have fused to form a large midfoot pad. The claws are retractable and rarely leave a print. They appear sharp and fine. The hind foot is narrower, more elongated and generally more like a vertical oval than the rounded front foot. The more elongated negative space between the midfoot and digital pads is another feature that can be used to differentiate between front foot and hind foot prints. The two outer lobes on the bottom edge of the midfoot pad are often further forward than the inner lobe.

Similar tracks
Domestic Cat tracks are much smaller. Hind foot prints can be mistaken for those of canids.

DIFFERENTIATING BETWEEN THE SEXES

Males tend to leave broader, stronger tracks. This characteristic is more noticeable in front foot prints than in hind foot prints. The differences between the front foot and hind foot tracks are more obvious in males than in females. Pardo & Galan (2013) use the size of the hind foot track to determine the sex of Iberian Lynx: W ≥ 5cm indicates a male animal.

GAITS

Both lynx species mainly move at a walk. They choose routes with plenty of cover that tend to run along the edge of the area they want to cross. If they do cross open areas, lynx usually prefer to trot. Lynx often stop to observe their surroundings. If they sneak up on prey, they usually approach in an understep walk. They bound and gallop when launching a surprise attack or to escape. Their legs can leave characteristic drag marks in deep snow.

Walk
Stride length: 56–121cm
Trail width: 10–25cm

Trot
Stride length: 80–160cm
Trail width: 10–19cm

Gallop
Group length: 45–140cm
Inter-group length: 50–164cm
Stride length: 100–304cm (up to 6m)
Trail width: 18–35cm

▼ *Iberian Lynx in overstep walk. Footfall from bottom to top: LF, LH, RF, RH. Coto de Doñana, Spain.*

▼ *Overstep walk.*

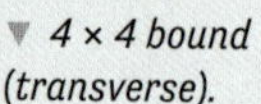

▼ *4 × 4 bound (transverse).*

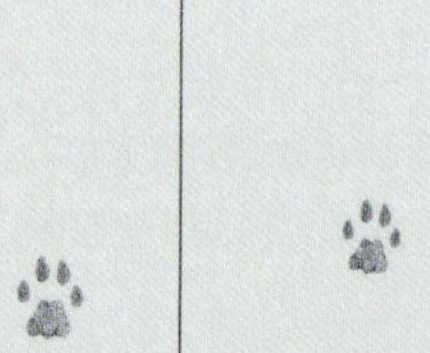

▲ *Eurasian Lynx in a rotary gallop. Jämtland, Sweden. Heide Ulrich.*

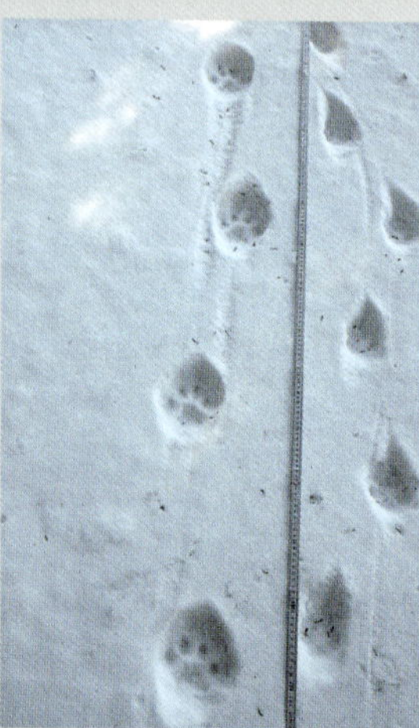

▲ *Eurasian Lynx in an overstep walk (left). Jämtland, Sweden. Laura Gärtner.*

◄ Lynx often bury their prey. Bavarian Forest, Germany. Markus Schwaiger, Lynx Project Bavaria.

▶ Heavy bleeding caused by the targeted throat bite is usually recognisable under the skin around the windpipe. 3–4 holes made by the pointed canine teeth are characteristic. Bavarian Forest, Germany. Markus Schwaiger, Lynx Project Bavaria.

SIGNS

SCRATCH MARKS   Lynx sharpen their claws on so-called scratching trees. They also make marks by scraping the ground and leaving scratch marks on trees. These marks are relatively rare and are more commonly found on the edge of a male's territory.

CARCASS FEATURES   Lynx often kill their prey with a targeted bite to the throat. A few (3–4) holes made by the pointed canine teeth can usually be recognised, leaving a clean, unfrayed edge. As lynx usually strangle their prey, clear bleeding can often be seen around the windpipe. Lynx sometimes also kill larger prey with a bite under the ear. They bite the neck of smaller animals, such as hares or deer fawns, or bite through their spine. Typical signs of predation by a lynx are few injuries, no severed body parts and, occasionally, skin that has been pulled off and over the head to create a kind of 'sack'. The stomach and intestines are untouched. Lynx often cover the carcass with available material such as leaves, grass, snow, moss and branches.

• **Distance between the canine teeth 2.7–3.2cm (2.5–3.6cm)**

EXCREMENT Typical cat droppings, usually segmented and with blunt ends. The droppings are generally more compact and contain fewer bone fragments and hairs than wolf droppings. The colour varies from brownish yellow, greenish brown and grey to dark brown or black. Like Domestic Cats, lynx can also bury and cover their droppings. However, scraping is more common, where scratch marks are made on the ground as a marking. Lynx often mark on or next to these scratch marks. Lynx droppings are mainly found at territory boundaries and prominent crossroads and have the pungent odour typical of carnivores.

• **L 5–25cm D 1.3–2.5cm**

▼ *Iberian Lynx droppings. Coto de Doñana, Spain. René Nauta.*

▲ *Eurasian Lynx dropping. The tapered end (left) comes out last. Bieszczady, Poland.*

EUROPEAN WILDCAT
Felis silvestris

The European Wildcat belongs to the genus of true cats (*Felis*). It is the only wild *Felis* species present in the region covered by this book. The European Wildcat is a highly specialised, stealthy hunter of mainly small rodents. It is crepuscular and nocturnal, but can also be seen during the day in secluded areas. It has excellent hearing and vision, while its sense of smell is more for close-range perception. It is a ground-based, stalk and ambush predator that reaches its prey in short bounds, before grabbing it with its claws and killing it with a bite to the throat or the back of the neck. It does not typically pursue prey over long distances. It is predated by wolves and lynx, but young animals are also taken by foxes, larger mustelids, Eurasian Eagle-owls, eagles and Eurasian Goshawks. Many European Wildcats are killed by vehicles on the road.

HTL 45–67cm
TL 21–35cm (up to 40cm)
W 2.5–8kg (up to 12kg)
Males are larger and heavier than females.

DISTINGUISHING FEATURES Short cat face. Faint stripes, not spotted. Narrow, almost black dorsal stripe from the shoulders to the base of the tail. Bushy, hairy tail, with blunt, black tip and 1–5 bands. Coat colour varies greatly.

TRACK

The asymmetrically arranged digital pads take up relatively little space in the entire track. The large midfoot pad can account for up to 50 per cent of the track, which roughly corresponds to the area of all four digital pads. The midfoot pad has two lobes on the front edge and three lobes on the back edge. Claw marks are not normally visible in the track. Exceptions may occur on slippery ground and at high speed.

▲ *Left front.*
Westerwald, Germany. Immo Meyer.

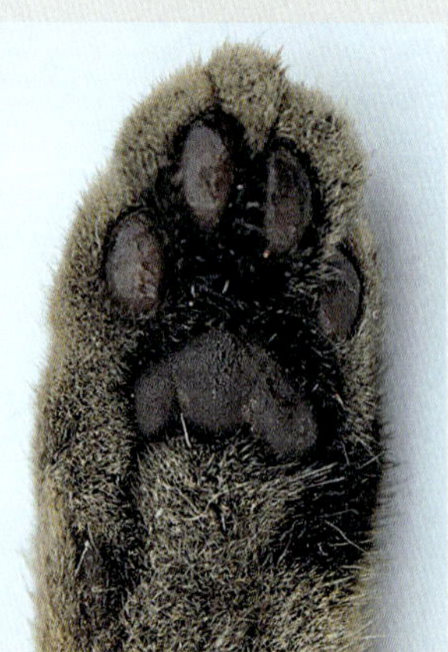

▲ *Left hind.*
Westerwald, Germany. Immo Meyer.

Front

L 3.5–4.9cm W 3–4.2cm

Medium-sized. Digitigrade. Asymmetrical. Five toes. Toe 1 is very short and sits further up on the inside of the foot. It rarely makes a print. The metacarpal pads have fused to form a large midfoot pad. A single thenar/hypothenar pad can leave a print at high speed or in deep ground. The claws are retractable and rarely leave a print. They appear fine and sharp and are higher up than the digital pads. The front foot is slightly larger and rounder than the hind foot.

▲ *Left front.*

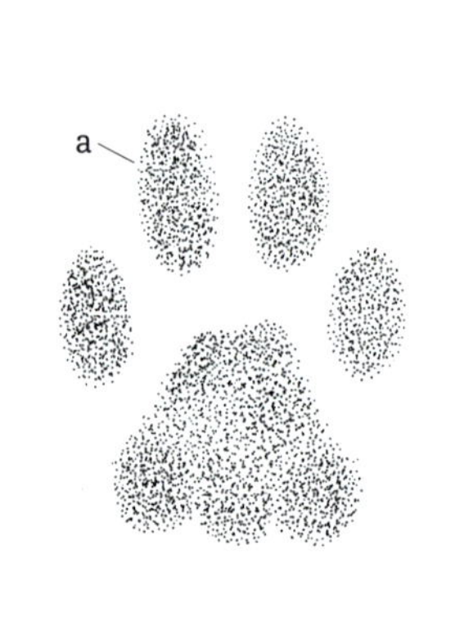

▲ *Left hind.*

Hind

L 3.2–4.5cm W 3–4cm

Medium-sized. Digitigrade. Slightly asymmetrical. Four toe prints, which often appear more elongated than on the front foot ⓐ. The metatarsal pads have fused to form a large midfoot pad. The claws are retractable and rarely leave a print. They appear sharp, fine and are higher than the digital pads. The hind foot is usually slightly narrower than the slightly rounder front foot.

▲ *Right front (bottom) and right hind (top). Huesca, Spain. Paloma Troya (SERAFO).*

▲ *The rare tracks of a European Wildcat. Westerwald, Germany. Immo Meyer.*

GAITS

European Wildcats mainly move at a walk. They prefer well-covered routes, which often lead along forest edges, hedgerows or similar cover. They mainly prefer to trot when crossing open spaces. European Wildcats often stop to observe their surroundings. When stalking prey, they usually approach it in an understep walk. They bound or gallop when launching a surprise attack or to escape from danger.

Walk
Stride length: 32–65cm
Trail width: 5–15cm

Trot
Stride length: 50–106cm
Trail width: 4–11cm

Data mainly from Domestic Cats.

▲ *European Wildcat in a slightly overstep walk. Footfall from bottom to top: RF, RH, LF, LH, RF, RH, LF, LH. Huesca, Spain. Paloma Troya (SERAFO).*

▼ *Direct register walk.*

▼ *Understep walk.*

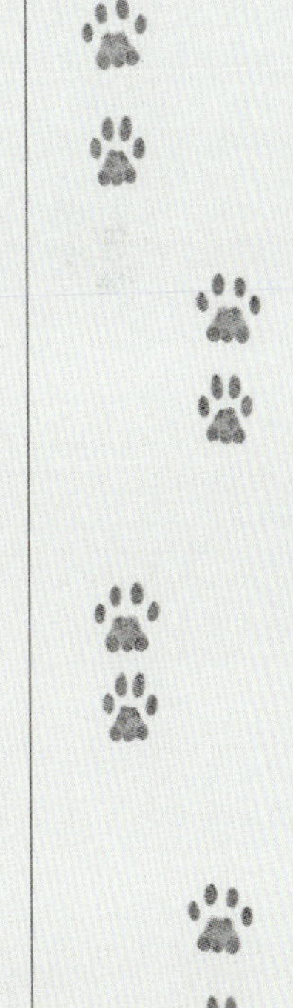

DISTRIBUTION AND HABITAT Iberian Peninsula, France, Italy, central Europe, Balkans and Carpathian Mountains. Northernmost occurrence in Scotland, absent from Scandinavia. The European Wildcat is thought to be restricted to large, unbroken areas of forest, although it can also be found in the immediate vicinity of settlements and in wetlands. Their historical range also includes the European lowlands and reintroduction is expected. European Wildcats also inhabit forest steppes in the south-east of their range. Warm, undisturbed places to rest, sleep and raise their young are important. European Wildcats are often found on steep, warm and dry south-facing slopes with a view.

DIET European Wildcats specialise in mice and especially voles but will also eat other small to hare-sized prey. More rarely, they will eat birds, insects, lizards and fish.

REPRODUCTION The mating season is from January to March. After a gestation period of 63–75 days, 1–3 (up to 6) young are born in a warm, dry, inaccessible hiding place and raised by their mother. The young leave the den for the first time after about a month. They are suckled for 3–4 months but begin to hunt with their mother after just two months. European Wildcats leave their mother's territory when they are around six months old. They reach sexual maturity at around 10 months.

Primitive form of the Domestic Cat

Following an extensive taxonomic revision in 2017, the closely related European Wildcat (*Felis silvestris*), Domestic Cat (*F. catus*) and African Wildcat (*F. lybica*), the latter being the ancestor of Domestic Cats, were recognised as separate species. The African Wildcat was probably domesticated around 9,000–10,000 years ago, during the emergence of agriculture in the Fertile Crescent in the Middle East. Its domestic descendants have since been spread worldwide by humans. The European Wildcat, the African Wildcat and the Domestic Cat can all mate with each other and give birth to fertile offspring. Just under 4 per cent of continental western European Wildcats have some Domestic Cat ancestry, with the wildcat apparently achieving genetic dominance in the long term. However, European Wildcats in Scotland may be functionally extinct because of hybridisation with Domestic Cats.

SIGNS

SCRATCH MARKS European Wildcats sharpen their claws on so-called scratching trees. They also mark territory by scraping the ground and leaving scratch marks on trees.

EXCREMENT Typical cat droppings, usually segmented and with blunt ends. The colour varies from brownish yellow, greenish brown and grey to dark brown or black. Like Domestic Cats, European Wildcats bury their excrement unless using it for marking. Scraping is also associated with passing droppings. European Wildcats often mark on or next to the scratch marks. Droppings are mainly found at territory boundaries and prominent crossroads and have the unpleasant odour typical of carnivores.

- L 4.5–18cm D 1.2–2.2cm

▶ *Domestic Cat scratching tree. Märkische Schweiz, Germany.*

▲ *Domestic Cats and European Wildcats tend to bury their excrement. These droppings are from a Domestic Cat. Märkische Schweiz, Germany.*

▲ *European Wildcat droppings are often very segmented with blunt ends. Lezáun, Spain. Paloma Troya (SERAFO).*

COMMON GENET
Genetta genetta

HTL 40–60cm
TL 40–55cm
(approximately the
same as body length)
W 1.2–2.5kg

The Common Genet belongs to the viverrid family (Viverridae) and is one of 14 species in the genet genus (*Genetta*). Only the Common Genet occurs in the area covered by this book. The predominantly crepuscular and nocturnal genet lives alone or in a family group. Family groups stay together for around a year. They are very agile, with well-developed hearing. Compared to cats, their sense of smell is highly developed. They are adept climbers and jumpers. They bed down in natural crevices, hollow trees, dense bushes or abandoned badger setts or rabbit burrows, and can release a strong musky odour if threatened. They are predated by lynx, foxes and large raptors.

DISTINGUISHING FEATURES   Similar to a Domestic Cat, but with an elongated, slimmer body with short legs, large eyes and large ears. Short, spotted coat. The tail usually has 9–10 wide black rings.

DISTRIBUTION AND HABITAT   South-west Europe. Found from Belgium and France to the Iberian Peninsula in Spain and Portugal.

Requires bushy terrain with vegetation that offers sufficient cover. Prefers dry areas with open terrain, but also occurs at the edge of woodland.

DIET The Common Genet is omnivorous and its diet consists mainly of insects, spiders, snakes, lizards and other small vertebrates. It occasionally also eats fruit and birds' eggs, as well as carrion.

REPRODUCTION Two litter periods with the main period from April to June and a secondary period from September to November. During these periods, there are 1–2 litters with 2–3 young, which are born almost hairless and blind, after a gestation period of 10–11 weeks. The animals are sexually mature from the age of two.

SIGNS

EXCREMENT Cat-like droppings, usually segmented and with blunt ends. Similar in shape to solid otter droppings, but usually contain more small bone fragments and hair. Colour varies from brown to grey, dark grey and black. Common Genets mark territory boundaries and prominent crossroads. They make latrines at these sites.

- L 5–20cm D 0.8–1.3cm

▼ *The more hairs the droppings contain, the more twisted they tend to be. Lezáun, Spain. Paloma Troya (SERAFO).*

▼ *Fernando Gomez finds a large genet latrine. Lezáun, Spain. Paloma Troya (SERAFO).*

◄ *The shape of the droppings depends greatly on the food ingested. Here, mainly crustaceans were eaten. Coto de Doñana National Park, Spain.*

TRACK

Front

L 3.1–4.2cm W 2.6–3.7cm

Medium-sized. Digitigrade. Asymmetrical, but generally more symmetrical than in cats. Five toes. Toe 1 is very short and sits further up on the inside of the foot. Rarely leaves a print but does so much more often than in cats. Toes 2–5 appear to radiate from the midfoot pad, while cats' toes tend to be parallel and point forwards. The front foot is larger and rounder than the hind foot.

Hind

L 2.8–3.8cm W 2.3–3.2cm

Medium-sized. Digitigrade. Slightly asymmetrical. Five toes. Toe 1 is very short and sits further up on the inside of the foot. It rarely leaves a visible print. Cats only have four toes on their hind foot. The toes often appear more elongated than on the front foot. Toes 2 and 5 are further back and further in than in cats. This gives the track a comparatively longer, narrower outline with a more compact negative space between the toes and midfoot pad. The hind foot is narrower, more elongated and generally more like an upright oval in shape than the rounded front foot.

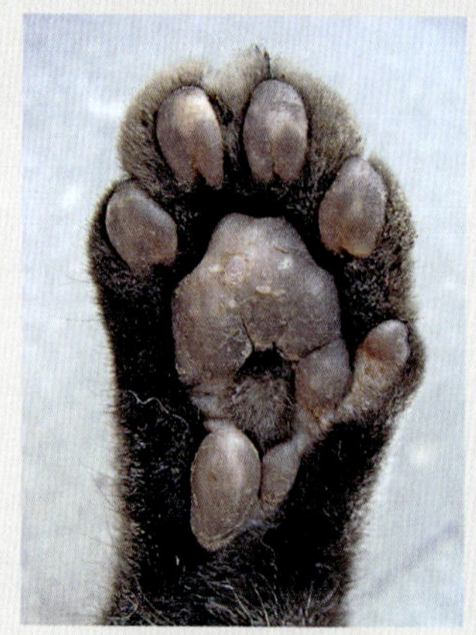

▲ *Right front. Lezáun, Spain. Paloma Troya (SERAFO).*

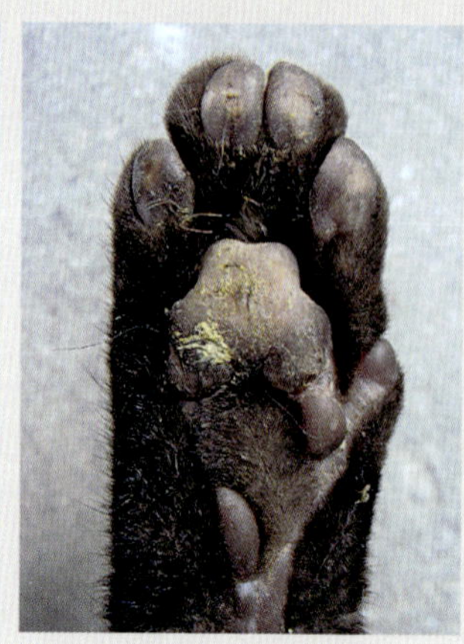

▲ *Right hind. Lezáun, Spain. Paloma Troya (SERAFO).*

▲ *Right front.*

▲ *Right hind.*

▲ *Right front, print from a dead specimen. Lezáun, Spain. Paloma Troya.*

▲ *Right hind. Coto de Doñana National Park, Spain.*

GAITS

The predominant gait is a brisk trot. The animals switch to a walk to stalk prey or to explore more intensively. They bound or gallop over short distances when fleeing or attacking. Like cats, genets prefer to move along border areas such as the edges of clumps of bushes, forest edges or other cover-rich zones.

Walk
Stride length: 21–43cm
Trail width: 5–8cm

3 × 4 bound (rotary) and 4 × 4 bound (transverse)
Group length: 17–22cm
Inter-group length: 15–60cm
Stride length: 32–82cm
Trail width: 5–9cm

Similar tracks

The tracks of the Common Genet can be confused with those of Domestic Cats and European Wildcats. Important differences are the more frequent print of toe 1 on the front foot and hind foot and the composition of the midfoot pad, where the interdigital pads have fused to form a large midfoot pad that tapers towards the outside back edge. The back edge of the metacarpal/metatarsal pad has a notch in the middle so it has two lobes, while cats have the characteristic three lobes. A single thenar/hypothenar pad can occasionally leave a print at high speed or in deep ground and is more frequently visible than in cats. In contrast to cats, the sharp, curved claws are only partially retractable and often leave visible marks.

▼ *Common Genet in an overstep trot. Coto de Doñana National Park, Spain.*

▲ *Direct register trot.*

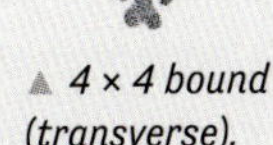

▲ *4 × 4 bound (transverse).*

▲ *Common Genet in a 4 × 4 bound (transverse). Footfall from bottom to top: RF, RH, LF, LH. Coto de Doñana National Park, Spain.*

EGYPTIAN MONGOOSE
Herpestes ichneumon

KRL 45–60cm
TL 35–55cm
W 1.9-4kg

The Egyptian Mongoose belongs to the mongoose family (Herpestidae), a family of mustelid-like carnivores with a slender body, pointed snout and short ears. There are three species in our region, two of which, the Small Indian Mongoose (*Herpestes auropunctatus*) and the Indian Grey Mongoose (*H. edwardsi*), occur in isolated populations as non-native species, so they are not discussed here. They usually live in family groups of up to five animals, but older males are more solitary. They typically walk in a line, with families crossing dense vegetation one behind the other. The nose of the rear animal touches the rear end of the front animal. Egyptian Mongooses exhibit highly developed playful behaviour. They are mainly active during the day, and are agile and very fast. They are known for attacking venomous snakes and overpowering them with a lightning-fast grab. Egyptian Mongooses are skilled jumpers and swimmers, and all their senses are well developed. As ground dwellers, they make their home under dense bushes, in self-dug burrows and in natural cavities. Their predators are larger carnivores and eagles.

DISTINGUISHING FEATURES Elongated body with short legs and a long tail.

DISTRIBUTION AND HABITAT The south of the Iberian Peninsula. Inhabits steppes, grasslands and Mediterranean maquis. A dense layer of vegetation is important for cover. They favour wetlands such as stream valleys.

DIET Insects, snakes, lizards, amphibians, birds and birds' eggs. Small mammals, mainly rodents, up to the size of rabbits. Occasionally they also eat fruit.

REPRODUCTION Reproduction has not yet been sufficiently researched. In Europe, most litters are normally born from May to June. After a gestation period of around 60–84 days, 2–4 young are usually born. They are sexually mature at the age of two.

Mongooses were revered in Ancient Egypt where their popularity was probably linked to their reputation as snake killers and rat exterminators.

SIGNS

EXCREMENT Varies in colour from grey to dark brown. Latrines are more common along territory boundaries because they serve as territory markings.
- L 4–12cm D 1.2–1.5cm

Similar tracks

The Small Indian Mongoose occurs as a non-native species in small populations in Croatia and Bosnia-Herzegovina. It is significantly smaller than the Egyptian Mongoose. Its track is about 2cm shorter. The Indian Grey Mongoose, also a non-native species in central Italy, is mentioned in the list of European mammals. In terms of size, it is between the Small Indian Mongoose and the Egyptian Mongoose. Both species only occur very locally and are not examined in detail. Egyptian Mongoose tracks can be mistaken for tracks of mustelids of a similar size.

TRACK

Front

L 4.3–6.2cm W 3.4–4.6cm

Medium-sized. Plantigrade. Asymmetrical. Five toes. Toe 1 is the smallest toe and cannot be reliably recognised. Toes 2–5 point forwards at a slight angle. The interdigital pads are fused but leave prints with individually recognisable bulges. Two further thenar/hypothenar pads, the T- and H-pads, can be visible at the back of the track. If the T-pad leaves a print, it is out of line and behind metacarpal pad IV ⓐ. If the T- and H-pads are not visible, the midfoot pad appears elongated to the back on the distal side ⓑ. As with mustelids, the H-pad is remarkably strong ⓒ. However, unlike mustelids, the digital and metacarpal pad prints appear stronger and the distinctive notch on the lower edge of the midfoot pad that is typical of mustelids is absent. Long, pointed claws are distinctive and usually leave prints. The front foot print is wider and more symmetrical than the hind foot print.

Hind

L 4–5.5cm W 3.2–4cm

Medium-sized. Plantigrade. Asymmetrical. Asymmetry clearer than in the front foot. Five toes Toe 1 is the smallest toe and sits further back than on the front foot. This makes the hind foot more asymmetrical. If toe 1 is not visible, the hind foot print

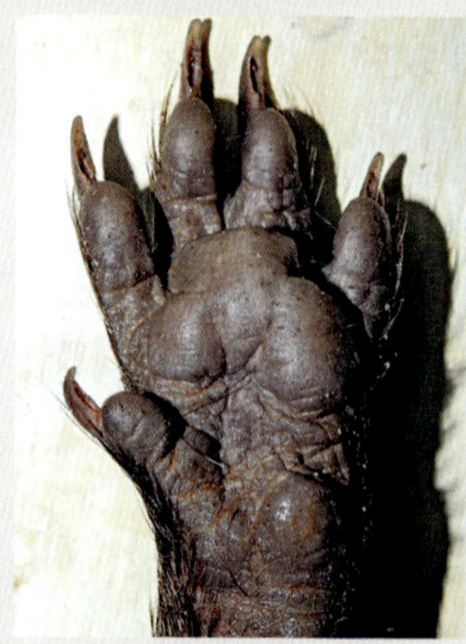

▲ *Left front. Spain. Paloma Troya (SERAFO).*

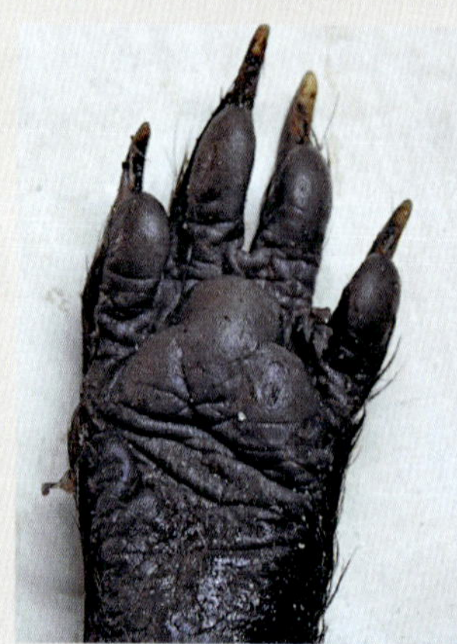

▲ *Left hind. Madrid, Spain. Paloma Troya (SERAFO).*

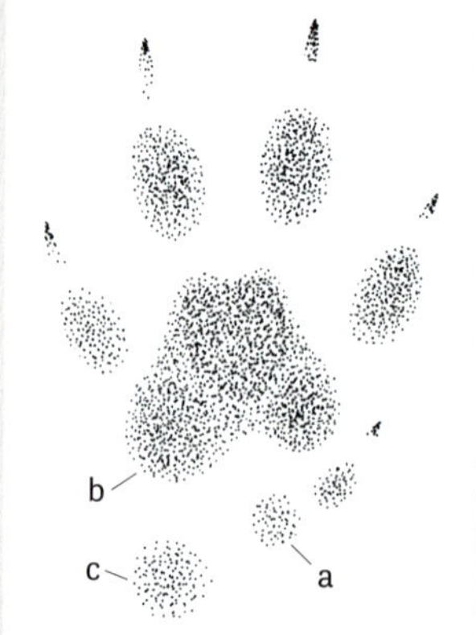

▲ *Left front.*

▲ *Left hind.*

▲ *Left front. Coto de Doñana National Park, Spain.*

▲ *Left hind. Coto de Doñana National Park, Spain.*

appears more symmetrical than the front foot print and may resemble the footprint of a small canid. Toe 1 is often not recognisable. Toes 2–5 point straighter forwards than on the front foot. Overall, the hind foot appears more elongated and slimmer. The interdigital pads are fused but leave prints with individually recognisable bulges. The long, pointed claws are only partially recognisable.

GAITS

The overstep walk and overstep trot are preferred gaits. Tail drag marks are often visible on the right surface.

Walk
Stride length: 28–50cm
Trail width: 7–11.5cm

Trot
Stride length: 42–86cm
Trail width: 4–9cm

▼ *Egyptian Mongoose tracks in the typical landscape. You can often find tail drag marks or drag marks from a snake carried in the mouth alongside the track. Coto de Doñana National Park, Spain.*

▼ *Overstep trot. Coto de Doñana National Park, Spain.*

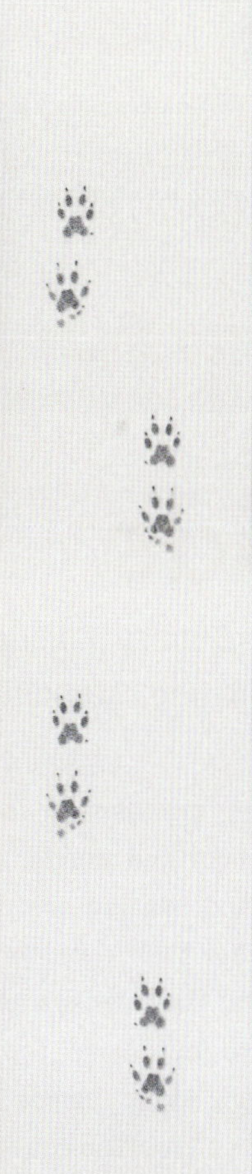

▲ *Overstep walk.* ▲ *Overstep trot.*

Canids

The dog or canid family (Canidae) is best known for its pronounced social behaviour. Canids live in strong pair bonds, family groups and large packs.

These carnivores are usually long-legged, with a pointed snout and slender body. Their sense of smell and hearing are particularly well developed, but their eyesight is weaker than that of cats. Their senses and physical conformation are optimised for life as a pursuit predator. Most canids are fast, tenacious hunters that often chase their prey over long distances. They mainly eat meat, but sometimes also plant-based food. Males are larger and stronger than females (sexual dimorphism).

There are three genera with five species in the area this book covers. Comparisons to help differentiate between certain species can be found at the end of the corresponding species portraits.

TRACKS AND SIGNS OF CANIDS

Digitigrade. Symmetrical. Five toes on the front foot and four toes on the hind foot. The first toe on the front foot rarely leaves a print as it does not carry any weight. As a rule, the four weight-bearing toes are relatively large and the fused midfoot pad appears rather small. Apart from a few exceptions, canids cannot retract their claws. They use their claws for a better grip when running rather than for killing prey. The claw prints are therefore generally blunter and can be recognised more reliably in the track than those of cats. However, canid and cat tracks can be confused with each other; the presence or absence of claw marks alone is not sufficient to tell them apart (see differentiation on page 362). Red Fox tracks differ in some respects from the typical canid characteristics and are described in detail in the corresponding portrait.

- Track formula: 4F × 4h + C

FRONT FOOT Symmetrical. Four symmetrically arranged toes that reliably leave a print. Toe 1 is very short and rarely leaves a print. The front feet are larger and stronger than the hind feet and have a larger midfoot pad in comparison to the hind feet. A single thenar/hypothenar pad may be visible in deep prints or at high speed.

HIND FOOT Symmetrical. Four symmetrically arranged toes that reliably leave a print. Toe 1 is absent. The hind feet are smaller than the front feet and their overall outline is more like an upright oval. The midfoot pad tends to be smaller and weaker than on the front foot and can look like the print of a single digital pad.

GAITS Canids are excellent long-distance runners and usually travel at a trot. They use all forms of trot. Moving in a side trot over longer distances is a unique characteristic of canines (see page 82). Canids also walk, bound and gallop, depending on the situation.

SCENT MARKING All canids mark their territory with excrement and urine. They usually deposit excrement in prominent places such as on hills, the edges of roads and paths. The urine of different species often has a characteristic odour and can enable us to distinguish between them. Domestic Dogs and Grey Wolves often scrape the ground with their front and/or hind feet after urinating or defecating. This spreads the scent mark over a larger area. The sweat glands on the paws also leave a scent mark and the scrape marks usually leave a conspicuous visual sign. Females come into heat between late winter and early spring. Blood can be found in female canids' urine shortly before they come into heat.

GOLDEN JACKAL
Canis aureus

HTL 60–100cm
TL 20–27cm
W 6.5–13.5kg (up
to 16.5kg)

The Golden Jackal is a medium-sized wild dog species that is mainly crepuscular and nocturnal. It belongs to the same genus as wolves (*Canis*), which comprises a total of about six species of typical long-limbed canids. Together with the Grey Wolf (*Canis lupus*), it is one of two species in the area. Golden Jackals are gregarious and seek lifelong partnerships. They live in small family groups from birth until the young move away. Golden Jackals in Europe mainly hunt alone but they may also hunt larger prey in a pack. They can sprint over long distances and are good swimmers with excellent hearing and sense of smell. Their way of life is similar to that of wolves in many respects. A slight decline in the population of Red Foxes has been observed in areas where Red Foxes and Golden Jackals occur together. Competition seems to occur without any direct attacks, but foxes avoid jackals. Apart from hunting by humans and being killed on the road, the Grey Wolf is the Golden Jackal's main threat. The absence of Grey Wolves can have a favourable effect on the spread of Golden Jackals. Eagles can also prey on Golden Jackals.

DISTINGUISHING FEATURES Smaller and slimmer than wolves but larger and leggier than foxes. Pointed muzzle and short, bushy tail with a black tip.

DISTRIBUTION AND HABITAT The Golden Jackal, originally from South Asia, mainly inhabits the Balkan Peninsula in Europe. It is currently spreading northwards and westwards further into central Europe. The core population lives in Bulgaria, Romania and Serbia, but it is now also increasingly found in Hungary. There are small populations in Italy and confirmed reports of first successful breeding in Austria and the Czech Republic. There is even isolated evidence of Golden Jackals in Germany and Switzerland. Golden Jackals are very adaptable in terms of habitat. They prefer open and dry terrain, rocks, fields and barren grassland. They are occasionally found near settlements. They tend to avoid dense forests.

DIET Opportunistic omnivore with a focus on animal-based food. Small to medium-sized mammals, insects, birds and reptiles. They also eat carrion, berries and fruit. Crops such as maize also form part of their diet, and they sometimes prey on livestock such as poultry and sheep.

REPRODUCTION The mating season lasts from January to February. After a gestation period of 60–63 days, there is usually a litter of 3–6 (up to 8) young in April or May. The young are born blind and raised by both parents. They usually reach sexual maturity after one year.

SIGNS

EXCREMENT The shape, colour and content of Golden Jackal droppings depend greatly on the food ingested. They are either regular and sausage-shaped, with one rounded end and one spindle-shaped end that tapers to a point, or broken up into individual parts. Typical colours are white (bone), green-brown, brown and black. It often contains hair or bone remnants. Golden Jackal droppings are often difficult to distinguish from those of Red Fox in the field.

- L 4.5–20cm D 1–2.5cm

Own data collection expanded by Jennifer Hatlauf and the database of the Golden Jackal Project in Austria.

▲ *Golden Jackal excrement.*
Jennifer Hatlauf, Golden Jackal
Project Austria.

TRACK

Front

L 4.9–7cm W 3.4–5cm

Small to medium sized. Digitigrade. Symmetrical. Five toes. Toe 1 is very short, is located higher up on the inside of the foot and rarely leaves a print. Toe prints 2–5 are reliably recognisable. Digital pads 3 and 4 are fused at the back, but this is rarely visible in the print. The interdigital pads have fused to form a triangular midfoot pad. A single thenar/hypothenar pad print can occasionally be visible at high speed or in deep ground. Sharp, pointed claw marks may be visible or absent. The front foot prints are larger and usually rounder than the hind foot prints.

Hind

L 4.8–6.4cm W 3–4.2cm

Small to medium sized. Digitigrade. Symmetrical. Four toes that reliably leave prints. Digital pads 3 and 4 are fused at the back, but this is rarely visible in the print. The interdigital pads have fused to form a triangular midfoot pad. Sharp, pointed claw marks may be visible or absent. The hind foot prints are smaller and more elongated than the front foot prints.

▲ *Right front. Blatino, Bulgaria. Paloma Troya (SERFAO).*

▲ *Right hind. Blatino, Bulgaria. Paloma Troya (SERFAO).*

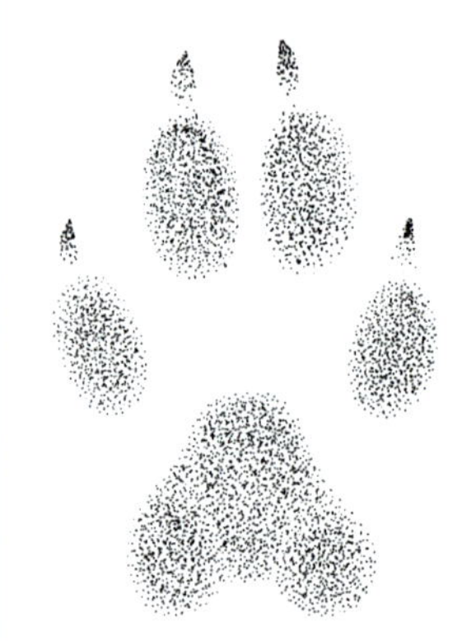

▲ *Right front.*

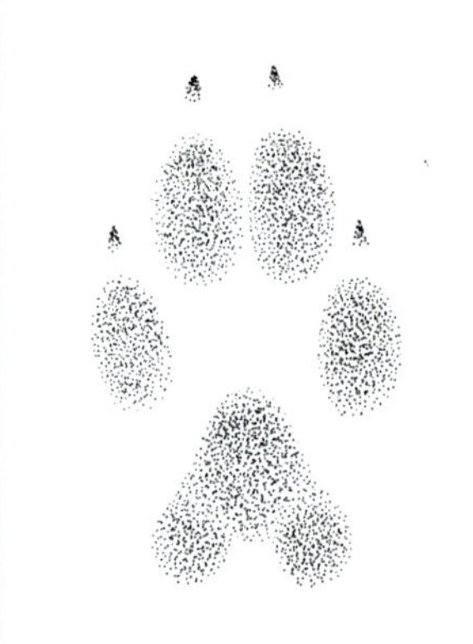

▲ *Right hind.*

▲ *Right front (right) and right hind (left). Sozopol, Bulgaria. Paloma Troya (SERFAO).*

GAITS

Walk
Stride length: 44–62cm
Trail width: 10–15cm

Trot
Stride length: 60–106cm
Trail width: 5–12cm

Based on small data sample.

Similar tracks
Red Fox, Domestic Dog and
Common Raccoon Dog.

▼ *Overstep walk. Footfall
from bottom to top: LF, LH, RF,
RH, LF, LH. Sozopol, Bulgaria.
Paloma Troya (SERFAO).*

▲ *Direct register trot.*

GREY WOLF
Canis lupus

HTL 100–170cm
TL 30–75cm
W 25–60kg (15–80kg)
Males are usually larger and heavier than females. Grey Wolves are largest in the north and smaller in southern Europe.

Grey Wolves are the largest species in the canid family and the ancestor of the Domestic Dog. They live in social family groups called packs. Packs generally consist of two parent animals, along with cubs and young from the last 1–3 years. Pack size varies throughout the year. They usually consist of 5–10 animals, but larger packs can also occur. When they reach sexual maturity at around 10–22 months, the young leave their parents' territory to start a family and establish their own territory. Young animals occasionally stay with their family until they are three or, more rarely, four or five years old. This is strongly influenced by the availability of food. Pack size depends on various factors, such as emigration, mortality, birth rate and, above all, food supply. Larger packs often form to hunt larger prey so more food is available at once. Grey Wolves mainly prey on wild even-toed ungulates and can sometimes cause a reduction in their population. However, it can be assumed that ungulates will adapt to Grey Wolves.

Grey Wolves generally prey on young, old and weak animals so the hunted herds are made up of stronger, healthier individuals.

Grey Wolves can travel far and sometimes cover considerable distances in a short time while searching for prey or a territory of their own. A young male Grey Wolf from Germany travelled 75km (as the crow flies) in one night. Grey Wolves are legendary for their stamina. They can trot at 8–10km/h and maintain their top speed of 64km/h for 20 minutes. Territory size is significantly, but not exclusively, influenced by the quantity and size of available prey. Latitude is another factor: the further north a pack lives, the larger its territory tends to be, as prey density is usually lower due to sparser vegetation. Refuges for rearing cubs and daytime dens also play an important role for Grey Wolves in cultivated areas. For example, research on Grey Wolves fitted with collar transmitters in Germany shows that they avoid settlements as much as possible and prefer forests (Reinhardt & Kluth 2016).

Grey Wolves are mainly crepuscular and nocturnal, although they are also active during the day. They are excellent swimmers that are capable of crossing rivers and lakes. They have very good hearing and an excellent sense of smell. Grey Wolves howl as a form of long-distance communication. They howl to call the family together, to search for a partner or to coordinate before a hunt. Howling also plays an important role in marking and maintaining territories and helps to strengthen social bonds. Grey Wolves howl most frequently during the mating season and while rearing cubs.

Grey Wolves are very social within their pack, for example older juveniles help to raise cubs. They play, ambush each other, stalk, catch and wrestle with each other. Play is crucial for physical and social development, especially for cubs. It offers the opportunity to practise physical skills for fighting without being exposed to the dangers of the real thing. Grey Wolves have no natural predators. Apart from the diseases typical of canids, the main causes of death are road traffic accidents and illegal and, in some countries, legal killings. Fights can break out as Grey Wolves defend their territory and these can sometimes result in death.

TRACK

Typical canid structure. The prints of adult males are usually larger than those of females. The footprints of cubs born in spring often reach the size of adult tracks as early as October/ November.

Front

L 8.5–12cm W 6–10cm

Large. Digitigrade. Symmetrical. Five toes. Toe 1 is very short, is located higher up on the inside of the foot and rarely leaves a print, even at high speed. Toe prints 2–5 are reliably recognisable and point forwards. The interdigital pads have fused to form a triangular midfoot pad. The negative space between the midfoot pad and the digital pads often has a kind of X- or H-shape. The H-shape is characteristic of the front foot. In rare cases, a single thenar/ hypothenar pad will leave a print at high speed. Prints made by the strong, sharp claws point forwards and are reliably recognisable. The front foot prints are larger and rounder than those of the hind foot.

Hind

L 7.5–10.5 cm W 5–8cm

Large. Digitigrade. Symmetrical. Four forward-pointing toes that reliably leave prints. The interdigital pads have fused to form a triangular midfoot pad that often leaves an incomplete print. In this track, the front

▲ *Right front. Lausitz, Germany. Helene Möslinger (LUPUS).*

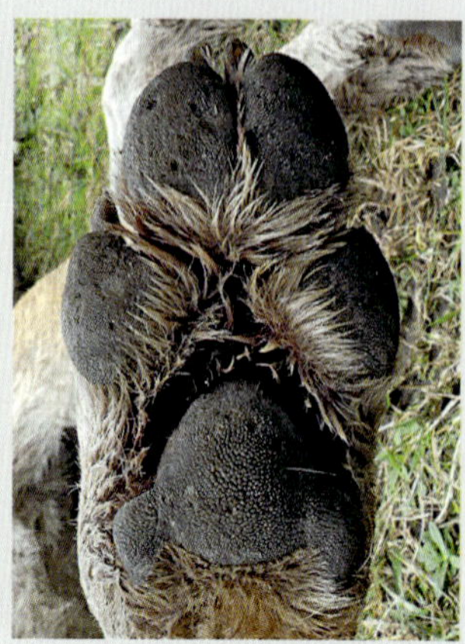

▲ *Right hind. Lausitz, Germany. Helene Möslinger (LUPUS).*

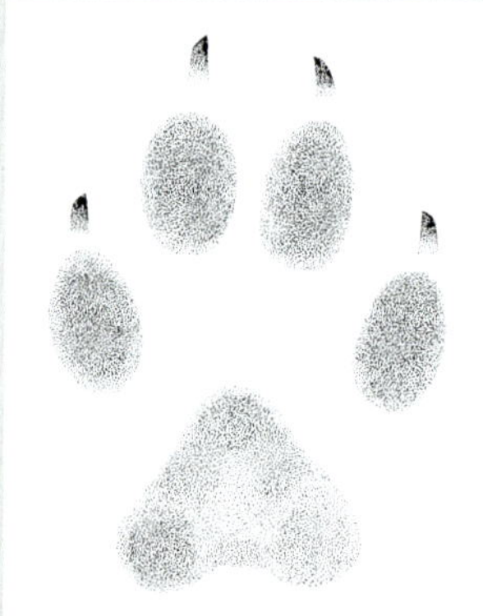

▲ *Right front.*

▲ *Right hind.*

▲ *Right front (bottom) and right hind (top) in comparison. Lausitz, Germany.*

▲ *Right hind. Lausitz, Germany. Helene Möslinger (LUPUS).*

area leaves a deeper print than the outer 'lobes' that are sometimes only faintly visible. The negative space between the midfoot pad and the digital pads has a kind of X-shape. The prints made by the strong, sharp claws are reliably recognisable and point forwards. The hind foot prints are longer and narrower than the front foot prints.

GAITS

To conserve energy, Grey Wolves choose easy routes such as footpaths, forestry tracks, tracks underneath power lines, lake shores and mountain passes. They also use frozen rivers and lakes, ski runs and cleared and snow-covered roads in winter. The animals mainly move at a trot. An even direct register trot with a relatively narrow trail width (called a 'straight-line' trot) over long distances is characteristic. Grey Wolves have an unusual and impressive ability to travel as a pack, one behind the other, leaving a track pattern that looks like the direct register track pattern of a single animal. Each animal steps exactly into the footprints of the one in front. The side trot is also a common gait that usually indicates a slightly higher speed than the direct register trot. Grey Wolves usually overstep in walk.

Walk
Direct register:
SL 60–118cm   TW 12–23cm
Overstep:
SL 74–122cm   TW 14–25cm

Trot
Direct register:
SL 105–158cm   TW 9–19cm
Side trot:
SL 120–174cm   TW 14–24cm

Gallop
Group length: 135–240cm
Inter-group length: 15–160cm
Stride length: 150–400cm
(larger exceptions are possible)
Trail width: 16–22cm

▲ A Grey Wolf in its favourite gait, the 'straight-line' trot. Lausitz, Germany. Helene Möslinger (LUPUS).

▲ The side trot is also a common gait. Lausitz, Germany. Helene Möslinger (LUPUS)

▼ Direct register trot.

▼ Side trot.

▼ Overstep walk.

▼ Grey Wolf tracks in the snow. Slovakia. Immo Meyer.

Similar tracks

Due to the wide variety of dog breeds and their close relationship to Grey Wolves, some Domestic Dog tracks can easily be mistaken for those of Grey Wolves. It is rarely possible to tell the difference between them based on a single track and accounts that claim otherwise are often simplistic and unreliable. Perfect tracks, precise examination of foot morphology, sufficient data and a lot of experience are necessary for reliable identification based on footprints. If possible, tracks and behaviour should also be considered (see page 396).

▲ Grey Wolf tracks in sand.
Lausitz, Germany. Helene Möslinger (LUPUS).

DIFFERENTIATING BETWEEN DOMESTIC DOGS AND GREY WOLVES BY THEIR TRACKS

Domestic Dog footprints come in all kinds of shapes and sizes. The diversity of their tracks can help to distinguish dogs from wolves. If you are familiar with the characteristic features of wolves, you can look for features that deviate from these to recognise dogs. The following information can help you to differentiate between individual prints:

Domestic Dog

a) More compact and wider basic outline.
b) Claws often point in different directions.
c) Distance between midfoot pad and hallux smaller.
d) Digital and metacarpal/metatarsal pads create prints of equal depth (flat).
e) Digital pads are often different shapes and point in different directions.
f) Outer toes usually clearly enclose the halluces.
g) Back edge of the midfoot pad is straight.

Grey Wolf

a) More elongated basic outline.
b) Claws tend to point straight forwards.
c) Distance between midfoot pad and hallux larger.
d) Digital pads normally leave deeper prints that the metacarpal/metatarsal pads.
e) The digital pads are evenly shaped and point forwards in parallel.
f) Outer toes only slightly enclose the halluces.
g) Back edge of the midfoot pad is notched.

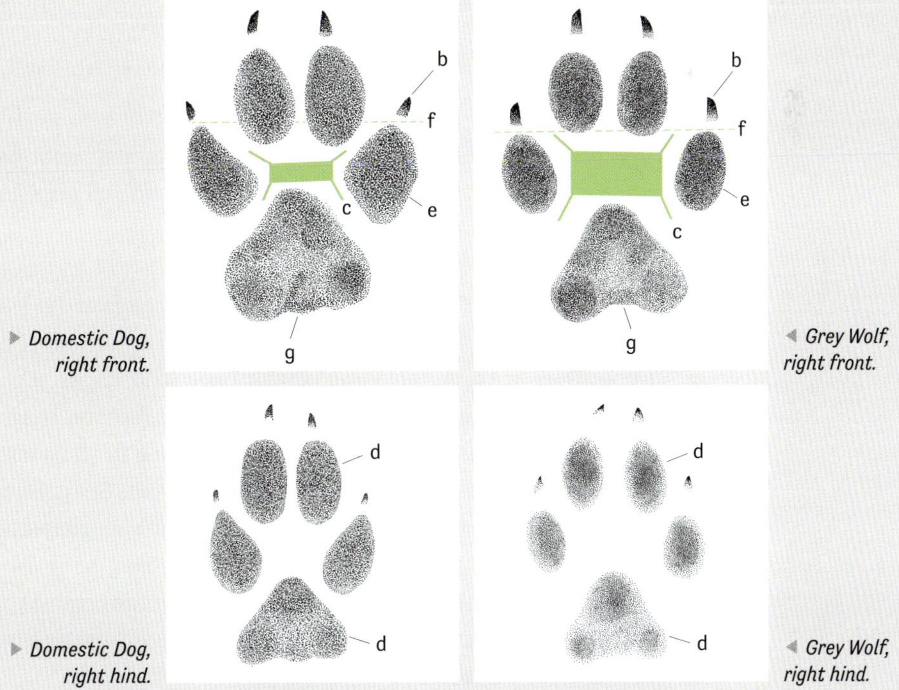

▶ *Domestic Dog, right front.*

◀ *Grey Wolf, right front.*

▶ *Domestic Dog, right hind.*

◀ *Grey Wolf, right hind.*

DIFFERENTIATING BETWEEN DOMESTIC DOGS AND GREY WOLVES BY THEIR TRACK PATTERN

Domestic Dogs and Grey Wolves use the same gaits, but you can sometimes tell the difference between them by their track patterns. The crucial factor here is the different behaviour that life as a domestic or wild animal entails. Grey Wolves generally try to conserve energy. They are purposeful, watchful and leave a correspondingly uniform track pattern. They tend to move purposefully and consistently in one gait over longer distances. Their characteristic gait is the direct register trot.

Domestic Dogs also use the direct register trot but less precisely. Their hind feet repeatedly land in different places, while Grey Wolves place their hind feet precisely in the prints of the front feet, even over longer distances. Overall, the track pattern of Domestic Dogs tends to be more irregular: it shows frequent changes of direction and gait and even continuous gaits are repeatedly interrupted by short bounds. In contrast to Grey Wolf tracks, Domestic Dog tracks appear playful and carefree.

There are also Domestic Dogs whose behaviour is like that of Grey Wolves. The Federal Agency for Nature Conservation in Germany therefore recommends examining a distance of 100–2,000m, depending on ground conditions and print depth, in order to determine the continuity of the features. Young animals make the situation more complicated. The playful behaviour of young Grey Wolves is like that of many Domestic Dogs, so their tracks can show the typical features of dog tracks. Context can be helpful here. Young Grey Wolves usually move in small groups, close to their rendezvous site. Tracks of the parent animals can often also be found in this area. If the tracks run over a longer distance next to human tracks that were created at the same time, it was probably a Domestic Dog.

DISTINGUISHING FEATURES   Like a large dog with a straight back, narrow chest, relatively long, strong legs, broad head and neck and a bushy tail with a black tip.

DISTRIBUTION AND HABITAT   Once found across almost the entire northern hemisphere, Grey Wolves have been wiped out in many parts of Europe. They are adaptable and inhabit open landscapes such as steppe, tundra and the edges of deserts, as well as forests. They can even cope in cultivated areas. Grey Wolves are found in all countries in Europe except the island states of Ireland, Iceland, Great Britain, Cyprus and Malta. We currently distinguish between nine populations (LCIE 2020). Some of these populations are spreading, so it can be assumed that Grey Wolves will continue to spread in the coming years and that genetic exchange between the populations will become more common again in the future.

DIET Grey Wolves are carnivores and mainly feed on even-toed ungulates such as Western Roe Deer, Wild Boar, Reindeer and Western Red Deer. They also occasionally eat small mammals, carrion, birds, insects and frogs. Grey Wolves can consume 1–3kg (up to 10kg) of meat in one meal but can also go hungry for up to two weeks. They build up food stocks when there is a surplus of food, mainly in summer, which they use when food is in short supply.

REPRODUCTION Grey Wolves form more or less constant pair bonds and these pairs often stay together for the rest of their life. The mating season is from December to March and usually peaks in February/March. After a gestation period of 61–64 days, 4–6 pups are usually born. Litters of eight cubs are possible. Grey Wolves usually reach sexual maturity at 1–2 years of age.

SIGNS

DWELLINGS Grey Wolves dig their own dens, renovate and extend fox dens and badger setts or live in natural hollows under rocks and tree roots. Wolf dens are normally well hidden in sandy or loose soil and located less than 1km from the nearest water source. There are often piles of bones and excrement in the immediate vicinity. The cubs are born in an enlarged living chamber at the end of the 1.5–4.5m long tunnel. A Grey Wolf den can have several entrances. The entrance holes can be different shapes and have diameters of around 35–55cm.

▶ *Entrance to a Grey Wolf den.*
Lausitz, Germany.
Helene Möslinger (LUPUS).

▼ The distance between the canine teeth in a bite wound provides important initial clues for identifying the predator. Lausitz, Germany. LUPUS Institute.

RENDEZVOUS SITES When cubs are young, they spend most of their day at the so-called rendezvous site. The parent animals return there repeatedly, especially before and after hunting. The rendezvous site is presumably mainly used for socialising and as a place to play and rest. The rendezvous site is close to the den until midsummer when the cubs become more mobile. There are often many tracks, droppings, bone remains and 'bite toys' for the cubs, for example deer antlers.

MARKINGS Urine and excrement are used as territory markings and for sharing information with other wolves. Droppings and urine markings mainly accumulate near the rendezvous site, around a carcass the animals have killed, near dens, along popular routes and at crossroads and territory boundaries. Occasionally these are accompanied by narrow, short scratch marks, usually with clear ejected material. Grey Wolves often only scrape with their hind legs and rarely their front legs. In a pack, the adult pair normally leaves the marks, usually with one leg raised. The other animals urinate by squatting or without lifting a leg. Oestrus blood may cause females' urine to look orange or reddish when they come into heat.

▶ A bite wound can often only be clearly identified after the fur has been removed. Lausitz, Germany. LUPUS Institute.

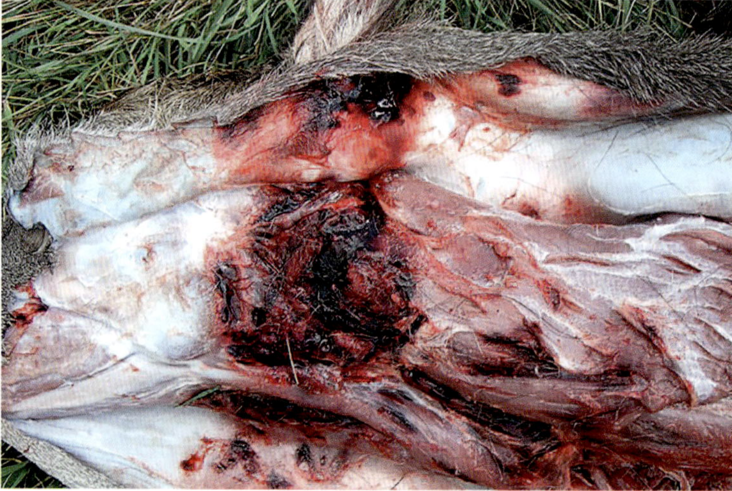

CARCASS FEATURES Grey Wolves hunt by chasing their prey for a few hundred metres until the opportunity arises to capture a single, usually weaker animal. Only about 10 per cent of all hunts end successfully. Grey Wolves usually kill with a bite to the throat called a choke bite. They often grab larger, more dangerous animals, such as Western Red Deer, Wild Boar, Reindeer and Elk, by the hind legs first and bring them down before biting their throat. This usually causes serious damage to the corresponding musculature and even larger bones can be shattered. Body parts, such as the chest, shoulders and flanks, can also suffer serious injuries. Grey Wolves occasionally bite their prey on the nose, causing them to suffocate. Choke bites and other bite wounds often only become visible after the fur or even skin of the dead animal has been removed.

◄ *Choke bite. Bites made by the canine teeth are often well hidden and difficult to see. Lausitz, Germany. LUPUS Institute.*

The carcass is usually opened on the abdomen and occasionally in other places, especially if several animals have eaten from the carcass. Grey Wolves eat the nutrient-rich entrails first, followed by the large haunch muscles and the tender meat on the back. If several animals are involved in a larger kill, they often tear off individual parts of the carcass and eat them at a distance, leaving the remains scattered around the surrounding area.

Unless there is an oversupply of food, Grey Wolves eat almost the entire animal, because the different parts and nutrients are important for their balanced diet. Even the hair provides nutrients and cushions larger bone fragments as they pass through the digestive tract. All that remains of smaller prey are usually scraps of skin, stomach contents and a few bones. If the prey is larger, Grey Wolves leave almost the complete skeleton and the fur. Unlike Brown Bears, Grey Wolves do not usually break open the skull of larger prey to eat the brain. If there is an excess of food, Grey Wolves can detach body parts, carry them away and put them in food stores. Food stores are holes dug in the ground where predators hide their prey by covering it over. Smaller prey is occasionally also hidden whole.

- **Distance between the canine teeth: 4–4.5cm (3.2–5.2cm)**

▶ *Grey Wolves often leave the almost complete skeleton and fur of larger prey. Lausitz, Germany. LUPUS Institute.*

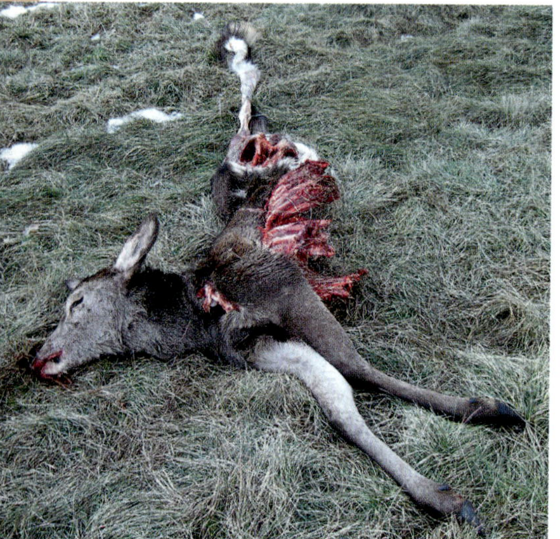

▲ *Grey Wolves eat the nutrient-rich entrails first, followed by the large haunch muscles and the meat on the back. Lausitz, Germany. LUPUS Institute.*

▲ *This typical Grey Wolf dropping is about 40cm long. On the left, you can see that the excrement contains hair. Lausitz, Germany.*

EXCREMENT Grey Wolf droppings are either regular and sausage-shaped with a rounded end and a spindle-shaped end that tapers to a point, or are broken up into individual pieces. It is typically white (bone), brown or black in colour. Black indicates nutrient-rich food, such as meat or organs. If the animal has eaten entrails, excrement may be a dark, shapeless pile. Distinguishing features of Grey Wolf droppings are their distinctive odour and the often high proportion of bone fragments, hair and hoof remains of ungulates (especially the small hooves of young Wild Boar).

The shape, odour and content of droppings greatly depend on the food eaten. Domestic Dog excrement is strongly dependent on the size of the animal and its diet. Shape, colour, size and odour vary much more than in Grey Wolves. Shop-bought dog food results in excrement with a pungent odour and mushy contents, that disintegrates easily. It rarely contains hair or bone fragments. Domestic Dogs rarely feed on wild animals in large enough quantities for their droppings to resemble wolf droppings in shape, colour, odour and contents. Nevertheless, if the diets are identical, the droppings of Domestic Dogs and Grey Wolves can look confusingly similar. In case of doubt, genetic analysis is necessary to identify the species with certainty.

- L 15–65cm, exceptionally 10–85cm (n = 5,085 droppings)
- D 1.3–4.8cm, exceptionally 1–5.1 cm (n = 4,147 droppings)

Own data collection supplemented by Paul Lippitsch and the database of the Senckenberg Museum of Natural Sciences, Görlitz.

REMARKS Grey Wolves were once found in forests across Europe. However, their reintroduction to countries such as Germany is new and has the potential to cause conflict. Supporters and opponents of reintroduction clash because of their polarised opinions. Grey Wolves are carnivores and it is in their nature to hunt but humans are not part of their classic prey pattern. Grey Wolf attacks on humans have not increased, even in areas where Grey Wolves are not hunted.

To make a well-founded assessment of wolf behaviour, it is important to gather as much information as possible, research specialist sources and familiarise yourself with the topic. We are often frightened by the unfamiliar. Eckhardt Grimmberger writes: 'even though around 12 people die in accidents on Germany's roads every day, we have become accustomed to this danger and keep getting in our cars. The probability of encountering a wolf once in your life and then the risk of being attacked by it is infinitely lower, but people are much more afraid of wolves than they are of their next car journey'.

Personally, I am in favour of peaceful coexistence between humans and wolves and I think that a regenerating population of native wild animals is worth striving for. In this context, I would like to emphasise the value of objective debate. The best possible approach for all stakeholders should be decided and continuously reviewed together with qualified experts, such as wildlife biologists, politicians and other scientists.

COMMON RACCOON DOG
Nyctereutes procyonoides

HTL 50–85cm

TL 13–25cm

W 3.5–6kg in
summer, 6–12kg
in winter
No sexual
dimorphism.

The Common Raccoon Dog is the only species of the raccoon dog genus (*Nyctereutes*) in the area this book covers. Common Raccoon Dogs live secretive lives and are rarely observed. They are mainly crepuscular and nocturnal but can also be observed during the day in undisturbed areas. Their sense of smell is important and they have good hearing. Common Raccoon Dogs can swim and dive. They rarely dig their own dwellings and often inhabit abandoned badger setts and fox dens. They occasionally also live in caves or make their home in dense reeds. As the only canids that hibernate, they build up fat reserves beforehand. The length of the hibernation period largely depends on the temperature and duration of the winter. Common Raccoon Dogs in Finland, for example, hibernate from November to March. In places with short, temperate winters, the hibernation period is correspondingly shorter and may even be broken up into intervals. Unlike most other canids, Common Raccoon Dogs do not bark. They are predated by wolves and lynx. Our conclusions about their numbers are mainly drawn from dead animals found at the roadside, and individuals killed by hunting.

DISTINGUISHING FEATURES   About the size of a fox with a shorter snout, short legs, small head and stocky build. In winter, they have noticeably long hair, with badger-like colouring and a raccoon-like face mask.

DISTRIBUTION AND HABITAT Native to East Asia, the species migrated to and colonised eastern Europe from 1928–1955. It now occurs from eastern Europe to France and from Finland to Bulgaria. As a very adaptable non-native species, it continues to spread. Common Raccoon Dogs prefer wetlands, such as moors, floodplains and damp forests with dense undergrowth, but they can also be found in cultivated landscapes and can come close to human settlements when foraging.

DIET Omnivore that rummages and gathers food like a European Badger rather than actively hunting like a Red Fox. On average, 50 per cent of its diet is made up of plant parts, such as acorns, chestnuts, leaves, berries and fruits, and crops, such as maize and potatoes. The proportion of plant-based food is particularly high in summer and autumn. In terms of animal-based food, they mainly eat insects, carrion and small mammals. Amphibians, reptiles, earthworms, snails, fish, birds and eggs make up a smaller proportion. The importance of food sources depends on their availability.

REPRODUCTION Common Raccoon Dogs mainly live in strong pair bonds. The pair raise their young, gather food and defend their territory together. The mating season is from mid-February until the end of April. After a gestation period of 59–64 days, around 7–9 (up to 16) young are born. These altricial young become sexually mature towards the end of their first year. The animals can respond to intense hunting by increasing their birth rate.

SIGNS

BROWSING MARKS Common Raccoon Dogs chew young trees and branches of shrubs in the immediate vicinity of their hibernation site. This is probably a form of territorial marking.

EXCREMENT Common Raccoon Dogs are omnivores, so the appearance of their droppings varies greatly. They may be whitish grey (bones and hair remains), dark green, brown or black (entrails and blood). Droppings are normally cigar-shaped with a thicker middle section, one blunt end and one end that tapers to a short point. Winter droppings are thinner, more kinked and have multiple, smaller segments. When berries and fruits are ripe, droppings can consist almost entirely of their seeds. Common Raccoon Dogs deposit their droppings individually or into hollows they have dug in the ground. These latrines serve as markers, are found in greater numbers near dwellings and are used by several animals. Common Raccoon Dogs eat more plant-based foods than European Badgers or Red Foxes, so the odour of their droppings is often vegetable-like and less pungent. The droppings can otherwise resemble those of a small European Badger.

- **L 2–11cm D 1.7–3cm**

Common Raccoon Dogs are one of the main carriers of rabies in Europe.

▲ *A Common Raccoon Dog latrine. Vledder, Netherlands. René Nauta.*

TRACK

Classic canid structure. Symmetrical. Digitigrade that, on average, puts more weight on the midfoot pad than the Red Fox. Strong digital pads. A notable feature is that toe 4 is often longer than toe 3. The negative space between the digital pads and the midfoot pad forms a wide X-shape or an H-shape. In contrast to the Red Fox, the hair on the soles of the feet is only found between the pads.

Front

L 4.3–6.9cm   W 4.3–7.2cm

Small to medium-sized. Five toes. Toe 1 is short. It sits slightly higher up on the foot and only leaves a print at very high speed or on slippery ground. Toes 3 and 4 are almost in line and have fused at their base (a). Toes 2 and 5 clearly enclose toes 3 and 4 (b). The interdigital pads have fused to form a triangular midfoot pad. A single thenar/hypothenar pad can occasionally be visible at high speed or in deep ground. The negative space between the digital pads and the midfoot pad is significantly smaller than in a Red Fox. Toes 2–4 appear splayed and are arranged in a fan shape from the midfoot pad. On the whole, wider and more asymmetrical than the hind foot print. The outline can look like a horizontal oval. Sharp, strong claws reliably leave prints.

▲ *Right front.*
Germany. Kerstin Krahwinkel.

▲ *Right hind.*
Germany. Kerstin Krahwinkel.

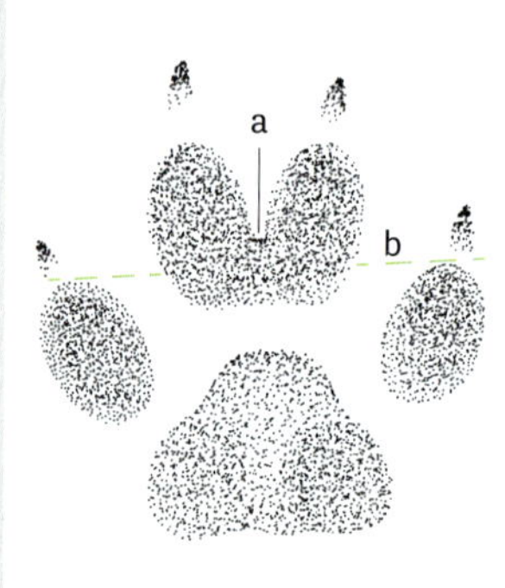

▲ *Right front.*

▲ *Right hind.*

▲ *Right front. Note the length of toe 4. Lausitz, Germany.*

▲ *Right hind.*

Hind

L 4.2–6cm W 3.8–5.8cm

Small to medium-sized. Four toes. Toes 3 and 4 are almost aligned and are slightly enclosed by toes 2 and 5. Both toes are further back than on the front foot, so the negative space between the digital pads and the midfoot pad is larger than on the front foot. Toes 2–4 are less splayed than on the front foot, but more so than in the Red Fox. The interdigital pads have fused to form a triangular midfoot pad. On the whole, the print is narrower (more like a vertical oval) and more symmetrical than the front foot print, but wider and coarser than in the Red Fox. Compared to the Red Fox, the midfoot pad leaves a much stronger print. The strong, sharp claws are reliably recognisable.

GAITS

Like the Red Fox, the Common Raccoon Dog explores at a direct register walk and switches to trot when it wants to cover distances. Because of its short legs and stocky build, it has a conspicuously short stride length and a relatively large trail width. The average stride length in walk and trot is only about half that of a Red Fox but the trail width is twice as wide on average. With a little experience, this distinctive track pattern can easily be distinguished from that of the Red Fox. The animals leave a wide furrow, especially in deep snow. As Common Raccoon Dogs usually travel in pairs, the tracks of two animals hunting together or resting together are often found. When alarmed, they flee by galloping or bounding.

Walk
Stride length: 35–54cm
Trail width: 11–29cm

Trot
Stride length: 52–75cm
Trail width: 7–15cm

Similar tracks
Domestic Dog, Red Fox and Golden Jackal.

▶ *A Common Raccoon Dog balanced along this snow-covered tree trunk. The stride length is short and the trail width is wide compared to the Red Fox. Märkische Schweiz, Germany.*

▲ *Direct register walk.*

▲ *Direct register trot.*

ARCTIC FOX
Vulpes lagopus

HTL 45–73cm
TL 25–50cm
W 3–8.5kg

The true fox genus (*Vulpes*) comprises 12 species, two of which occur in the area. They are canids with pointed snouts, relatively short limbs, large ears and a long, bushy tail. The Arctic Fox is active during the day and at night and has a very well-developed sense of smell and good hearing. It is the only wild dog whose coat colour changes with the seasons. The coat is mainly brownish in summer but normally pure white in winter. Arctic Foxes live in crevices between boulders or in a den they have made themselves. The animals look for ground on riverbanks, by lakes and in elevated areas not affected by permafrost to make their dwellings. Suitable sites can be hard to find, so once established, dwellings are often used for many generations, creating a complex, impressively large tunnel system with many entrances and exits.

Arctic Foxes have a complex social structure. The animals are solitary for six months until they meet their mate in spring. They establish and defend a territory together for about six months before temporarily dissolving their pair bond and driving away their young. Arctic Foxes are usually monogamous and the same pair will come together in spring over several years. There are exceptions where one male mates with two females. In this case, both females suckle the young together. Two pairs have also been observed to share a den and raise their young together. Family groups of up to six animals can occur, especially when food supply is ideal. Arctic Foxes are predated by Grey Wolves, Wolverines and Red Foxes.

DISTINGUISHING FEATURES Smaller than a Red Fox. The shorter legs and muzzle and short, rounded ears reduce heat loss and the risk of frostbite to the extremities. The animals build up fat reserves until autumn, which can increase their weight by up to 50 per cent. The fat serves both as an energy reserve and as insulation through winter.

DISTRIBUTION AND HABITAT The Arctic Fox is mainly found north of the Arctic Circle in Scandinavia, Iceland and Greenland. It inhabits open tundra and mountains, preferably near the coast.

DIET The Arctic Fox is an omnivore. It feeds mainly on small mammals such as voles and, especially, lemmings. If these are available in abundance, Arctic Foxes will hardly eat anything else. They also eat Mountain Hares, ground-nesting birds, eggs, mussels, insects, carrion, waste and berries. Near the coast, seabird and goose colonies can be important sources of food. Arctic Foxes tend to take eggs rather than birds, sometimes in very large numbers. Their impact on populations of bird colonies during the breeding season has been analysed in detail and found to be drastic at times. Arctic Foxes follow Polar Bears to benefit from leftovers of their prey. The animals stockpile food when there is a surplus. Arctic Foxes can play dead to outwit potential prey.

The fur of the Arctic Fox has the best insulating properties of all European land mammals.

TRACK

Typical canid structure. Very large tracks in comparison with other canids and in relation to body size. The scientific name means 'hare-footed fox' and refers to the thick hair on the soles of the feet, which is characteristic of hares. This thick hair often makes the prints of the digital and midfoot pad appear unclear. The individual pads can be difficult to make out in winter due to the thick covering of hair on the soles of the feet. The negative space between the toes and the midfoot pad usually appears very large.

Front

L 5.7–7cm   W 4.4–5.4cm
Medium-sized. Digitigrade. Symmetrical. Five toes. Toe 1 is very short and sits further up on the inside of the foot. The interdigital pads have fused to form a triangular midfoot pad. The negative space between the midfoot pad and the digital pads forms an 'X'. In rare cases, another thenar/hypothenar pad can leave a print at the back of the track. Sharp, pointed claw prints may or may not be visible. The front foot prints are larger and often rounder than those of the hind feet.

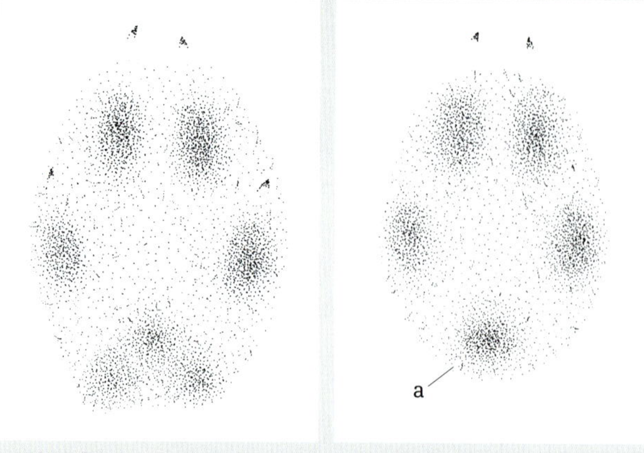

▲ *Right front.*   ▲ *Right hind.*

▲ *Front foot print (top) and hind foot print (bottom). Iceland. Heide Ulrich.*

Hind

L 4–6.3cm W 3–4.7cm

Medium-sized. Digitigrade. Symmetrical. Four toes that reliably leave prints. The interdigital pads have fused to form a triangular midfoot pad that often leaves an incomplete print and looks like another, misplaced digital pad (a). Sharp, pointed claw marks may or may not be visible. The hind foot prints are smaller and narrower than the front foot prints.

GAITS

Arctic Foxes mainly travel in bounds. Over longer distances, they often use the typical canid side trot or direct-register trot. If they are particularly interested in something, they slow down to a walk.

Walk
Direct register:
SL 30–40cm TW 9.5–11.5cm

Trot
Direct register:
SL 46–73cm TW 5.5–10cm

3 × 4 bound (rotary) and 4 × 4 bound (transverse)
Group length: 66–80cm
Inter-group length: 15–25.5cm
Stride length: 81–105.5cm

Similar tracks
Red Fox and Common Raccoon Dog.

▶ *Direct register trot of an Arctic Fox in its preferred habitat.*

▲ *Direct register trot.*

▲ *Side trot.*

REPRODUCTION   The mating season is in spring (March/April). After a gestation period of 49–57 days, 3–9 young are born in a thickly lined underground living chamber. The male provides the mother with food after the birth and also shares the task of feeding the young later on. The young reach sexual maturity after around 10 months. If food supplies are plentiful, young Arctic Foxes remain with their parents even after reaching sexual maturity and form family groups.

SIGNS

DWELLING   The continuous accumulation of decomposing prey remains, excrement and urine around the den results in nutrient-rich soil and lush vegetation. This makes Arctic Fox dens visible from far away, especially in open tundra. The entrance holes are usually 10–20cm high and 15–25cm wide.

DIGGING MARKS   The Arctic Fox digs up rodent nests and tunnels to get to its prey. The digging marks are similar to those made by Red Foxes (page 417).

MARKINGS   Arctic Foxes use excrement and urine to mark their territory, especially at boundaries. They mark conspicuous trees, fence posts, elevations and similar. They regularly visit the same marking sites, which sometimes results in large piles of droppings.

CARCASS FEATURES   See Red Fox (page 418).
- Distance between the canine teeth: 1.9–2.2cm (1.7–2.4cm)

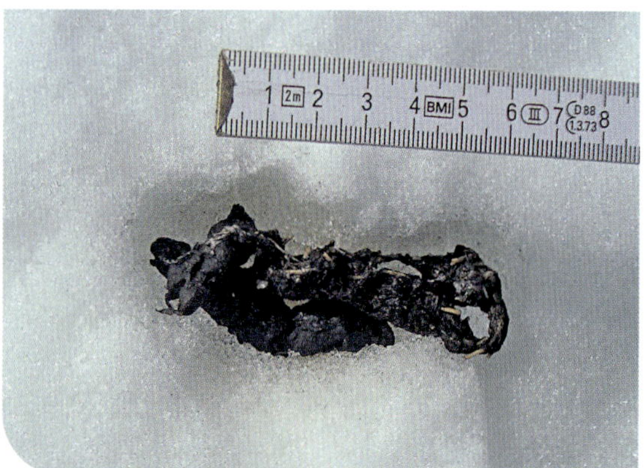

EXCREMENT Arctic Fox droppings are similar to those of the Red Fox, but slightly smaller with less variety of colour and shape due to a less varied diet.
- L 5–11cm D 0.8–1.6cm

◄ *Arctic Fox droppings are often much smaller than those of Red Fox. Helagsfjäll, Sweden. Heide Ulrich.*

RED FOX
Vulpes vulpes

The Red Fox is one of the best-known wild animals in Europe. Many tales tell of their cunning. Phrases such as 'sly old fox' are well founded, as the animals behave cautiously, learn quickly and develop remarkable problem solving and innovative hunting strategies. An example of this is the clever strategy of moving curled-up hedgehogs into water to make them open up. Another equally impressive example is playing dead to attract crows and then killing them in a surprise attack. Red Foxes are mainly crepuscular and nocturnal but can also be observed during the day. They have an excellent sense of smell and hearing and a very good sense of touch. Red Foxes are solitary and pair up during the breeding season. They dig their own den or move into abandoned parts of a badger sett. They are predated by wolves, as well as lynx, Wolverines, and eagles. However, most deaths are caused by road traffic accidents, diseases and parasites.

HTL 50–90cm

TL 30–50cm

W 4–10kg (up to 14.5kg)

Measurements vary greatly depending on season and geographical location.

TRACK

Some distinctive deviations from the typical canid structure. Large, very hairy feet in relation to body size. The prints of the digital pads and the midfoot pad often appear unclear due to the thick hair. The negative space between the toes and the midfoot pad appears very large. An imaginary, horizontal line runs between the inner and outer toes. In contrast to most other canids, this imaginary line, which connects the front edges of the outer toes, is clearly behind the inner toes (a).

Front

L 4.8–7.3cm   W 3.5–5.7cm

Small to medium sized. Digitigrade. Symmetrical. Five toes. Toe 1 is very short and sits further up on the inside of the foot. The interdigital pads have fused to form a triangular midfoot pad. The midfoot pad often appears angular or like a narrow, horizontal line (b). This is a clear distinguishing feature of the species. The negative space between the midfoot pad and the digital pads forms an 'X'. In rare cases, an individual thenar/hypothenar pad can leave a print at the back of the track. Sharp, pointed claw marks may or may not be visible. The front foot prints are larger and usually rounder than the hind foot prints.

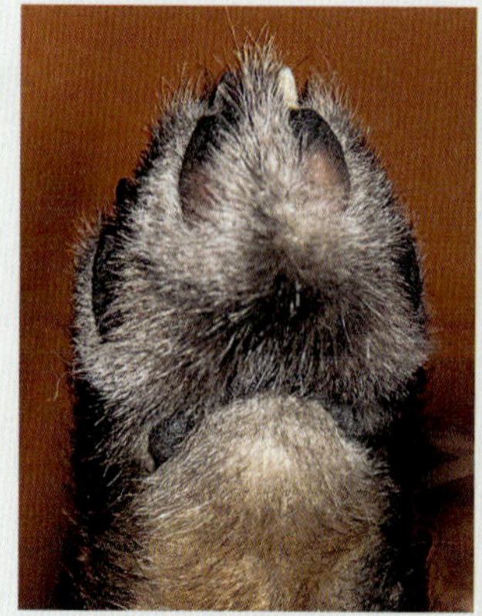

▲ *Right front. Großglockner, Austria.*

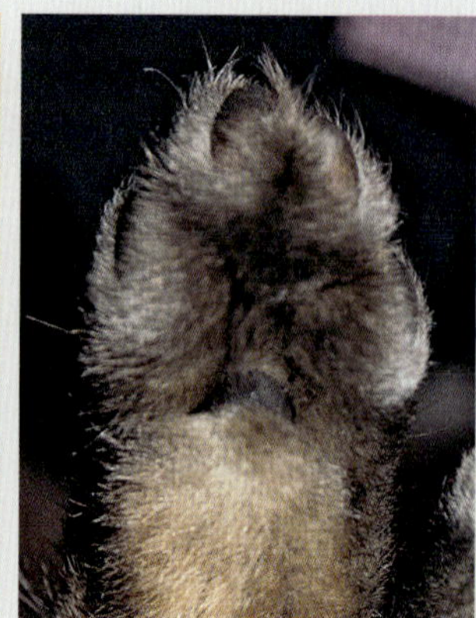

▲ *Right hind. Großglockner, Austria.*

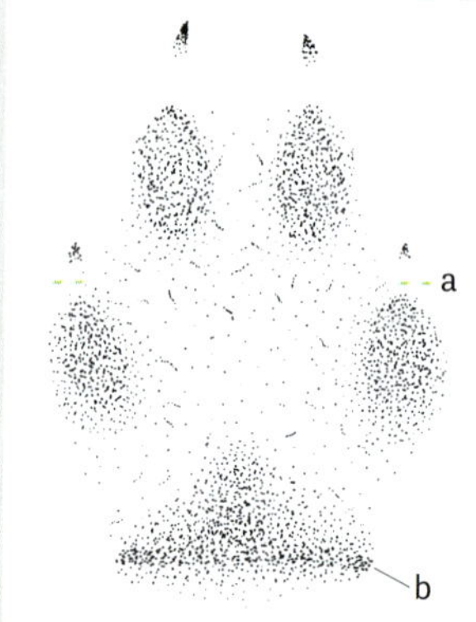

▲ *Right front.*

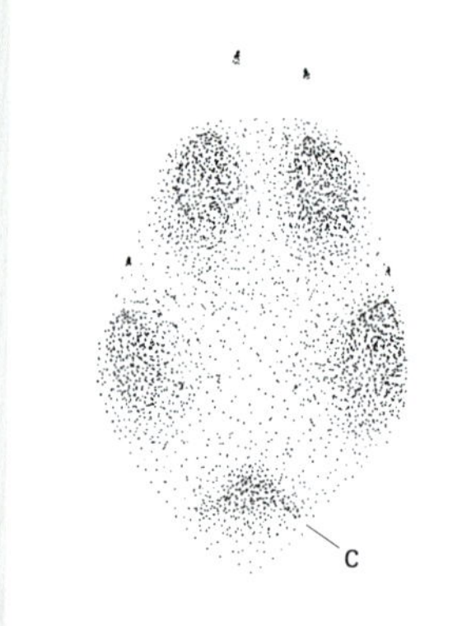

▲ *Right hind.*

▲ *Right front. A perfect print that even shows the rarely visible toe 1. Märkische Schweiz, Germany.*

▲ *Right hind. The midfoot pad can resemble another digital pad. Märkische Schweiz, Germany.*

Hind

L 4–6.3cm   W 3–4.7cm

Small to medium sized. Digitigrade. Symmetrical. Four toes that reliably leave prints. The interdigital pads have fused to form a triangular midfoot pad that often leaves an incomplete print and looks like another, misplaced digital pad Ⓒ. The sharp, pointed claws may or may not leave a print. The hind foot prints are smaller and more elongated than the front foot prints.

GAITS

Red Foxes are known for their 'straight-line' direct register trot. In this track pattern, the double prints are in an almost perfectly straight line because of the narrow trail width. The straight-line trot is the most common gait, but the typical canid side trot is often used for longer distances. The straddle trot is often used over short distances and especially during changes of gait (see page 82). Red Foxes bound and gallop to reach high speed. They switch to a walk to investigate something more closely or to stalk prey. Red Foxes favour the direct register walk in deep snow. They can easily balance along fallen trees, walls and other narrow, raised objects. Their supple flexibility can be reminiscent of cats.

Walk
Direct register: SL 42–61cm   TW 7–15cm
Overstep: SL 45–66.5cm   TW 8.3–12.9cm

Trot
Direct register: SL 58–99cm   TW 5–11.5cm
Side trot: SL 70–124cm   TW 6.5–16cm
Straddle trot: SL 92–100cm   TW 10–16cm

3 × 4 bound (rotary) and 4 × 4 bound (transverse)
Group length: 46–65cm
Inter-group length: 23–46cm
Stride length: 74–111cm
Trail width: 8–19cm

Gallop
Group length: 61.5–188cm
Inter-group length: 25–117cm
Stride length: 74.5–295cm
Trail width: 8.5–26cm

▶ *The Red Fox switches from a direct register trot to a direct register walk in deep snow. This is a straight-line walk along a fallen tree. Märkische Schweiz, Germany.*

◀ *Red Fox changing from a straddle trot to a side trot. Coto de Doñana National Park, Spain.*

▲ *Red Fox in a 4 × 4 bound (transverse). Footfall from bottom to top: LF, RF, LH, RH. Märkische Schweiz, Germany.*

▲ *The side trot is a distinctive gait where both hind feet overtake the front feet on the same side. Märkische Schweiz, Germany.*

Similar tracks

Domestic Dog, Common Raccoon Dog, Arctic Fox and Golden Jackal.

DIFFERENTIATING BETWEEN THE RED FOX AND THE ARCTIC FOX

Red Fox

a) Clear footprints with hair on the soles of the feet.
b) Track outline tends to be more elongated.
c) Angled midfoot pad.
d) Large negative space. The digital pads can appear small in proportion.

Arctic Fox

a) Footprints often appear unclear because of the hairy soles.
b) Track outline tends to be slightly rounder.
c) Midfoot pad not angled but prints of the back outside edge of the midfoot pad can appear as a short, horizontal line.
d) Large negative space. The digital pads can appear very small or even invisible in proportion, especially in winter and due to the thick hair on the soles of the feet.

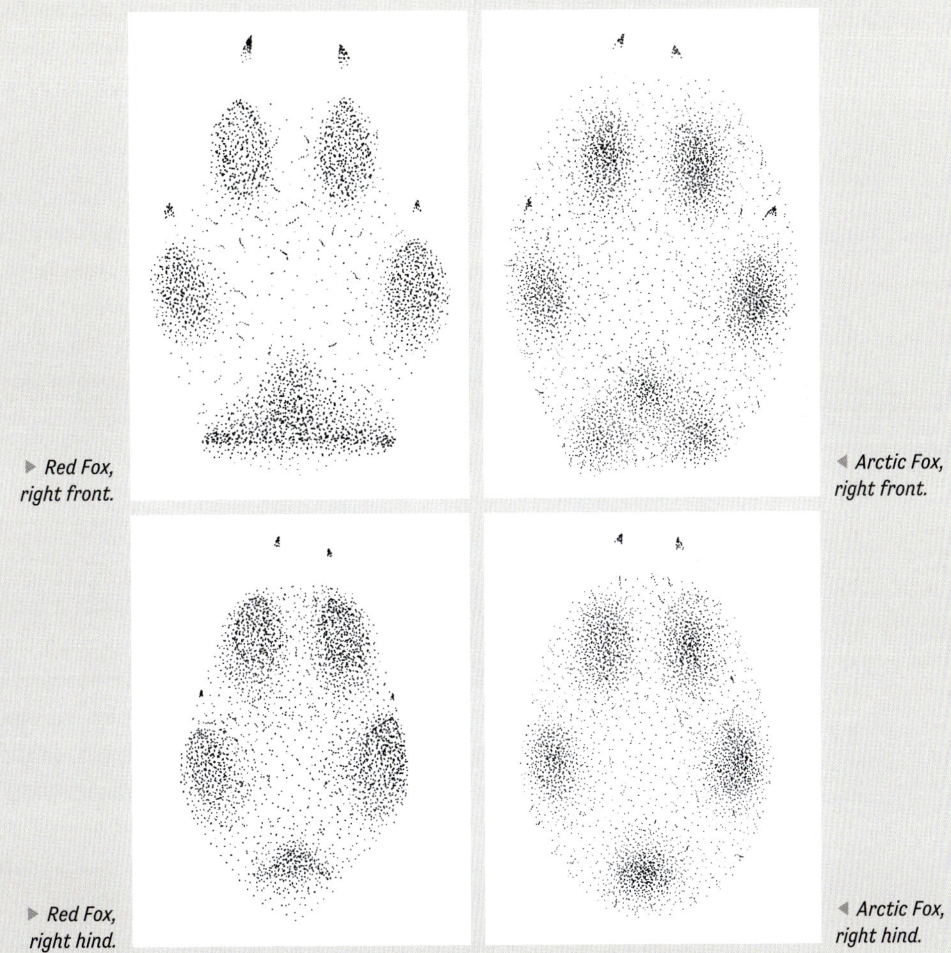

▶ Red Fox, right front.

◀ Arctic Fox, right front.

▶ Red Fox, right hind.

◀ Arctic Fox, right hind.

DISTINGUISHING FEATURES Long, bushy tail. Distinctive pointed ears that are darker on the back.

DISTRIBUTION AND HABITAT The Red Fox is the most geographically widespread of all carnivores and is found throughout the whole of Europe with the exception of Iceland. It is an extremely adaptable species, often associated with human activity, and inhabits forests, meadows, tundra, moors, cultivated areas and cities. Urban Red Foxes prefer gardens, parks, cemeteries and railway embankments.

DIET The Red Fox is an opportunistic omnivore that primarily eats animal-based food. Foxes in many areas specialise in catching voles. Mice often make up more than 50 per cent of their diet. Depending on season and availability, they also eat insects, earthworms, beetles, snails, ground-nesting birds, young birds, eggs, berries, fruit, mushrooms, grass and potatoes, and raid rubbish bins. Red Foxes also prey on deer fawns and domestic fowl. They are unable to keep up with adult hares. The animals stockpile food when there is a surplus.

REPRODUCTION Males make a loud, hoarse-sounding bark during the mating season from January to February. After a gestation period of 51–54 days, 4–5 (up to 13) young are born, usually from April to May. They are born in a well-lined underground living chamber and are raised by both parents. The young reach sexual maturity after around 10 months. When hunting pressure is high, more females take part in reproduction, which often results in an increased birth rate.

SIGNS

DWELLING Red Fox dens are usually located at the edge of a forest, on a south-facing slope or other sunny spot. They usually have several entrances and exits, as well as a lined living chamber, and are recognisable by the bones that pile up outside the den. Unlike European Badgers, Red Foxes cannot turn their paws sideways to dig, so there are no horizontal claw marks from digging out the entrance. Another difference between fox dens and badger setts is the 18–28cm vertical oval entrance hole, which is more of a horizontal oval in badger setts. Fox dens also lack the 'channel' that is typical of badger setts.

▶ *The slightly sloping position and the vertical oval shape of the entrance hole are characteristic of a Red Fox den. Offenbach, Germany. Simone Roters.*

▼ *Unlike Red Foxes, European Badgers can turn their paws outwards to dig sideways. This is the straight digging mark of a Red Fox. Lausitz, Germany.*

▲ *The entrance hole of a Red Fox den in a field. Maintal, Germany. Simone Roters.*

▲ *Red Foxes cache eggs in the ground to eat at a later time. West Sussex, England.*

DIGGING MARKS   Like typical canids, Red Foxes dig rapidly with their front feet to get to food, such as the nests of small rodents. Due to the parallel position of their front legs, the predominantly straight pile of ejected soil is located centrally underneath the animal's body. The resulting sign is a pile of soil with the length of an approximately 10–15cm deep hole. The hole in the ground and the ejected material are long and narrow compared with European Badger digging marks. The precise way Red Foxes dig for prey is reflected in the overall impression of the digging marks, which tend to be small and neat.

▲ *A classic sign: a Red Fox marking. The short 'detour' on an otherwise straight path is often a sign of a marker. Märkische Schweiz, Germany.*

MARKINGS Red Foxes mainly mark territory boundaries with excrement and urine. They seek out conspicuous trees, fence posts, elevations and similar that they regularly visit to renew the markings. The resulting curved and irregular track pattern is distinctive and can often give a clue to a marking site even from a distance. Note: Males do not always lift a leg and females do not necessarily squat to urinate. Fox researcher J. David Henry (1993) describes how Red Foxes can adopt 12 different positions for urinating, depending on where the scent mark is to be left. Males generally urinate clearly towards the centre of the body, far in front of their hind feet. Females, on the other hand, urinate almost exactly between their hind foot prints.

CARCASS FEATURES Red Foxes stalk their prey and kill it after a few bounds of up to 5m. They do not chase down their prey. They repeatedly bite an easily accessible part to bring a larger animal to the ground, usually the legs, flanks and abdomen. Once the animal is on the ground, the fox kills it with numerous bites to the neck area. These leave many small puncture wounds in a sieve-like pattern in the skin, resembling holes from a shotgun. Red Foxes usually open the carcass at the belly and eat the digestive tract and entrails first.

This is an important distinguishing feature from cats which do not normally eat entrails. They typically also tear off body parts such as the head and legs, carry them away and hide them in food stores. Food stores are holes dug in the ground, where they hide their prey by covering it over.

- **Distance between the canine teeth: 1.9–2.2cm (1.6–2.7cm)**

EXCREMENT Red Fox droppings are either sausage-shaped with one rounded end and one spindle-shaped end that tapers to a point, or broken up into individual pieces. Shape and colour vary greatly depending on the contents. Typical colours are white (bone), reddish, brown and black. A black colour indicates nutrient-rich food such as flesh or entrails. Eating entrails often results in a dark, shapeless pile of excrement. In summer and autumn, fox droppings can contain the remains of fruit and lots of fruit seeds. The remains of insects, such as beetle wing cases, and waste are also not uncommon. Fresh Red Fox droppings smell very strong and unpleasantly pungent to humans (see box on page 137). The droppings can occur in atypical shapes and sizes. If it is not clearly identifiable, search for any surrounding signs for reliable identification, such as the placement of droppings or tracks. Red Foxes often leave droppings on exposed areas as scent markers. These can be tree stumps and other raised sites, but they also defecate in the middle of tracks or paths.

- **L 3–12cm (up to 20cm) D 1–2cm (up to 3cm)**

Caution!

Red Fox excrement can contain the fox tapeworm (*Echinococcus*), which is also dangerous for humans. Infections are rare and the risk of infection is low. In Europe, it is mainly found in southern Germany, northern Switzerland, western Austria and eastern France.

▼ *Red Fox droppings are a favourite food for dung beetles. The distinctive placement next to a tuft of grass is characteristic. Märkische Schweiz, Germany.*

▲ *Red Fox droppings vary greatly and can have many different components. Here, we can make out bones, hair, insect remains and seeds. Spessart, Germany.*

Bears

The bear family (Ursidae) consists of five genera with eight species, of which only two species of the genus *Ursus*, the Brown Bear (*Ursus arctos*) and the Polar Bear (*Ursus maritimus*), occur in Europe.

We will not be discussing the Polar Bear further here due to its low numbers in Europe. Bears are the largest land carnivores in Europe. They have a compact build with strong limbs, dense fur, small, rounded ears, small eyes and a barely visible stumpy tail. Bears have an excellent sense of smell and hearing, but their sight is relatively poor.

Bears are omnivores. Adult males are much larger and stronger than adult females – an expression of sexual dimorphism. Due to their eating habits, large size and species-typical territorial behaviour, bears can leave large numbers of impressive and unmistakable tracks and signs.

TRACKS AND SIGNS OF BEARS

Plantigrade. Asymmetrical. Five toes on the front and hind feet. Bear tracks are usually noticeably large and evenly flat. In comparison, digitigrades tend to put more weight on their toes, so the toe prints are deeper. Bear tracks can resemble human footprints. In bears, the smallest toe is the inner toe 1, whereas in humans the outer toe 5 is the smallest. Claw marks are usually visible.

- **Track formula: 5F × 5h + C**

FRONT FOOT Asymmetrical. Toes 2–4 are arranged symmetrically and reliably leave a print. Toe 1 is not reliably visible. The front feet are proportionally wider and generally appear stronger than the hind feet. A single thenar/hypothenar pad may be visible in deep prints or at high speed.

HIND FOOT Asymmetrical. If the whole foot leaves a print, the hind foot print is much longer than the front foot print.

GAITS The most common gait is a ponderous, overstep walk. Other gaits are used according to the situation.

SIGNS Bears usually leave many conspicuous signs. They eat sedges, grasses and shrubs, as well as insects, larvae and meat. This results in various signs associated with foraging, such as digging marks, broken deadwood trunks or the remains of raided wasp nests. Conspicuously wide bear paths are common. The relatively low belly often flattens down the vegetation.

BROWN BEAR
Ursus arctos

HTL 150–300cm

TL 8–14cm

W 100–250kg (up to 340kg)

Weight varies greatly depending on range and before and after hibernation. Males are much larger and heavier than females.

Brown Bears have an excellent sense of smell and good hearing. Their rather poor eyesight plays a secondary role. They are good swimmers and can run at speeds of up to 50km/h over short distances. These shy, solitary animals are mainly crepuscular and nocturnal, but they can also be observed during the day in undisturbed areas. Brown Bears build up considerable fat reserves before retreating into underground dwellings or caves for a 2–6 month hibernation period between October and March. The exact duration of hibernation is mainly determined by the ambient temperature. In contrast to true hibernation, body temperature only drops slightly and there can be frequent interruptions. By the time hibernation ends in spring, bears can have lost up to 40 per cent of their body weight. Brown Bears have no predators.

DISTINGUISHING FEATURES Unmistakable strong build with a distinctive hump on the back of the neck and a broad head with a long muzzle, short ears and small eyes.

DISTRIBUTION AND HABITAT Originally widespread throughout almost all of Europe. European Brown Bears avoid human activity and are now mainly found in densely forested low-mountain regions, in isolated populations.

DIET Opportunistic omnivore that mainly feeds on plant foods. Their diet is highly dependent on season and food supply. Berries, fruits, roots, seeds, nuts, fungi and grains, as well as insects, invertebrates, rodents, birds and their eggs, can make up a large part of their diet. Brown Bears like to fish, and honey is also a popular food. They turn over large stones and tree trunks to get to ants and insect larvae. Like European Badgers, Brown Bears dig out bee and wasp nests. They eat more grasses, leaves and herbs in spring. However, if the opportunity arises, they will also kill the young of deer and other large mammals. Brown Bears may also prey on domestic animals like sheep.

REPRODUCTION The mating season usually lasts from April to July. After fertilisation, an embryonic diapause of 4–5 months sets in. Active pregnancy lasts around 6–8 weeks. Young are born during the hibernation period, with usually two (very rarely one or 3–5) cubs born between January and March. The extremely altricial cubs are born blind and almost hairless in the female bear's lined hibernation den. It takes 3–4 years for the young to reach sexual maturity.

Brown Bears are strictly protected. When they come into proximity with humans, conflicts can arise. These are partly due to a lack of experience of dealing with each other. Brown Bear attacks on people are rare in European countries where bears and people have lived alongside each other for many years.

SIGNS

MARKINGS Brown Bears mark by rubbing against trees, scratching the bark with their long claws and biting bark and thinner branches. Alternatively, they may also use wooden telegraph poles, fence posts or similar. The characteristic wavy hairs with silver tip are often found there. Marking sites are usually found near resting places, along popular routes or near favourite feeding grounds. Bears use these markings to communicate information, for example about their physical condition, territory boundaries and sex.

TRACK

Front

L 13–31cm   W 10–21cm

Very large. Plantigrade. Asymmetrical. Hairless sole. The outline is usually short and wide. Five toes. Their prints are very close together and arranged in a slight curve in front of the midfoot pad. The negative space between the midfoot pad and the digital pad can look like a distinctive wall (a). Toe 1 is the smallest toe and cannot be reliably recognised. The interdigital pads have fused to form a large midfoot pad that is much wider than it is long. A roundish thenar/hypothenar pad can be visible at the back of the track. The large claws almost always leave a print and are sometimes very long. Claw length varies.

Hind

L 18–35cm   W 11.8–19cm

Very large. Plantigrade. Asymmetrical. Hairless sole. Five toes. Their prints are very close together and arranged in a slight curve in front of the midfoot pad. Toe 1 is the smallest toe. The interdigital pads have fused to form a large midfoot pad. It is connected to a thenar/hypothenar pad by a thick layer of skin (b).

Strong, usually visible claw marks, which are significantly shorter than on the front foot. Without claw marks, the track resembles a large human footprint. In humans, however, toe 5 is the smallest and toe 1 is the 'big toe'.

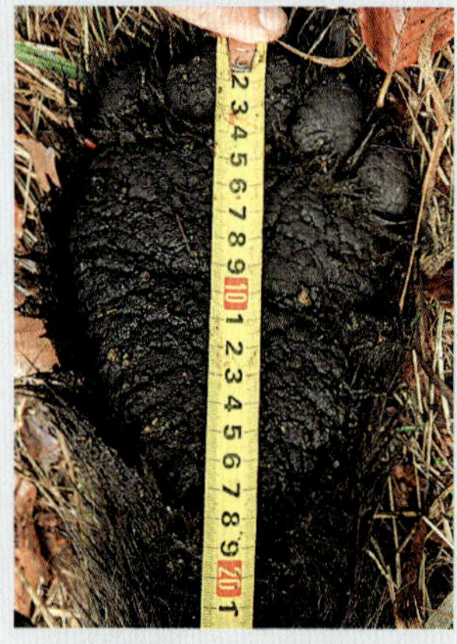

▲ *Right front. High Tatras, Slovakia. Michaela Skuban.*

▲ *Brown Bear, left hind. High Tatras, Slovakia. Michaela Skuban.*

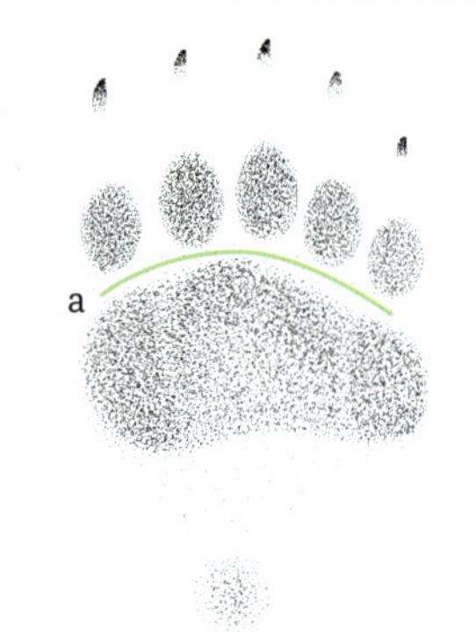

▲ *Left front.*

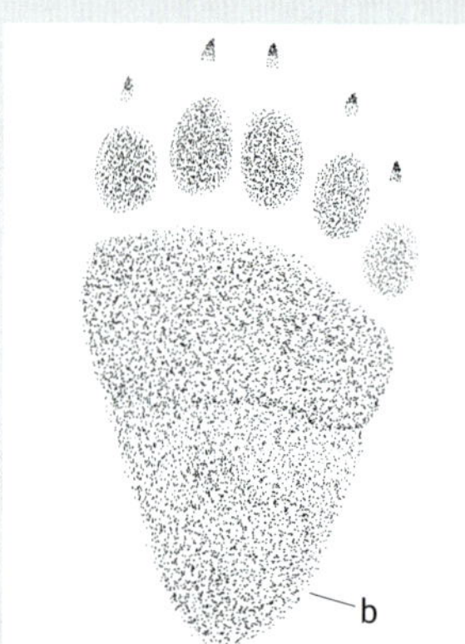

▲ *Left hind.*

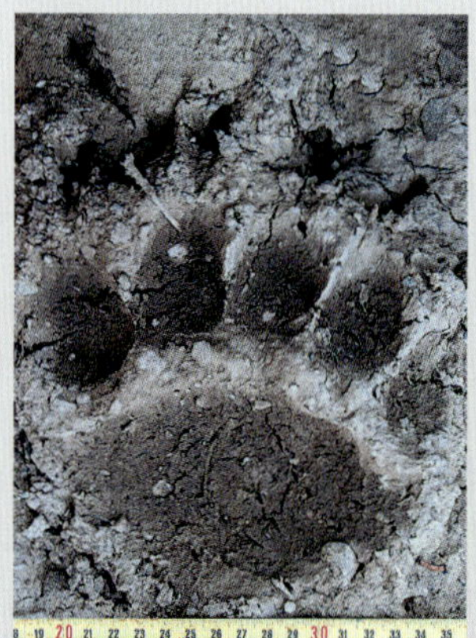

▲ *Left front. Bieszczady, Poland.*

▲ *Left hind foot print, the front foot print beneath is slightly covered. Bieszczady, Poland.*

GAITS

Brown Bears have a leisurely walk where they move both legs on one side almost simultaneously (amble). They often overstep and their feet typically turn in. They prefer the more energy-efficient direct register walk in deep snow. The animals move in rapid bounds or at a gallop when alarmed or in pursuit of prey.

Walk
Direct register:
SL 90–140cm
TW 35–50cm
Overstep:
SL 128–195cm
TW 22–45cm

Gallop
Group length:
200–232cm
Inter-group length:
72–90cm
Stride length: 271–322cm
Trail width: 30–50cm

Similar tracks

Brown Bear footprints are unmistakable in Europe. The tracks of young animals can be mistaken for European Badger or Wolverine tracks.

▶ *Brown Bear in its preferred gait, the slightly overstepping walk. Bieszczady, Poland.*

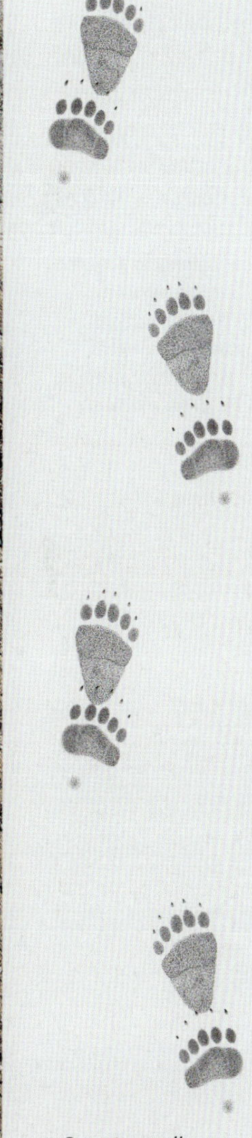

▲ *Overstep walk.*

▲ *Similar to European Badgers, Brown Bears dig out and raid wasps' nests. High Tatras, Slovakia. Michaela Skuban.*

DIGGING AND FORAGING MARKS   Brown Bears use their long claws to dig for roots, tubers, small rodents or insects. Digging tracks are often found in meadows, where Brown Bears have lifted and dug over clumps of turf about the size of a front foot. Brown Bears revisit preferred digging sites over years, making the grassy landscape look churned up and uneven. They dig over larger clumps of turf in comparison with Wild Boar.

Brown Bears can break open and hollow out rotten tree trunks with their claws when searching for beetle larvae and insects. Fan-shaped ejected material in the direction of the animal is usually recognisable. Foraging Black Woodpeckers leave signs that can be mistaken for those of Brown Bears (page 713) and European Badgers can show similar behaviour (page 158). Unlike European Badgers and Black Woodpeckers, Brown Bears can roll or turn over even massive logs to reach interesting feeding sites. If the site shows signs that such strength has been used, Black Woodpeckers and, depending on the weight of the tree trunk, European Badgers can be ruled out.

CARCASS FEATURES   Brown Bears kill their prey in their typical way, by seemingly random blows with their paw to the nose, neck and back or by bites to the neck and back. Massive injuries and large claw marks are usually clearly visible. When Brown Bears have killed a larger herbivore or carried off a carcass, they usually begin by eating the inside of the ruminant's stomach, entrails and stomach contents. One explanation for this behaviour, which is otherwise not typical of carnivores, is that ingesting the fermenting bacteria present in ruminant stomachs helps Brown Bears to digest fibre-rich plant food.

Brown Bears will occasionally tear apart a carcass and individual parts can be spread over a larger area. They avoid eating skin and bones. The fur is usually found intact and inside out next to the skeleton or connected to the nose. If there is a surplus of food in autumn, Brown Bears sometimes only eat nutrient-rich body parts, such as the brain and entrails, and leave the rest of the prey behind. If Brown Bears want to store a carcass for later, they cover it with leaves and surrounding material. They have also been seen to store carcasses in running water.

- **Distance between the canine teeth: 5.7–7cm (5–8.5cm)**

EXCREMENT   As is typical for omnivores, the appearance of droppings depends greatly on the food consumed. Brown Bear droppings usually have blunt ends and break apart easily. Excrement that consists mainly of entrails and meat is usually a shapeless, dark pile that disintegrates quickly. When fruits and berries are ripe, droppings can consist almost exclusively of their remains. In this case, excrement is usually either a shapeless pile of poorly digested fruit, or sausage-shaped droppings consisting of tightly packed seeds. They also leave large piles of droppings containing the remains of plants, such as beechnuts, seeds and leaves. Animal remains such as fur and bones are rarer, in which case the individual segments decompose much more slowly. Brown Bears defecate frequently and seemingly indiscriminately. Piles of droppings are found near resting places and preferred feeding sites.

- **L 15–50cm D 3–7cm**

▶ *Cherry stones are clearly visible in these Brown Bear droppings. Bieszczady, Poland.*

Mustelids

The mustelid family (Mustelidae) comprises 22 genera with 59 species, 12 of which occur in the area this book covers. These very small to medium-sized carnivores usually have a slender, elongated body with short legs, and a short snout and small ears.

The European Badger and the Wolverine are the more sturdily built exceptions, with a longer snout. Most mustelids have scent glands in the anal area, which they use for territory marking, communication and sometimes for defence. Mustelids are energetic and curious. Their sense of smell and hearing is well developed but their sight is rather weak. They are predominantly carnivores, but some can also eat plant-based food, depending on the season. Adult males can be up to 60 per cent larger than adult females (pronounced sexual dimorphism).

TRACKS AND SIGNS OF MUSTELIDS

Plantigrade. Asymmetrical. Five toes on the front foot and hind foot. The claws cannot be retracted so claw marks are usually recognisable. The toes protrude asymmetrically from an arched midfoot pad that is notched on the proximal side. The print of the midfoot pad resembles a croissant in shape and usually consists of four clearly separate interdigital pads. The outer metacarpal/metatarsal pad (IV) is stronger than the tapered, inwardly curved part of the midfoot pad (metacarpal/metatarsal pad I). It is located slightly further back in the track and is often faint or does not leave a print. The two highly specialised species – Eurasian Otter (swimming) and European Badger (digging) – deviate from the typical foot morphology described here. Dimensions can vary greatly due to pronounced sexual dimorphism. The maximum values for a species indicate a male. Foot morphology needs to be analysed in detail to eliminate confusion with the next largest species. Clear identification can be difficult.

- Track formula: 5F × 5h + C

FRONT FOOT The front feet are proportionally wider and generally appear stronger than the hind feet. Another thenar/hypothenar pad (H-pad) can leave a print at the back of the track. Toe 1 does not reliably leave a print but does so more often than on the hind foot.

HIND FOOT The hind foot is more asymmetrical than the front foot. Toe 1 and metatarsal pad I are often not visible, in which case the hind foot appears more symmetrical than the front foot. Hind foot prints without toe 1 may resemble the hind foot print of a small canid.

GAITS Most mustelids mainly move in various forms of bound. The preferred gait depends greatly on the situation and the species. In principle, all mustelids can use all gaits. The use of a constant gait or frequent gait changes can provide important clues for species identification. Many mustelids explore at a walk. European Badgers often trot. The 2 × 2 bound is characteristic of mustelids, except for European Badgers. This leaves a track pattern that looks like a series of slightly offset track pairs. However, these are actually double prints of the front and hind pair of feet, which have landed on top of each other (see page 87). The trail width of a 2 × 2 bound track pattern can give a further clue to the type of mustelid.

EXCREMENT Mustelid droppings are sausage-shaped and usually kinked with pointed ends. However, the shape strongly depends on the animal's diet. If they consume a lot of liquid and little fibre (e.g., entrails or earthworms) or a lot of fruit, the droppings are often quite loose or shapeless. Droppings containing lots of small fruit stones are a typical characteristic of Pine and Beech Martens (see pages 456 and 463), as they are the only mustelid species to consume large quantities of fruit.

EURASIAN OTTER
Lutra lutra

HTL 50–95cm
TL 28–55cm
W 5.5–15kg
Sexual dimorphism
much less
pronounced than
in other mustelids.
Males slightly larger
than females.

The otter genus (*Lutra*) comprises two or three species, one of which occurs in the region. Otters are excellent, agile swimmers. In clear waters, they use their excellent eyesight to hunt under water. In poor visibility, otters use their many long, stiff whiskers (vibrissae) to track down prey. Otters are both diurnal and nocturnal. In areas close to humans, they shift their activity from sunset to morning. When undisturbed, the animals tend to be active during the day. Otters sleep in above-ground daytime hiding places under dense vegetation, in caves or under tree roots. To give birth to and rear their young, otters retreat into dwellings that they dig themselves or take over from Red Foxes, Muskrats and Eurasian Beavers. The entrance to these dwellings is usually above the surface of the water and the living chamber is lined with plant material. Otters are usually solitary. However, communal territories can also occur on seashores with a high food supply. The maternal family can stay together for just over a year. Otters have virtually no predators. In rare cases, they are preyed upon by wolves, lynx, White-tailed Eagles or Wolverines. Mortality is largely due to habitat destruction by humans.

DISTINGUISHING FEATURES   Slender, elongated body and short legs. Flat, broad head. Tail is thick at the root and tapers evenly towards the tip.

DISTRIBUTION AND HABITAT   Apart from a few exceptions, such as the Mediterranean islands and Iceland, Eurasian Otters can be found throughout Europe, although usually only in isolated populations. Eurasian Otters need clean water with naturally overgrown, wooded and undisturbed banks. They prefer fast flowing water to stagnant water but can also be found in still lakes in the forest, marshy areas or on the seashore. They will readily travel long distances over land to reach a suitable habitat. Their dependence on water makes Eurasian Otters vulnerable to changes in water availability and quality, disturbance from water sports, and agricultural pollution.

DIET   Eurasian Otters are mainly opportunistic carnivores whose diet varies according to the season. If available, they will almost exclusively eat fish. In winter, when waters freeze over and fish are difficult to catch, frogs, toads, crayfish and mussels can make up a large part of the diet. In rare cases, they will eat smaller mammals such as water voles, shrews or Muskrats, as well as ground-nesting birds (mostly waterfowl) and their eggs.

REPRODUCTION   Eurasian Otters are not restricted to a fixed mating season, but in Europe they usually mate from February to April. Their mating behaviour appears playful. The gestation period is 58–66 days. As a rule, 2–3 (up to 5) young are born in March/April. Females give birth once a year or every other year. The altricial young are born in a lined dwelling and do not start swimming until they are 2–3 months old. They reach sexual maturity after 17–22 months at the earliest.

REMARKS   Eurasian Otters compete for food with the American Mink (*Neovison vison*). They hunt, kill and eat this smaller relative and destroy or cover up its scent markings. Eurasian Otter populations declined significantly in the twentieth century due to pollution and hunting. The species is listed as Near Threatened on the IUCN Red List.

TRACK

Front

L 4.9–9.5cm W 5–8.2cm
(W < 4.5cm are young animals)
Medium-sized. Plantigrade.
Asymmetrical. Five toes, which
usually appear strong, oval and
teardrop-shaped in the track.
Toe 1, the smallest, is not always
easy to recognise and is almost
opposite toe 5 (a).

The webbing between the toes
only occasionally leaves a print.
The interdigital pads have fused
to form a large midfoot pad.
Another thenar/hypothenar pad
that is slightly offset to the back
can leave a print at the back of
the track. In contrast to other
mustelids, the front foot is smaller
than the hind foot, yet it usually
appears stronger and wider.
The claws do not regularly leave
prints.

Hind

L 4.8–11cm W 5–8.5cm
Medium-sized. Plantigrade.
Asymmetrical. Five toes, which
usually appear strong, oval and
teardrop-shaped in the track.
Toe 1 is well developed and sits
much further back than the front
foot (b). This makes the hind
foot more asymmetrical. The
webbing between the toes only
rarely leaves a print. If toe 1 is
not visible, the hind foot print
appears more symmetrical than
the front foot print and may
resemble the footprint of a canid.
The interdigital pads have fused

▲ *Left front.*
Heide Ulrich, Germany.

▲ *Left hind.*
Heide Ulrich, Germany.

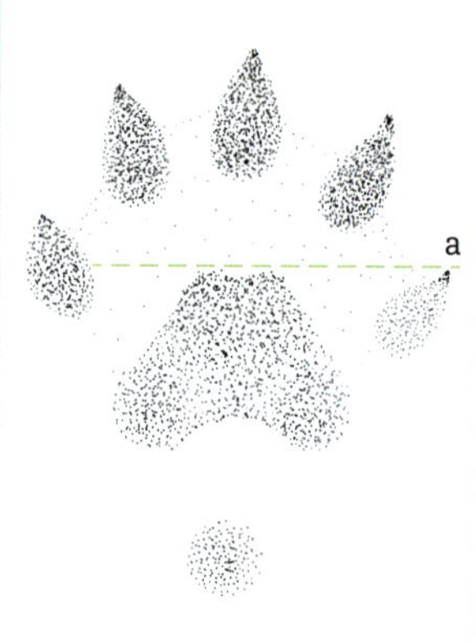

▲ *Left front.*

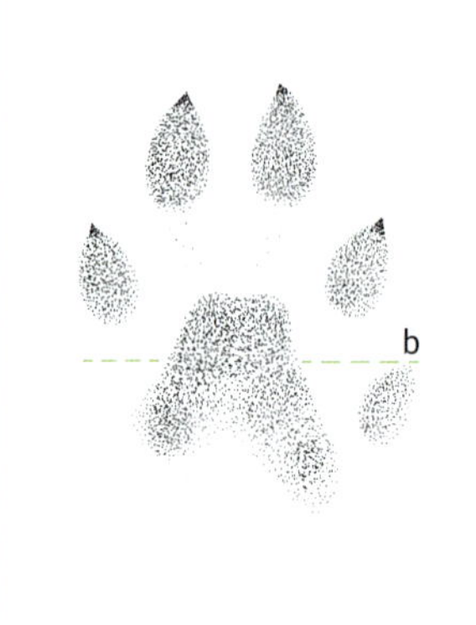

▲ *Left hind.*

▲ *Left front. Lausitz, Germany.*

▲ *Left hind, toe 1 is clearly offset
to the rear. Lausitz, Germany.*

to form a large midfoot pad, which extends further back on the inside of the track in a complete print. No thenar/hypothenar pad. If the thenar/hypothenar pad of the front foot is not visible, the hind foot print will be larger than the front foot print. Nevertheless, the hind foot print usually appears longer and narrower. The claws do not regularly leave prints.

GAITS

Eurasian Otters move at a walk when exploring. When travelling, they prefer the 2 × 2 bound, 3 × 4 bound (rotary) and 4 × 4 bound (transverse) or a lope. The Eurasian Otter is one of the few mammals capable of travelling long distances in a 3 × 4 bound (rotary). They gallop if they need to move at higher speed.

Walk
Stride length: 31–66cm
Trail width: 12–21cm

2 × 2 bound
Group length: 11–24cm
Inter-group length: 27–94cm
Stride length: 39–115cm
Trail width: 12–18cm

3 × 4 bound (rotary) and 4 × 4 bound (transverse)
Group length: 22–55cm
Inter-group length: 27–65cm
Stride length: 51–120cm
Trail width: 10–26cm

Gallop
Group length: 23–65cm
Inter-group length: 37–66cm
Stride length: 71–133cm
Trail width: 15–25cm

Similar tracks
Large Beech Marten tracks and footprints of Northern Raccoons in which the digital pads have not printed all the way to the midfoot pad.

◀ *A Eurasian Otter in a direct register walk. Kreba-Neudorf, Germany.*

◀ *The 4 × 4 bound (transverse) is a classic Eurasian Otter track pattern. Footfall from bottom to top: RF, RH, LF, LH. Kreba-Neudorf, Germany.*

▼ *Parallel bound. Footfall from bottom to top: LF, RF, LH, RH. Kreba-Neudorf, Germany.*

▼ *Lope.*

▼ *Direct register walk.*

3 × 4 bound (rotary).

▲ This distinctive track pattern is typical of Eurasian Otters and indicates a lope or 4 × 4 bound (transverse). Märkische Schweiz, Germany.

▲ Eurasian Otters often cover long distances in a mix of 3 × 4 bound (rotary) and 4 × 4 bound (transverse). The angled position of the groups of four is distinctive and characteristic. Märkische Schweiz, Germany.

SIGNS

DUST BATHS   Eurasian Otters roll on the ground to groom their fur (see page 177). This cleans and dries the coat and helps to maintain its insulating properties. They often leave scent marks and scratch marks when rolling.

SCENT MARKING   Eurasian Otters use a gelatinous, strong-smelling anal secretion and droppings to scent-mark distinctive sites. They mainly leave scent marks in centres of activity, on elevations such as large rocks or tree stumps, next to entrances and exits on banks, under bridges and at intersections of important paths.

▼ Eurasian Otters roll and scratch the ground to groom their fur and leave a scent mark. Oderbruch, Germany.

▼ Eurasian Otters use a mixture of excrement and anal secretions for scent marking. Spreetal, Germany.

▶ Eurasian Otters scrape together mounds of earth in prominent places for scent marking. They then leave remains of excrement, urine and anal secretions on the mound to emphasise the odour. Kreba-Neudorf, Germany.

▼ *With a little practice, you can distinguish Eurasian Otter paths from the paths of other animals. Lausitz, Germany.*

▼ *Entry and exit point of a Eurasian Otter. Märkische Schweiz, Germany.*

▶ *Feeding sites like this are typical of Eurasian Otters. The head of a fish was the meal here. Lausitz, Germany. Heide Ulrich.*

MARKING MOUNDS Eurasian Otters scrape together earth and surrounding material to make marking mounds. They leave droppings, urine and/or anal secretions as territory markings.

PATHS Distinctive paths can be created between bodies of water. Eurasian Otters tunnel through higher vegetation, while geese and ducks trample the plants. Their paths measure about 15–25cm in width. Eurasian Otters also leave conspicuous entry and exit points at the edges of riverbanks, where they often place territory markings.

FEEDING MARKS Eurasian Otters generally eat fish whole so feeding marks are a rare find. However, you can occasionally find piles of mussels and fish remains on the banks along an otter territory. If there are no additional signs, these feeding sites can be difficult to distinguish from the feeding sites of other species.

SLIDES Eurasian Otters slide downhill on steep, grassy slopes or in snow, creating deep furrows (otter slide). The animals can sometimes slide for several dozen metres. Sliding is an efficient form of locomotion and possibly also a form of play.

EXCREMENT Droppings can be a shapeless, black to grey pile with scales and bones or a solid, sausage-like form, with one blunt end and one end that tapers to a point. Regardless of the initial form, the droppings gradually decompose into a pile of scales and bones. Animals that have eaten more mammals and hair will produce firmer, more sausage-shaped droppings. Droppings are occasionally accompanied by yellowish, brown or green anal secretions.

- L 7–16cm D 1.2–2.2cm

▼ *Eurasian Otter excrement varies greatly in size and shape. Longer, sausage-shaped droppings can be mistaken for fox or dog droppings at first glance. Scales and bones can be important clues for differentiation. Märkische Schweiz, Germany.*

◄ *Eurasian Otters mark important points, such as entry and exit points or crossroads, by creating latrines there. Recurring contamination with uric acid turns the vegetation brownish ('brown-out') in these places. Lausitz, Germany.*

▲ *Wolverines mark their territory with specially placed scratch marks.*
Sweden. Frank Jermis.

▲ *A Wolverine has scratched this birch trunk in search of larvae.*
Ljungdalen, Sweden. Heide Ulrich.

SIGNS

DWELLING Wolverine dwellings are important for resting and for giving birth to and rearing young. The animals use existing, hard-to-reach gaps between rocks and fallen trees. If no suitable structures are available, they dig their own dwelling. In higher areas with lots of snow, they sometimes create complex tunnel systems. In areas with less snow, Wolverines make use of natural crevices. They prefer places with an above-average amount of snow and where snow cover lasts longer. Dwellings have one or more entrances. The entrance holes are on average 25–35cm wide.

TERRITORY MARKINGS Wolverines often mark saplings and large, solitary, prominent trees along the route of their territory. To do this, they climb the lower part of tree trunks and make scratch and bite marks in the bark and on the branches. Occasionally they even climb head-down to mark the tree trunk and the surrounding ground, usually by producing an anal secretion. This serves as a scent mark for intraspecies communication and territory marking.

FEEDING MARKS Wolverines create winter food stores by carrying prey far away from the kill site and then burying it in the snow. The stores can vary in size. They usually only visit small food stores once but will return to larger stores repeatedly throughout winter. In this case, the paths they make can radiate from the store like the spokes of a wheel. Like Brown Bears and European Badgers, Wolverines are known to turn over stones and pull apart rotting tree trunks in summer to get at insects or larvae.

- **Distance between the canine teeth: 2.6–3.3cm (2.4–3.5cm)**

EXCREMENT Wolverine excrement often contains a lot of hair and bone remains. If it does, it will be kinked and taper at the ends to a long point, which is typical of mustelids. If more nutritious food, such as entrails, is eaten, the droppings will be shapeless and darker in colour. They often leave droppings on stones, tree trunks and other exposed places, which also serves to mark their territory. Unlike some other mustelid species, Wolverines do not create latrines.

- **L 7.5–20cm D 1–2.5cm**

▶ *Partially snow-covered Wolverine droppings. Jämtland, Sweden. Laura Gärtner.*

EUROPEAN BADGER
Meles meles

HTL 64–90cm
TL 11–20cm
W 7–18kg
Body weight can
increase by up to 6kg
before hibernation.

The genus of badgers (*Meles*) comprises three or four species, of which the European Badger is the only one found in the region. With its shovel-like front feet and long claws, the European Badger is an excellent, effective digger. Badgers dig their own dwellings consisting of tunnels and chambers, which they line with dry grasses, moss, leaves and ferns. They tuck the nesting material between their chin and front feet and carry it backwards into the dwelling. A European Badger sett may be used for several generations for more than 100 years and can be constantly extended. Larger setts can be multi-storeyed and reach an incredible size. One particularly large sett in England (Cotswolds) measured over 35 × 15m and had 12 entrances, a total tunnel length of 310m and a volume of 15m³.

European Badgers are mainly crepuscular and nocturnal. Their dominant senses are smell and hearing and their eyesight is weak. They live gregariously in families consisting on average of 6 (2–20) adults. European Badger families occasionally visit each other and spend the night in each other's sett. They hibernate in colder regions. In Europe, they are only predated by wolves, dogs and humans.

DISTINGUISHING FEATURES   Powerful build with short, sturdy legs and a relatively small head with a long, pointed snout. The black and white, longitudinally striped face mask is an instantly recognisable feature. Strong, short neck and short tail.

DISTRIBUTION AND HABITAT   The European Badger is widespread throughout Europe, except for northern Scandinavia, Iceland and some Mediterranean islands, and is found in various habitats. Mixed deciduous forests with a high density of earthworms and a mild, humid climate are ideal for them. They are usually only absent from low-lying areas such as marshes, wetlands with a high risk of flooding, and at high altitudes.

DIET   As an adaptable, opportunistic omnivore, the European Badger's diet is one of the most varied and plant-rich of European carnivores. Earthworms are often an important part of the diet. Other important food sources include large insects, small mammals, carrion (especially in winter), cereals, fruit and roots. Prey (mainly young animals) includes rabbits, rats, mice, voles, shrews, moles and hedgehogs. Their preferred insect prey are beetles and caterpillars as well as the contents of bee and wasp nests. They occasionally also eat frogs, toads, snakes, lizards, snails, fungi, roots, berries, fruit and acorns. European Badgers also seek out surrounding farmland with cereal crops as a seasonal food source. The diet can therefore be very broad, but also very limited at certain times – this is highly dependent on season and availability. For example, they may temporarily feed almost exclusively on earthworms, frogs or young mice.

REPRODUCTION   European Badgers can mate all year round, with a peak from February to May. The gestation period is extended by embryonic diapause. Implantation usually occurs in December and the embryo develops rapidly. Active pregnancy lasts 42–49 days. An average of 2–3 (1–5) altricial young are born underground in the sett, mainly in February/March. The young emerge from their burrow for the first time after around eight weeks, usually in April. Related animals help each other to raise their young. The young generally reach sexual maturity after 12–18 months, but this can be as early as 8–9 months in exceptional cases.

Rabbits, foxes and small rodents sometimes live in their own areas of the sett, where they are tolerated by the badgers.

TRACK

Front

L 5.3–11.5cm W 4.4–7cm

Medium-sized. Plantigrade. Asymmetrical. Five toes. Toe 1 is the smallest toe. It is further back and does not reliably leave a print. Toes 2–5 form an even, wide arc which, together with the short, wide negative space between the digital pads and the midfoot pad, is a characteristic identifying feature (a). The more or less parallel digital pads are rarely splayed. The interdigital pads have fused to form a large, strong midfoot pad that is wider than it is long. A thenar/hypothenar pad that is slightly offset to the outside occasionally leaves a print and can make the track appear longer. Long, strong claws, which are used for digging, usually leave a print and are a characteristic feature.

Hind

L 4.5–9cm W 3.5–5.7cm

Small to medium sized. Plantigrade. Asymmetrical. Five toes. Toe 1 is the smallest toe and does not regularly leave a print. The interdigital pads have fused to form a large midfoot pad, which is narrower and more elongated than on the front foot. The strong claws are considerably shorter than on the front foot and are not always visible (b). The hind foot print is smaller and narrower than the front foot print.

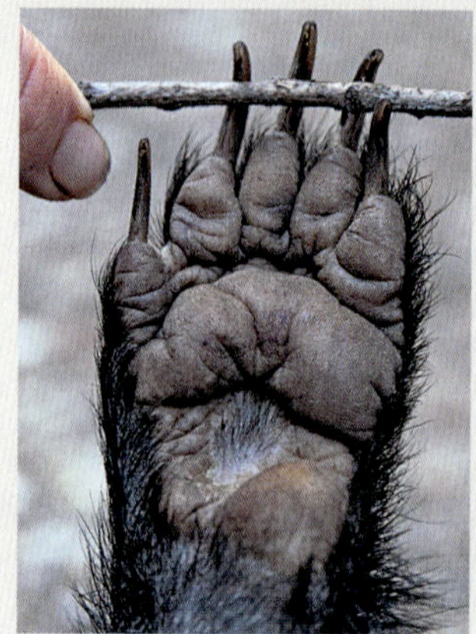

▲ *Left front.*
Märkische Schweiz, Germany.

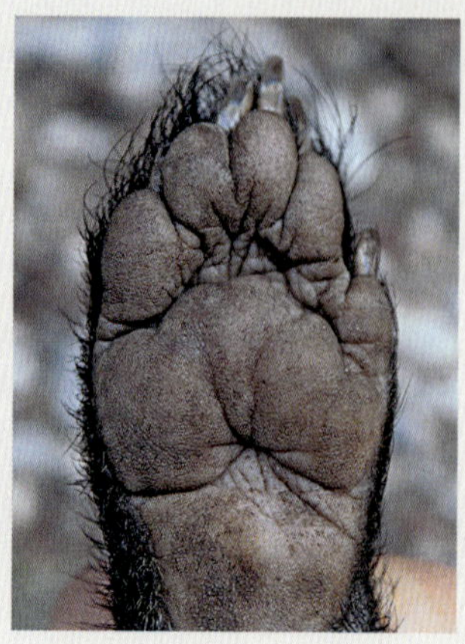

▲ *Right hind.*
Märkische Schweiz, Germany.

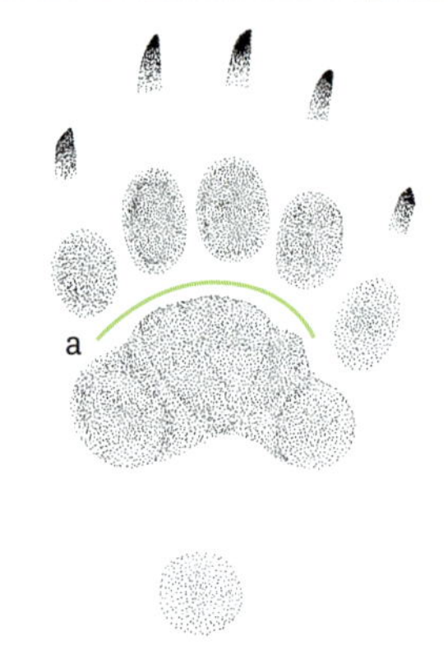

▲ *Left front.*

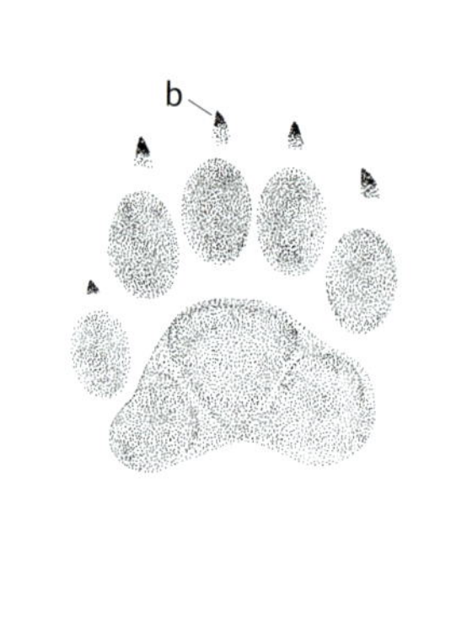

▲ *Right hind.*

▲ *Left front.*
Märkische Schweiz, Germany.

▲ *Right hind.*
Märkische Schweiz, Germany.

GAITS

European Badgers usually travel at a brisk trot and forage at a slow walk. The hind feet often noticeably turn in at an angle. When alarmed or threatened, European Badgers can gallop at speeds of up to 30km/h over short distances. Regularly used paths are distinctive and clearly visible. They are remarkably wide relative to their low height, and connect different burrows, favourite feeding sites and latrines. Roadsides and forest tracks are frequently used.

Walk
Stride length: 42–70cm
Trail width: 11.5–17.5cm

Trot
Stride length: 56–115cm
Trail width: 6–15cm

3 × 4 bound (rotary) and 4 × 4 bound (transverse)
Group length: 38–72cm
Inter-group length: 17–34cm
Stride length: 55–103cm
Trail width: 10–16cm

Gallop
Group length: 43–110cm
Inter-group length: 31–68cm
Stride length: 78–156cm
Trail width: 10–18cm

Similar tracks
Unmistakable if prints are clear. Can be mistaken for Crested Porcupine tracks, although their front foot claws are much shorter.

▼ *The classic track pattern of a European Badger at a gallop. Märkische Schweiz, Germany.*

▼ *Lope.*

▼ *Direct register trot.*

◄ *At first glance, this looks like a direct-register trot. But if you look closely, the thenar/hypothenar pads of the two bottom tracks stand out, which the European Badger only has on its front feet. Footfall from bottom to top: RF, LF, RH, LH. This is actually a transverse gallop. Märkische Schweiz, Germany.*

SIGNS

DWELLINGS Preferred sites for setts are south-facing slopes, usually not more than 500m from the edge of the forest. Badger setts have an average of 3–10 large entrances with sometimes metres-wide, fan-shaped piles of ejected sand outside them. Unlike Red Fox dens, European Badger setts usually have a clearly recognisable 'channel' and a horizontal oval shape. A well-trodden network of paths is often visible between the individual entrances, which is not the case with Red Fox dens. In cooler or more humid parts of Europe, European Badgers prefer to orientate the entrance holes of their sett towards the south. Entrance holes have a diameter of over 25cm, but they are often much larger.

◄ *Entrance to a European Badger sett. The wide, well-trodden channel is a distinguishing feature that is absent from a Red Fox den. Märkische Schweiz, Germany.*

▲ *European Badger sett with fresh ejected sand in front of the entrance. Märkische Schweiz, Germany.*

◀ European Badger scratching trees are conspicuous markers and can indicate proximity to a sett. West Sussex, England.

▲ Nesting material. European Badgers do a kind of spring clean. Märkische Schweiz, Germany.

SCRATCHING TREES   European Badgers scratch trees, mainly elder trees near their sett, to mark their territory. Claw and mud marks can be found on trees at a height of up to 1m.

NESTING MATERIAL European Badgers gather fresh nesting material and get rid of old nesting material, mainly in spring and autumn. Old bedding, such as dry grass, moss, leaves and ferns, as well as associated drag marks, can be found around their sett. On sunny winter days, they carry old nesting material out of their burrow, put it near the entrance to dry and later carry it back into their burrow.

▶ *Here, a European Badger has dug up and raided a wasp nest. Remains that provide clues to the prey can often be found in the sett. Vledder, Netherlands. René Nauta.*

▲ *A hole in the ground dug by a European Badger in search of food. If you look closely, you will see the claw marks on the wall and the track in the fresh ejected material. Lausitz, Germany.*

DIGGING MARKS AND FORAGING European Badgers make conspicuous holes in the ground when foraging for earthworms, insect larvae and roots. It is important to distinguish whether an animal has dug the hole with its front feet or 'stabbed' it with its pointed snout. Holes dug by European Badgers can be very deep and usually have cone-shaped ejected material and claw marks on the side walls. These marks tend to be horizontal, because European Badgers, like Beech Martens but unlike Red Foxes, can turn their front feet sideways. Compared with holes that have been dug with the claws, holes made with the snout are much smaller: they have a diameter of 5–20cm, are a maximum of 15cm deep and the soil is pushed rather than dug out. These holes can be numerous in meadows.

Badgers can break open and hollow out rotten tree trunks to get at beetle larvae and other insects. Fan-shaped ejected material is usually visible. Black Woodpeckers in search of food leave signs that can be mistaken for those of European Badgers (page 158) and Brown Bears also show similar behaviour (page 426).

EXCREMENT European Badger excrement ranges from loose to mushy and may appear muddy if the animal has eaten many earthworms. Consistency depends heavily on diet. A strong-smelling, orange-coloured anal gland secretion is sometimes left behind with the droppings. European Badgers prefer to dig small hollows in the earth, into which they repeatedly defecate. These latrines are mainly established near setts, at preferred feeding sites or along territory boundaries. European Badgers sometimes create large latrines that you can smell before you come across them. In addition to latrines, they may also leave their droppings en route. In this case, they usually only defecate once on or near the path.

- L 4.2–12.5cm   D 1.2–2.2cm

▲ *European Badger droppings after raiding a mining bee colony. Neustadt, Germany.*

▼ *A recently used latrine. Latrines are sometimes created next to each other and spread to cover an area of several square metres over the years. West Sussex, England.*

EUROPEAN PINE MARTEN
Martes martes

HTL 36–56cm
TL 17–28cm
W 0.8–2.2kg

The European Pine Marten belongs to the genus *Martes*, which has a total of eight species of which two occur in the area. European Pine Martens have the typical marten shape with a bushy tail and short legs. They are highly adept climbers. These are solitary animals, which are mainly active at dawn, dusk and at night, but in summer they can also be seen during the day. They can run down tree trunks headfirst, jump up to 4m from branch to branch and are usually so nimble that you only spot a movement in the branches, rather than seeing the animal itself. There are reports of European Pine Martens jumping out of trees from a height of 20m with their front legs outstretched and then landing safely on all fours, like a flying squirrel. They prefer to shelter in trees, using Black Woodpecker holes, bird and squirrel nests or bushy growths in the treetop. For shelter on the ground, they use hollows between tree roots, piles of branches or abandoned fox dens. European Pine Martens are predated by foxes, wolves and lynx, as well as eagles and Eurasian Eagle-owls.

DISTINGUISHING FEATURES   Long, slender body with a long, bushy tail. Their legs are longer and their body slimmer than a Beech Marten. The normally yellowish throat patch varies greatly in colour and shape and in some individuals is indistinguishable from that of a Beech Marten. Compared to Beech Martens, European Pine Martens have longer ears with a yellowish-white edge.

DISTRIBUTION AND HABITAT   The European Pine Marten is found throughout almost the whole of Europe, apart from the southern Iberian Peninsula. It tends to avoid human activity more than the Beech Marten. It prefers large, unbroken deciduous and mixed forests over coniferous forests as they offer a richer food supply. In rare cases, the animals can also be found in human settlements. Unlike Beech Martens, European Pine Marten populations are declining in some parts of Europe due to infrastructure development and the associated fragmentation of woodland by busy roads.

DIET   Opportunistic generalist with a varied diet, similar to the Beech Marten, but mainly eats prey from the forest. These include mammals (murids, voles and shrews, squirrels and rabbits), birds and birds' eggs (wren, thrushes, jays and owls), berries and fruit (forest berries, cherries, rowan fruit) and insects, as well as frogs, toads, fungi, nuts and honey. The movement of birds in a hen house can trigger a killing instinct that occasionally provokes European Pine Martens into wiping out the entire flock.

REPRODUCTION   Reproduction and development of young identical to Beech Martens (page 461).

REMARKS   European Pine Martens are endangered according to the Red List. Paul Marchesi *et al.* (2010) write: 'Knowledge of wildlife migration routes between forest areas, as well as their preservation, and where necessary, improvement and restoration (wildlife bridges), is important for European Pine Martens and many other forest-dwelling species.' These migration axes can be located with tracking, which is a minimally invasive method of wildlife monitoring.

TRACK

Typical mustelid arrangement. Remarkably large feet in relation to their size. European Pine Martens have very hairy soles, so their digital and metacarpal/metatarsal pads are covered and the tracks appear unclear, depending on how hairy the soles are. The thickness of the hair depends on the time of year and individual factors – it may even be largely absent in some individuals. If there is evidence of a lot of hair in a track, you can rule out the Beech Marten. Clear footprints are rare.

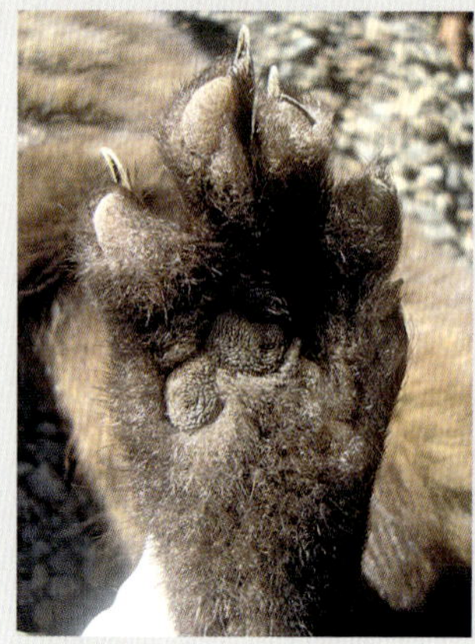

▲ *Right front.*
Göttingen, Germany. Heide Ulrich.

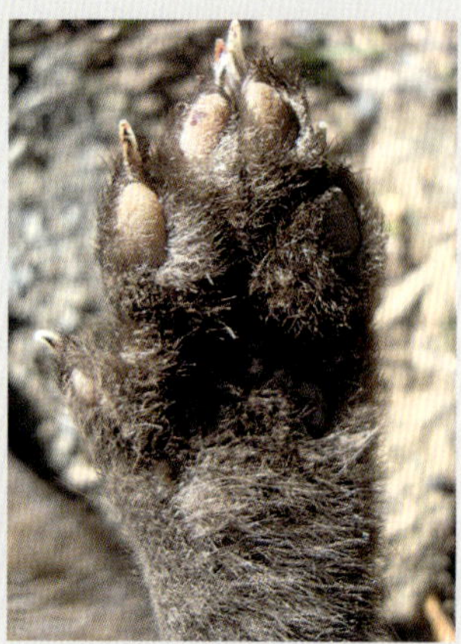

▲ *Right hind.*
Göttingen, Germany. Heide Ulrich.

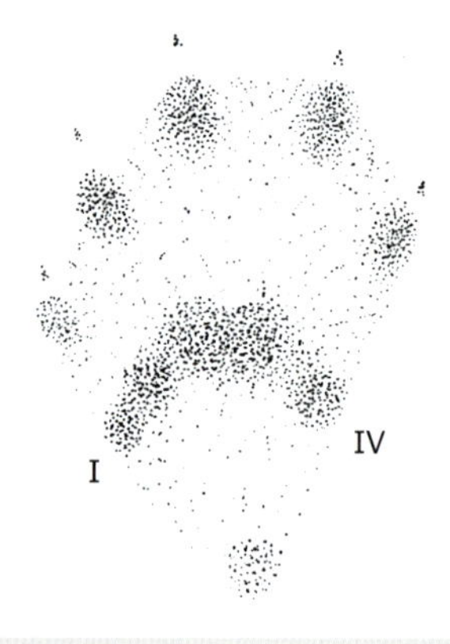

▲ *Right front.*

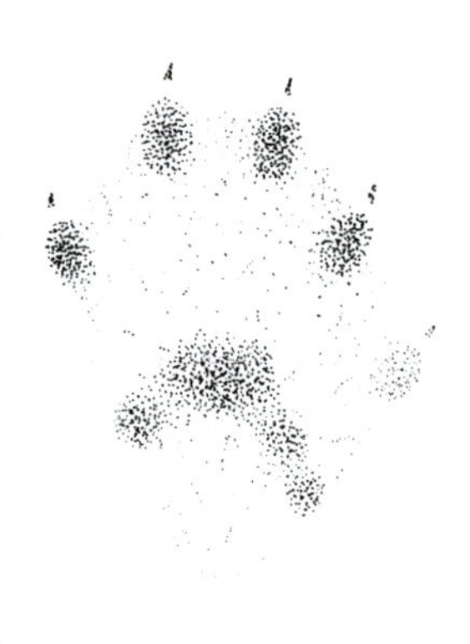

▲ *Right hind.*

Front

L 4–6.5cm W 4.5–6.5cm
Small to medium-sized. Plantigrade. Asymmetrical. Broader overall and more symmetrical than the hind foot print. Five toes: Toe 1 is the smallest toe and cannot be reliably recognised. The interdigital pads are fused, but clear bulges are individually recognisable. If metacarpal pad I is recognisable, it leaves an individual print that is further back than metacarpal pad IV. The complete midfoot pad is rarely visible. A single thenar/hypothenar pad may be visible at the back outside edge of the track. The fine, pointed claws do not always leave marks.

Hind

L 4.2–6.4cm W 4.2–5.5cm
Small to medium-sized. Plantigrade. Asymmetrical. Five toes: Toe 1 is the smallest toe and is further back than on the front foot. This makes the hind foot more asymmetrical. However, toe 1 is often not recognisable, so the hind foot print can be more symmetrical than the front foot print. The interdigital pads have fused to form a midfoot pad but with individually recognisable bulges. If metatarsal pad I is visible, it leaves an individual print and is much further back than metatarsal pad IV. The fine, pointed claws are sometimes recognisable.

GAITS

European Pine Martens explore and hunt in 2 × 2, 3 × 4 (rotary) and 4 × 4 (transverse) bounds. If inspecting something more closely, they switch to the slower walk. They can gallop when fleeing or when threatened. They usually favour the 2 × 2 bound in deep snow.

2 × 2 bound
Group length: 6–19cm
Inter-group length: 16–107cm
Stride length: 22–132cm
Trail width: 6–12cm

3 × 4 bound (rotary) and 4 × 4 bound (transverse)
Group length: 14–45cm
Inter-group length: 15–38cm
Stride length: 38–85cm
Trail width: 8–16cm

Similar tracks

Larger on average than European Polecat, mink or Stoat tracks, but can easily be confused with a Beech Marten. European Pine Marten digital pads are slightly more delicate compared to the size of the footprint than those of Beech Martens. They are more or less parallel to the orientation of the foot and make up a smaller proportion of the track surface. Nevertheless, the digital pads are much larger and stronger than in Stoats or European Polecats. Compared to mink, the digital pads and metacarpal/metatarsal pads leave more even prints and the digital pads are less splayed. European Pine Marten metacarpal/metatarsal pads are roundish and usually difficult to make out because of the very hairy soles.

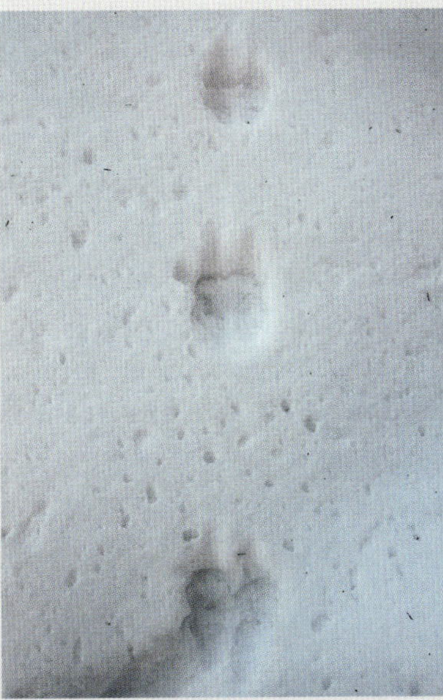

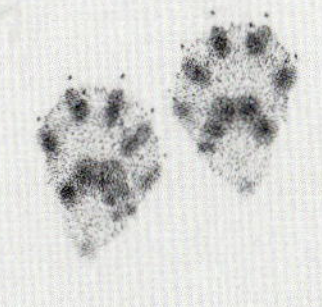

◀ A classic track pattern: the European Pine Marten in a 2 × 2 bound. Bavarian Forest, Germany.

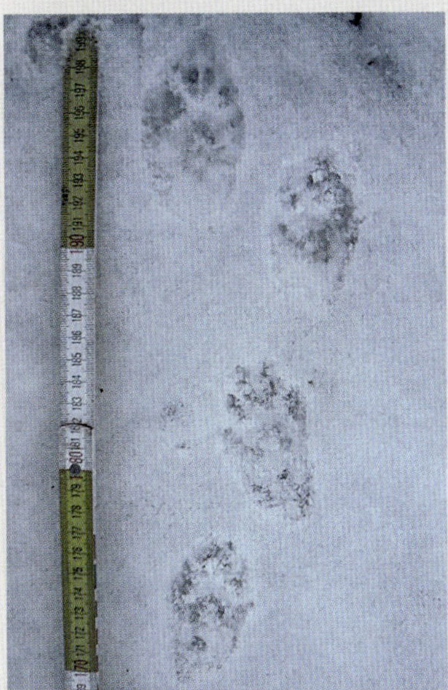

▲ Parallel bound track of a European Pine Marten in the snow. Sequence of footfalls from bottom to top: LF, RF, RH, LH. Ljungdalen, Sweden. Heide Ulrich.

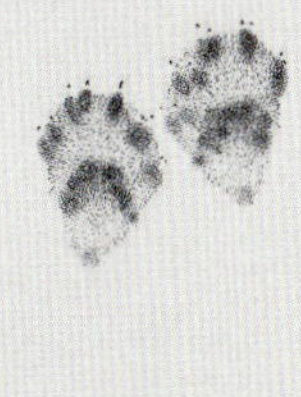

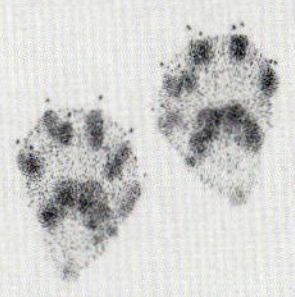

▲ 2 × 2 bound.

SIGNS

FOOD STORES Like many mustelids, European Pine Martens store food. Finding stores, however, is rare as they are usually hidden in hollows in trees or abandoned bird nests.

EXCREMENT Droppings are usually long, kinked and pointed at one end. A high proportion of small mammal hairs give the droppings the classic mustelid kinking and shape. However, the tip is usually less pronounced than in Stoats or Least Weasels, which feed almost exclusively on small mammals. The droppings are usually grey or greyish-black and tend to be slightly lighter in colour than Beech Marten droppings. Appearance and contents can vary greatly depending on the season and the food available. Towards the end of summer and in autumn, the droppings can consist almost entirely of fruit stones. There is usually a pile of droppings near the shelter. Like most mustelids, European Pine Martens prefer to deposit their droppings in elevated places. The odour of European Pine Marten droppings is associated with a pleasant fruity, musky smell, while Beech Marten droppings smell unpleasantly putrid. It is not possible to distinguish between the droppings of the two species by appearance alone.
- L 5.1–12.7cm D 0.5–1.6cm

◄ *European Pine Marten droppings. On closer inspection, you can make out small mammal bones. Märkische Schweiz, Germany.*

BEECH MARTEN
Martes foina

Beech Martens are mainly active at dawn, dusk and night. They have an excellent sense of smell and very good hearing and eyesight. The tactile hairs on their face enable them to move safely both in the dark and through narrow passages. The animals can even climb vertically up seemingly smooth walls while holding prey in their mouth. They often nest inside buildings, especially in attics, where they occasionally damage the insulation. Unlike Beech Martens, European Polecats prefer to nest on the ground floor. Beech Martens can also be found in cellars or garden sheds. They are both known and feared for their fondness for damaging cars. Away from human settlements, Beech Martens shelter in burrows, piles of stones and similar places that offer protection. Their main predators are foxes and other canids, but also eagles, Eurasian Eagle-owls and lynx. Persecution by humans has the greatest impact on numbers.

HTL 38–55cm

TL 19–30cm

W 1–2.3kg

TRACK

Typical mustelid arrangement. Remarkably large feet in relation to the body size. Beech Martens have strong, hairless digital pads that are more or less parallel to the orientation of the foot and make up a relatively high proportion of the track. This is a clear difference to Stoats or European Polecats. Compared to mink footprints, the digital pads and midfoot pad leave a more even print and the digital pads are less splayed. Beech Marten interdigital pads are roundish and have a robust appearance. They are usually easily recognisable because Beech Martens have less hairy soles than European Pine Martens. This can be a reliable distinguishing feature in winter. The difference is less pronounced in summer and only tends to be visible in very clear prints.

Front

L 4–6.9cm W 3–6.5cm

Small to medium-sized. Plantigrade. Asymmetrical. Generally wider and more symmetrical than the hind foot print. Five toes: Toe 1 is the smallest toe and cannot be reliably recognised. The front centre pads are fused, but with individually recognisable bulges. If metacarpal pad I is visible, it usually leaves an individual print and is further back than metacarpal pad IV. If the entire midfoot pad is visible, it appears to be elongated towards the back on the side closest to the body (a).

▲ Left front.

▲ Left hind.

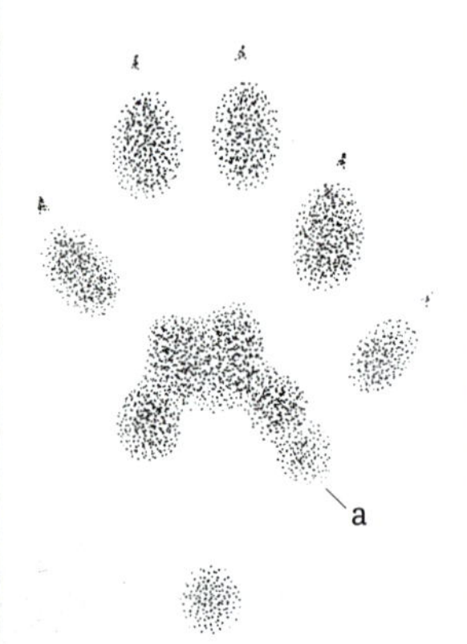

▲ Left front.

▲ Left hind.

▲ Left front.
Lausitz, Germany.

▲ Left hind.
Lausitz, Germany.

A single thenar/hypothenar pad may be visible at the back of the track. Sharp, pointed claw marks are sometimes visible. The front foot print is generally, but not always, larger than the hind foot print. The hind foot print appears larger when the front foot leaves a print without the thenar/hypothenar pad and the hind foot leaves a complete print.

Hind

L 3.2–6.5cm W 3–5.6cm
Small to medium-sized. Plantigrade. Asymmetrical. Asymmetry clearer than in the front foot. Five toes: Toe 1 is the smallest toe and is further back than on the front foot. This makes the hind foot more asymmetrical. If toe 1 is not visible, the hind foot print appears more symmetrical than the front foot print and may resemble the footprint of a small canid. Toe 1 is often unrecognisable. The hind foot generally appears more elongated. The interdigital pads are fused, but there are usually individually recognisable bulges. If metatarsal pad I is visible, it usually leaves an individual print and is further back than metatarsal pad IV. If the entire midfoot pad is visible, it appears to be elongated towards the back on the side closest to the body. Sharp, pointed claw marks are sometimes visible.

GAITS

Beech Martens explore and hunt in 2 × 2 bounds, 3 × 4 bounds (rotary) and 4 × 4 bounds (transverse). If they need to slow down to inspect something more closely, they switch to a slower walk. They gallop when fleeing or threatened.

2 × 2 bound
Group length: 8.5–21cm
Inter-group length: 8–95cm
Stride length: 15–116cm
Trail width: 9.5–13.5cm

3 × 4 bound (rotary) and 4 × 4 bound (transverse)
Group length: 12–41cm
Inter-group length: 17–34cm
Stride length: 42–83cm
Trail width: 6–17cm

Gallop
Group length: 10–81cm
Inter-group length: 16–91cm
Stride length: 25–135cm
Trail width: 7–17.5cm

Similar tracks
European Pine Marten, European Polecat.

◀ *Beech Martens often favour the 2 × 2 bound in snow. Bavarian Forest, Germany.*

◀ *Beech Marten in 3 × 4 bound (rotary) 4 × 4 bound (transverse). Footfall from bottom to top: LF, RF, LH, RH. Lausitz, Germany.*

▲ *3 × 4 bound (rotary) and 4 × 4 bound (transverse).*

▲ *Parallel bound.*

DISTINGUISHING FEATURES Long, slender body with a tail that is about half the length of the body. Shorter legs and a slightly stockier build than a European Pine Marten. Throat patch varies greatly in shape and colour in each individual; it is generally whitish, but occasionally also yellowish, in which case it is indistinguishable from that of European Pine Marten. The edges of the ears are white in Beech Martens and yellowish in European Pine Martens.

DISTRIBUTION AND HABITAT Beech Martens are widespread in Europe. They are absent from Great Britain, Ireland, Iceland and Scandinavia. Beech Martens are an adaptable species associated with human activity and are found in villages, cities and woodland areas. They can also be found far away from human dwellings and can change their habitat. In forests, they often occur together with European Pine Martens. Unlike European Pine Martens, European Polecats, Stoats and Least Weasels, Beech Marten populations are increasing in many places in Europe. This is presumably due to the animal's marked ability to adapt to resources close to human settlements.

DIET The Beech Marten is an omnivorous generalist with a remarkably varied diet. Depending on availability, Beech Martens feed on mammals (from small rodents to young rabbits), sometimes in significant quantities, berries and fruits (cherries, apples, rose hips, rowan berries, forest berries and similar), birds and birds' eggs, as well as insects (beetles, bee and wasp nests), amphibians and earthworms. The movement of birds in a hen house can trigger a predatory instinct that occasionally provokes Beech Martens into killing the entire flock. Beech Martens also eat other food depending on availability.

REPRODUCTION The mating season lasts from mid-June until the end of August. The female is in heat for about 10 days during this time. Delayed implantation results in a prolonged gestation period of nine months. Active pregnancy lasts four weeks. An average of 3–5 young are born in March/April. The young are altricial and only able to open their eyes after about a month. This is very late compared to other similar mammals.

Beech Martens benefit from being near humans because of the availability of food, shelter and warmth. They are unpopular because they destroy the cables in car engines. However, reliable methods to protect cars against Beech Martens are now available. Trapping and killing only helps in the short term, as vacated territories are soon reoccupied.

SIGNS

FOOD STORES Beech Martens are well known for carrying off large eggs and storing them when there is a surplus of food. Food stores can contain lots of eggshells and other remains of prey, such as individual body parts.

DIGGING MARKS Beech Martens may dig holes when hunting for voles or bee nests. These deep holes are almost circular and correspond to the animal's long, slender build. Badgers and foxes make equally deep holes but they are wider because these animals have wider skulls.

▶ A Beech Marten has raided a wasps' nest. Vledder, Netherlands. René Nauta.

▲ *Only the tubular body of a mustelid could fit into this almost circular, deep hole. Digging marks made by a Beech Marten on the hunt for voles. Lausitz, Germany.*

EXCREMENT Beech Marten excrement is usually long, with a relatively large diameter, kinked and pointed at one end. The high proportion of small mammal hairs give the droppings the classic mustelid kinking and shape. The tip is usually less pronounced than in Stoats or Least Weasels, which feed almost exclusively on small mammals. Beech Marten droppings are usually grey, brown or black and often darker in colour than those of European Pine Martens.

Appearance and contents can vary greatly depending on the season and the food available. Towards the end of summer and in autumn when cherries, for example, are ripe, the droppings can consist almost entirely of their stones. Like most mustelids, Beech Martens prefer to deposit their droppings in elevated places. They make foul-smelling latrines near their shelter. European Pine Marten droppings have a pleasant, musky odour. It is not possible to distinguish between the droppings of the two species by appearance alone.

- L 4–12cm D 1–1.5cm

▼ *The contents of Beech Marten excrement can vary because of their mixed diet. Cherry stones and mouse hair are visible here. Hoyerswerda, Germany.*

EUROPEAN POLECAT
Mustela putorius

HTL 30–50cm

TL 8–19cm

W 0.65–2kg

Very pronounced sexual dimorphism. Males significantly (up to a third) larger than females. Exceptionally small feet in relation to height.

The European Polecat belongs to the genus of weasels (*Mustela*). Weasels are very small to small carnivores with an elongated, agile body and short legs. There are 17 species, six of which occur in the area. This book does not cover the Steppe Polecat (*Mustela eversmannii*) or the European Mink (*Mustela lutreola*) due to insufficient data or very limited distribution. The European Polecat is solitary, crepuscular and nocturnal. It has an excellent sense of smell and very good hearing. It lives in the dwellings of its prey, such as rabbits and Common Hamsters, and also uses old fox dens. If unable to find a suitable dwelling, the European Polecat digs its own, usually a simple burrow. In rare cases, it is predated by foxes and diurnal birds of prey. Most deaths are caused by humans and their dogs.

DISTINGUISHING FEATURES   Typical long, weasel body with relatively short legs. Dark face mask, edges of ears white. American Mink do not have either of these features (page 476).

DISTRIBUTION AND HABITAT   One of the most common native mustelids, found throughout almost the whole of Europe and only absent from northern Scandinavia and Ireland. The European Polecat is often, but not exclusively, associated with wetlands. It usually lives in forests and copses on the edges of swamps and reedbeds, in structurally rich wetlands and near houses. It can even be found in towns and cities. Altitudinal distribution in the mountains up to around 2,000m. In some regions of Europe, evidence of European Polecat populations is declining. This is presumably related to loss of wetlands. The Steppe Polecat replaces the European Polecat in eastern Europe.

DIET   Mainly small mammals, such as smaller rodents, up to the size of a rabbit, and amphibians. Earthworms, fish, birds' eggs and insects make up a small part of their diet. Compared to other mustelids, the European Polecat's diet falls somewhere between that of Least Weasels, which specialise in small mammals, and generalists, such as European Pine Martens and Beech Martens. It is most often a generalist but, if food supplies allow, it can also specialise in certain food groups, especially amphibians, rats and rabbits. It eats rats, Muskrats and European Rabbits more often than do other mustelids. It pursues them into their burrows. European Polecats have been shown to control rabbit and rodent populations. If there is a surplus of food, European Polecats build up stocks. Their food stores have been found to contain several rabbits and sometimes more than 100 frogs and toads.

REPRODUCTION   Mating season lasts from the end of February until the end of June. After a gestation period of 35–43 days, an average of 4–8 (up to 12) young are born, mainly in May and June. Exceptionally, young can be born in early March, or late October. Although there is usually only one litter per year, a second litter is possible if the first one is lost, which could explain late births. The altricial young reach sexual maturity towards the end of their first year.

European Polecats used to be coveted for their fur. They are the wild ancestors of the Ferret, although the genetic influence of the Steppe Polecat is unclear.

TRACK

Typical mustelid arrangement. The soles of the feet are hairy, but the relatively strong, rounded digital pads and the rather delicate interdigital pads are usually clearly recognisable.

Front

L 2.5–4.7cm W 2.2–4cm
Small. Plantigrade. Asymmetrical. Five toes. Toe 1 is the smallest toe and not always easy to recognise. The digital pads usually point straight forward from the midfoot pad. Toes 3 and 4 are usually less than 4mm apart. The rather narrow interdigital pads have fused, but they usually leave a print with individually recognisable bulges. Metacarpal pad III forms a horizontal rectangle, which can help to distinguish the prints from those of Stoat or mink (a). If metacarpal pad I is visible, it leaves an individual print. A thenar/hypothenar pad that is offset to the outside may be visible at the back of the track. European Polecats have characteristic long, sharp, powerful digging claws that usually leave a clear print.

Hind

L 2.5–3.9cm W 2–3.4cm
Small. Plantigrade. Asymmetrical. Five toes. As the smallest toe, toe 1 is often hard to make out. If toe 1 does not leave a print, the other four toes have an arrangement similar to the toes of a small canid.

▲ *Right front.*
Nahrendorf, Germany.

▲ *Right hind.*
Nahrendorf, Germany.

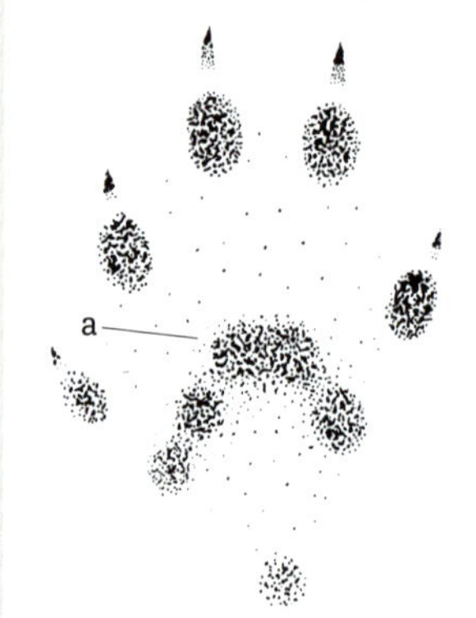

▲ *Right front.*

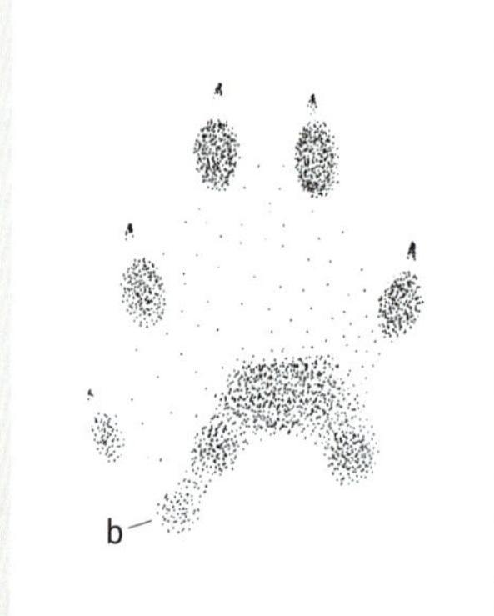

▲ *Right hind.*

▲ *Right front. The rectangular metacarpal pad is a characteristic feature. Spessart, Germany.*

▲ *Right front.*
Vledder, Netherlands. René Nauta.

The interdigital pads have fused, but there are usually individually recognisable bulges. If metatarsal pad I is visible, it leaves an individual print ⓑ. The sharp claw marks are often visible.

GAITS

The European Polecat's preferred gaits are 2 × 2, 3 × 4 bound (rotary) and 4 × 4 bound (transverse). European Polecat paths are often found along richly structured field and forest edges where they forage for food.

2 × 2, 3 × 4 bound (rotary) and 4 × 4 bound (transverse)
Group length: 5–30cm
Inter-group length: 15–75cm
Stride length: 30–105cm
Trail width: 4.5–11cm

▲ *The 4 × 4 bound (transverse) of a European Polecat. Bieszczady, Poland.*

Similar tracks

Beech and European Pine Marten footprints are larger. Mink tracks are similar in size, but are usually clearly identifiable by their characteristic otter-like appearance. Some of the dimensions overlap with those of Stoats. Although it is sometimes possible to make a clear statement based on foot morphology, this requires almost perfect footprints (see 'Differentiating between Stoats, European Polecats and American Mink', page 481).

SIGNS

AMPHIBIAN REMAINS In spring, European Polecats hunt frogs and toads as they migrate to their spawning waters to lay their eggs. To avoid the poisonous parotoid gland behind the ears of toads, they do not eat the head or the front half of the body. They also do not touch the spawning pouches of females. Without further clues, amphibian remains are difficult to distinguish from those left by mink, otters or buzzards.

FOOD STORES Like many other mustelids, European Polecats hide surplus food in food stores in natural chambers between stones, in hollow tree trunks or in holes in the ground.

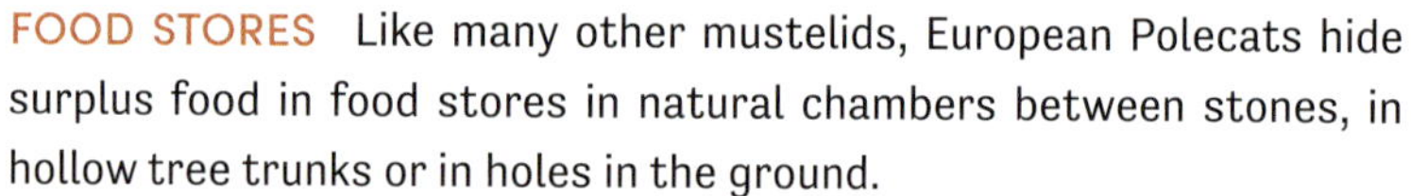

EXCREMENT Long, cylindrical, kinked and pointed at one end. Eating amphibians or earthworms results in looser excrement. Colouring is highly dependent on diet but is usually dark brown to black and with a light covering of mucus. Droppings often contain bone remains from frogs as well as bones and hair from small rodents and shrews. European Polecats use latrines near the dwelling. Sometimes difficult to distinguish from mink droppings, unless they have mostly eaten fish. Generally much larger than Least Weasel or Stoat droppings.

- L 3–7cm D 0.5–0.9cm

▼ Cylindrical, kinked European Polecat droppings that taper to a point at one end. Midhurst, England. John Rhyder.

▶ The remains of spawn from a female amphibian predated by a European Polecat. Jetzendorf, Germany.

LEAST WEASEL AND STOAT

Mustela nivalis, Mustela erminea

Stoats (known as ermine in their white winter pelage) and Least Weasels are small, energetic carnivores. They can move incredibly fast and their reaction speed is exceptional. Their hearing and sense of touch are very well developed and they can climb and swim well. Due to their similar way of life and similar tracks and signs, we will consider them together. Both species have long, slender bodies that are ideal for hunting in confined spaces, such as in hollow trees or between stones. The smaller females, in particular, pursue and hunt rodents deep into their dwellings and tunnels. In areas that are permanently covered with snow in winter, Stoats and Least Weasels also frequently hunt under the snow. Stoats and Least Weasels do not dig dwellings themselves but take over tunnels dug by their prey. Least Weasels often spend time underground or under snow. The relatively small size of Stoats and Least Weasels makes them potential prey for many other carnivores. Raptors can sometimes have a major impact on Stoat and Least Weasel populations.

Least Weasel
HTL 12–25cm
TL 3–8cm
W 40–210g

Stoat
HTL 17–33cm
TL 7–14cm
W 110–470g

SIZE Stoats and Least Weasels can differ considerably in size depending on range. In European populations, the animals are smallest in the north-east and become larger towards the south and west. In areas such as the Mediterranean region, where Stoats are absent, Least Weasels can reach the size of Stoats. Due to pronounced sexual

The Least Weasel is the smallest carnivore on earth. Stoat winter pelts (or ermine) have been used as fur for royal robes since the late Middle Ages.

dimorphism (males can reach 1.5 to 2 times the weight of females), the sizes of small female Stoats and large male Least Weasels may overlap in areas where both species occur. It is therefore normally only possible to reliably differentiate between tracks outside these areas.

DISTINGUISHING FEATURES   The Stoat has a long, slender body, short legs and long tail with a black tip. Least Weasels are usually smaller, with a shorter tail and no black tip. The brownish coat turns white in winter in some populations, though in mild climates they remain brown all year round. The tip of the Stoat's tail remains black even in its winter coat, while the Least Weasel's winter coat is completely white.

DISTRIBUTION AND HABITAT   Both are found throughout almost the whole of Europe. Stoats are absent from the Mediterranean region and the Balkans. After European Badgers, Least Weasels are the most widespread mustelid species and are only absent from Ireland and Iceland. The habitat of both species covers a wide range of different ecosystems. They can occur in almost any biotope and at any altitude, as long as sufficient prey and cover are available. They prefer varied, structurally rich landscapes with hedges, dense undergrowth and dry stone walls and tend to avoid open terrain. Stoats are less common than Least Weasels at the centre of larger forests.

DIET   Specialist feeders that mainly focus on the various vole species – Common Vole, Short-tailed Field Vole, Water Vole and Bank Vole – as well as other small mammals. Their prey also includes murids, lemmings, squirrels, hamsters, sousliks, rabbits and shrews, with Stoats more regularly taking the larger species. Small birds, birds' eggs, amphibians, insects and earthworms make up a much smaller part of their diet. Stoats and Least Weasels need to eat 25–40 per cent of their body weight every day, which is equivalent to about two mice a day. If there is a surplus of food, Stoats and Least Weasels build up stocks of food.

REPRODUCTION   The mating season lasts from April to July. In Stoats, delayed implantation results in a prolonged gestation period of 9–10 months. Active pregnancy lasts four weeks. An average of 4–9 altricial young are born in spring, with a peak in April. Female Stoats may mate and conceive at the remarkably young age of 3–6 weeks. Embryonic diapause delays development of the embryo so the animals are fully grown by the time they give birth. Least Weasels have a gestation period of 34–37 days with no embryonic diapause. They can give birth at almost any time of year. If food supplies are good, they can produce two litters in a year. In Least Weasels, both sexes reach sexual maturity after 3–4 months.

NOTE Stoats employ an unusual hunting strategy where they 'hypnotise' European Rabbits. They run back and forth, dart sideways and leap in the air, seemingly at random, confusing the rabbit and causing it to stop instead of running away. The Stoat systematically approaches the rabbit in a series of random jumps until it is close enough to overpower it.

SIGNS

FOOD STORES If there is a surplus of food, Stoats and Least Weasels can build up large food stores, for example under tree trunks. Carefully stacked carcasses of field mice and voles near the dwelling are a characteristic sign.

EXCREMENT Excrement is long, thin, kinked and tapers to a long point at one end. The kinking and shape are typical of mustelids and caused by the high proportion of small mammal hairs. Bone remains are rarely found. Least Weasel droppings are around 50 per cent thinner than Stoat's and usually shorter. Being able to distinguish between the two species based on droppings alone requires experience and knowledge of the individual animals present. Least Weasel can generally be ruled out if L > 6cm and/or D > 0.3cm. Droppings are usually greyish brown to black in colour and tend to be deposited in clearly visible raised sites such as on tree stumps or stones.

- **Least Weasel** L 2.5 – 6cm D 0.2–0.3cm
- **Stoat** L 4 – 8cm D 0.5cm

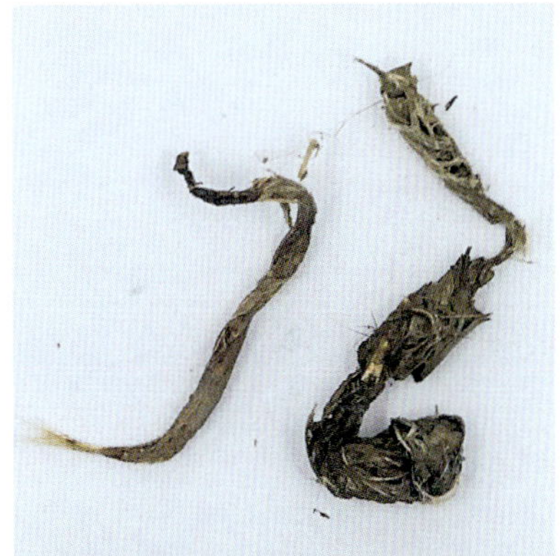

▼ *Least Weasel droppings. Apart from by size, it is not possible to reliably distinguish between the two species based on their droppings. John Rhyder, England.*

▶ *Stoat excrement (bottom), and Least Weasel excrement (top right). Lausitz, Germany. Heide Ulrich.*

TRACK

Typical mustelid arrangement. The soles of the feet are hairy, which can make the tracks appear larger and less clear. The foot morphology of Stoats and Least Weasels is almost identical. Apart from by size, it is nearly impossible to distinguish the species from each other. In areas where both species occur, the Stoat is always the larger species. However, in areas where Stoats are absent, Least Weasels can be much larger, reaching typical Stoat size. Here we have only used measurements from areas where both species occur and the Least Weasel is correspondingly smaller than the Stoat.

Front

Least Weasel
L 0.8–2cm   W 0.7–1.7cm

Stoat
L 1.6–3.8cm   W 1.2–3cm

Complete track with L < 1.4cm = Least Weasel
Complete track with L > 2.4cm = Stoat

Tiny to very small (Least Weasel). Very small to small (Stoat). Plantigrade. Asymmetrical. Five toes. Toe 1 is the smallest toe and cannot be reliably recognised. The interdigital pads have fused, but there are usually individually recognisable bulges. If metacarpal pad I is visible, it leaves an individual print. Another thenar/hypothenar pad may be

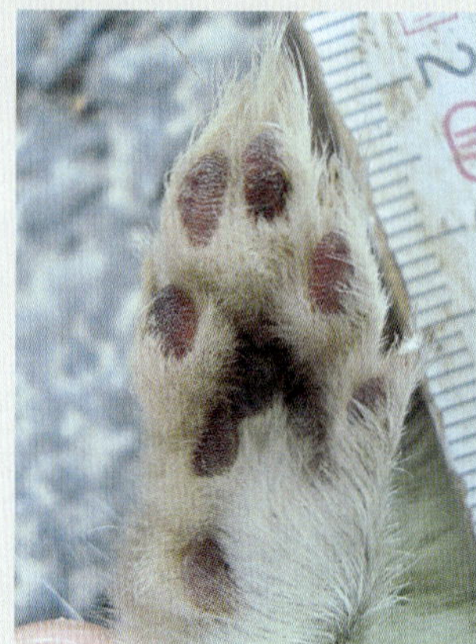

▲ *Stoat, front right. Göttingen, Germany. Heide Ulrich.*

▲ *Least Weasel, right hind. Germany. Heide Ulrich.*

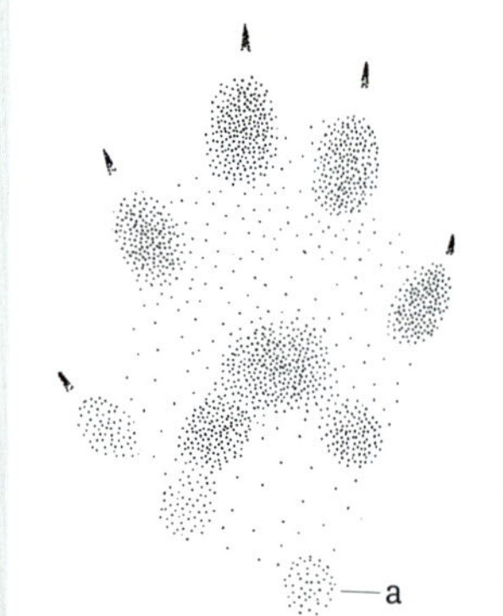

▲ *Stoat, front right.*

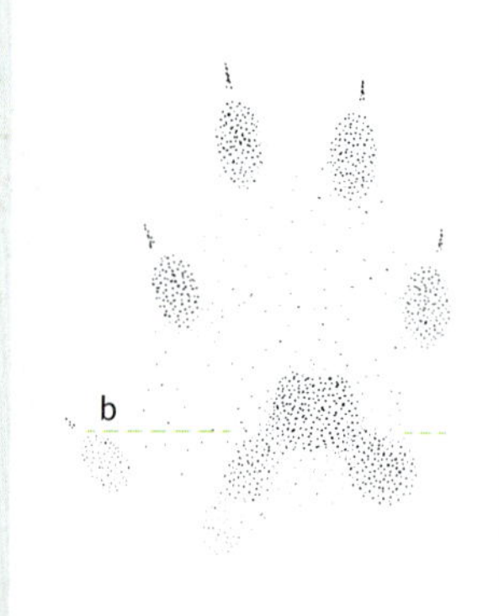

▲ *Stoat, front hind.*

▲ *Stoat, front right. A complete print is a rare find. Nahrendorf, Germany.*

▲ *Stoat, right hind. Lower Oder Valley, Germany.*

visible at the back of the track (a). The fine, sharp and relatively short claw marks are sometimes visible. The front foot print is generally larger and, most importantly, wider and stronger than that of the hind foot. However, the hind foot print can occasionally appear larger. This is mainly the case when the front foot leaves a print without the thenar/hypothenar pad and the hind foot leaves a complete print.

Hind

Least Weasel
L 0.8–1.9cm   W 0.6–1.4cm

Stoat
L 1.6–3.2cm   W 1.3–2.5cm

Tiny to very small (Least Weasel). Very small to small (Stoat). Plantigrade. Asymmetrical. Asymmetry clearer than in the front foot. Five toes. Toe 1 is the smallest toe and is situated much further back than on the front foot (b). This makes the hind foot more asymmetrical. If toe 1 is not visible, the hind foot print appears more symmetrical than the front foot print and may resemble the footprint of a small canid. Toe 1 is often unrecognisable. The interdigital pads have fused, but there are usually individually recognisable bulges. If metatarsal pad I is visible, it leaves an individual print. The hind foot print generally appears narrower and more elongated than the front foot print. Fine, sharp and relatively short claw marks are sometimes visible.

▼ *Left front (top) and left hind (bottom). Clear Least Weasel* (Mustela nivalis) *tracks are rarely found. Welzow, Germany.*

GAITS

Stoats and Least Weasels explore and hunt in bounds. When they investigate something more closely, they switch to the much slower walk. They can gallop when fleeing or when threatened. Least Weasel bounds often appear irregular. Stride length measurements can vary greatly from bound to bound. Least Weasels can make very sharp turns, so a 360° turn from one group of four to the next is possible. When searching for prey, Least Weasels and Stoats prefer to move along covered border areas such as hedges, dry stone walls and the banks of small streams. Paul Marchesi called these paths in his research area 'stoat highways' because they were so frequently used (Marchesi *et al.* 2010, page 54). Border areas with dense cover offer the tiny Least Weasel both food and protection from predators. Both species favour the 2 × 2 bound, especially in deep snow, and often tunnel in soft snow.

2 × 2 bound
Group length: 1.2–3.8cm
Inter-group length: 6–70cm
Stride length: 7.5–72cm
(Least Weasels usually under 50cm)
Trail width: 2–5.5cm

3 × 4 bound (rotary) and 4 × 4 bound (transverse)
Group length: 3.5–18cm
Inter-group length: 5–65cm
Stride length: 8.5–84cm
Trail width: 2.5–7.5cm

A Least Weasel's stride length is usually less than 50cm. The trail width of a Least Weasel is usually between 2–4cm and not larger than 4.5cm. The trail width of a Stoat is usually between 2.8–5cm and not larger than 5.5cm.

▼ *Stoat, 4 × 4 bound (transverse). Footfall from bottom to top: LF, RF, LH, RH. Bieszczady, Poland.*

▲ *Stoat, 4 × 4 bound (transverse). Lausitz, Germany.*

▼ *Stoat, 2 × 2 bound.*

▼ *Stoat, 4 × 4 bound (transverse).*

Similar tracks

Least Weasel tracks are normally smaller than tracks, and in some cases can be identified by the size of the track alone. However, large Least Weasel tracks can be difficult or even impossible to distinguish from those of small Stoats.

Beech and European Pine Marten footprints are much larger and the tracks of American Mink are usually clearly identifiable by their characteristic otter-like appearance. Stoat and European Polecat tracks can look very similar. If track sizes are similar, they are difficult to tell apart. Although it is sometimes possible to make a clear identification based on foot morphology, this requires almost perfect footprints (see 'Differentiating between Stoats, European Polecats and American Mink' page 481).

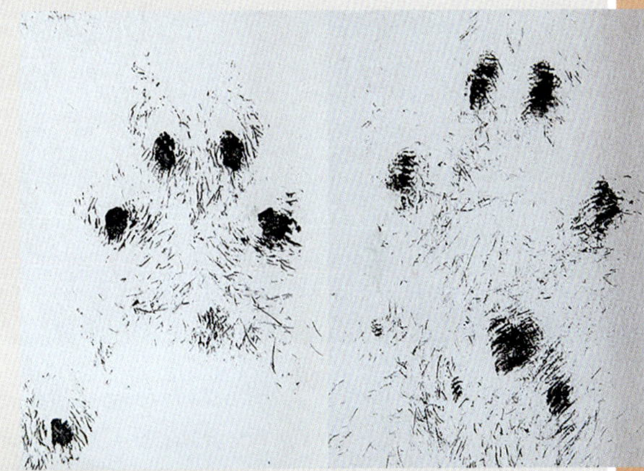

▲ *Ink prints of a Least Weasel track (left) compared to those of a Stoat (right). While metatarsal pads II–IV are often recognisable in Stoats, often only metatarsal pad III is seen in Least Weasels.*

▼ *Left front foot print of a Least Weasel (left) and a Stoat (right) in size comparison. Germany. Aaron Tiedemann.*

AMERICAN MINK
Neovison vison

HTL 30–55cm

TL 13–23cm

W 0.45–2.3kg

The American Mink is the only species of the genus *Neovison* that occurs in the area covered by this book (note that some taxonomists place it in the genus *Neogale*). It is solitary, territorial and mainly crepuscular and nocturnal. However, it can also be seen during the day. Sense of sight, hearing and touch play an important role for mink. They can swim, dive and climb well. American Mink are vulnerable to almost all larger predators. Humans and hunting dogs cause the greatest losses.

DISTINGUISHING FEATURES   Similar size to a European Polecat, but more uniformly coloured (no face mask) and normally darker. Dense, shiny coat. Unlike European Polecats, American Mink do not have white-rimmed ears. Much smaller than a Eurasian Otter.

DISTRIBUTION AND HABITAT This species is native to North America. Its fur was once in great demand, leading to the spread of fur farms all over the world. Escapes and deliberate releases of farmed mink have allowed permanent wild populations to become established in Europe since the 1950s. American Mink are highly dependent on water. However, they are also found in other habitats with sufficient prey, such as large rabbit populations. They generally live near streams, rivers and lakes where there is sufficient bushy vegetation, overhanging banks, tree roots and similar structures. These provide cover and contribute to a plentiful food supply. In coastal regions, American Mink favour stony habitats with good cover.

DIET The American Mink is an opportunistic carnivore that feeds almost exclusively on animal-based food. Its varied diet includes mammals, fish, crustaceans, amphibians, reptiles, birds and invertebrates. Main food sources vary greatly depending on season and availability of prey. Small to medium-sized mammals, such as Muskrats, rabbits, squirrels, voles and murids, are particularly important. They enter Muskrat burrows to prey on young animals. They also hunt waterfowl, such as ducks, Moorhens and coots on rivers and lakes, as well as fish. They catch, kill and eat adult ducks and terns, their young and, occasionally, their eggs. This can lead to significant population losses. If American Mink fish, they mainly catch small, slow-swimming fish such as perch (*Perca*). In rare cases, they will eat amphibians, snakes, beetles and earthworms. They store surplus food. If Eurasian Otters and American Mink share a habitat, the latter's diet changes in favour of more land-based food and less aquatic food, evidence that American Mink spend less time near water when Eurasian Otters are present in the area.

REPRODUCTION The breeding season lasts 3–4 weeks between February and April. Implantation of the fertilised egg can be delayed by 11–48 days, resulting in a prolonged gestation period of 39–76 days. Active pregnancy lasts 28–30 days. An average litter of 3–6 young is born during the period from April to June, but mainly in May. The altricial young become sexually mature at around one year of age.

TRACK

Typical mustelid arrangement. Proximal webbing is occasionally visible, especially in deep prints. The claws often merge with the digital pads in the print, creating a pointed impression. The delicate digital pads have a teardrop shape (a). Tracks often appear deep and splayed in the digital pad area, especially on soft ground. The midfoot pad is relatively strong, but usually leaves a faint print. The soles of the feet are less hairy than in other mustelids.

Front

L 3.2–4.9cm W 2.4–4.3cm

Small. Plantigrade. Asymmetrical. More symmetrical overall than the hind foot. Five toes. Toe 1 is the smallest toe and cannot be reliably recognised. The relatively delicate digital pads extend in a fan shape from the midfoot pad, which often results in characteristic splaying. The distance between toes 3 and 4 is usually greater than 4mm. The rather strong interdigital pads have fused, but usually leave prints with individually recognisable bulges. If metacarpal pad I is visible, it leaves an individual print. A further thenar/hypothenar pad that is offset to the outside can be visible at the back of the track. The fine, pointed claw marks are often visible. The front foot print usually appears wider and stronger than the hind foot print.

▲ *Right front.*
Märkische Schweiz, Germany.

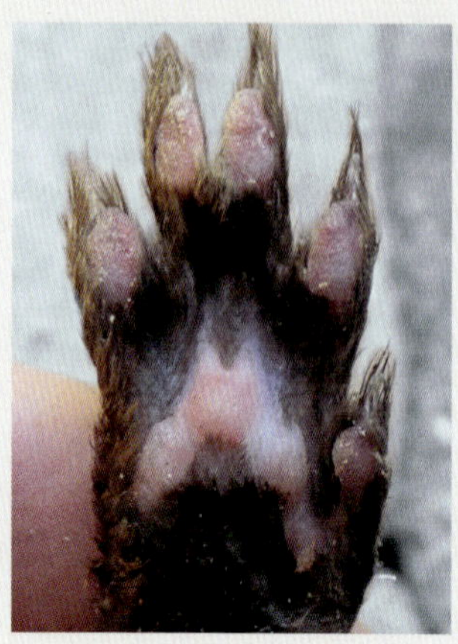

▲ *Right hind.*
Vledder, Netherlands. René Nauta.

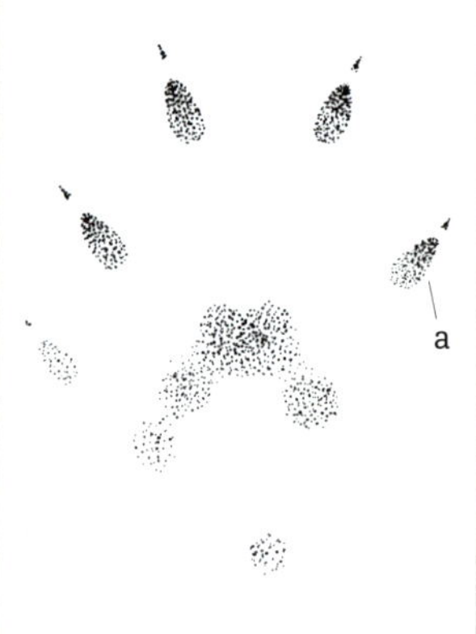

▲ *Right front.*

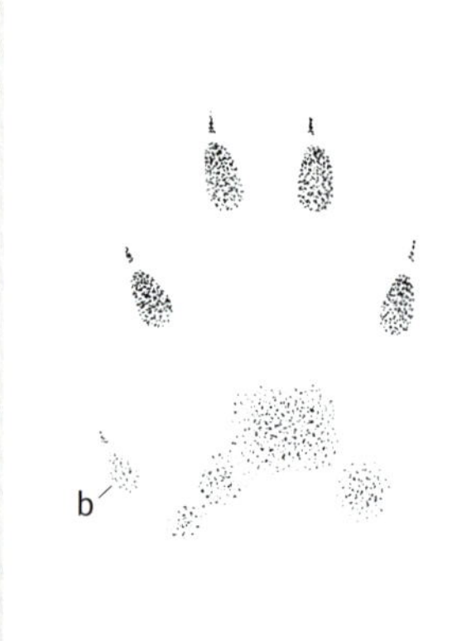

▲ *Right hind.*

▲ *Right front.*
Lausitz, Germany.

▲ *Right hind.*
Lausitz, Germany.

Hind

L 2.5–4.2cm W 2.1–3.8cm

Small. Plantigrade. Asymmetrical. Asymmetry clearer than in the front foot. Five toes. Toe 1 is the smallest toe and is situated much further back than on the front foot ⓑ. If toe 1 is not visible, the track may appear more symmetrical. The interdigital pads are fused, but there are usually individually recognisable bulges. If metatarsal pad I is visible, it leaves an individual print. The digital pads splay less than on the front feet. Fine, pointed claw marks are often visible. The hind foot print usually appears narrower and more elongated than the front foot print.

GAITS

American Mink mainly explore and hunt in 2 × 2 bounds and 3 × 4 bounds (rotary). When slowing to investigate something more closely, they switch to a walk. They gallop when fleeing or when threatened. They favour the 2 × 2 bound in deep snow. Compared to Stoats and Least Weasels, American Mink have a more constant and relatively shorter stride length in the 2 × 2 bound. Similar to Eurasian Otters, the track groups of the 3 × 4 (rotary) and 4 × 4 bound (transverse) are often diagonal to the direction of travel. This feature can help to distinguish American Mink from Stoats and European Polecats, whose track groups tend to be parallel to the direction of travel.

Walk
Stride length: 23–36cm
Trail width: 8–13cm

2 × 2 bound
Group length: 6–18.5cm
Inter-group length: 19–79cm
Stride length: 25–97.5cm
Trail width: 5–11.5cm

3 × 4 bound (rotary) and 4 × 4 bound (transverse)
Group length: 13–35cm
Inter-group length: 16–85cm
Stride length: 29–110cm
Trail width: 5–11cm

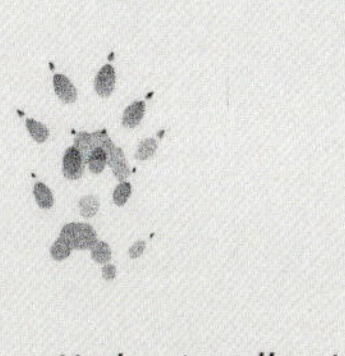

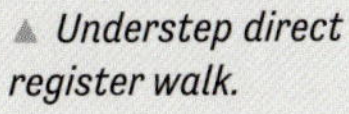

▲ *Understep direct register walk.*

▲ *3 × 4 bound (rotary).*

◀ *American Mink in a walk. American Mink mainly walk when they investigate something more closely. Lausitz, Germany.*

▲ *In the 2 × 2 bound typical of martens, both hind feet step into the prints of the front feet. Here the left hind foot print is offset to the inside. Lausitz, Germany.*

▲ *American Mink in a 4 × 4 bound (transverse). Footfall from bottom to top: LF, RF, LH, RH. Kreba-Neudorf, Germany.*

Similar tracks

European Pine Marten and Beech Marten tracks are larger. European Polecat tracks are a similar size. It can be difficult to tell them apart, but it is usually possible if the footprints are suitable.

▼ *Ink prints of a Stoat (left) and European Polecat (right) for comparison. Lützel, Germany.*

DIFFERENTIATING BETWEEN STOATS, EUROPEAN POLECATS AND AMERICAN MINK

Stoat

a) Rounded digital pads, usually slightly splayed and coming straight from the midfoot pad.

b) Relatively short, fine claw marks.

c) Metacarpal/metatarsal pad III roundish in shape.

d) Tracks tend to be smaller and narrower than European Polecat tracks.

e) Thick hair on the soles of the feet. Metacarpal/metatarsal pad prints often appear unclear or are absent altogether.

European Polecat

a) Relatively strong, roundish digital pads, usually slightly splayed and coming straight from the midfoot pad. The distance between toes 3 and 4 is usually less than 4mm.

b) Relatively long, strong claw marks, which are usually more than 6mm in front of the respective digital pad (Harrington 2008).

c) Interdigital pads rather fine compared to those of American Mink. Metacarpal/metatarsal pad III is noticeably rectangular in shape.

d) Tracks quite small in relation to body size and tend to be smaller and more elongated than American Mink tracks.

e) The soles of the feet are hairy, but the digital and metacarpal/metatarsal pads are usually clearly visible.

American Mink

a) Relatively fine, teardrop-shaped digital pads, usually clearly splayed from the midfoot pad. The distance between toes 3 and 4 is usually greater than 4mm.

b) Relatively short, fine claw marks, usually less than 6mm in front of the respective digital pad (Harrington 2008).

c) Interdigital pads rather strong. Metacarpal/metatarsal pad III is roundish in shape.

d) Tracks large in relation to body size and tend to be larger and rounder than European Polecat tracks.

e) The soles of the feet are comparatively lacking in hair. The digital and metacarpal/metatarsal pads are more clearly recognisable.

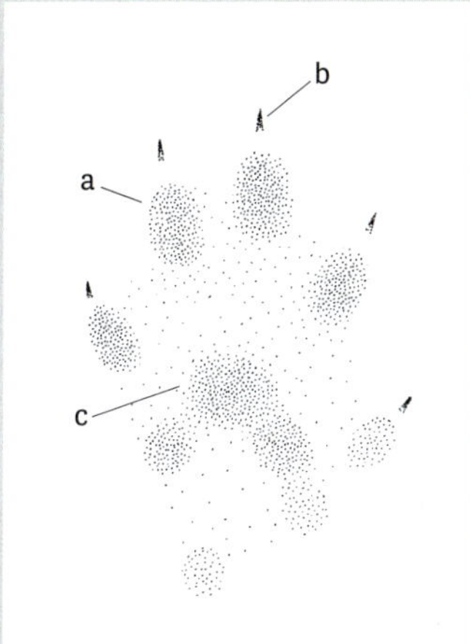

▲ Stoat, left front.

▲ European Polecat, left front.

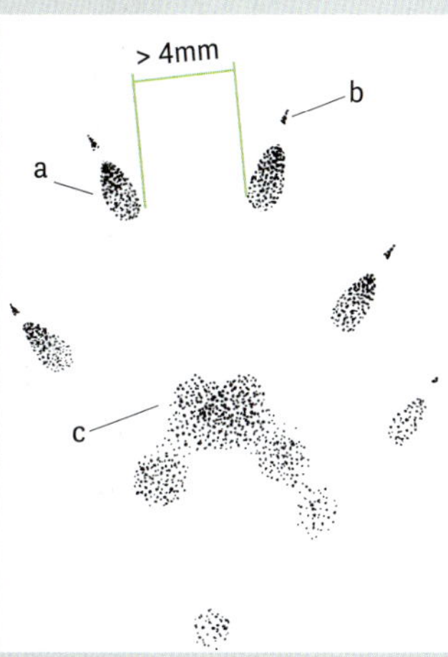

▲ American Mink, left front.

REMARKS Competes with the Eurasian Otter and European Mink. If both mink species occur together, the American Mink displaces the European Mink. Not protected as a non-native species, and globally it is classed as Least Concern.

SIGNS

▲ *This American Mink food store contained over a dozen small fish. Lausitz, Germany.*

FOOD STORES American Mink build up food stores, mainly in autumn and early winter. Their 'larder' can contain frogs, fish, ducks and smaller mammals up to the size of a European Rabbit or Muskrat. American Mink hide their food supplies in natural chambers between stones and in hollow tree trunks or holes in the ground. One particularly large food store was found to contain 13 Muskrats, two Mallards and a Eurasian Coot.

▼ *An American Mink dwelling. Witzenhausen, Germany. Heide Ulrich.*

DWELLINGS American Mink take over and sometimes extend hollows in trees, existing burrows or Muskrat and Eurasian Beaver dwellings. Their tunnel system featuring several chambers, entrances and exits covers around 1–3m². The dwellings are usually less than 2m away from water. The entrance holes have a diameter of 5–9cm.

PATHS As with Eurasian Otters, conspicuous paths can occur along bodies of water. The paths are about 10–15cm wide, so they are much smaller than otter paths. However, the minimum width of an otter path can overlap with the maximum width of an American Mink path.

CARCASS FEATURES Like Eurasian Otters, American Mink usually eat whole fish, so feeding marks are rare.
- Distance between the canine teeth: 0.8–1cm (0.6–1.2cm)

EXCREMENT Long, cylindrical, kinked and pointed at one end. If many fish have been eaten, excrement will contain large amounts of scales and bones. Droppings are generally dark green, brown or black in colour, depending on diet. American Mink use droppings to mark the entrance to their dwelling, elevated sites (such as stones and tree roots), as well as entrances and exits, bridges and crossroads. They deposit more droppings at territory boundaries. Repeated use of these distinctive places results in larger latrines. Eurasian Otter droppings can be found in similar places. They smell slightly more pleasant (sweetish) and are normally much larger and often softer. European Polecat droppings are very similar to American Mink droppings, although less likely to be mostly made up of fish remains.
- L 4–9.2cm D 0.5–1.1cm

▶ *American Mink excrement. The droppings resemble otter droppings but are usually smaller and contain finer fish remains.*
Lausitz, Germany.

NORTHERN RACCOON
Procyon lotor

HTL 41–70cm
TL 19–30cm
W 4–15kg

Raccoons belong to the New World carnivore family Procyonidae. In terms of shape, they are somewhere between a marten and a bear. They have a stocky build with a rather short, broad head and usually a longer tail. Procyonids are mainly nocturnal omnivores that are adept climbers. In Europe, the Northern Raccoon is the only procyonid that has successfully become established as a non-native species.

Northern Raccoons are clever and dexterous. Their sense of touch is particularly well-developed, which is why they are said to be able to 'see with their hands'. They swim and climb well, and can turn their hind feet 180°, enabling them to climb upside down. Northern Raccoons usually spend the day in the safety of trees or in their dwelling. They mainly forage on the ground at night. Northern Raccoons are territorial, but females can form small groups with their daughters and males sometimes form groups of 3–4. Depending on the climate, they may go into hibernation. They are predated by wolves and other canids.

DISTINGUISHING FEATURES Barely the size of a fox, with a stocky, bear-like body and a black face mask. Short, pointed muzzle and striped tail.

DISTRIBUTION AND HABITAT Native to North and Central America, the animals escaped or were released from fur farms in various parts of Europe during the Second World War. Northern Racooons have become established as a non-native species in many areas of Europe and continue to spread. Today they are found in Austria, Poland, Germany, the Netherlands, France and Luxembourg. Northern Raccoons favour mixed deciduous forests with a high proportion of dead wood and many wet areas. As dwellings, they often use hollow trees or abandoned badger setts or fox dens, as well as old buildings and attics. The animals are very adaptable and can even colonise less suitable habitats, such as steppes. As a species associated with human activities, they are also found in cities. The only places they avoid are very dry areas and mountains above 2,500m.

DIET Opportunistic omnivores that eat more plants than animals. Fruit, berries, grass, maize, acorns and beechnuts make up the majority of their plant-based diet. Insects, worms and snails, as well as crayfish, fish and mice, play a major role as animal food sources. Northern Raccoons pull the skin off toads before eating them. They also eat human rubbish and prey on birds and their eggs.

REPRODUCTION The mating season lasts from January to March and is the only time when males and females live together. If a female loses her first litter or does not become pregnant, a second ovulation can occur in May–July. After a gestation period of 60–73 days, 2–4 (up to 7) young are born from March to September. The young are almost independent by the end of autumn, but they often hibernate with their mother and become sexually mature after 1–2 years.

SIGNS

SLEEPING PLACES Northern Raccoons prefer hollows in old standing trees, often oaks. Their sleeping places are normally located at a height of 3–12m and no further than 100m from the nearest body of water (Gehrt 2003). They often sleep in abandoned fox dens or badger setts or in attics. Hair, latrines or the sounds of young animals in spring can draw attention to the main sleeping places.

▶ *Northern Raccoon hair like this can be found near day shelters and main sleeping areas. Märkische Schweiz, Germany.*

TRACK

Front

L 4.1–8cm W 4–7cm

Small to medium sized. Plantigrade. Symmetrical. Five finger-like toes that can leave prints like a small human hand. The toes usually leave a print all the way to the midfoot pad and are arranged in a fan shape. This usually makes the toes appear more spread out than on the hind foot. The interdigital pads have fused to form a larger midfoot pad, which can have a clearly curved C-shape (a). The largest interdigital pad is metacarpal pad IV which is located on the outside of the foot (b). A further thenar/hypothenar pad may also be visible. The small, pointed claws leave marks. The front foot print usually appears shorter and rounder than the hind foot print.

Hind

L 4.8–10cm W 3.6–6.8cm

Medium to large. Plantigrade. Asymmetrical. Five toes. Toe 1 is the smallest toe and is clearly further back in the print. The toes are slightly shorter and more parallel than on the front foot. They usually leave a print that extends to the midfoot pad. The interdigital pads have fused to form a large midfoot pad that is considerably longer than it is wide. The triangular-shaped back of the midfoot pad usually leaves a fainter print and may be completely absent (c). The small, pointed claws can leave marks.

▲ *Left front.*
Märkische Schweiz, Germany.

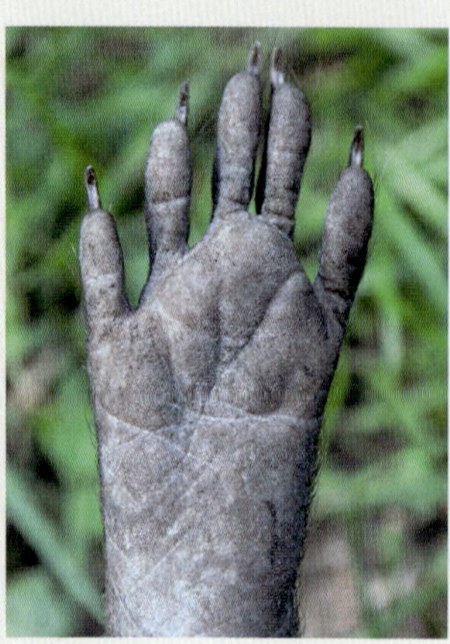

▲ *Left hind.*
Märkische Schweiz, Germany.

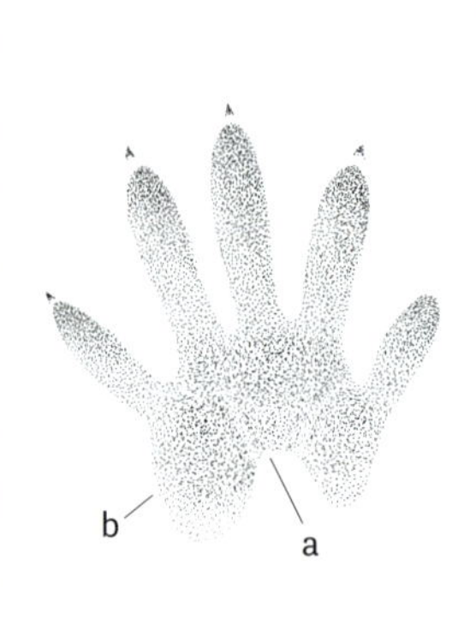

▲ *Left front.*

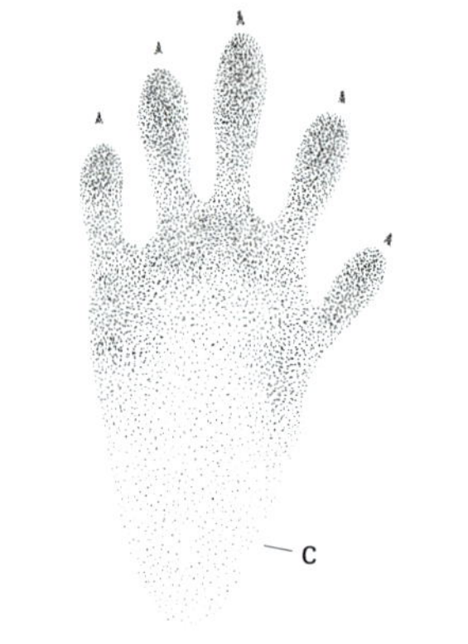

▲ *Left hind.*

▲ *Left front.*
Witzenhausen, Germany.

▲ *Left hind.*
Witzenhausen, Germany.

GAITS

Northern Raccoons leave one of the most characteristic track patterns of all European mammals: the extreme overstep walk. This track pattern indicates that the raccoon's favoured gait is the amble. The hind foot of one side of the body lands next to the front foot on the other side. The extent to which the hind feet overstep indicates the speed of the amble (page 80). Raccoons prefer a direct register walk in deep snow. They bound and gallop when alarmed or threatened.

Amble
Stride length: 30–113cm
Trail width: 7.5–18cm

Direct register walk
Stride length: 42–65cm
Trail width: 9–16cm

3 × 4 bound (rotary) and 4 × 4 bound (transverse)
Group length: 25–67cm
Inter-group length: 30–54cm
Stride length: 55–106cm
Trail width: 11–16cm

Parallel bound
Group length: 31–68cm
Inter-group length: 28–75cm
Stride length: 59–135cm
Trail width: 15–25cm

Similar tracks
If the toes do not leave a print all the way to the midfoot pad, Northern Raccoon tracks can be mistaken for Eurasian Otter footprints.

▲ Amble.   ▲ Overstep amble.   ▲ Parallel bound and gallop.

▶ Northern Raccoon in an offset bound, from bottom to top: RF, LF, LH, RH. Oderbruch, Germany.

◀ *In this characteristic track pattern, the front foot print of one side of the body and the hind foot print of the opposite side alternate with each other. Witzenhausen, Germany.*

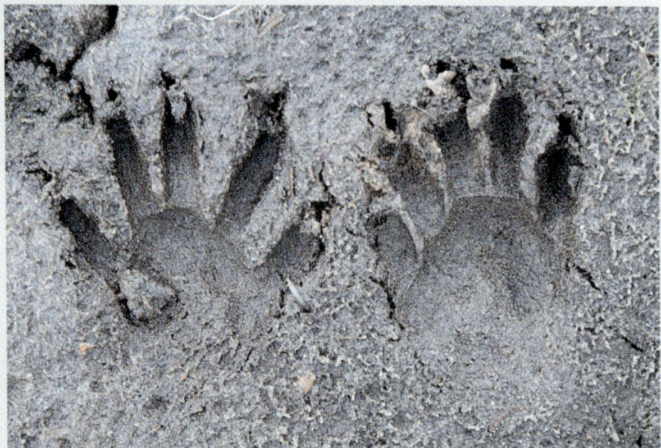

◀ *Left front (left) and right hind (right). Oderbruch, Germany.*

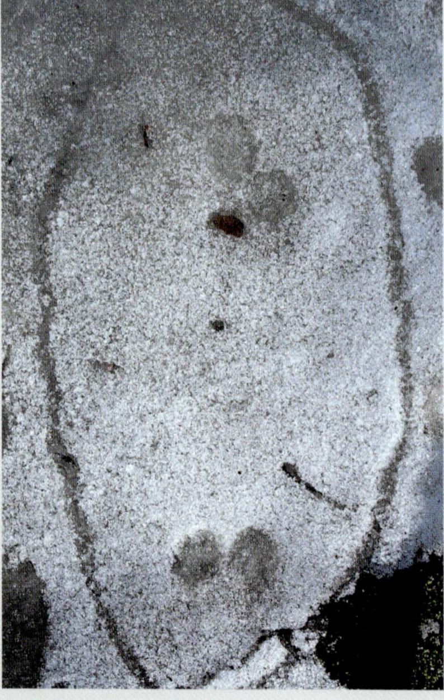

▲ *Even when no details are visible, Northern Raccoons are often identifiable by their distinctive track pattern alone. The roundish tracks come from the front foot and the elongated ones from the hind foot. Märkische Schweiz, Germany.*

DIGGING MARKS AND SCRATCH MARKS Northern Raccoons mainly dig softer surfaces and near water to get to prey like crayfish or insect larvae. Their marks are relatively short and boxy compared to Red Fox digging marks. Spoil heaps from European Badger digging activity are larger and more cone-shaped in comparison. Northern Raccoons also dig in harder ground on the hunt for bee or wasp nests. This results in tunnel-like entrances that are narrower than European Badger digging marks and usually do not expose the nest. Northern Raccoons also make scratch marks on the bark of trees when they climb them. They often use the same route and noticeably even, almost parallel scratch marks can appear over time. Marten scratch marks are finer and more irregular and erratic by comparison.

CARCASS FEATURES
- **Distance between the canine teeth: 2–2.4cm (1.7–2.8cm)**

EXCREMENT Droppings are usually sausage-shaped and blunt-ended. They break up readily and resemble the excrement of small bears. Because Northern Raccoons eat such a wide range of food, the appearance of their excrement varies greatly depending on diet and season. Droppings can be a shapeless pile or, more rarely, kinked. Northern Raccoons create large communal latrines at the base of distinctive trees near water or under rocky outcrops. They usually use the side furthest from the water.
- L 5.5–15.2cm D 0.8–2.4cm

▲ *Northern Raccoons have lived in the hollow of this old oak tree in the marshland of Märkische Schweiz for over a decade. Märkische Schweiz, Germany.*

▼ *Northern Raccoon latrine. Spessart, Germany.*

▲ *Typical Northern Raccoon excrement in autumn, here with cherry stones. Lausitz, Germany.*

Caution!

The raccoon roundworm (*Baylisascaris procyonis*), which is harmful to humans, can be picked up when inspecting Northern Raccoon excrement. Although the probability of infection is extremely low, never sniff dry raccoon excrement.

WALRUS AND TRUE SEALS
Odobenidae, Phocidae

Tracks of five species of pinnipeds – the Walrus and four species of true seals – can be found on Europe's coasts. The legs have evolved into flippers with very strong bones and a shape optimally adapted to life in water. The upper canine teeth of Walruses have developed into long tusks. Seals do not have these. Walruses use their front and hind flippers under their body for locomotion. On land, seals move by using their front flippers to drag their body along. This is known as as 'galumphing' or 'hobbling'. Differentiating the species is difficult but as a group their signs are distinctive and characteristic.

TRACKS AND SIGNS OF WALRUSES AND SEALS

Plantigrade. Asymmetrical. Five toes on the front foot and hind foot.

- Track formula: 5F × 5h + C

FRONT FLIPPERS   The front flippers leave stronger, wider prints than the hind flippers. Large, strong claw marks are usually visible.

BACK FLIPPERS   Smaller claw marks than on the front flippers but usually do not leave prints.

TRACK PATTERN   The front flippers support the body as the animal pulls itself along on its front. The body leaves a clear drag mark with regular, successive front flipper prints at the sides. The five claw marks are almost parallel to the print of the body, but all five claws do not always leave marks. This track pattern is so distinctive that it cannot be confused with any other animal group.

▼ *The characteristic track of a Grey Seal. The decreasing stride length as the animal approaches the water indicates deceleration.*

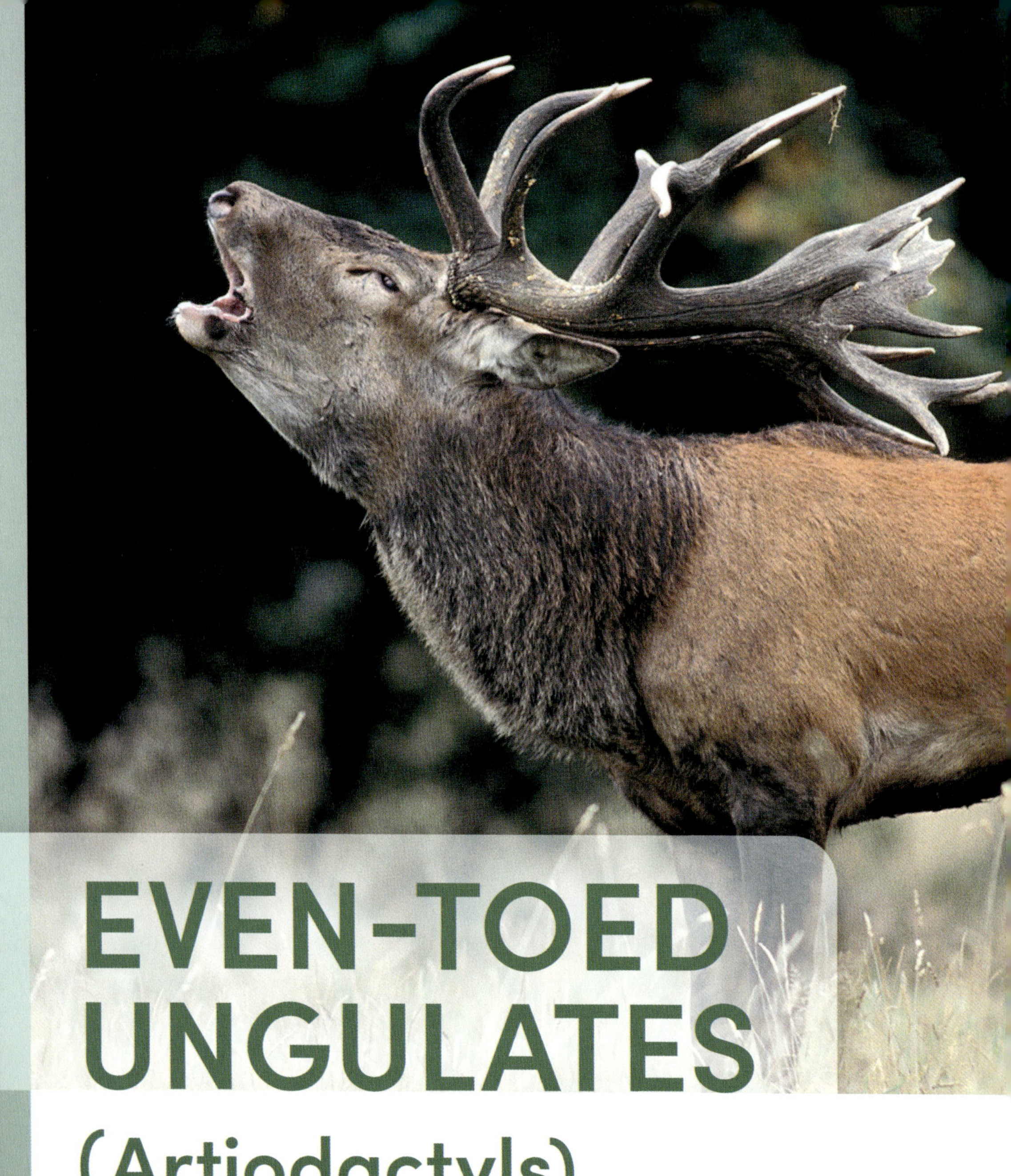

EVEN-TOED UNGULATES
(Artiodactyls)

Almost all even-toed ungulates (artiodactyls) that occur in Europe are pure herbivores and ruminants (Ruminantia) that are polygastric (they have multiple stomachs). This enables them to break down indigestible plant food, such as cellulose, by regurgitating predigested food when resting, so they can chew it again before swallowing the finer plant matter. Ruminating causes the fermenting bacteria in the stomach to multiply. These bacteria can digest tough cellulose. At the same time, chewing also softens the rough surfaces, for example of grasses. The pig family (Suidae) are an

exception because they are omnivorous and monogastric. Monogastric animals have a full set of teeth, while ruminants have a bony upper dental pad in their upper jaw instead of incisors. The males of many of the species in this order have developed horns or antlers that are mainly used for rutting behaviour, such as territory marking and fights with rivals, or for defence. In deer (Cervidae), they are known as antlers. They are made of bone and are shed and regrown every year. The horns of bovids (Bovidae) normally grow for the whole of the animal's life. Use of horns or antlers can leave distinctive signs, such as marks made by striking with the antlers or rubbing the horns on trees and shrubs.

All even-toed ungulates have excellent hearing and a well-developed sense of smell. Their sight is especially sensitive to movement. They tend to only see the outline of motionless objects or not see them at all. Typical of prey species, the eyes are positioned on the side of the head to give a wide field of vision that enables the animals to spot threats as early as possible. Their stereoscopic vision is limited because there is a limited overlap between the field of vision of each eye. Three families with a total of 19 species occur in Europe. We have grouped together the tracks and signs of these families according to their similarities.

TRACKS AND SIGNS OF EVEN-TOED UNGULATES

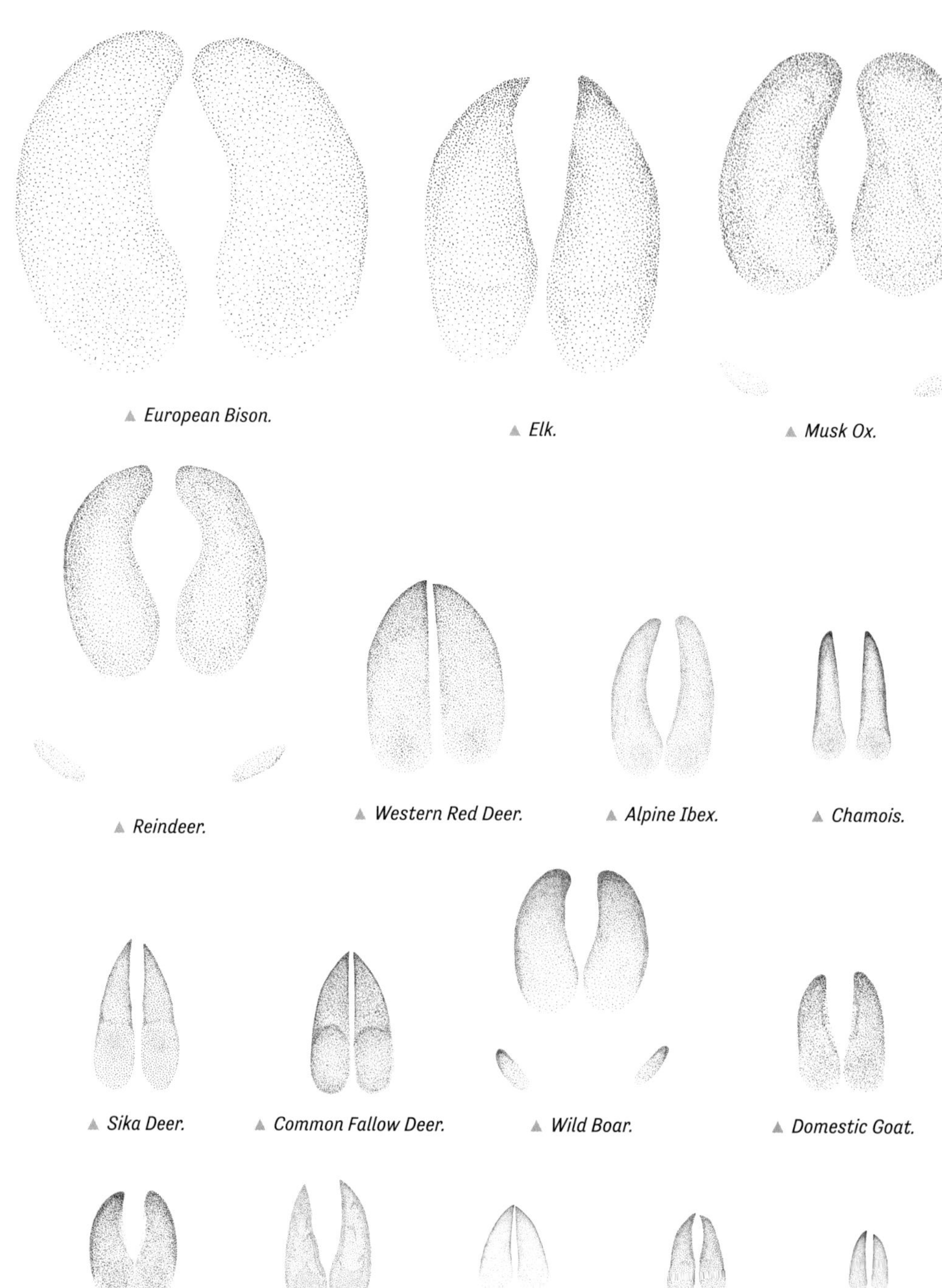

▲ *European Bison.*

▲ *Elk.*

▲ *Musk Ox.*

▲ *Reindeer.*

▲ *Western Red Deer.*

▲ *Alpine Ibex.*

▲ *Chamois.*

▲ *Sika Deer.*

▲ *Common Fallow Deer.*

▲ *Wild Boar.*

▲ *Domestic Goat.*

▲ *Domestic Sheep.*

▲ *Mouflon*

▲ *Western Roe Deer.*

▲ *Water Deer.*

▲ *Reeves's Muntjac.*

Clear Artiodactyl tracks are usually easily recognisable as such. Even-toed ungulates are referred to as 'cloven-hoofed' because their print looks like a single hoof with two halves (Henrichs 2004). The hooves are strong structures that contain keratin and enclose the ends of the limbs. Toes 3 and 4 have developed into a pair of hooves with horny shells. They are weight-bearing and always leave a print. The tip and lateral parts form the unguis and are harder than the underside, the subunguis (Hildebrand 2004). In contrast to most other land mammals, toe 3 is usually slightly smaller than toe 4. Toes 2 and 5 are very short, made of horn and referred to as dew claws. They are higher up on the back of the foot and are usually only visible in deeper prints or at high speed. Exceptions are Wild Boar and Reindeer, where the dew claws are lower down on the foot and leave prints with corresponding regularity. Toe 1 is absent altogether. It can be difficult to identify artiodactyl prints because the foot

▲ *Front foot (left) and hind foot (right) of a Western Red Deer. Note the different position of the dew claws. Märkische Schweiz, Germany.*

structure of many species are very similar. It is important to pay particular attention to the shape of the hooves, the distance between them and the size of the subunguis. Track pattern, social behaviour and biotope can also give important clues for species identification.

- **Track formula: 2F × 2h**

FRONT FOOT Slightly asymmetrical, larger, wider and stronger than the hind foot. The dew claws are closer to toes 3 and 4 and larger than on the hind foot. They can leave a print that is almost horizontal to the direction of travel. This is often intensified at high speed.

HIND FOOT Slightly asymmetrical, narrower and smaller than the front foot. The dew claws are further from toes 3 and 4 and narrower than on the front foot. They usually leave a print that is more parallel to the direction of travel. The dew claws can splay out more at high speed, which makes it difficult to tell them from the front foot dew claws. However, the degree of splaying is anatomically limited and normally less than in the front foot.

GAITS The most common gait is a walk. Apart from Wild Boar, all species have long legs and a correspondingly long stride. The relationship between stride length and trail width can be a useful clue for telling the difference between Wild Boar and other artiodactyls. At higher speeds, they usually use a form of trot. Even-toed ungulates bound and gallop to escape predators.

FEEDING MARKS Apart from Wild Boar, even-toed ungulates do not have any incisors in their upper jaw. As a result, they leave characteristic feeding marks on plants and fungi. Vegetation, like grasses, shrubs or woody plants, that have been eaten by artiodactyls normally have an uneven, frayed edge, because the animals clamp the plant material between their lower incisors and the dental pad in their upper jaw and then tear, rather than bite, it off. By contrast, rodents and lagomorphs leave very clean edges because they have sharp incisors in their upper and lower jaw. Artiodactyl browsing can cause damage to shrubs and young trees, giving them a short, deformed appearance. Heavy browsing gives larger trees a noticeable and even, horizontal browse line on the underside of the crown. This conspicuous sign occurs when the animals have eaten all the leaves, shoots, buds and twigs within their reach. Many even-toed ungulates use their lower incisors to peel bark off trees. Bark peeling is a form of cambium feeding (see page 154).

EXCREMENT With the exception of Wild Boar, all even-toed ungulates produce large quantities of droppings because they eat a lot of undigestible plant material. The colour of the faecal pellets normally varies between yellowish brown, green and black. Ruminants deposit large quantities of faecal pellets at once, while lagomorphs deposit each faecal pellet individually. This results in conspicuous, often rounded piles of these faecal pellets. Faecal pellets in all species are cylindrical if dry food has been eaten. A typical shape has one pointed and one blunt or indented end. Another possible form is acorn-shaped with two rounded ends. Wetter food in spring often results in cowpat-like droppings. Summer droppings often consist of fully or incompletely formed faecal pellets that are stuck together in lumps or in a sausage shape, because of the moister diet. It is possible to identify some species by their droppings, but in other cases the area of overlap is too great.

DETERMINING SEX BASED ON TRACKS AND SIGNS

There are different theories about how to determine the sex of even-toed ungulates from their tracks. However, they are rarely reliably applicable in practice. The following is a summary of information that can be used for sex identification. It should be understood more as a set of guidelines than universally applicable rules.

TRACKS In the case of all even-toed ungulates, the tracks of adult males are larger than the tracks of adult females. Therefore tracks in the maximum range of measurements probably come from male animals, but the measurements of a younger male can overlap with those of an adult female. In male Western Red Deer, the size difference between the front feet and hind feet can be more marked, especially in terms of width, than in females and young animals (Halfpenny 1999). My own measurements of 50 Western Red Deer have demonstrated this. Louis Liebenberg writes about similar observations of even-toed ungulates in Africa (Liebenberg 1990). Hypothetically, this size difference could be down to the need to support the extra weight of the head, neck, thorax and horns or antlers. Mark Elbroch, Brian McConnell and David Moskowitz describe how this size difference also occurs in other cervids in the northwest of the US and in Bighorn Sheep (Elbroch 2003, Moskowitz 2010). However, it relates more to the length of the track or foot than the width. This criterion can be used as a means of differentiation for some species, including Western Red and Common Fallow Deer, but not for some other species, such as Western Roe Deer.

Measuring tracks in the field

We compare the size of the front foot and hind foot in the field by measuring the circumference of both tracks. A piece of string can be a useful tool for this. A clear size difference indicates a male animal.

SOCIAL STRUCTURE In some species, different social structures between the sexes can also result in clear differences in track patterns. Many even-toed ungulates move in groups that consist of adult females, their young from the current year, and young from the previous year. Adult males live apart from adult females and young animals for most of the year. If a group has a high variance of track sizes it is probably made up of adult females with young of both sexes. Males, especially younger animals, form so-called 'bachelor herds'. Similarly sized tracks from a group of adult animals were probably made by males. Older males are typically solitary in some species. Their usually noticeably large tracks are often found in isolation over longer distances, because adult males normally only associate with other animals during the annual rut.

SCENT MARKING AND RUTTING BEHAVIOUR Male even-toed ungulates of the species covered in this book leave conspicuous signs when they engage in scent marking and rutting behaviour. They cause distinctive damage to trees by striking them with their antlers or horns, and rubbing their body on the bark. Scrape marks on the ground, as well as mud baths and rutting hollows, are equally distinctive (page 176).

WILD BOAR
Sus scrofa

HTL 120–185cm

TL 15–30cm

W 50–200kg (up to 350kg)

Males are much stronger and heavier than females. Weight also fluctuates greatly depending on region and season. Weight and size increase from the southwest to the northeast.

In Europe, the Wild Boar is the only wild representative of the pig family (Suidae). Unlike ruminants, Wild Boar have upper incisors. Both jaws have elongated canine teeth.

Wild Boar live gregariously in family groups that consist of a female leader (the sow), her adult daughters, and young animals. Old males (boars) live alone, while young boars form bachelor groups in their second year. Wild Boar are active during the day and at night. If they are intensively hunted, they shift their activity cycles to dawn, dusk and night. They have excellent hearing and sense of smell. They are good swimmers and can run fast over long distances. Their eyesight is relatively weak. They are mainly predated by wolves and bears. Young animals are also taken by lynx and Eurasian Eagle-owls.

DISTINGUISHING FEATURES   Solid-looking, stocky body with relatively short legs and a short tail. The wedge-shaped skull extends into a snout with a wide, circular end made of hard cartilage that the animal can use to break up even frozen ground. Adult males, in particular, have large, protruding, upwardly curving canine teeth (tusks). Wild Boar have bristly fur, with hairs that are usually split at the ends.

DISTRIBUTION AND HABITAT   Wild Boar are found throughout the whole of Eurasia, except for western Scandinavia. They prefer deciduous and mixed forests with plenty of cover and reserves in bogs, dense spruce thickets or reed beds and marshy areas. As an adaptable species associated with human activity, Wild Boar occurs in cultivated landscapes and, increasingly, in cities.

DIET   Omnivore that mainly eats plants. Important plant-based food sources are fruits and seeds of trees, such as beechnuts, acorns and nuts. During mast years, almost the entire stomach contents of Wild Boar can consist of acorns and beechnuts. Roots and foliage, as well as cereals, potatoes, maize and beet, are other important plant foods. Wild Boar also eat fungi. Animal-based food mainly consists of grubs, earthworms, voles and snails. If the opportunity presents itself, they will catch frogs, lizards, young mammals such as Western Roe Deer fawns, and also young birds and eggs. They also eat carrion and rubbish.

REPRODUCTION   Wild Boar are polyoestrous and can have fertile periods throughout the year. The mating season in Europe mainly lasts from November to February, with a peak in December. After a gestation period of 112–121 days, 4–8 (up to 13) young are born from March to April. They are born with fur and can see. If a female loses her litter, there can be a second litter, which is why young Wild Boar may also be seen in January. Young females reach sexual maturity after 7–12 months and males after 18 months. The animals appear to respond to intensive hunting by increasing their reproduction rate.

The Wild Boar is the ancestor of Domestic Pigs.

TRACK

Front
L 5.1–6.5cm   W 4–7.5cm

Hind
L 4.9–6.2cm   W 4.2–7cm

Medium-sized. Unguligrade. Slightly asymmetrical. Four toes. Toes 2 and 5 (the dew claws) are very short and slightly higher up on the back of the foot. The dew claw prints are usually clearly visible and are further apart than the hoofprints of toes 3 and 4 (a), which gives the overall outline of the track a triangular appearance. Toes 3 and 4 are wide. There is a gap between the tips of the hooves at the front, even if they are not splayed. The hooves are rounded in comparison to the hooves of other even-toed ungulates. The negative space between toes 3 and 4 can be large with a wedge-shaped appearance. The hooves of young animals are more pointed at the front. The outer hoof walls are wide and often leave a solid print (b). The inner hoof walls are slightly concave. The outline of each hoof often has a square appearance. The front foot is larger and wider than the hind foot.

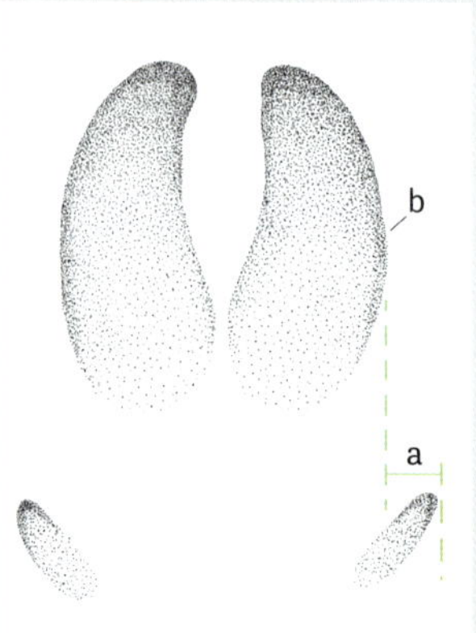

▲ *Left front.*

▲ *Double print. The hind foot print partly covers the front foot print. The difference in the dew claws is clearly recognisable: the dew claws on the hind foot are much finer and almost parallel to the direction of travel.*
Märkische Schweiz, Germany.

GAITS

Wild Boar mainly move at a walk. They eat on the move and their snouts leave distinctive rooting marks. They trot to cover longer distances. The animals flee in bounds or at a gallop when alarmed.

Walk
Stride length: 59–96cm
Trail width: 14–22cm

Trot
Stride length: 102–163cm
Trail width: 7–15cm
Straddle trot SL: 155–165cm
Straddle trot TW: 18–22cm

Bound/gallop
Group length: 82–130cm
Inter-group length: 51–175cm
Stride length: 160–303cm
Trail width: 16–24cm

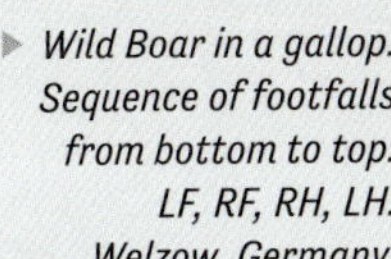
▶ *Wild Boar in a gallop. Sequence of footfalls from bottom to top: LF, RF, RH, LH. Welzow, Germany.*

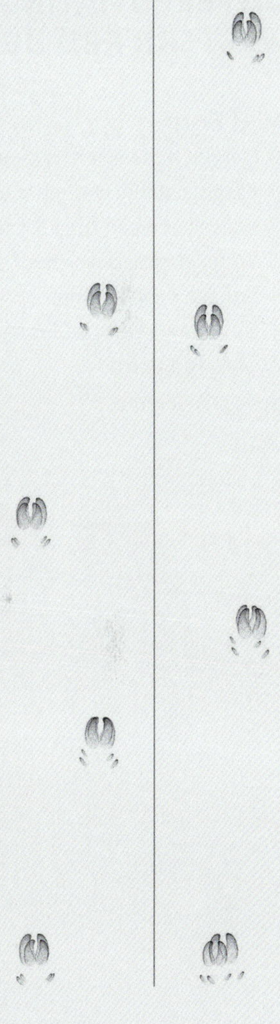
▼ *Direct register walk.* ▼ *Direct register trot.*

◀ *The wide, well-trodden paths and the choice of route are characteristic: Wild Boar often break through undergrowth. Coto de Doñana, Spain.*

Similar tracks

Dew claw prints mean clear footprints are unmistakable. Young Wild Boar tracks can be mistaken for Western Roe Deer footprints but are much smaller and squarer. If the dew claws do not leave prints, Wild Boar tracks can be confused with Western Red Deer tracks. Even trained trackers or hunters can confuse the two in these circumstances. Avoid jumping to conclusions and pay attention to the following foot morphology features:

DIFFERENTIATING BETWEEN WILD BOAR AND WESTERN RED DEER

Wild Boar

a) Curved hooves with rounded tips.
b) Large, usually wedge-shaped negative space between toes 3 and 4.
c) Wide, strong outer hoof walls.
d) Square hoof outline.
e) Dew claw prints wider than the hoof prints of toes 3 and 4.

Western Red Deer

a) Straight hooves with angular tips.
b) Relatively small negative space between toes 3 and 4.
c) Narrow, fine outer hoof walls.
d) Elongated hoof outline.
e) Dew claw prints aligned behind toes 3 and 4.

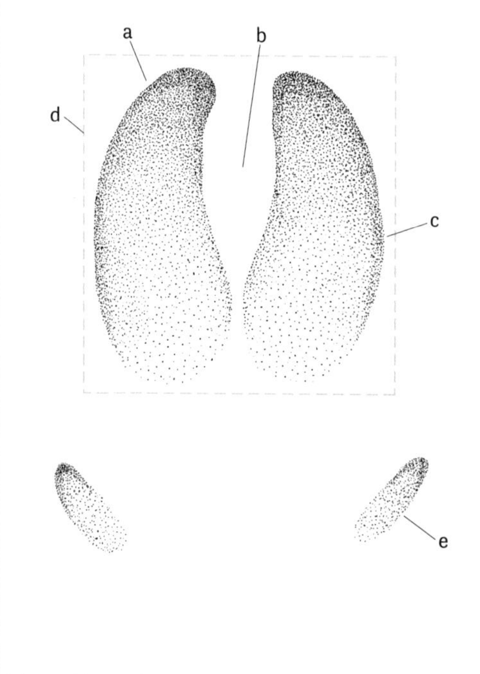

▲ *Wild Boar, left front.*

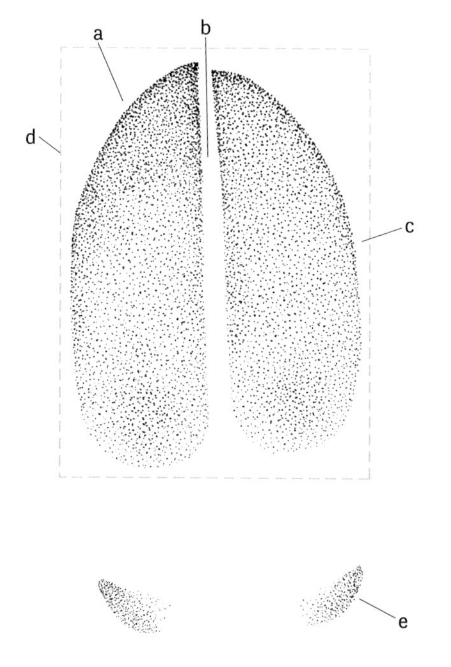

▲ *Western Red Deer, left front.*

▲ Double prints of Wild Boar (left) and Western Red Deer (right) in comparison. Wild Boar tracks are much shorter if the dew claws do not leave a print. Aaron Tiedemann.

▲ The curved shape of the hooves and the thick outer hoof wall help to distinguish the prints from those of Western Red Deer. Lausitz, Germany.

▶ This track looks too small for a Wild Boar but the dew claw marks at the sides help to solve the mystery: This is the track of a young Wild Boar. Welzow, Germany.

SIGNS

MUD BATHS Wild Boar take regular mud baths (wallowing) to protect against insects and parasites. Animals repeatedly visit popular wallowing sites, so the surrounding vegetation is usually heavily trampled, often with clear, wide paths leading to the muddy pool.

RUBBING TREES After wallowing, Wild Boar rub on trees. Some of the mud is rubbed off, leaving muddy patches on the tree. Wild Boar like to repeatedly rub on the same tree so large areas of bark are rubbed off over the years. Their bristly hairs are often found at the edges of the rubbed patches. Rubbed patches normally extend from the ground to a

◀ *Wild Boar mud bath, close-up. Spessart, Germany.*

▲ *Wild Boar mud bath, the 'bathroom' of a Wild Boar herd. Spessart, Germany.*

▲ *A classic Wild Boar track leading through a muddy marsh area. Märkische Schweiz, Germany.*

▲ *Tree where a Wild Boar rubs. The deep furrows left by the tusks of a Wild Boar are visible at the top edge. Märkische Schweiz, Germany.*

height of 60–100cm (up to 120cm). Rubbed patches made by Western Red Deer are higher up and can be up to 150cm high. When male animals rub, they often bite the bark and damage it with their tusks.

BED/NEST Wild Boar beds are usually in secluded locations, for example in a spruce thicket or dense reedbed. Wild Boars are the only even-toed ungulates to occasionally line their bed with green coniferous branches or grasses that they snip off with their teeth. Like Western Roe Deer, they scrape a depression in humus-rich ground. It is normally much larger than those made by other even-toed ungulates and can be mistaken for the rut pit of a Common Fallow Deer.

FEEDING AND DIGGING MARKS Wild Boar root in the ground with their snout when searching for voles, roots, insect larvae, acorns and worms, and leave long flat rooting marks. They churn up the soil with their snout and push it forward and to the sides. This behaviour is known as rooting. A herd of Wild Boar can damage fields and completely destroy a garden overnight. Feeding marks on fungi can be found but they are rarer because Wild Boar bite off the whole fungus.

▼ *The distinctive bed of a Wild Boar. The secluded location under overhanging branches is typical. Offenbach, Germany. Simone Roters.*

▼ *The end of their snout is reinforced with cartilage, which enables deep rooting. Offenbach, Germany. Simone Roters.*

▲ *As they search for food, Wild Boar can break up large expanses of earth, leaving distinctive signs. Offenbach, Germany. Simone Roters.*

Unlike when other even-toed ungulates eat fungi, the toothmarks of the upper and lower incisors are visible.

EXCREMENT   Wild Boar are omnivores, so the appearance of their droppings varies greatly according to diet and season. The droppings differ greatly from those of ruminant even-toed ungulates. Droppings are usually black in colour, but occasionally brown to yellowish. They mostly consist of compressed, short, flat faecal pellets that can form a thick, sausage-shaped cylinder. As it decomposes, this compressed cylinder breaks down into individual segments and the individual faecal pellets can be confused with Western Red Deer droppings. However, Wild Boar pellets are normally much flatter and their surface is rougher because of the higher coarse fibre content.
- L 10–28cm D 3–7cm

▼ *Wild Boar excrement can resemble Western Red Deer excrement but it is more often compressed into a sausage shape and contains more coarse fibre. Märkische Schweiz, Germany.*

▼ *When Wild Boar 'root', they push the soil forwards and to the sides with their snout. Here, you can even make out the direction of movement (from right to left). Märkische Schweiz, Germany.*

Deer

The deer family (Cervidae) includes Common Fallow, Western Red, Western Roe, Water and Sika Deer, as well as Reeves's Muntjac, Elk and Reindeer. The Chital or Axis Deer (*Axis axis*), which also occurs in the area, is not covered because it is very rare.

Deer are characterised by their usually branched antlers that are formed every year at mating season and then shed. The bony antlers sit on the pedicles of the frontal bone. Normally only male deer grow antlers. Reindeer are the exception as the females also have antlers, although they are much smaller than the male's. When they start growing, antlers are covered in 'velvet'. The velvet supplies the bony substance of the antlers with blood during the growth phase. When growth is completed, deer rub or strike off the velvet on small trees and shrubs.

This is known as 'fraying'. Cervids mainly use their antlers for rutting behaviour, such as territory marking and fighting with rivals, during the mating season, after which they are shed. You may occasionally find discarded antlers in the field. Water Deer do not have antlers. Instead, the males have very large upper canine teeth.

TRACKS AND SIGNS OF DEER

The tracks and signs of deer have the characteristic features of even-toed ungulates (page 494) and can be confused with those of bovids.

STRIKING Striking is a common and distinctive sign caused by deer striking trees with their antlers and leaving both visual and scent marks. Deer often strike trees where paths intersect and along the territory boundaries of male animals. However, signs of striking are also found at the centre of a territory. Immediately before and during the rutting season, it can sometimes be possible to precisely identify the territory of male animals through marks caused by striking.

The terms fraying and striking are sometimes used synonymously, especially with reference to Western Roe Deer. In both cases, deer leave clear signs on trees and shrubs with their antlers, but it is not possible to tell whether these marks are caused by fraying or striking. In rare cases, the remains of velvet provide a clear indication of fraying. However, finding velvet is rare because of its popularity as a nutrient-rich food source for other species.

ELK
Alces alces

HTL 250–280cm

TL 5–13cm

W 240–550kg (up to 850kg)

Males (bulls) are much heavier than females (cows).

This is the only species in the genus *Alces*. Elk (or Moose) are mainly solitary animals that are active at dawn and dusk and during the day. They come together during rutting, and may form small groups of 3–5 animals in winter. They normally spend the rest of the year alone. Elk have well-developed hearing and an excellent sense of smell. They are good swimmers and the only deer able to eat underwater. Another special feature an approximately 7cm-long fold of skin between the two parts of the hoof. When the hooves are splayed, this skin stretches and acts like webbing that enables the animals to cross muddy ground without sinking too deeply. This gives them a crucial advantage in marshy areas. Adult animals do not have any predators because of their size. Calves are taken by wolves, Wolverines, bears and lynx.

DISTINGUISHING FEATURES   Larger than a horse. Long head with a convex 'Roman nose', and an overhanging upper lip. Very long legs in proportion to the body. Large ears and small eyes. Elk bulls usually have palmate antlers (similar in shape to those of Common Fallow Deer). However, in Scandinavia their antlers are more commonly branch-like in shape.

DISTRIBUTION AND HABITAT   Larger populations in Sweden, Finland, Norway and the Baltic states. Also found in Poland, Czech Republic and Belarus. Individual Elk occasionally enter Germany from Poland. The animals prefer large, damp deciduous and mixed forests with marshes, bogs and bodies of water.

DIET   Elk mainly eat nutrient-rich foods such as buds, shoots and aquatic plants. The animals can dive down to 5m for more than 60 seconds to reach aquatic plants. Other food sources are leaves, tree bark, fungi, lichen, thin twigs from deciduous and coniferous trees, as well as foliage and grasses. Elk eat to build up fat reserves to sustain them through the winter season.

REPRODUCTION   Rutting lasts from September to November. Violent fights can break out between males before and during mating season. After a gestation period of around eight months (224–266 days), 1–3 (often 2) young are born. They reach sexual maturity after 1–2 years. Males often do not reproduce until their third or fourth year.

The Elk is also found in North America, where it is known as the Moose. A different large deer species (*Cervus canadensis*, a close relative of the Western Red Deer), is known as the Elk in North America.

SIGNS

BEDS   Shaped like a Western Red Deer bed (page 546), but larger.
- **Length 130–190cm**

MUD BATHS AND RUBBING TREES   Like Western Red Deer (page 547), Elk wallow in mud. Elk often leave rubbed patches on tree trunks as a form of marking. These rubbed patches resemble those made by Western Red Deer. However, patches made by Elk are usually further up the trunk (up to 2.5m high).
- **Length of a mud bath: 90–300cm**

TRACK

Front
L 11–17.5cm W 10–15cm

Hind
L 10–16.5cm W 8–14cm

Large to very large. Unguligrade. Slightly asymmetrical. Four toes. Toes 2 and 5 (the dew claws) are very short and higher up on the back of the foot, though they are lower than those of Western Red Deer and often leave a print. Their prints are the same width as the outer edges of toes 3 and 4. Toe 3 is slightly smaller than toe 4. The front half of the inside hoof wall is slightly concave. The pads stand out clearly from the hard edge of the hoof and the relatively narrow subunguis. The front foot is larger and rounder than the hind foot. The hooves become very splayed, especially on slippery ground and at high speed. Even when not splayed, the individual hooves are relatively far apart compared with other deer ⓐ. The hoof outline of the hind foot can appear heart-shaped.

Similar tracks
European Bison, Musk Ox, Cattle.

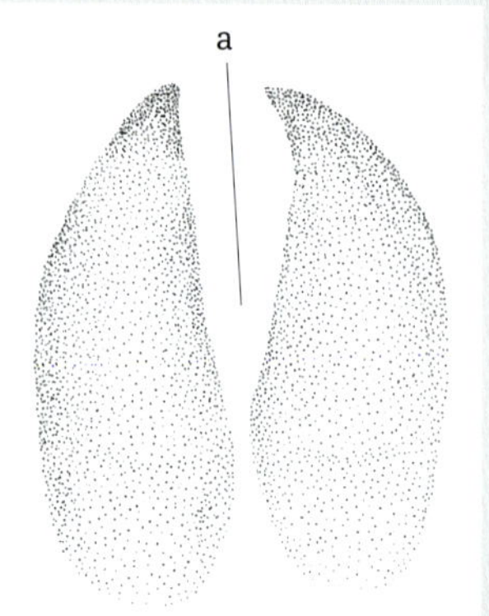

▲ Left front.

▲ Left front.
Värmland, Sweden. Heide Ulrich.

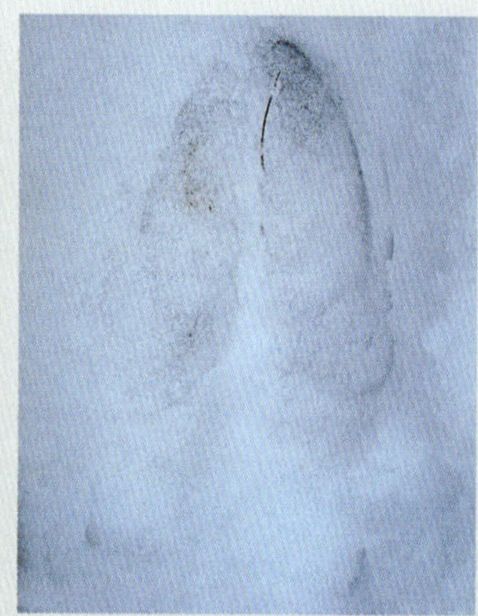

▲ Left hind.
Värmland, Sweden. Heide Ulrich.

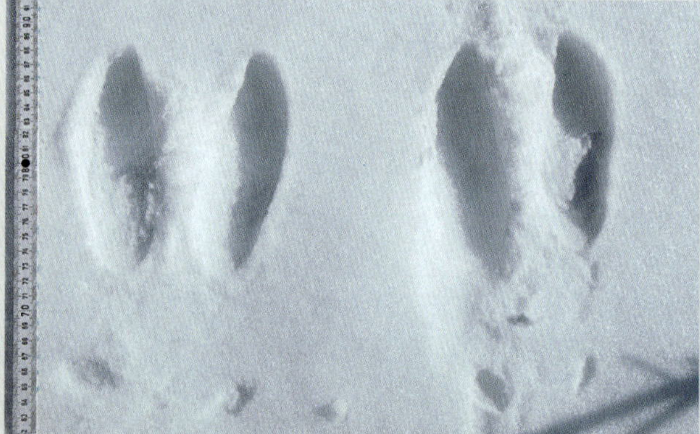

▲ Front foot print (left) and hind foot print (right) in comparison. The dew claws of the front foot are further forward and point more to the sides. The dew claws of the hind foot are further back and more forward-pointing.

GAITS

Elk mainly move at a walk. If they are alarmed or threatened, they will bound and gallop to flee. They use their own established paths that become heavily trampled over time and can be very distinctive. The direct register trot is the most common faster gait. Elk are less hindered by deep snow than other even-toed ungulates and they normally stay in a direct register walk or trot.

Walk
Direct register: SL 119–198cm   TW 23–52cm

Trot
Direct register: SL 180–298cm   TW 8–31cm
Straddle trot: SL 250–370cm   TW 30–62cm

Gallop
Group length: 145–300cm
Inter-group length: 120–432cm
Stride length: 265–703cm

▲ *Direct register walk.*

▲ *Straddle trot.*

◄ *Several Elk paths lead through this wintry landscape. Jämtland, Sweden. Heide Ulrich.*

▲ *Elk beds are like Western Red Deer beds but much larger. Jämtland, Sweden. Laura Gärtner.*

RUT PITS During mating season, males use their front feet to dig hollows in the ground called 'rut pits'. They urinate in the pit and then lie in it. A fresh rut pit gives off a very strong odour (see also Common Fallow Deer, page 538).

STRIKING Striking bushes and shrubs with the antlers is another method of marking during rutting. Shrubs that have been struck will have many broken branches and scratched bark.

FRAYING From July to August, male Elk fray the velvet from their newly grown antlers. The highly nutritious velvet is a rare find as it is a food source for other animals.

BARK PEELING Elk peel bark, making two wide furrows in the tree with their incisors. Because of their considerable size, Elk can cause significant damage to commercial forests, occasionally knocking over whole trees.

- Max. height of peeling: 2.5m
- Single toothmark width: 0.8–1.3cm

▲ The unmistakably wide toothmarks. Jämtland, Sweden. Laura Gärtner.

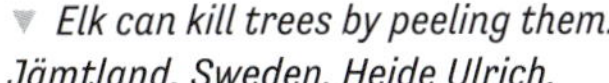

▼ Elk can kill trees by peeling them. Jämtland, Sweden. Heide Ulrich.

BITE MARKS Elk pull foliage and buds towards them with their mouth, often breaking branches. The branches break above or slightly below the height of the animal's head. Elk feeding marks are massive and easily recognisable. They can even break off strong terminal shoots and change the appearance of the landscape. Elk have another method for reaching food above head-height; they push over smaller trees by standing on the trunk and gradually moving forward until the tree bends so far that it breaks.

▶ *Elk bite mark. Elk feeding marks can have a lasting impact on the landscape where the animals live. Jämtland, Sweden. Heide Ulrich.*

▲ *Summer droppings. Elk droppings contain more water during summer. These ruminants can fully digest cellulose. Jämtland, Sweden. Heide Ulrich.*

▼ *Winter droppings. Faecal pellets are drier and more fibrous in winter. Värmland, Sweden. Heide Ulrich.*

▲ *Individual Elk droppings are much larger than those of Western Red Deer. Jämtland, Sweden. Laura Gärtner.*

EXCREMENT Elk excrement has two different main forms. Cowpat-like piles mainly appear in spring or after excessive consumption of aquatic plants. Faecal pellets are more typical of winter and are like those of Western Red Deer but much larger. Many different variations are possible between those two main forms. Fresh excrement is normally dark green to black. Old, dried out faecal pellets have a brownish colour.

- L 2.2–4.5cm D 1.2–2.2cm

Using droppings to determine sex

As in Western Red Deer, the difference between cylindrical and acorn-shaped excrement is a reliable way of identifying sex. MacCraken and Van Bellenberghe (1987) researched whether it would be possible to identify the sex of Elk using their droppings. They concluded that a diameter of less than 1.63cm cannot come from a male and a diameter of over 1.64cm cannot come from a female or a calf. Both researchers also observed that dried faecal pellets that weigh more than 2g come from males. They also claimed that faecal pellets from males are square while faecal pellets from females are more roundish. Mark Elbroch points out that Western Red Deer and Musk Ox droppings can also be square or roundish. Whether there is a link to sex remains to be proven.

WESTERN ROE DEER
Capreolus capreolus

HTL 90–135cm

TL 2–5cm

W 15–35kg

Weight is highly dependent on food supply, altitude and climate. It tends to increase from the southwest to the northeast and from low to high altitudes.

The Western Roe Deer is Europe's most common deer species and the only species from the roe deer genus (*Capreolus*) to occur in the area. Western Roe Deer are active both during the day and at night. They mainly find their way using their excellent sense of smell. They can hear well. They have a wide field of vision and are especially attuned to motion. However, if a human stands completely motionless, the animals will not see them. During the summer, Western Roe Deer live alone or in small groups consisting of a mother and her fawns. The animals can come together in large herds in winter. 'Field' Roe Deer can form herds of up to 100 animals. Western Roe Deer are mainly predated by wolves, lynx and foxes. Red Foxes mainly prey on fawns but they can also take adults if deep snow prevents the deer from fleeing. More rarely, Western Roe Deer fall prey to Golden Eagles, Wolverines, Wild Boar, European Wildcats and Domestic Dogs. Road collisions and accidents with agricultural machinery are a common cause of death in some areas.

DISTINGUISHING FEATURES Approximately goat-sized, slim body, long neck, long legs and a very short tail. Coat colour changes from the reddish-brown summer coat to the greyish-brown winter coat with a distinctive white patch on the rump. Males have small antlers.

DISTRIBUTION AND HABITAT Western Roe Deer are present throughout the whole of Europe, apart from some islands including Corsica, Sardinia and Sicily. They occur in any suitable habitats. These mainly include deciduous and coniferous forests with good cover and access to fields and open agricultural land. Patchwork landscapes with lots of adjoining woodland areas are the ideal conditions for the dietary requirements of Western Roe Deer. As a highly adaptable species associated with human activity, Western Roe Deer also occur in the borders of settlements, gardens, parks or in cemeteries. 'Field' Roe Deer live almost exclusively on agricultural crops and remain in the fields all year round.

DIET Western Roe Deer are specialist feeders that are referred to as 'concentrate selectors'. Unlike Western Red Deer, for example, they consume a variety of carefully chosen food types instead of eating large quantities of one food type at once. They prefer easily digestible, high-energy, high-protein food. Their diet is mainly made up of shoots, buds and leaves of trees and shrubs, such as brambles, raspberries, hazel, elder, fir, ash, maple, oak and beech, as well as fruits, beechnuts, acorns and chestnuts. Plants like dandelion, plantain, chervil or grasses, as well as fungi constitute a smaller part of their diet.

REPRODUCTION The summer mating season (July/August) and the embryonic diapause that can last up to 150 days are unusual for cervids. Embryonic diapause can increase the gestation period to 273–294 days. There can be another rut in November/December. If there is, there is no embryonic diapause so the 1–2 (up to 4) young are born in May/June. The mothers give birth in a hidden location and remain an average of 50–150m away. The fawns reach sexual maturity at 7–8 months old.

TRACK

Front
L 3.1–6.2cm W 2.5–5.5cm

Hind
L 3–6cm W 2.4–5.2cm

Medium-sized. Unguligrade. Slightly asymmetrical. Four toes. Toes 2 and 5 (the dew claws) are very short and higher up on the back of the foot. The dew claws are rarely visible in prints. If they are, they leave a print behind the hoof and are the same width as the outer edges of toes 3 and 4. Toe 3 is slightly smaller than toe 4. The hoof outline can be heart-shaped. The individual hooves are narrow with tapered ends. The front half of the inside hoof wall is slightly concave. The pad makes up 25 per cent to one third of the total length of the track. The front foot leaves a slightly larger and wider print than the hind foot. The hooves can spread apart at the front at high speed and on slippery ground.

▲ *Right front.*
Vledder, Netherlands. René Nauta.

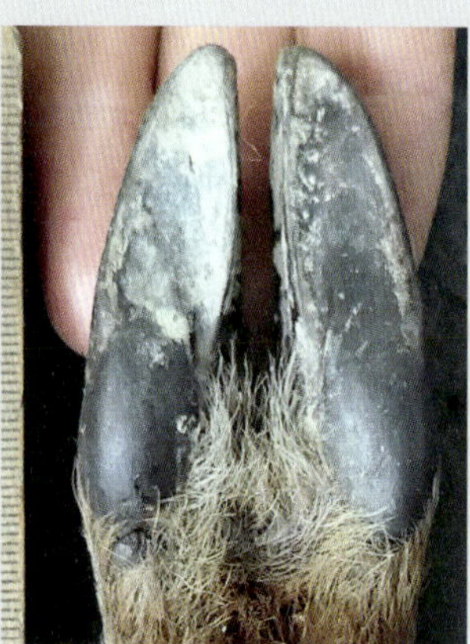

▲ *Right hind.*
Vledder, Netherlands. René Nauta.

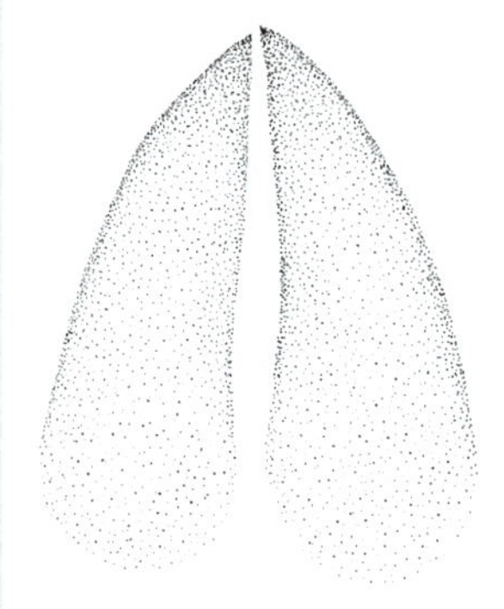

▲ *Right hind.*

▲ *Right hind.*
Bielefeld, Germany. Ulrike Quartier.

▲ *Double print of a Western Roe Deer. The characteristic heart shape is clearly visible. Märkische Schweiz, Germany.*

Similar tracks
Reeves's Muntjac, Water Deer, Common Fallow Deer and Sika Deer, young Wild Boar.

GAITS

Western Roe Deer mainly move at a walk. Their tracks often turn out slightly. An understep walk can indicate an older animal, while overstepping is often a sign of the more spirited behaviour of young animals. However, other factors can also cause overstepping, such as the transition from trot into gallop. Western Roe Deer use tracks and minor roads but prefer their own paths. The direct register trot is the most common faster gait. Western Roe Deer flee by bounding and galloping when alarmed. They can make considerable bounds of up to 7m.

Walk
Direct register:
SL 53–98m   TW 7–23.5cm
Understep:
SL 67–76cm   TW 12–16cm

Trot
Direct register:
SL 89–136cm   TW 6.3–13cm
Overstep:
SL 104–160cm   TW 8–20cm
Straddle trot:
SL 138–143cm   TW 22cm

3 × 4 bound (rotary) and 4 × 4 bound (transverse)
Group length: 48–128cm
Inter-group length: 84–120cm
Stride length: 115–248cm
Trail width: 11–20cm

Gallop
Group length: 58–257cm
Inter-group length: 50–323cm
Stride length: 150–440cm
Trail width: 10–25cm

▼ The direct register walk is the preferred gait of the Western Roe Deer. A Northern Raccoon has crossed the path from left to right. Märkische Schweiz, Germany.

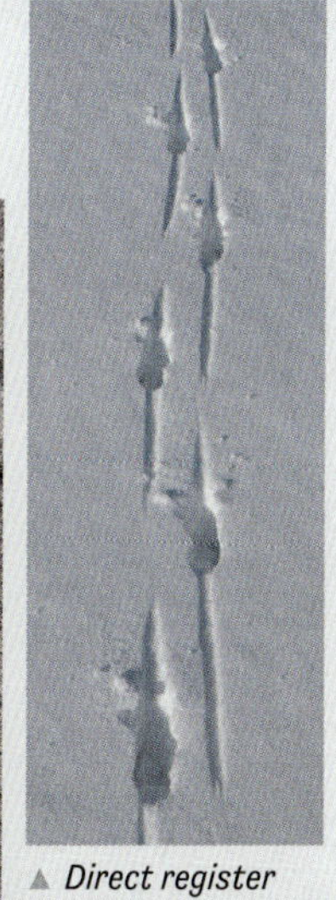

▲ Direct register walk. Drag marks are often left in snow. Jämtland, Sweden. Laura Gärtner.

▲ Western Roe Deer in a straddle trot. Märkische Schweiz, Germany.

▲ Rotary gallop. The cycle of footfalls from bottom to top is: LF, RF, RH, LH. Märkische Schweiz, Germany.

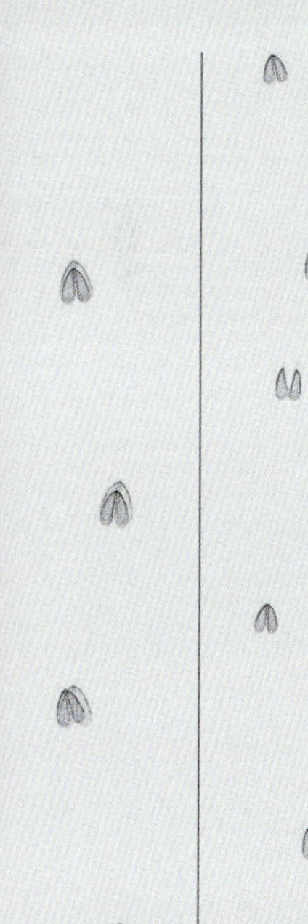

▲ Direct register walk.   ▲ Rotary gallop.

SIGNS

On the left, the wide, rounded shape of the abdomen and on the right, the tapered front of a Western Roe Deer bed. Spessart, Germany.

BED Unlike other deer, Western Roe Deer scrape leaves away from their bed before lying down. Their beds have the shape that is typical of all even-toed ungulates (page 174).

- Length 50–65cm

FRAYING Western Roe Deer grow new, velvet-covered antlers from March to June. Bucks rub or 'fray' off this velvet on small trees and shrubs. The resulting damage to vegetation can range from minimal to significant. They also leave behind the remains of the velvet but it is rarely found because it is eaten by either the deer themselves or by birds. Adult males tend to fray earlier in the year than young males.

Western Roe Deer bed in the snow. The head was to the left. Jämtland, Sweden. Laura Gärtner.

▲ *The classic combination of scraping and striking is typical of Western Roe Deer. Bavarian Forest, Germany.*

STRIKING Western Roe Deer make strike marks by hitting trees with their antlers and rubbing their forehead and cheeks against them ('forelock rubbing'). This results in bark being removed about 80cm above ground, at the height of the animal's head, leaving a smooth, pale, shiny surface. Repeated striking of surrounding vegetation is a clear form of visual marking. The glands on the animal's forehead and other parts of its head also leave scent marks. Striking and scraping (see below) often occur together. Western Roe Deer bucks begin this type of marking in February. They continue marking all through summer, with particular intensity immediately before and during rutting. Hunters refer to roebuck striking as 'fraying'.

SCRAPING Scraping is another form of marking. The animals use their front hooves to scrape undergrowth, leaves and needles back under their body. Their hooves leave parallel scrape marks and the glands between the toes leave scent marks. Scrape marks are often found at the base of trees. They frequently occur in conjunction with strike marks, but roebucks can also scrape without leaving strike marks. Western Roe Deer scrape from around the beginning of March to the end of September.

▼ *The distinctive strike marks of a roebuck. We often find these signs together with an area that has been scraped. Lausitz, Germany.*

▲ *Striking or fraying marks like these are often found at distinctive places such as crossroads. Märkische Schweiz, Germany.*

▲ *Roe ring. A roebuck has pursued a doe in a circle for a long time. West Sussex, England.*

▲ *Unlike marks made by striking, bark peeling marks usually have clear toothmarks. West Sussex, England.*

FAIRY RING During the mating season, roebucks drive the female along in front of them. They sometimes drive her around in a circle for so long that they create a ring of trampled vegetation. These are known as 'roe rings'.

BARK PEELING Western Roe Deer peel off pieces of bark so they can eat the cambium below. The incisors scrape from bottom to top like carving tools, leaving two furrows. The toothmarks are narrower than those of Common Fallow Deer. At the bottom of the peeled area, there will be a straight cut, while the top edge will be frayed. The animals peel off strips of bark from the bottom up. Compared to Western Red Deer, Western Roe Deer rarely graze and mainly do so when there is more snow or less food available.

- Max. height of peeling: 1.2m
- Single toothmark width: 0.2–0.4cm

BUD BROWSING   Instead of incisors, cervids have a dental pad in their upper jaw. They clamp a stem or stems between the incisors of the lower jaw and the dental pad and tear it off. This creates a frayed end that is very different from the clean cut made by hares, rabbits and rodents.

BROWSE LINE   The typical height of the browse line on shrubs, trees and bushes is around 120–150cm.

EXCREMENT   Colour varies between yellowish brown, greenish black and black. The faecal pellets are either rounded at both ends or have an indentation at one end and a small point at the other. As ruminants, Western Roe Deer deposit large quantities of faecal pellets at once. Summer droppings often stick together because the food eaten is richer and contains more moisture. Western Roe Deer produce 17–23 piles of pellets a day.

- L 1–1.7cm   D 0.7–1.1cm

▼ The frayed, untidy appearance of deer bite marks is characteristic of all deer. Märkische Schweiz, Germany.

▼ Western Roe Deer excrement. The individual faecal pellets can vary in size. The average size is crucial for identification. Märkische Schweiz, Germany.

REINDEER
Rangifer tarandus

HTL 120–220m
TL 7–18cm
W 60–220kg (up to 300kg)
Females are smaller than males.

Reindeer is the only deer species in Europe in which the females have antlers. The males shed their antlers after rutting in winter, but females only shed theirs after giving birth to their young in spring. Reindeer have a well-developed sense of smell and can see UV light. They are gregarious and can form herds of several hundred animals, depending on their range. Females and young males live together in large herds, while older males form smaller herds or roam alone. Considerable numbers of tundra-dwelling animals come together for seasonal migrations. In some areas, thousands of Reindeer travel together to the southern taiga. After migration, they disperse into smaller groups of 10–100 individuals. Mountain-dwelling animals do not migrate but roam around the region where they live.

Reindeer are mainly active during the day. They have a well-developed sense of smell; they also swim well and can run for long distances. Their thick coat protects them from temperatures as low as -50°C. They have distinctive, large, wide hooves that enable them to move safely on ice and snow. Their toe joints make an audible cracking sound when they walk. They are predated by wolves, Wolverines and bears. Young animals can also be hunted by lynx, foxes and eagles.

Many people in northern Europe, especially indigenous peoples such as the Sámi, base their livelihood around Reindeer.

DISTINGUISHING FEATURES   Powerfully built, medium-sized deer. Female Reindeer are the only female deer that have antlers.

DISTRIBUTION AND HABITAT   Norway, Finland, Svalbard, Greenland, Scotland and Iceland. The large Reindeer herds in Lapland are semi-domesticated. Reindeer inhabit the Arctic and subarctic tundra, the taiga, northern mountains and dense, damp coniferous forests.

DIET   In summer, Reindeer mainly eat grasses and herbs, as well as other plant foods such as buds, twigs, leaves, tree bark and fungi. In winter, mosses and lichens are an important food source, for example the Reindeer Lichen (*Cladonia rangiferina*).

▼ *Five Reindeer head towards the horizon. Helagsfjäll, Sweden. Heide Ulrich.*

TRACK

Front
L 8–11.5cm   W 9.5–14.5cm

Hind
L 7.5–10cm   W 9–13cm

Medium to large. Unguligrade. Slightly asymmetrical. Four toes. Toes 2 and 5 (the dew claws) are very short and higher up on the back of the foot, but are located further down on the foot than in Western Red Deer, so they frequently leave prints. On the front foot, they are situated horizontally to the direction of travel and, on the hind foot, they are positioned vertically to the direction of travel. At higher speeds, they spread apart, so the vertical dew claws of the hind foot can leave prints that are almost horizontal to the direction of travel. The dew claws of the front foot usually leave a print and, like Wild Boar, protrude laterally beyond the outer hoof walls ⓐ, which can give the track a triangular appearance. The dew claws of the hind foot do not reliably leave a print.

Without the dew claws, the track has a distinctively roundish appearance. Toe 3 is slightly smaller than toe 4. The outer hoof walls are very convex, while the inner hoof walls are obviously concave. The negative space between toes 3 and 4 is large. The tips of the hooves are curved and become rounder as they wear

▲ *The front feet of a reindeer. Jämtland, Sweden. Heide Ulrich.*

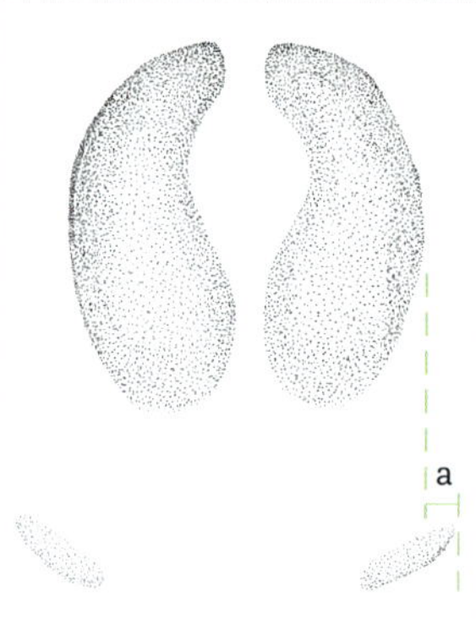

▲ *Left front.*

▲ *Left front. Jämtland, Sweden. Laura Gärtner.*

▲ *Left front (left) and left hind (right). Comparison shows how much longer and narrower the hind foot print is. Note the different position of the dew claws. Jämtland, Sweden. Heide Ulrich.*

down. The front feet are larger, wider and rounder than the hind feet. The concave indentation of the inner hoof wall is more pronounced in the front foot than in the hind foot. The hooves spread apart on slippery ground and at high speed.

GAITS

Reindeer move mainly at a walk and often in an overstep walk. Their hooves leave characteristic drag marks in the snow. They bound and gallop to flee if they are alarmed or feel threatened. The animals usually travel in groups. Their favourite paths become heavily trampled over time and can be very conspicuous.

Walk
Stride length: 88–172cm
Trail width: 18–40cm

Trot
Stride length: 200–306cm
Trail width: 15–29cm

Gallop
Group length: 149–350cm
Inter-group length: 91–166cm
Stride length: 250–469cm
Trail width: 21– 43cm

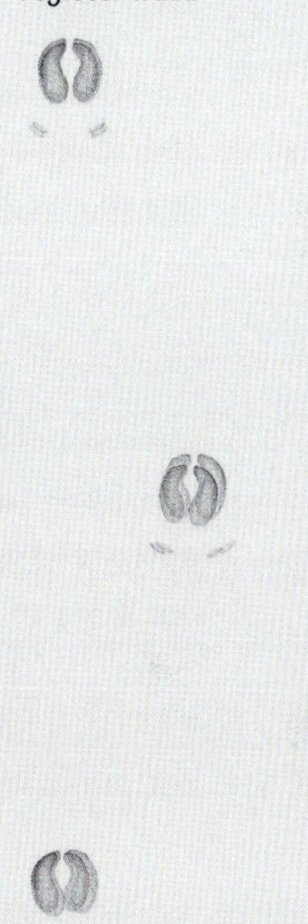

▲ The direct register walk of a Reindeer. Helagsfjäll, Sweden. Laura Gärtner.

▲ Reindeer in a rotary gallop. Jämtland, Sweden. Laura Gärtner.

▼ *Direct register walk.*

▼ *Overstep walk.*

Similar tracks
Unmistakable.

REPRODUCTION The mating season lasts from mid-September until the end of October. Males attract a harem and mate with as many females as possible. Fights often take place between rival males. Normally, one calf is born in May/June after a gestation period of 7–8 months. Calves reach sexual maturity after 1.5–2.5 years.

SIGNS

BED Similar shape to a Western Red Deer bed (page 546), but slightly smaller.
- Length 60–115cm

MUD BATHS AND RUBBING TREES Like Western Red Deer, Reindeer also wallow. The rubbed areas on trees can serve as markings and are like those of Western Red Deer (page 547).
- Length 90–230cm

STRIKING Reindeer also mark during rutting by striking their antlers against bushes and shrubs. Shrubs that have been struck by Reindeer will have many broken branches and scratched bark. The highest antler strike marks of a male animal are about 2m above the ground, but most of the damage will be at a height of about 90–120cm.

FRAYING Male Reindeer fray the velvet from their newly grown antlers from August to September.

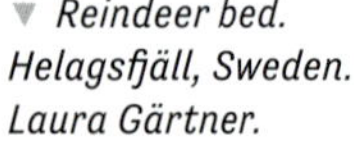
▼ *Reindeer bed. Helagsfjäll, Sweden. Laura Gärtner.*

▲ *Large areas of lichen and moss that have been eaten away and cleared by hooves are a typical winter feeding mark. Jämtland, Sweden. Laura Gärtner.*

FEEDING MARKS Extensive areas where lichen and moss have been eaten away can occasionally be found in places where Reindeer stay for longer. This conspicuous sign indicates a Reindeer herd's winter grazing.

EXCREMENT Reindeer droppings are like those of Common Fallow Deer and Western Roe Deer. They are slightly smaller than sheep excrement and consist of irregularly shaped, short faecal pellets which, as in Wild Boar, can also be stuck together in a sausage shape. If food in winter mainly consists of lichen and moss, winter droppings can be small tar-like blobs.
- L 0.9–2cm D 0.8–1.5cm

In a study in Svalbard, Norway, Morden *et al.* (2011) were able to reliably identify different age groups of female Reindeer based on the size of their faecal pellets:

- Adult animals:
  L > 8mm D > 7.5mm
- One-year-old animals:
  L = 7.3–7.7mm D = 6.8–7.2mm
- Calves: L < 7.2mm D < 6.8mm

▼ *Reindeer excrement. Helagsfjäll, Sweden. Laura Gärtner.*

▲ *Reindeer excrement, close-up. Helagsfjäll, Sweden. Laura Gärtner.*

COMMON FALLOW DEER
Dama dama

HTL 120–150m

TL 12–24cm

W 35–90kg (up to 110kg)

Males are larger and much heavier than females. Males, in particular, lay down fat in summer and lose it again during the rut.

Common Fallow Deer are adaptable deer that are less sensitive to disturbance than Western Red Deer and Sika Deer. The animals can even be observed resting and grazing in open spaces near busy roads. They can be active during the day and at night and their activity cycles do not appear to be affected by hunting or other disturbances. Their eyesight is better than that of other deer. They have a good sense of smell and hearing and they swim and jump strongly. Like Western Red Deer, Common Fallow Deer mainly live in small- to medium-sized groups in summer. In winter, they sometimes form considerable herds of more than 100 males and females. They are mainly predated by wolves and, in rare cases, by lynx and hunting dogs.

DISTINGUISHING FEATURES Smaller than Western Red Deer but larger than Western Roe Deer. Males have distinctive palmate antlers. Their summer coat has characteristic light patches and a black dorsal stripe along the middle of their back. White and melanistic forms can be frequent.

DISTRIBUTION AND HABITAT Native to the Mediterranean region, Common Fallow Deer were brought to central Europe by the Romans. They are found throughout the whole of Europe, except for a few islands. They prefer flat terrain and generally stay below 1,000m in altitude. They like areas with a mix of cover and open spaces. Other typical habitats include sparse deciduous and mixed forests next to fields, meadows or park-like landscapes. Common Fallow Deer use the forest as cover but tend to feed in open areas. They generally avoid dense, coniferous forests.

DIET Grasses, herbs, leaves, buds, shoots, tree bark and crops, such as maize, rape, cereals and beet. From March to September, over 60 per cent of their diet can be made up of grasses. In autumn, when the mast is ripe, they prefer acorns, beechnuts and chestnuts. Common Fallow Deer are less discriminating in their choice of food than Western Roe Deer.

REPRODUCTION Reproduction peaks between October and November. The males use their antlers to dig pits in the ground (rut pits) and then lie down in them and attract females with their scent and vocalisations. The males give their mating call to indicate their readiness to mate. Dominant stags can vocalise 25,000 to 30,000 times in 24 hours (Petrak 1987). In contrast to rutting Western Red Deer stags who follow females, a Common Fallow Deer buck gathers a harem around him. The females move to the rutting ground, which the male defends against other bucks. The rut pits of several bucks can be close together and a dominant buck may return to his rutting ground repeatedly over several years. After a gestation period of 31–33 weeks, 1–2 fawns are born, mainly in May–July. Both sexes become sexually mature after 18 months, but the males usually only start reproducing after the age of three. During the rut, the males do not eat, losing up to 27 per cent of their body weight.

Common Fallow Deer have been kept in enclosures and farms for meat production since the early 1980s.

TRACK

Front

L 5.5–7.5cm W 3.5–6.6cm

For adult animals, the rule of thumb is L < 5.5cm = female and L > 6.5cm = male. Additional information on foot morphology or social behaviour is required to tell the difference between one-year-old males and adult females.

Hind

L 5–7cm W 3–6cm

Medium-sized. Unguligrade. Slightly asymmetrical. Four toes. Toes 2 and 5 (the dew claws) are very short and higher up on the back of the foot, and their prints are rarely visible. They are located behind the hoof prints and are the same width as the outer edges of toes 3 and 4. Toe 3 is slightly smaller than toe 4. The inner hoof walls are slightly concave. The pad prints make up around 50 per cent of the total length of the track ⓐ. A ridge is often visible between the hard unguis and the soft subunguis. The pads and hooves are at different heights and appear out of line ⓑ. The front foot is larger and wider than the hind foot. The hooves can spread apart at the front at high speed and on slippery ground.

▲ *Left front.*
Vledder, Netherlands. René Nauta.

▲ *Left hind.*
Vledder, Netherlands. René Nauta.

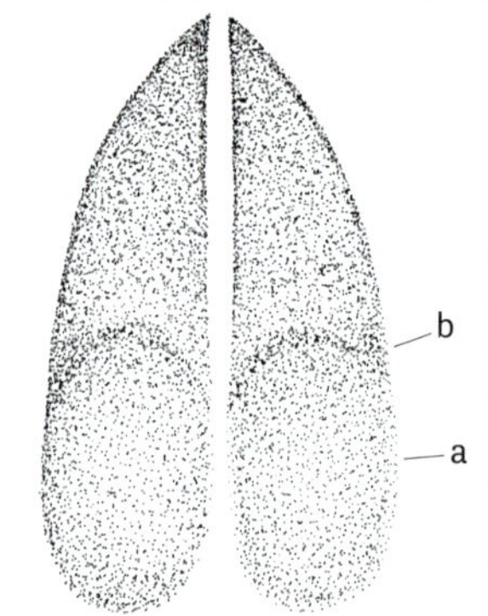

▲ *Left front.*

▲ *Left hind.*
Marianka, Slovakia. Heide Ulrich.

▲ *Left hind (left) and left front (right). The characteristic difference in height between pads and hooves is clearly recognisable. Steyerberg, Germany.*

GAITS

Common Fallow Deer mainly move at a walk. The direct register trot is their most common fast gait. When alarmed, the animals bound and gallop to flee. One notable feature of Common Fallow Deer is the relatively frequent use of the pronk, which helps the animals to quickly get a good view in open spaces.

Direct register walk
Stride length: 60–114cm
Trail width: 8.5–22cm

Direct register trot
Stride length: 96–189cm
Trail width: 6–16.5cm

Straddle trot
Stride length: 145–195cm
Trail width: 13–29cm

3 × 4 bound (rotary) and 4 × 4 bound (transverse)
Group length: 77–158cm
Inter-group length: 108–185cm
Stride length: 166–294cm
Trail width: 10–30cm

Gallop
Group length: 121–195cm
Inter-group length: 95–345cm
Stride length: 186–534cm
Trail width: 10–30cm

▶ *Common Fallow Deer in its preferred gait, the direct register walk. Hainburg, Germany. Simone Roters.*

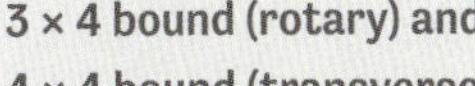

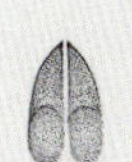
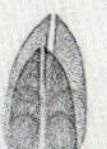

▲ *Direct register walk.*

▲ *Pronk.*

▲ *Common Fallow Deer in a 4 × 4 bound (transverse). Footfall from bottom to top: LF, LH, RF, RH. Lebensgarten Steyerberg, Germany.*

DIFFERENTIATING BETWEEN COMMON FALLOW DEER AND WESTERN RED DEER

Common Fallow Deer
a) Slim hooves that taper to a point.
b) Pads make up about 50 per cent of the track.
c) Elongated hoof outline.

Western Red Deer
a) Strong hooves that become flatter towards the tip.
b) Pads make up about 25–30 per cent of the track.
c) Proportionally wider hoof outline.

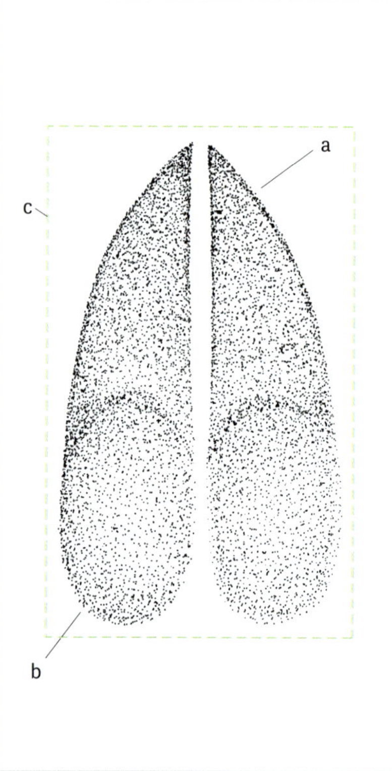

▲ *Common Fallow Deer, left front.*

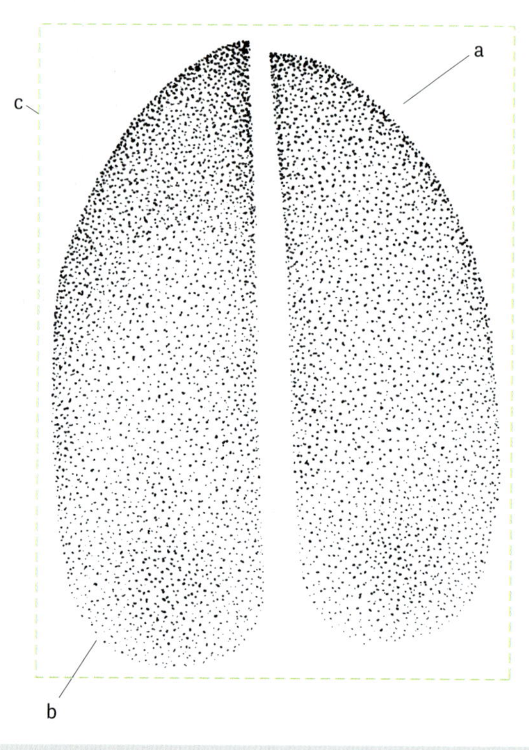

▲ *Western Red Deer, left front.*

Similar tracks
The tracks of Common Fallow Deer calves can easily be mistaken for Western Roe Deer tracks. If the pad prints are not visible, it is even possible to mix up the tracks of adult animals. However, Common Fallow Deer hooves are more parallel than Western Roe Deer hooves and the outline of the track is less heart-shaped. Western Red Deer tracks are much larger, but the tracks of small Western Red Deer can be mistaken for those of larger and heavier Common Fallow Deer. Sika Deer footprints (page 556) can also be difficult to distinguish from those of Common Fallow Deer.

▲ *Common Fallow Deer (left) and Western Red Deer (right) in comparison. Aaron Tiedemann.*

SIGNS

Unlike Western Red Deer, Common Fallow Deer do not wallow or rub. Otherwise, their signs are sometimes so similar to those of Western Red Deer that it can be difficult to tell them apart.

PATHS   Like Western Red Deer, Common Fallow Deer use established, often clearly recognisable paths within their territories. These paths usually link up grazing and resting places.

BEDS   Unlike Western Red Deer, Common Fallow Deer often lie down in the grass right next to footpaths and country roads. The beds of entire herds can often be found in the tall grass at the edges of fields, where vegetation provides adequate visual protection. The bed has the shape typical of all ruminant even-toed ungulates (page 174) and is smaller than that of a Western Red Deer.
• **Length 65–90cm**

▲ *Common Fallow Deer bed. Droppings are often found in them. Midhurst, England. John Rhyder.*

RUT PITS   During mating season, males use their front feet to dig hollows in the ground called rut pits. They urinate in the pit and strike the surrounding ground with their palmate antlers to stir it up. They then lie down and smear their oily secretions into the pit as a scent marking. A fresh rut pit gives off a strong odour that is typical of Common Fallow Deer during this time.

STRIKING   Like Western Red Deer, males show display and marking behaviour. They strike small trees and shrubs with their antlers, leaving broken branches and scratched bark. Strike marks build up before the rut.

FRAYING   From August to September, the males fray to remove the velvet from their new antlers. Young males tend to fray a little earlier than adult males.

BARK PEELING Common Fallow Deer tear off pieces of bark to feed on the cambium underneath. The incisors scrape from bottom to top like carving tools, leaving two furrows. The tooth furrows are narrower than in Western Red Deer. The bottom of the peeled area is bitten off cleanly, while the upper edge is frayed. The animals peel off strips of bark from the bottom up. It is not always possible to distinguish the peeling marks from those of Sika Deer.

- Max. height of peeling: 1.8m
- Single toothmark width: 0.3–0.7cm

MARKINGS Common Fallow Deer bite into the bark and rub a small, isolated area with their body, usually with the head and neck area. Surrounding cuts in the bark caused by striking with the antlers are often conspicuous visual markings.

▶ *Fraying at the upper and lower ends of the bark as well as furrows made by the lower incisors are important criteria for distinguishing bark peeling from markings. West Sussex, England.*

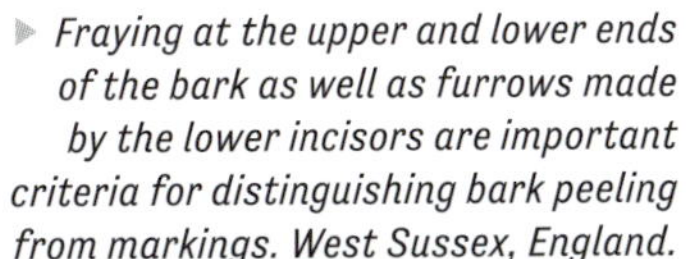

▲ *Visual markings like these are distinctive signs that can often be found in an area with Western Red or Common Fallow Deer. West Sussex, England.*

▲ *Common Fallow Deer browse line. The horizontal edges can look as precise as if they have been made with garden tools. West Sussex, England. John Rhyder.*

BUD BROWSING   Common Fallow Deer clamp their food between the incisors of the lower jaw and the dental pad of the upper jaw and tear it from the rest of the plant. This gives the remaining plant a frayed appearance. This feeding mark cannot be distinguished from that of Western Roe or Western Red Deer without additional clues. Like Western Red, Common Fallow Deer can leave a conspicuous feeding mark by snapping off branches of young trees to reach the terminal buds. They bite into the branch and then break it off by twisting their head. The dangling branch forms a kind of upside-down 'V' with the standing part of the tree. The breaks are located at about the animal's head height. This clue can help us distinguish it from terminal bud browsing by Western Red Deer.

BROWSE LINE The typical height of the browse line on trees is around 140–160cm.

EXCREMENT Common Fallow Deer excrement is like that of Western Red Deer, but usually much smaller. The colour varies between yellowish brown, green and black. The cylindrical and acorn-shaped faecal pellets are either rounded at both ends or have an indentation at one end and a small point at the other. As ruminants, Common Fallow Deer pass a large amount of faeces at a time. Richer and juicier food eaten in summer often causes droppings to stick together. Common Fallow Deer deposit 22–28 piles of droppings every day. In winter, the animals generally eat drier food and defecate less frequently (14–18 times a day).

- L 1–2cm D 0.8–1.5cm

▶ *Common Fallow Deer droppings are like those of Western Red Deer, but generally much smaller. Klein Auheim, Germany. Simone Roters.*

◀ *A small point may be apparent at one end. Klein Auheim, Germany. Simone Roters.*

WESTERN RED DEER
Cervus elaphus

HTL 160–250cm

TL 10–15cm

W 70–350kg

Males grow large antlers, but females do not have any. Males are much larger than females.

There are five species in the genus *Cervus* in all, of which only two occur in the area covered by this book. Western Red Deer have an excellent sense of smell. They can see and hear well, can jump high and wide and are good swimmers. Although they are naturally both diurnal and nocturnal, these animals now tend to be more active at dawn, dusk and at night in many areas due to human activity. Females live in herds, together with their calves. These herds also join up with other herds, especially in winter. Herds of more than 50 animals are possible. Young males form bachelor groups with other males of the same age, while adult stags are often solitary. In rare cases, healthy adult Western Red Deer are predated by wolves and bears. Wolverines or lynx can take young, ill or weak animals.

DISTINGUISHING FEATURES   Largest deer species in Europe. Long, slender legs and large ears.

DISTRIBUTION AND HABITAT   Occurs throughout almost all of Europe but absent from Iceland and northern Scandinavia. Western Red Deer were originally steppe dwellers but their high degree of adaptability means they can be found in many habitats from sea level to above the treeline. In many parts of Europe, Western Red Deer have retreated into forests to avoid humans. They favour deciduous and mixed forests with adjacent open spaces such as meadows and pastures. They are also found in wetlands with reedbeds, as well as moorland (Scotland) and open grassland.

DIET   Herbs, grasses, leaves, shoots, buds, tree bark, fungi, acorns, beechnuts and chestnuts, as well as lichens, wild berries and arable crops. As ruminants, the animals' daily rhythm is defined by alternating between active feeding and rumination. Every day, they move between their resting places in thickets and their grazing grounds.

REPRODUCTION   During mating season (rut) from September to October, a dominant stag gathers a herd of up to 20 hinds on the rutting ground. Violent fights can break out as they defend the herd against other stags. The dominant stag marks his territory by making a roaring sound that can be heard for miles. He will attempt to mate with all the females in the rutting herd. Stags hardly eat during this time, so they are severely weakened by the end of the rut. In very cold winters, this can mean death. After a gestation period of 230–240 days, 1–2 calves are born. The hind gives birth in a secluded location and initially only visits the calf to suckle. The animals reach sexual maturity after 1.5–2.5 years, but Western Red Deer stags usually do not start mating until the age of 4–6 years.

Their impressive antlers make Western Red Deer popular trophies for hunters. At least 12 subspecies occur in Europe.

SIGNS

PATHS   Western Red Deer keep to fixed, often clearly recognisable paths within their territories. These paths normally link up grazing and resting places. People can usually follow their paths without major difficulty as they are relatively free from vegetation.

TRACK

Front
L 7.5–10.5cm   W 5.2–8cm

Hind
L 7–9.5cm   W 5–7.5cm

Medium to large. Unguligrade. Slightly asymmetrical. Four toes: toes 2 and 5 (the dew claws) are very short and higher up on the back of the foot. They are rarely visible and their outer edges do not protrude beyond the lateral outer edges of the hooves. Toe 3 is slightly smaller than toe 4. The hoof outline is an upright rectangle. The individual hooves are broad and have wide, rounded tips. The inner hoof walls are slightly concave. The pad print makes up around 25 per cent of the total length of the front foot track (a) and around 30 per cent of the hind foot track. The front foot is larger and wider than the hind foot. The hooves can spread apart at the front at high speed and on slippery ground.

▲ *Left front.*
Märkische Schweiz, Germany.

▲ *Right hind.*
Märkische Schweiz, Germany.

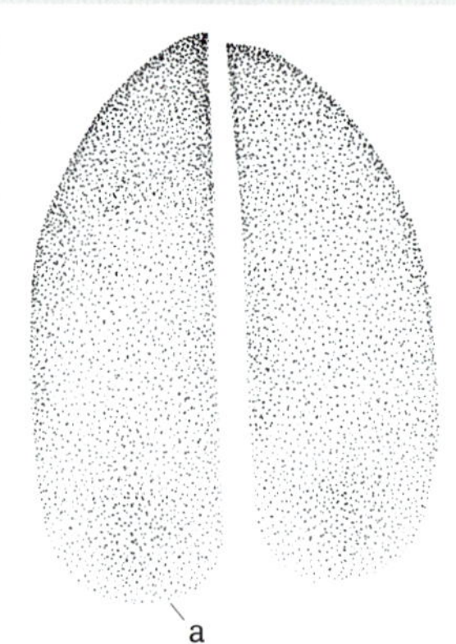

▲ *Left front.*

▲ *Western Red Deer and Western Roe Deer tracks.*

▲ *Left hind over left front.*
Kreba-Neudorf, Germany.

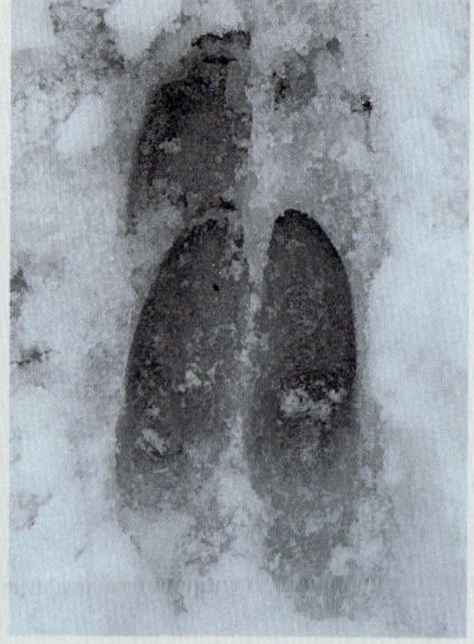

▲ *Left hind and left front.*
Bavarian Forest, Germany.

GAITS

Western Red Deer mainly move at a walk. The direct register trot is their most common fast gait. When alarmed, the animals bound and gallop to flee. Straddle trot is typically used for short distances and when they change from a trot to a gallop or from a gallop to a trot.

Walk
SL 72–173cm   TW 15–35cm

Trot
Direct register: SL 122–240cm
Overstep:
SL 218–258cm   TW 8–24cm
Straddle trot:
SL 192–266cm   TW 34–42cm

3 × 4 bound (rotary) and 4 × 4 bound (transverse)
Group length: 135–230cm
Inter-group length: 33–55cm
SL 245–272cm   TW 21–35cm

Gallop
Group length: 166–325cm
Inter-group length: 55–117cm
SL 186–347cm   TW 19–33cm

Similar tracks
Adult Wild Boar footprints can be mistaken for Western Red Deer footprints if dew claw prints are not visible. Even then, it is usually possible to tell them apart by considering the morphology of the foot (see page 502). Smaller Western Red Deer tracks are sometimes confused with the tracks of larger and heavier Common Fallow Deer (for differentiation, see page 536).

▼ *Rotary gallop, the two smaller hind feet overstep the two larger front feet. Märkische Schweiz, Germany.*

◄ *Western Red Deer in its most common gait, the direct register walk. Märkische Schweiz, Germany.*

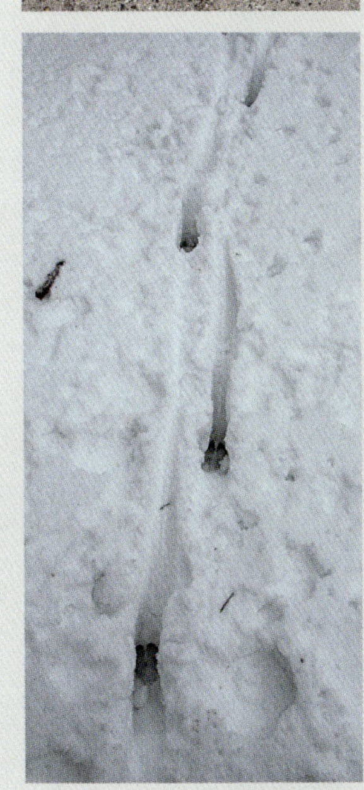

▲ *Direct register walk. The hooves often leave drag marks in deep snow. Bavarian Forest, Germany.*

▲ *Direct register walk.*

▲ *Direct register trot.*

▼ *Tracks can even be found on dry leaves. Can you make out the footprint? Märkische Schweiz, Germany.*

▲ *The almost perfect print of a Western Red Deer bed. The straight edges at the bottom are the prints of the front and hind legs. The clear curve at the top left is the print of the rump and back. The head was to the right. Bavarian Forest, Germany.*

▲ *A heavily trodden Western Red Deer path near the Großglockner. Großglockner High Alpine Road, Austria.*

BEDS   Unlike Western Roe Deer, who scrape shallow hollows in the ground before lying down, Western Red Deer lie down without making any preparations. Their beds have the typical shape of all even-toed ungulate beds (page 174). If you find the beds of several animals, it is worth comparing the different sizes and positions to glean information about the social structure. Beds are often found in well-covered border areas that offer protection and a good view. They often contain hairs.

- Length 100–135cm

MUD BATHS   Western Red Deer take regular mud baths (wallowing) to protect against insects and parasites. Animals repeatedly visit popular wallowing sites, so the surrounding vegetation is usually heavily trampled, often with clear, wide paths leading to the muddy pool.

- Length 90–250cm

RUBBING TREES Western Red Deer rub on trees and remove large areas of bark over time. Their hair is often found on the edges of the place where the bark has been rubbed off. Rubbed patches begin at least 40cm above ground and can be up to 150cm high. In comparison, rubbed patches made by Wild Boar usually begin at ground level and can be a maximum of 120cm high.

▶ Tree where a Western Red Deer has rubbed. Areas rubbed by Wild Boar usually reach the ground. Lake District National Park, England.

▲ This Western Red Deer wallowing site has been visited repeatedly over the years. Bieszczady, Poland.

In contrast to bark peeling, rubbing to mark territory leaves a smooth surface without toothmarks. Lausitz, Germany.

Trees that Western Red Deer use for striking, like this fir, usually have a combination of different types of damage, such as strike or rub marks from the antlers. Bieszczady, Poland.

MARKINGS Western Red Deer leave markings by biting tree bark and rubbing a small, isolated area with their body, usually the head and neck area. Marks where their antlers have struck the bark are often visible.

STRIKING Striking is a form of display and marking behaviour during the rut. Males strike small trees and bushes with their antlers, leaving both clearly visible strike marks and olfactory territory markings. It can be hard to tell the difference between bark damage caused by striking, and feeding marks such as bark peeling.

FRAYING Western Red Deer stags fray to remove the velvet from their newly grown antlers, approximately between July and August.

BARK PEELING Western Red Deer tear off pieces of bark to feed on the cambium beneath. The incisors leave two clear furrows that plane from bottom to top like carving tools. There is often a straight cut at the bottom of the peeled surface, while the top edge is heavily frayed. In contrast to territory markings, which tend to be strategically and individually positioned, bark peeling marks are usually found in large numbers on several trees, as Western Red Deer normally eat large amounts of cambium. This can cause considerable damage to commercial forests.

- Max. height of peeling: 2m
- Single toothmark width: 0.6–1.1cm

▶ *Western Red Deer bark peeling normally begins at the animal's head height. The deer only sampled this part, but they can extensively debark trees. Bavarian Forest, Germany.*

◀ *Close-up. Bavarian Forest, Germany.*

▶ *A Western Red Deer has snapped the branch off this young tree to get to the terminal bud. Bavarian Forest, Germany.*

▼ *Close-up. Elk leave similar signs. Bavarian Forest, Germany.*

▲ *Bud browsing can often only be associated with a specific species if other tracks and signs are present. Coto de Doñana National Park, Spain.*

▲ *Clear toothmarks left by the lower incisors help distinguish this from the feeding marks of a Wild Boar. Lausitz, Germany.*

BUD BROWSING Western Red Deer clamp a stem or stems between the incisors of the lower jaw and the dental pad of the upper jaw and tear it off. This gives the remaining plant a frayed appearance. These feeding marks can only be differentiated from those of Western Roe Deer or Common Fallow Deer by their height. Western Red Deer leave typical feeding marks by snapping the branches off younger trees to get to the terminal buds. They bite into the branch and then break it off by twisting their head. Together with the standing part of the tree, the dangling branch forms a kind of upside-down 'V'. The breaks are located at approximately the animal's head height.

BROWSE LINE The typical height of the browse line on trees is around 160–200cm.

FEEDING MARKS ON FUNGI The width of the lower incisors gives a clue as to which species was feeding. Western Red Deer toothmarks are much wider than those of Western Roe Deer. Wild Boar also have incisors in their upper jaw, so the fungi will be damaged from both sides.

EXCREMENT Colour varies between yellowish brown, green and black. The cylindrical or acorn-shaped faecal pellets are either rounded at both ends or have an indentation at one end and a small point at the other. It is sometimes suggested that this is a sex difference, but this has not been confirmed. Western Red Deer deposit large quantities of faecal pellets at once, whereas rabbits and hares deposit each pellet individually. Summer droppings often stick together due to the moister and richer food available in the summer months. The animals deposit 24–29 piles of droppings a day. They generally eat drier food in winter and defecate less frequently (19–25 piles of droppings).

* L 1.3–3.1cm D 0.9–1.9cm

▼ *The wetter summer droppings often clump together. Märkische Schweiz, Germany.*

▶ *The faecal pellets of winter droppings are clearly separate from each other. Bavarian Forest, Germany.*

SIKA DEER

Cervus nippon

HTL 95–160cm

TL 7.5–20cm

W 30–65kg (up to 100kg)

Pronounced sexual dimorphism, with males 30–40 per cent larger and heavier than females.

This mainly crepuscular and nocturnal species is roughly the size of a Common Fallow Deer. Sika Deer have a very good sense of smell and hearing. They are good swimmers and, like Water Deer, may flee into water when threatened. The animals mainly live alone or in small groups made up of mothers and their young. Groups of both sexes can be observed more frequently during the mating season and the following winter. However, these parties rarely consist of more than five animals. Sika Deer usually only form larger groups in a more open habitat. In Europe, they are predated by wolves, lynx and hunting dogs.

DISTINGUISHING FEATURES About the size of a Common Fallow Deer. Like Common Fallow Deer, their summer coat has pale spots. They also have characteristic dark stripes above their eyes, a faint dark stripe down the tail centre (less pronounced than that of Common Fallow Deer), and a circular pale gland halfway up the hind leg. The stag's antlers are branched, not palmate.

DISTRIBUTION AND HABITAT Introduced to Europe from their native East Asia in the nineteenth and twentieth centuries, isolated populations now occur in large parts of Europe. Sika Deer are very adaptable and prefer deciduous and coniferous forests near water, with dense undergrowth and relatively acidic soil. They appear to be more dependent than Western Red Deer on densely covered landscapes.

DIET Grasses, herbs, aquatic plants, leaves, needles, shoots, buds and bark. Sikas peel the bark of deciduous and coniferous trees. High population densities can cause considerable damage to forests. If they live near agricultural land, Sika Deer also eat crops, such as oilseed rape and beet, and can cause major damage.

REPRODUCTION Mating season lasts from the end of September until the end of December, with a peak in November. Like Common Fallow Deer, males attract a harem and defend their territory against other stags. After a gestation period of 210–223 days, one calf is usually born in May–June. Twin births occur in rare cases. The young reach sexual maturity after 18–24 months, but the males often do not mate until later.

Sika Deer can mate with Western Red Deer and give birth to fertile offspring.

TRACK

Front
L 5.5–7.9cm   W 3.5–6.5cm

Hind
L 5.4–6.8cm   W 3–6cm

Medium-sized. Unguligrade. Similar tracks to Common Fallow Deer, but more asymmetrical. Four toes: toes 2 and 5 (the dew claws) are very short and higher up on the back of the foot. Dew claw prints are rarely visible. They are located behind the hooves and the same width as the outer edges of toes 3 and 4. Toe 3 is always smaller than toe 4. The hoof outline is elongated and the individual hooves are slender and taper to a point. The inner hoof walls are clearly concave. The outer hoof walls taper slightly between the front end of the pad and the start of the hoof (a). The pad prints make up around 50 per cent of the total length of the track (b). A ridge is often visible between the hard unguis and the soft subunguis. The pads and hooves are at different heights and appear out of line. The front foot print is larger and wider than the hind foot print. The hind foot can appear heart-shaped, like a Western Roe Deer. The hooves can spread far apart at the front when the animal is moving at high speed or on slippery ground.

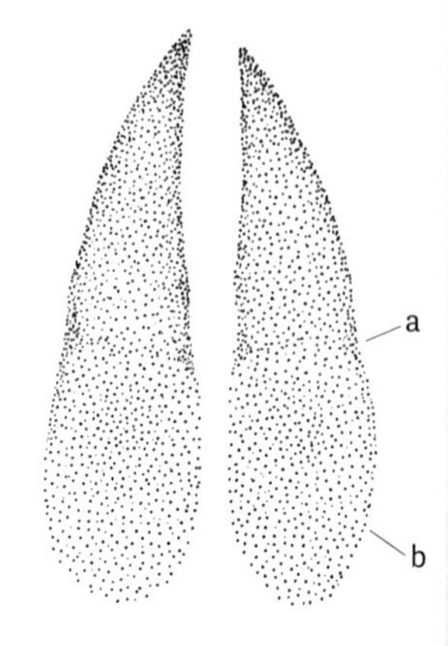

▲ *Left hind.*

▲ *Double print, left hind covering left front. Note the large pad that makes up almost 50 per cent of the print. Donau-Auen, Austria. Andreas Wenger.*

▲ *Sika Deer, left hind. Donau-Auen, Austria. Andreas Wenger.*

GAITS

Sika Deer mainly move at a walk. The direct register trot is their most common fast gait. When alarmed, the animals bound and gallop to flee.

Walk
Stride length: 75–115cm
Trail width: 10–25cm

Limited data available.

▼ *The preferred gait of the Sika Deer: the direct register walk. Donau-Auen, Austria. Andreas Wenger.*

▼ *Direct register walk.*

▼ *Direct register trot.*

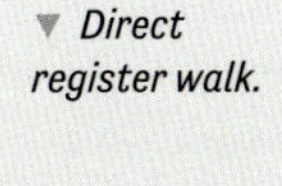

Similar tracks

Western Red Deer are larger. The tracks of Sika Deer calves can easily be mistaken for Western Roe Deer tracks, especially if there is no pad print. It can be difficult to distinguish them from Common Fallow Deer.

DIFFERENTIATING BETWEEN SIKA DEER AND COMMON FALLOW DEER

Sika Deer

a) Relatively slender, more pointed hooves.
b) Inner hoof walls more concave in comparison, with a wide negative space between the hooves.
c) Hoof outline more elongated and asymmetrical.
d) Clearly tapered area.

Common Fallow Deer

a) Relatively wide, less pointed hooves.
b) Inner hoof walls more parallel, with a narrow negative space between the hooves.
c) Hoof outline wider and more symmetrical.
d) Only slightly tapered area.

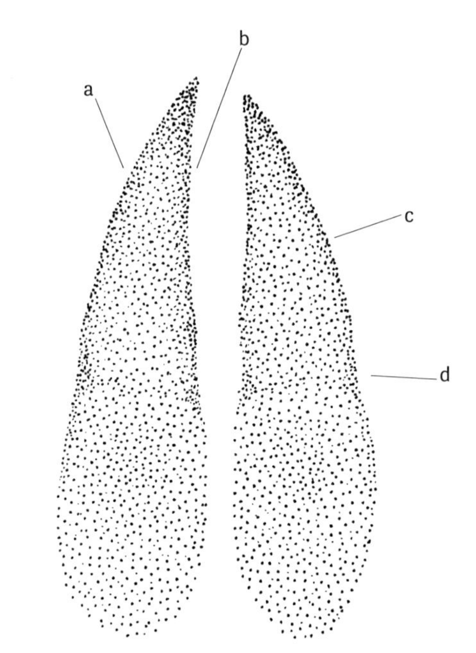

▲ *Sika Deer, left front.*

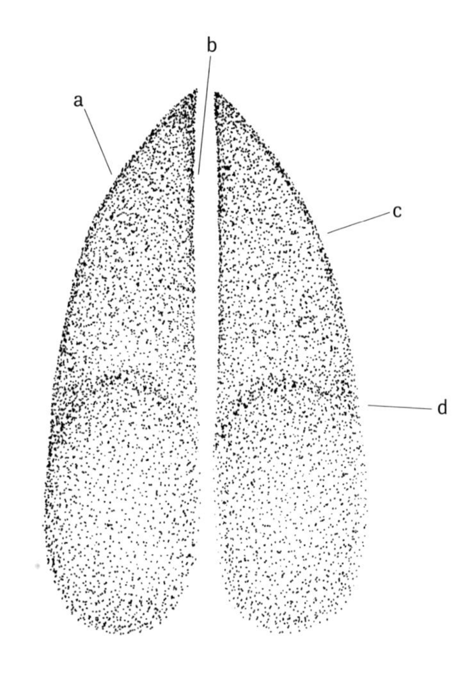

▲ *Common Fallow Deer, left front.*

SIGNS

Many of the signs are similar to those of Common Fallow Deer, so they are not always clearly identifiable.

BED   As for Common Fallow Deer (page 537).

RUT PITS   Like Common Fallow Deer (page 538), Sika stags make rut pits during mating season by scraping hollows in the ground with their front legs.

MUD BATHS   Unlike Common Fallow Deer (page 547), Sika Deer wallow in the same way as Western Red Deer.

FRAYING   When the new antlers emerge in August–September, they are covered in soft velvet that the animals remove by rubbing them on small trees, which is known as 'fraying'.

▼ Sika Deer excrement. Donau-Auen, Austria. Andreas Wenger.

STRIKING   Sika Deer stags mark their territory, especially during the rut, by 'bole scoring', i.e. gouging deep, vertical furrows in the tree bark with their antlers. This sign is like the deep furrowed markings that male Wild Boars leave on trees with their tusks (page 505), but the furrows are longer and found higher up on the tree. This behaviour is considered characteristic of this species.

BARK PEELING   Sika Deer can cause extensive damage by bark peeling.

BUD BROWSING   Not reliably distinguishable from Common Fallow Deer (page 540).

EXCREMENT   Droppings are similar to those of Common Fallow Deer.
- L 1.1–1.8cm D 0.8–1.4cm

▶ Close-up. Donau-Auen, Austria. Andreas Wenger.

REEVES'S MUNTJAC
Muntiacus reevesi

HTL 70–100cm
TL 10–16cm
W 9–16kg (up to 22kg)

The genus of muntjacs (*Muntiacus*) includes 11 species, one of which occurs in the area this book covers. Reeves's Muntjacs are Europe's smallest deer and are active both during the day and at night. Peak activity occurs at dawn and dusk. They have a well-developed sense of smell and can hear and swim well. They are secretive, and may be solitary or live in small groups consisting of a mother and her offspring. Occasionally, several can be observed feeding together, but apart from this, groups of four or more animals are rare. Like Western Roe Deer, they alternate short grazing periods with rest periods several times a day, during which the animals lie down to ruminate. Red Foxes kill and eat fawns and, more rarely, adult animals if they can catch them. Along with long, snowy winters, humans and hunting dogs, and traffic accidents, cause the greatest losses.

DISTINGUISHING FEATURES   About the size of a fox and considerably smaller than a Western Roe Deer. The males (bucks) have tusk-like canine teeth in their upper jaw, which can grow up to 2.5cm long. Very short antlers (< 15cm) and V-shaped, dark facial markings.

DISTRIBUTION AND HABITAT   Native to southeastern China, it was released in southern England and France, where stable populations have since become established. Reeves's Muntjacs prefer deciduous and coniferous forests with dense undergrowth and diverse vegetation. They also live in parks and cemeteries.

DIET   Like Western Roe Deer, Reeves's Muntjacs are concentrate selectors. They specifically select energy-rich, easily digestible food sources instead of eating large quantities of roughage at once. Shoots, buds and leaves of various deciduous trees and shrubs, as well as fruits, beechnuts and acorns make up a large part of their diet. They also eat flowers, such as primroses, orchids and bellflowers, as well as eggs, small animals and carrion. Grasses are only an important food source for a short time in spring or winter, when preferred food sources are scarce or difficult to access.

REPRODUCTION   Reeves's Muntjacs are able to reproduce all year round and females can become pregnant again immediately after giving birth (postpartum oestrus). Females can therefore be pregnant almost continuously for many years. After a gestation period of around 210 days, 1–2 young are born. They reach sexual maturity after 6–12 months. The average reproduction rate is around 1.6 young per female, per year.

In areas where they occur in high densities, muntjacs can cause serious damage to broad-leaved trees. They never or hardly ever damage coniferous trees. Muntjac meat is considered a delicacy in Asia.

SIGNS

Reeves's Muntjac signs are often only distinguishable from Western Roe Deer signs by their size or height. Even if the measurements point to a Reeves's Muntjac, you should also consider young Western Roe Deer. Knowledge about differences in behaviour and about the region are often essential for identification.

BED   Reeves's Muntjac beds resemble those of Western Roe Deer but are slightly smaller.

FRAYING   When the new antlers emerge in August–October, they are covered in soft velvet. Reeves's Muntjac bucks rub off the velvet on trees and bushes.

TRACK

Front
L 2.5–3cm W 2–2.4cm

Hind
L 2.3–2.5cm W 1.9–2.2cm

Small. Unguligrade. Slightly asymmetrical. Four toes: toes 2 and 5 (the dew claws) are very short and higher up on the back of the foot. Dew claw prints are rarely visible. They are located behind the prints of the hooves and are the same width as the lateral outer edges of toes 3 and 4. These toes always leave a print, with toe 3 being smaller than toe 4. The individual hooves are narrow and the outer hoof walls slightly convex. The front half of the inside hoof wall is slightly concave. The pad makes up 25 per cent to one third of the total length of the track and is rarely clearly recognisable. The front foot print is larger and wider than the hind foot print. The hooves can spread apart at the front at high speed or on slippery ground.

Similar tracks
Western Roe Deer and Water Deer.

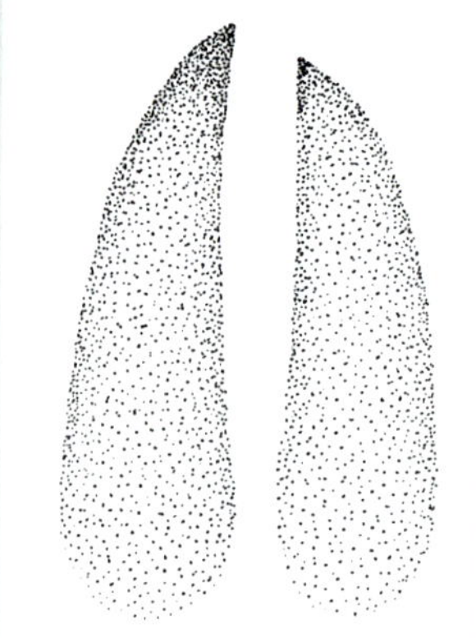

▲ *Left front.*

▲ *Front foot print, partially covered by a hind foot print. West Sussex, England. John Rhyder.*

GAITS

When undisturbed, Reeves's Muntjacs mainly move at a walk. Their paths usually lead through dense undergrowth and are difficult for humans to follow. The trot is the most common faster gait and is mainly used to reach a destination quickly. When alarmed, animals flee short distances in bounds or gallop before stopping to investigate the cause of the disturbance.

Walk
Stride length: 22–68cm
Trail width: 8–12cm

Trot
Stride length: 60–105cm
Trail width: 6–10cm

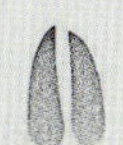

▶ *A Reeves's Muntjac in a walk. The step is shorter than the width of an adult human hand. West Sussex, England. John Rhyder.*

▲ *Direct register walk.*

DIFFERENTIATING BETWEEN REEVES'S MUNTJAC AND WESTERN ROE DEER

The difference in size between toes 3 and 4 is more pronounced than in Western Roe Deer. The individual hooves of Reeves's Muntjac are longer and narrower compared to Western Roe Deer. The tracks are rarely heart-shaped and smaller than those of an adult Western Roe Deer.

▲ Strike marks on a pine tree, about 10–20cm above the ground. The height indicates a Reeves's Muntjac. West Sussex, England.

STRIKING Reeves's Muntjacs strike in a similar way to Western Roe Deer. They mark areas about 10–40cm above the ground, so the strike points are usually lower than a Western Roe Deer's. Tusk marks made by Reeves's Muntjacs are another important distinguishing feature. Like Wild Boar, their enlarged canine teeth damage the bark and sometimes the wood beneath.

SCRAPING Reeves's Muntjacs scrape ground vegetation as well as leaves and needles backwards under their body, using their front hooves. Like Western Roe Deer, they leave scent secretions from their interdigital glands, as well as a clear visual marking. Scrape marks are rarer than in Western Roe Deer and are mainly found in areas with higher muntjac population densities, where territory disputes are more common. As Western Roe Deer occasionally only scrape very small patches and Reeves's Muntjacs can gradually enlarge a patch, it is not possible to reliably tell them apart (see Western Roe Deer, page 523).

BARK PEELING Reeves's Muntjacs usually peel bark from close to ground level up to a height of around 50cm. They peel bark with their enlarged canine teeth (tusks), sometimes leaving deep gouges in the wood. Reeves's Muntjacs peel bark all year round, but usually only on very thin shoots.

BUD BROWSING Reeves's Muntjacs typically browse blackberry bushes as well as seedlings and fresh plant shoots close to the ground. Like Western Red Deer, they snap branches off small trees to reach the terminal buds. They bite into the branch and then break it off by twisting their head. The breaks will be at a height of between 50–70cm. Together with the standing part of the tree, the dangling branch forms a kind of upside-down 'V'. Other traces of browsing cannot be distinguished from Western Roe Deer browsing (see Western Roe Deer, page 525).

BROWSE LINE   The typical height of the browse line on shrubs, trees and bushes is around 60–115cm.

EXCREMENT   Almost roundish, usually shiny black faecal pellets, which are either rounded at both ends or flat to concave at one end with a small point at the other. As ruminants, muntjacs produce large quantities of droppings at a time (20–120 faecal pellets). In areas with high population density, the animals create latrines to mark their territory.

- L 0.5–1.5cm D 0.5–1.3cm

▶ *Reeves's Muntjac droppings. West Sussex, England.*

◀ *CyberTracker evaluator John Rhyder demonstrates a field test for distinguishing between Reeves's Muntjac, Western Roe Deer and Common Fallow Deer pellets, using the hand size of an average adult male. If the faecal pellet is smaller than the fingernail of the little finger, this indicates a Reeves's Muntjac. If it is about the size of the fingernail, this points to a Western Roe Deer. If it is larger than the fingernail, this indicates a Common Fallow Deer. West Sussex, England.*

WATER DEER
Hydropotes inermis

Water Deer is the only species of the genus *Hydropotes* in the area and the only deer species in Europe without antlers. These shy, solitary animals are mainly active at dawn and dusk due to hunting, although they would naturally be active during the day and at night. They have a good sense of both hearing and smell and are excellent swimmers. When threatened, Water Deer may flee to water and can wade or swim across larger stretches. Red Foxes sometimes prey on fawns. Cold, wet winters can be fatal for exhausted males, especially after the rut.

HTL 70–100cm
TL 4–9cm
W 9–16kg

DISTINGUISHING FEATURES Slightly larger than a Reeves's Muntjac but smaller than a Western Roe Deer. The males have tusk-like canine teeth in their upper jaw that can grow up to 8cm long. Noticeably large and rounded ears.

DISTRIBUTION AND HABITAT Water Deer, along with Reeves's Muntjac, were introduced from their native east Asia to England in 1929 by the Duke of Bedford. Today, Water Deer can be found in Bedfordshire, the Cambridgeshire Fens and Norfolk Broads. Another feral population, introduced in 1954, lives in France. The animals prefer marshes, high, open grasslands and cultivated areas near water.

DIET Like Western Roe Deer, Water Deer are concentrate selectors, but to a lesser extent. Their diet includes aquatic plants, buds, beechnuts, acorns, sedges and grasses. They also eat crops such as carrots and leftover potato and winter wheat crops. Unlike most other deer species in Europe, Water Deer do not peel bark.

REPRODUCTION   Mating season lasts from November to January, with a peak in December. The males use their canine teeth in fights with rivals. Many males have scars on their necks and ears from previous fights. In England, 1–2 young are usually born in May–June after a gestation period of 165–210 days. The young reach sexual maturity after 5–7 months.

SIGNS

Unlike many other deer species, Water Deer do not peel bark. As the males do not have antlers, they do not leave the fraying or striking marks typical of other deer.

BED   See Western Roe Deer bed, page 522.

PATHS   Water Deer paths often lead through dense riverbank vegetation and usually end in a marsh or body of water.

EXCREMENT   Usually dark brown or black faecal pellets, which are rounded at one end and have a small point at the other. To mark their territory during the mating season, the animals create latrines at the boundaries.
• L 1–1.7cm D 0.5–1cm

▼ *Water Deer droppings are found in damp places that Western Roe Deer tend to avoid. Hickling Broad, Norfolk, England.*

▲ *Here, Water Deer have been moving between water and land. Hickling Broad, Norfolk, England.*

▲ *Close-up. Hickling Broad, Norfolk, England.*

TRACK

Front
L 4–5cm W 3–4cm

Hind
L 3.5–4.2cm W 2.3–3.2cm

Small. Unguligrade. Slightly asymmetrical. Four toes: toes 2 and 5 (the dew claws) are very short and higher up on the back of the foot. Dew claw prints are rarely visible. They are located behind the hoofprints and their outer edges make prints as wide as the lateral outer edges of toes 3 and 4. The print of toe 3 is slightly smaller than that of toe 4. The pad is enlarged and takes up about 35–40 per cent of the total area of the track (a). A tapered area may be visible where the pad meets the hoof (b). The front foot print is larger and wider than the hind foot print. The hooves can spread apart at the front at high speed or on slippery ground.

Similar tracks
Western Roe Deer, Reeves's Muntjac.

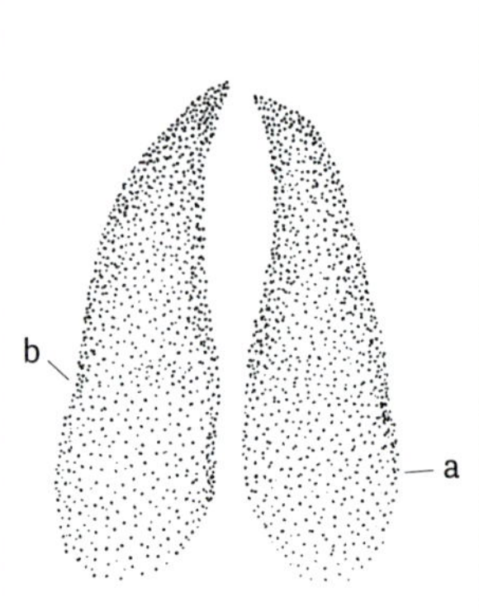

▲ *Left hind.*

▲ *Left hind. Note the tapered area between the large pad and the left hoof. Hickling Broad, Norfolk, England.*

GAITS

Water Deer mainly move at a walk. When alarmed, they flee over short distances in bounds or at a gallop. Like Western Roe Deer, Water Deer are physically unable to cover longer distances when fleeing, and usually seek cover in a few bounds.

Walk
Stride length: 50–80cm
Trail width: 8–15cm

Trot
Stride length: 60–100cm
Trail width: 5–12cm

Based on limited data.

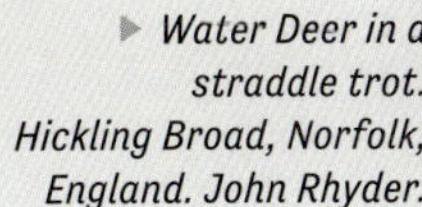
▶ *Water Deer in a straddle trot. Hickling Broad, Norfolk, England. John Rhyder.*

▲ *Direct register walk.*

◀ *Water Deer prefer to inhabit open grasslands with marshes and various bodies of water. Hickling Broad, Norfolk, England.*

Bovids

The bovid family (Bovidae) includes a total of some 50 genera with about 143 species. They are the most species-rich group of even-toed ungulates and include domestic livestock such as sheep and goats. There are nine species in the area this book covers.

The most obvious difference to deer is the hollow horns. They are covered with horny tissue and do not branch, in contrast to antlers. Horns keep growing throughout the animal's life, have a blood supply and are not cast like antlers. The shape and structure of the horns vary according to the species. Both sexes generally have horns, but males' horns are usually larger than females'. The animals mainly use their horns during mating season, for fighting with rivals or for territory marking.

TRACKS AND SIGNS OF BOVIDS

The tracks and signs of bovids essentially have the characteristic features of even-toed ungulates (page 494) and can be confused with those of deer.

PATHS Many bovids keep to fixed, well-trodden and clearly recognisable paths within their home range. These can be wider than deer paths.

DAMAGE CAUSED BY HORNS Rubbing and striking with the horns, usually against smaller trees, leave visual and scent markings that are a common distinctive sign. Striking or rubbing with the horns can leave deep furrows in the wood that can reach a considerable size. For this reason, some domestic goat breeds are used specifically in landscape conservation to curb heavy scrub encroachment.

EUROPEAN BISON
Bos bonasus

HTL 210–350cm
TL 30–80cm
W 400–1,000kg
Bulls are much larger and heavier than cows, but this pronounced sexual dimorphism only becomes apparent from the third year.

The European Bison is the only representative of the genus *Bos* (true cattle) in the region. They are the largest land mammals in Europe. Adult females live in herds of 8–20 animals led by an experienced lead cow. These groups generally consist of several adult cows, their calves and sometimes also young males. Bulls live together in small groups but become more solitary as they get older. Bulls usually only go to the female herd during the rut. European Bison are mainly active during the day. They mostly remain in a particular home range, but also wander in search of food. They spend about 18 hours a day ruminating and resting and about six hours foraging. Their sense of smell and hearing are particularly well developed, but their eyesight is weak. They can swim well and jump across a distance of up to 3m. Adults have no predators. Calves and weak animals can be prey for wolves.

DISTINGUISHING FEATURES Europe's largest ungulate. Massive, heavy bovine with a short head, small eyes and strong legs.

Conspicuous hump in the shoulder region and short, curved horns that point upwards. Dense coat, mane-like on the chest.

DISTRIBUTION AND HABITAT European Bison were formerly widespread in large parts of Europe. Their population began to decline around 6,000 years ago and they were extinct in the wild by the beginning of the twentieth century. Thanks to conservation breeding in zoos and game reserves, they have been successfully reintroduced and there are now small, stable populations in Poland, Lithuania, Germany and Slovakia. The total population comprises more than 3,000 animals. The first successful reintroduction was in 1952 in Białowieża National Park in Poland. In terms of habitat, European Bison need large, undisturbed deciduous and mixed forests with thick undergrowth, marshy areas and grassy clearings.

DIET Herbs, grasses, leaves and buds, tree bark from softwoods and thin branches. European Bison also eat mosses and lichens, as well as seeds like acorns and beechnuts. They need to eat between 30 and 60kg of food a day.

REPRODUCTION Bulls visit the herd during the mating season (rut) in August–October, but fights between bulls are rare compared to other even-toed ungulates. The males become very aroused during the rut. They churn up the ground with their hooves, destroy young trees with their horns and mark the ground with urine before rolling in it. Pregnant cows isolate themselves from the herd before giving birth. They seek a sheltered place to have their calf and then rejoin the herd with the calf after giving birth. The gestation period is 254–279 days. European Bison usually have one calf per year; twins are rare. The young are precocial and can stand and walk a few hours after birth. Females become sexually mature at 3 years of age and males at 2–4 years. However, due to competition from older, stronger bulls, young males often do not start reproducing until they are 6–12 years old.

▼ *Dust bath. Loose soil around exposed root plates of fallen trees is a favourite for dust baths. Nalibotskaya Pushcha, Belarus. René Nauta.*

TRACK

Front
L 11–18cm   W 10.5–15cm

Hind
L 10–15cm   W 10–13cm

Large to very large. Unguligrade. Slightly asymmetrical. Four toes: toes 2 and 5 (the dew claws). are very short and higher up on the back of the foot. Dew claw prints are rarely visible. They are located behind the hoofprints and their outer edges make prints as wide as the lateral outer edges of toes 3 and 4. Toe 3 is slightly smaller than toe 4. The outline of the front foot is rounded, while the outline of the hind foot is a vertical oval and rounder than in a Western Red Deer. The outer hoof walls of the front foot are convex along their entire length, while the inner hoof walls are concave. This creates the rounded impression. The negative space between the hooves is at

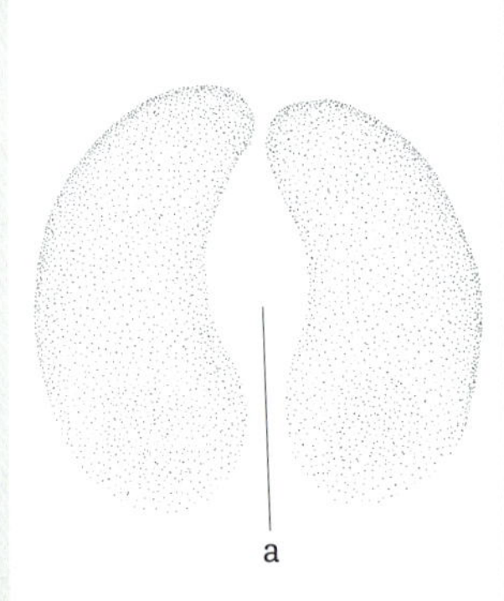

▲ *Left front.*

▲ *Left hind. Bieszczady, Poland.*

▲ *Left front. Bieszczady, Poland.*

▲ *Left hind. Bieszczady, Poland.*

the centre and most noticeable in the front foot ⓐ. There is less space between the hooves of the hind foot. The individual hooves are wide, with broad, rounded tips. The pad print makes up about 20 per cent of the total length of the track. The negative space between the back of the pads is shaped like an inverted V. The front foot print is larger and wider than the hind foot print. The prints of the cows tend to be slimmer, smaller and more pointed than the chunkier, rounded hoofprints of the bulls. The hooves can spread apart at the front at high speed or on slippery ground.

GAITS

European Bison mainly move at a steady walk. When alarmed, the animals flee by bounding and galloping and can reach speeds of up to 60km/h over short distances (around 100 metres).

Walk
Stride length: 75–165cm
Trail width: 30–49cm

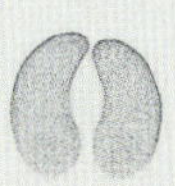

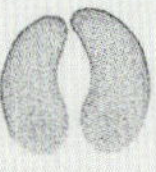

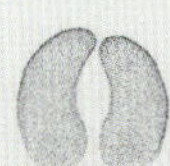

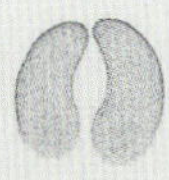

▲ *Direct register walk.*

▼ *The wide, well-trodden paths of European Bison are distinctive and reminiscent of Cattle trails in the mountains. Bieszczady, Poland.*

Similar tracks

The footprints can be mistaken for Elk and Domestic Cattle footprints. It is generally possible to clearly differentiate between the two by considering the morphology of the foot.

▲ A European Bison dust bath can cover an impressively large area. Nalibotskaya Pushcha, Belarus. René Nauta.

SIGNS

PATHS   European Bison keep to fixed, often clearly recognisable paths within their territory. These paths usually link up grazing and resting places. People can normally follow the paths without great difficulty, as they tend to be clear of vegetation. The distinctive cowpats that European Bison leave along these paths usually allow clear identification. Well-worn paths are an obvious sign of European Bison.

MUD BATHS   Like Western Red Deer (page 546), European Bison take mud baths, which are correspondingly large and conspicuous.

DUST BATHS   European Bison regularly take dust baths and sunbathe. They often make extensive changes to the vegetation in these places. European Bison dust baths have a diameter of 1.5–4m.

RUBBING TREES AND DAMAGE CAUSED BY HORNS   The males strike small trees with their horns, leaving characteristic marks. They prefer small trees with a diameter of 12–25cm. European Bison peel strips of bark off trees at a height of 15–170cm above the ground. The bark is often found in conspicuous piles at the foot of the tree. Their horns leave deep furrows in the wood. European Bison usually strike

▲ *European Bison excrement can look like Domestic Cattle dung. Insect activity indicates that the droppings are fresh. Bieszczady, Poland.*

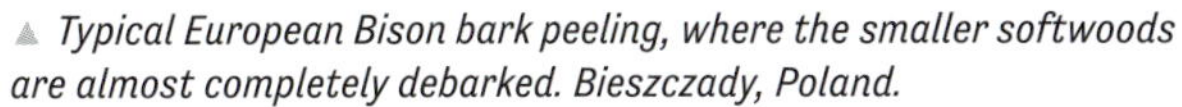

▲ *Typical European Bison bark peeling, where the smaller softwoods are almost completely debarked. Bieszczady, Poland.*

with their horns during mating season from August to October. This behaviour is often accompanied by rubbing the neck and shoulders on trees. Their wool often sticks to the resin of the trees they rub on.

BARK PEELING   European Bison tear off pieces of bark to feed on the cambium underneath. When peeling the bark, the lower incisors leave two distinct furrows as they scrape from bottom to top like carving tools. There is often a straight cut at the bottom of the peeled surface, while the top edge is heavily frayed. The animals peel off strips of bark from the bottom up. Bark peelings tend to pile up around neighbouring trees. European Bison usually eat cambium in larger quantities. Unlike deer, they prefer to peel bark from smaller softwoods.

EXCREMENT   Varies in colour between dark green and black. Generally very similar to Domestic Cattle droppings. In summer, European Bison droppings mainly consist of herbaceous plants, which gives them a thinner consistency, resulting in a cowpat-like pile. European Bison eat drier food in winter, resulting in more compact, solid excrement. It resembles the faecal pellets of Wild Boar that stick together in a sausage shape.

• **D 25–40cm**

MUSK OX
Ovibos moschatus

HTL 180–245cm
TL 6–14cm
W 180–200kg /
265–400kg
The bulls are
bigger and
heavier than
the cows.

These mostly diurnal and sedentary animals are fast runners with good vision. They live in small groups in summer and form large herds, sometimes with over 100 animals, in winter. Older males are often solitary. Musk Oxen are very well adapted to extreme cold. They lay down fat reserves that they use to sustain themselves during the long Arctic winter. If they do not manage to build up adequate fat reserves before winter, the most common causes of death are starvation and hypothermia. If threatened by snowstorms or attacks by predators, the herd closes together in a ring with their horns pointing out and their calves in the centre. Musk Oxen hardly ever flee. They are predated by Grey Wolves and Polar Bears.

DISTINGUISHING FEATURES   Massive body and a long, brown, shaggy coat. Pointed horns that first curve down and then up. The bulls' horns are larger than the cows' and can grow up to 70cm long.

DISTRIBUTION AND HABITAT   Small herds in Sweden and Norway. They prefer tundra with low precipitation and low snow cover.

DIET Mainly herbs and grasses, such as sedges. They also eat flowering plants and leaves of dwarf shrubs as well as lichens, ferns and mosses.

REPRODUCTION   The mating season lasts from July to October, with a peak in July and August. During this time, the bulls take part in ritualised fights where they repeatedly crash head-on into each other until one of the two opponents gives up. Dominant bulls gather a harem around them for mating. After a gestation period of around nine months, one calf is born. Twin births occur in rare cases. The males reach sexual maturity after 5–6 years, the females after 3–4 years.

▼ *Bark peeling.*
Tännäs, Sweden.
Laura Gärtner.

TRACK

Front
L 10.8–14.5cm W 11.1–15.2cm

Hind
L 9.5–11.4cm W 10.2–11.4cm

Large. Unguligrade. Slightly asymmetrical. Four toes: toes 2 and 5 (the dew claws) are very short and higher up on the back of the foot. The dew claws are not regularly visible. If they are, they leave a print behind the hoof almost horizontal to the direction of travel and are the same width as the outer edges of toes 3 and 4. Dew claw prints are closer to the hooves than in Reindeer. Without the dew claws, the track has a distinctively roundish appearance. Toe 3 is slightly smaller than toe 4. The outer hoof walls are very convex, while the inner ones are clearly concave along their entire length. Each individual hoofprint appears kidney-shaped. The front foot print is larger and wider than the hind foot print.

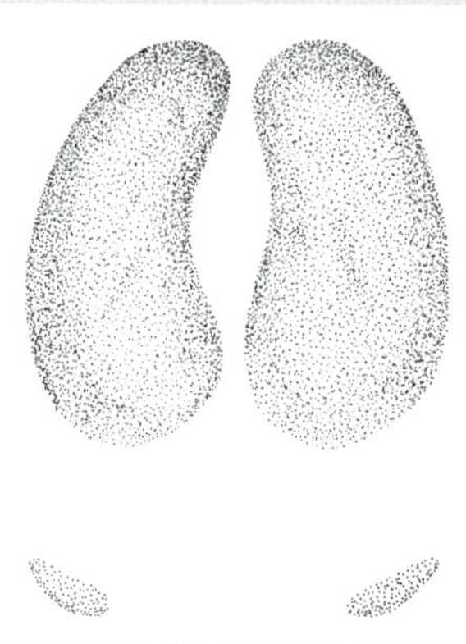

▲ *Left front.*

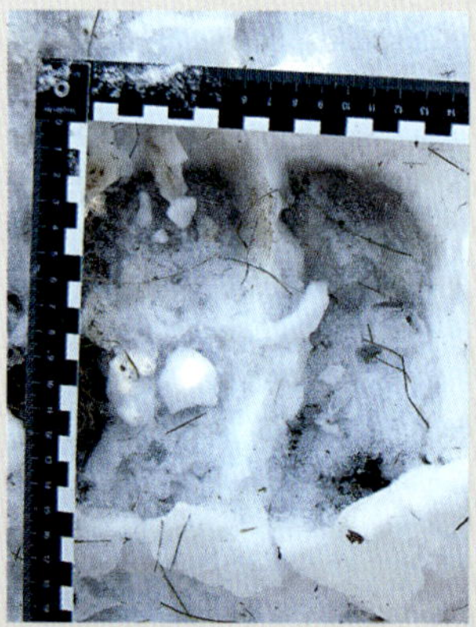

▲ *Musk Ox, left front.*
Tännäs, Sweden. Laura Gärtner.

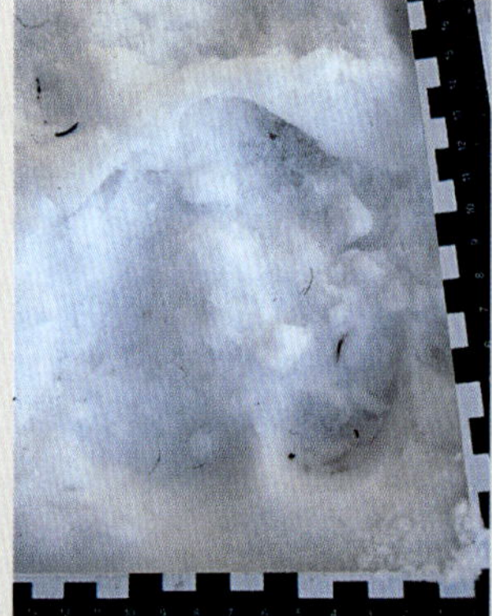

▲ *Right hind.*
Tännäs, Sweden. Laura Gärtner.

Similar tracks
The prints can be confused with Reindeer prints, but the negative space between the hooves of the Musk Ox is much smaller.

GAITS

Musk Oxen mainly move at a walk. The stride length is short in relation to the trail width. They usually trot to cover longer distances at higher speeds. The animals flee in bounds and at a gallop when alarmed.

Walk
Stride length: 80–130cm
Trail width: 25–40cm

Based on limited data.

◀ *A Musk Ox walks towards the camera. Tännäs, Sweden. Laura Gärtner.*

▲ *Direct register walk.*

◀ *Musk Oxen leave conspicuously wide, well-worn paths in deep snow. Tännäs, Sweden. Laura Gärtner.*

SIGNS

FEEDING MARKS As for other bovids.

BARK PEELING As for other bovids.

BEDS As for other bovids. Musk Oxen often lie close together to rest and sleep.

EXCREMENT Musk Ox excrement is like Reindeer excrement but it is slightly larger and consists of more oval faecal pellets. If Musk Oxen eat food with a high water content in summer, their excrement often clumps together or forms a lighter-coloured, mushy mass. Long tufts of shaggy fur are often found with the droppings.
- L 1.1–2.1cm D 0.8–1.3cm

◀ *Musk Ox excrement. Tännäs, Sweden. Laura Gärtner.*

▲ *Resting places. Tännäs, Sweden. Laura Gärtner.*

ALPINE IBEX AND IBERIAN IBEX

Capra ibex, Capra pyrenaica

Ibexes are members of the goat genus (*Capra*). There are nine species within this genus, two of which occur in the area. Due to the similarity of their tracks and signs, we discuss them together here. Ibexes are excellent rock climbers who can jump, hear, see and smell very well. Adult and young females (does) and males (bucks) that have not reached sexual maturity live in herds of up to 30 or more animals. Bucks form their own buck pack from age two or three. Very old bucks can be solitary. Ibexes are mainly active during the day and are not particularly timid. In case of danger, they retreat to steep, inaccessible areas. They do not have any predators, but young animals can be attacked by Golden Eagles.

TRACK

Front
L 6.2–10cm   W 4.2–6.4cm

Hind
L 5.5–8cm   W 3.8–5cm

Medium to large. Unguligrade. Slightly asymmetrical. Four toes: toes 2 and 5 (the dew claws) are very short and higher up on the back of the foot. The dew claws are occasionally visible. They leave prints behind the hooves at the same width as the outer edge of toes 3 and 4. Toe 3 is slightly smaller than toe 4. The hoof outline is elongated and noticeably irregular. In males, the outer edges of the hoof are slightly convex, giving them a bulky, rounded outline. The females' hooves are much smaller and their outer edges are straight or slightly concave. The inner hoof walls have a conspicuous, slightly concave indentation in the centre, making the negative space quite wide at this point. This bulge between the hooves is most obvious in the male front foot but barely recognisable in the female hind foot. The hoof is wider in the pad area, so the gap between the two hooves is narrower. The individual hooves are tapered and slightly splayed, which is usually visible at the front of the track. The shape of the tracks can vary greatly as ibex hooves are flexible so they can adapt to the ground. The front foot print is larger and wider than the hind foot print.

▲ *Iberian Ibex, right front. SERAFO, Spain. Paloma Troya & Fernando Gómez.*

▲ *Iberian Ibex, right hind. SERAFO, Spain. Paloma Troya & Fernando Gómez.*

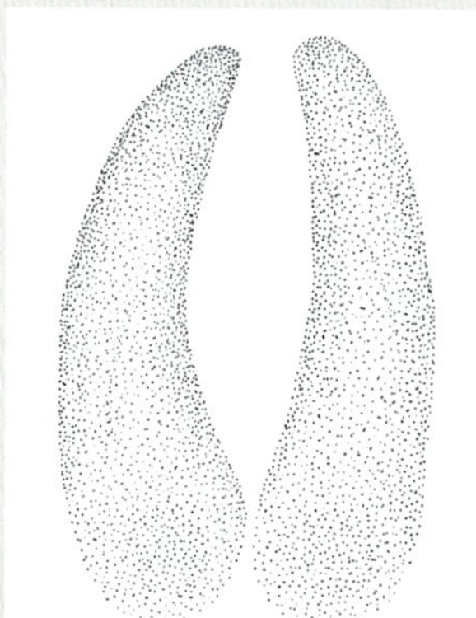

▲ *Right hind.*

▲ *Iberian Ibex, right hind. SERAFO, Spain. Paloma Troya & Fernando Gómez.*

GAITS

Ibexes usually move at a walk. They mainly cover longer distances at a trot. They bound and gallop to escape when alarmed. Because the animals usually move in rocky terrain, it is rarely possible to recognise gaits.

Walk
Stride length: 60–128cm
Trail width: 13–25cm

Based on limited data.

◀ *Gaits are rarely discernible on the rocky terrain inhabited by ibexes. Iberian Ibex droppings are visible at the bottom of the image. SERAFO, Spain. Paloma Troya & Fernando Gómez.*

Similar tracks
Chamois tracks are similar, but smaller, slimmer and more parallel. The prints can be mistaken for Domestic Goat prints. Domestic Goat prints are comparatively wider, more irregular and usually smaller overall (L ≤ 6.5cm, W ≤ 4.5cm), but as with most domesticated animals, the size and shape of the print can vary greatly depending on the breed.

▲ *Direct register walk.*

Alpine Ibex

♀ HTL 75–115cm

♂ HTL 140–170cm

♀ TL 10–20cm

♂ TL 10–20cm

♀ W 40–50kg

♂ W 70–120kg
(up to 150kg)

Iberian Ibex

♀ HTL 105–120cm

♂ HTL 130–150cm

♀ TL 10–15cm

♂ TL 10–15cm

♀ W 25–35kg

♂ W 60–80kg

Pronounced sexual dimorphism. Males much larger and heavier than females. Bucks' horns can measure slightly more than 1m, doe horns up to 30cm.

DISTINGUISHING FEATURES Goat-like shape with a bulky, strong build and relatively short legs. Characteristic sabre-shaped, backward-curving horns. Bucks' horns have thick transverse ridges on the front. The number of notches on the back of the horn corresponds to the age of the buck.

DISTRIBUTION AND HABITAT The Alpine Ibex lives in the Alps. It used to be widespread throughout the region until intensive hunting around 1820 left only around 100 animals in the Italian Alps. When the species was given legal protection, its population gradually recovered, and stock was later relocated to other Alpine regions. All the Alpine Ibexes alive today are descended from this population. The Iberian Ibex lives on the Iberian Peninsula. Its original range covered the Pyrenees and various mountain ranges in Portugal and Spain. There are four subspecies of Iberian Ibex, two of which are extinct.

Ibexes inhabit rock and scree regions above the tree line at an altitude of 2,100–3,500m. Changing location according to season is typical of Iberian Ibex: in winter, they seek out places with less snow, such as south and south-west-facing slopes or low-lying areas where food is more readily available. Unlike chamois, ibexes even remain above the tree line in winter.

DIET Mainly herbs and grasses, but also leaves, buds and, more rarely, shoots of woody plants. They also eat mosses, lichens and bark.

REPRODUCTION Fights break out between older bucks during the mating season (rut) from December to January. Fights over mating rights follow a certain ritual. Fighting bucks rear up on their hind legs and clash their horns together at full force. Only superior bucks mate with the does. After a gestation period of 161–175 days, one calf is born. A second calf is rare. The typical precocial calf is born fully furred and can see and walk immediately after birth. The animals are sexually mature after their second or third year.

SIGNS

DAMAGE CAUSED BY HORNS Males rub and strike their horns against small trees such as firs, spruces and pines. This leaves a scent mark as well as a visual sign that serves as a territory marker. This characteristic sign is like the Western Roe Deer's fraying or strike marks but involves more severe damage to the vegetation. Ibexes mainly rub with their horns immediately before and during mating season.

▼ Alpine Ibex excrement. Heiligenblut, Austria.

EXCREMENT Cylindrical to round, mostly irregularly shaped faecal pellets that are deposited in piles. One end may have a slightly extended point. The individual faecal pellets are generally furrowed and often have one or more flattened or indented side(s). The colour varies from brownish or greenish black to black. Smaller faecal pellets can be mistaken for chamois droppings but tend to be slightly larger and more irregular in shape.

- L 1.1–2cm   D 0.8–1.5cm (usually ≤ 1.1cm)

▲ Iberian Ibex, territory marking. SERAFO, Spain. Paloma Troya & Fernando Gómez.

ALPINE CHAMOIS AND PYRENEAN CHAMOIS

Rupicapra rupicapra, Rupicapra pyrenaica

HTL 90–140cm

TL 3–10cm

W 14–60kg

Males are larger
and heavier than
females.

There are between two and six species within the genus *Rupicapra*,
two of which occur in the area. Chamois are diurnal, usually timid
animals. However, they are often less timid in areas where they are
not hunted. They are excellent rock climbers and good at running and
jumping. They can see well and have a good sense of smell. Chamois
form large herds of up to 100 animals in autumn and winter. As is
normal for bovids, chamois do not shed their horns but grow them all
year round. The inconspicuous growth rings on the back of the horns

give an indication of an individual's age. When threatened, chamois stamp their front feet on the ground as a warning signal. Martin Görner writes that they also heed the whistling sounds emitted by marmots as a warning (Görner & Hackethal 1987, page 346). When chamois take flight, they usually flee uphill to inaccessible rocky areas and scree slopes. They are predated by wolf, lynx and bear. Young animals may also be killed by Common Ravens and eagles.

DISTINGUISHING FEATURES Goat-like shape. Two broad, dark stripes on the face, which run from the muzzle to the base of the ears, are striking and characteristic features. Both sexes have relatively thin, backwards-curving hooked horns that can reach a length of 20–32cm.

DISTRIBUTION AND HABITAT Various low and high mountain ranges in central, south-eastern and southern Europe. Chamois prefer rocky areas near the tree line. When chamois are found in lower-lying habitats, this is often due to the weather, especially heavy snow. However, chamois can also be found in forests all year round, although they avoid very dense woodland.

DIET Herbs, grasses, leaves, shoots of deciduous and coniferous trees (e.g., spruce), tree bark, mosses and lichens.

REPRODUCTION Fights often break out between older bucks during the mating season (rut) from October to January. Fights over mating rights can be dangerous when they occur in high rocky areas and on steep rock faces. The bucks pursue each other energetically, sometimes leading to fatal falls. After a gestation period of 170–190 days, 1–3 young are born. The young can stand and walk just a few hours after birth. They stay with their mother until the next year's young are born. Young animals form their own small groups within the herd and play energetically with each other. Males become sexually mature at 3–4 years, females at 1–3 years.

TRACK

Front
L 5.5–7.5cm   W 3.8–5.5cm

Hind
L 5–6.8cm   W 3.2–5.1cm

Medium-sized. Unguligrade. Slightly asymmetrical. Four toes: toes 2 and 5 (the dew claws) are very short and situated very high up on the back of the foot. Dew claw prints are rare and usually only visible in deep snow. Dew claws leave prints behind the hooves at the same width as the outer edges of toes 3 and 4. Toe 3 is slightly smaller than toe 4. The hoof outline is elongated and the hooves are an even width. The individual hooves are slender with rounded, narrow tips. The overall outline is less pointed than a Western Roe Deer hoof. The outer edges of the hind foot are slightly concave, with a convex bulge at the back, where the noticeably short pad is located ⓐ. The outer edges of the front foot are almost parallel up to the beginning of the pad, then widen. The pads are noticeably weak on the front foot and on the hind foot ⓑ. The inner hoof walls are parallel along almost their entire length. The negative space between the hooves is correspondingly straight and very wide, especially in the front foot. This can give the impression that the hooves are always slightly splayed.

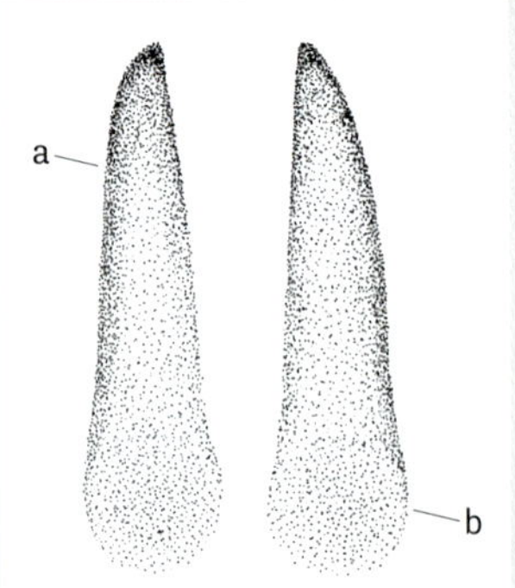
▲ *Right front.*

▲ *Right front. Monte Baldo, Italy. Andreas Wenger.*

▲ *Right hind. Monte Baldo, Italy. Andreas Wenger.*

▲ *Chamois, right front (right) and right hind (left). Note the much more concave outer hoof wall on the hind foot and the relatively short front foot pad. Monte Baldo, Italy. Andreas Wenger.*

The hooves can spread wide to give the animals a better grip on rocky terrain and to create a kind of 'snowshoe effect' that prevents them from sinking into the snow. The front foot print is larger and slightly wider than the hind foot print.

GAITS

Chamois mainly move at a walk. They bound and gallop to escape when alarmed.

Walk
Stride length: 52–115cm
Trail width: 8–19cm

▼ *Direct register walk.*

▲ *Direct register walk. The first double print of the right side is at the bottom right. The left double print and another double print on the right follow from bottom to top. Monte Baldo, Italy. Andreas Wenger.*

Similar tracks
The tracks can be confused with ibex, sheep and goat footprints. Differentiation is usually possible if the foot morphology is considered.

SIGNS

PATHS Chamois often make recognisable paths within a territory. These paths usually link up grazing and resting places. Chamois paths are often above the tree line. The animals can easily cross scree fields, and steep slopes with extreme drops are their natural habitat.

▲ *Chamois tracks are often well trodden and clearly recognisable. Monte Baldo, Italy. Andreas Wenger.*

DAMAGE CAUSED BY HORNS Males rub and strike with their horns against small fir, spruce, pine or alder trees. This leaves a scent secretion and a visual sign to mark their territory. This characteristic behaviour is like that of Western Roe Deer (page 523) and mainly occurs immediately before and during the rut.

EXCREMENT The acorn-shaped to roundish faecal pellets vary in colour from brownish to black and are either rounded at both ends or have one rounded and one pointed end. A pile of droppings will often contain both acorn-shaped and flattened faecal pellets with one blunt and one pointed end. This is typical of chamois. Chamois droppings are like Western Roe Deer droppings but tend to be slightly more narrow. Summer droppings often stick together because the food eaten has a higher moisture content. Chamois generally eat drier food in winter, resulting in firmer, more compact and generally larger droppings that stick together less.

- L 0.9–2cm (usually ≤ 1.5cm) D 0.6–1.5cm (usually ≤ 1cm)

▶ *Chamois winter droppings are usually slightly larger and have a rougher surface than their summer droppings. Monte Baldo, Italy. Andreas Wenger.*

◀ *Summer droppings often stick together due to the lusher food. Monte Baldo, Italy. Andreas Wenger.*

MOUFLON

Ovis gmelini musimon

HTL 105–130cm

TL 7–11cm

W 20–55kg

Rams about 35–55kg

Ewes about 20–35kg

Pronounced sexual dimorphism: rams are larger and heavier than ewes.

Mouflon remain in their habitat and are usually active during the day. They have excellent hearing and eyesight. They can run fast and jump and climb well. In summer, the females (ewes) live in herds with their young. The males (rams) form their own herds during this time, but older rams can also be solitary. Large, mixed herds of up to 300 or more animals can form in winter. They are mainly predated by wolves, lynx and hunting dogs.

DISTINGUISHING FEATURES   Built like short-haired Domestic Sheep. The characteristic spiral horns can grow to over 1m long in rams. The ewes rarely have horns, but if they do, they are a maximum of 20cm long.

DISTRIBUTION AND HABITAT   Wild populations in Corsica, Sardinia and Cyprus, and the mainland further east. From there, the animals have spread throughout most of Europe. Some populations derive from deliberate introductions for hunting. Mouflons prefer dry, rocky mountain landscapes but they are very adaptable. Populations also occur in deciduous and mixed forests with meadows and other open spaces, as well as in lowland areas. Hoof diseases are more common in wetlands, which makes them unsuitable as a habitat.

DIET   Mainly herbs, grasses and leaves. Mouflons also eat buds, ferns, moss and tree bark. They eat crops if available.

REPRODUCTION   Ritualised fights may break out between older rams during the mating season (rut) from October to December. The males headbutt each other and clash horns at full force. Unlike ibexes, they do not rear up on their hind legs. After a gestation period of around five months, 1–2 young are born. The animals reach sexual maturity after 1–1.5 years. However, because they are no match for the older and more dominant rams, young males do not start to reproduce until the age of four.

The domestication of wild sheep dates back at least 11,000 years. Kept for their meat, milk and wool, and valued for their self-sufficient nature, sheep are important livestock in many countries. Mouflons can breed with all domestic sheep.

SIGNS

FEEDING MARKS   Bud browsing as for Western Roe Deer (page 525).

BARK PEELING   Like Western Roe and Western Red Deer, Mouflons eat the cambium under the bark of trees. They strip some trees almost completely of their bark, especially in winter. The bark peelings are indistinguishable from those of the Western Roe Deer (page 524).

EXCREMENT   Egg-shaped or spherical faecal pellets. The individual faecal pellets are often sausage-shaped or clumped together, which can cause them to lose their rounded shape and appear more angular. The colour varies from brownish or greenish black to black. If the animals have eaten new grass, their droppings often clump together or form a lighter-coloured, mushy mass. Chamois droppings are similar, but much larger. Not clearly distinguishable from Domestic Sheep excrement.

- L 0.9–1.2cm   D 0.6–1.2cm (usually ≤ 0.9cm)

TRACK

Front
L 4.5–6.5cm   W 3.2–5.5cm

Hind
L 4.2–6cm   W 3–5cm

Medium to large. Unguligrade. Slightly asymmetrical. Four toes: toes 2 and 5 (the dew claws) are very short and higher up on the back of the foot. The dew claws are rarely visible. They leave prints behind the hooves at the same width as the outer edges of toes 3 and 4. Toe 3 is slightly smaller than toe 4. The hoof outline is elongated, but proportionally wider than that of the Alpine Ibex. The outer edges of the hoof are clearly convex. The inner hoof walls are usually parallel. The strong pads make the hooves wider and the area between the two hooves much narrower (a). The individual hooves become narrower towards the front. They are elongated and pointed. The outer edges of the hoof usually leave clear prints. Mouflon hooves characteristically spread apart at the front of the track, something that is noticeable even at slower gaits (b). The front foot print is larger and wider than the hind foot print. The hind foot pads are usually less well developed. It is not possible to distinguish clearly between rams and ewes.

▲ *Right front.*
Vledder, Netherlands. René Nauta.

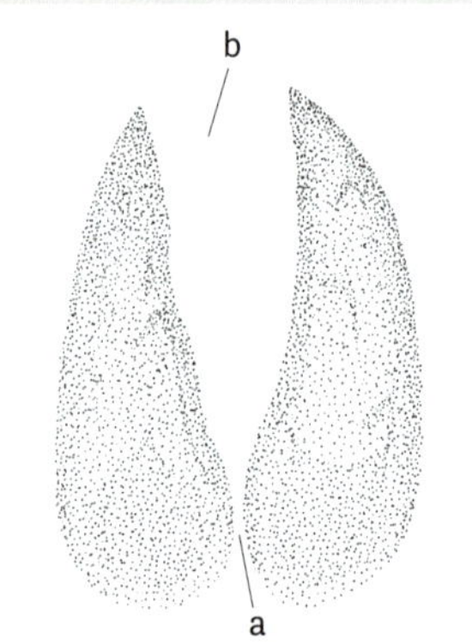

▲ *Right front.*

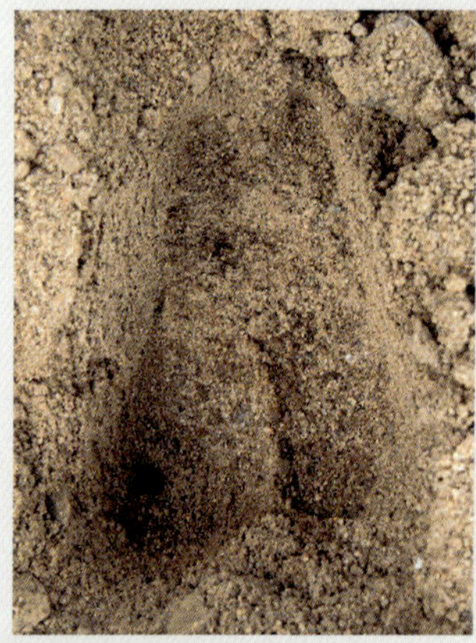

▲ *Right front.*
Austria. Andreas Wenger.

▲ *Right front.*
Vledder, Netherlands. René Nauta.

▲ *Right hind.*
Vledder, Netherlands. René Nauta.

GAITS

Mouflons mainly move at a walk. They usually trot to cover longer distances at higher speeds. Their track pattern is characterised by a relatively wide trail width in comparison to the stride length. They bound and gallop to flee when alarmed. Mouflons have an extraordinary ability to jump.

Walk
Stride length: 60–120cm
Trail width: 10–15cm

Based on limited data.

▼ *Overstep walk. Footfall from bottom to top: RF, RH, LF, LH, RF, RH. You can make out the tracks of a Red Fox in side trot at the left edge of the image. Vledder, Netherlands. René Nauta.*

▲ *Direct register walk.*

Similar tracks
Domestic Goats. The prints can also easily be mistaken for those of Domestic Sheep.

DOMESTIC SHEEP AND DOMESTIC GOAT

Ovis gmelini aries, Capra aegagrus hircus

White Polled Heath sheep

♀ W approx. 40–50kg

♂ W approx 60–70kg

Merinolandschaf sheep

♀ W approx. 70–100kg

♂ W approx. 120–160kg

Pygmy Goat

♀ W approx. 40–50kg

♂ W approx 60–70kg

German White Noble Goat

♀ W approx. 55–75kg

♂ W about 70–100kg

Pronounced sexual dimorphism, males larger and heavier than females.

Domestic Sheep and Goats are among the oldest domestic animal breeds. Their domestication probably began 11,000–13,000 years ago in the Fertile Crescent. Whether the Domestic Sheep is descended from the wild form of the Mouflon (*Ovis gmelini*) is disputed, as there is also a theory that the Mouflon is a feral form of the Anatolian Domestic Sheep. The wild ancestor of the Domestic Goat is the Wild Goat (*Capra aegagrus*), which inhabits large parts of western Asia. We consider Domestic Sheep and Goats together here because of their domestication history, similar tracks and signs, and similarities as farm animals.

Today, Domestic Sheep and Goats are farmed all over the world for their milk and meat. In economic terms, wool production plays a subordinate role in modern European sheep farming. Both species are also used in landscape conservation to graze extensive green areas. Thanks to their climbing skills, goats can graze on steep and rocky slopes. Most sheep in Europe are kept in Great Britain. Goat farming plays a secondary role in European agriculture.

Many different sheep and goat breeds have developed because of spreading and trading by humans and associated local breeding. However, numerous regional breeds are currently at risk of extinction due to the dominance of more economically viable breeds such as the Merinolandschaf or German White-headed Mutton. Due to the enormous variety in appearance, weight and size and the equally different forms of husbandry, we have only described a few representative breeds in more detail below. The descriptions focus on tracks and signs, as these findings can be confused with the tracks and signs of wild animal species.

SIGNS

BUD BROWSING Similar to Western Roe Deer (page 525).

DAMAGE CAUSED BY HORNS Both male and female goats strike small trees with their horns, leaving characteristic marks. They prefer trees with a diameter of 5–25cm. They remove the bark at a height of 15–150cm above the ground. Goats peel the bark off trees in strips and conspicuous piles of it are often found at the foot of the tree. Their horns usually leave deep furrows in the wood. Some breeds can cause such extensive damage with their horns that they are sometimes intentionally used in landscape conservation to curb heavy scrub encroachment. Only a few Domestic Sheep breeds have horns, and they strike less frequently than Domestic Goats do.

TRACK

Due to the wide variety of Domestic Sheep and Goat breeds, their tracks show great variation in size. Large sheep, such as Merino or German Black-headed Mutton, can usually be distinguished from smaller sheep, such as moorland sheep or the Alpines Steinschaf. Pronounced sexual dimorphism means that large tracks within a breed generally come from males, while relatively small tracks tend to come from females. Note, however, that the track sizes of a large male of a small breed and the track sizes of a small female of a larger breed can overlap, although the tracks of the fine-limbed hardy sheep breeds tend to be finer, even if they are similar in size. It is usually possible to distinguish large goat breeds, such as the Boer or German White Noble, from smaller breeds, such as Pygmy Goats. Sexual dimorphism is less evident in Domestic Goats.

Medium to large. Unguligrade. Slightly asymmetrical. Four toes: toes 2 and 5 (the dew claws) are very short and higher up on the back of the foot. The dew claws are rarely visible. They leave prints behind the hooves at the same width as the outer edges of toes 3 and 4. Toe 3 is slightly smaller than toe 4. The track outline is longitudinally rectangular and has a distinctive boxy shape, especially in the front foot print. This is a

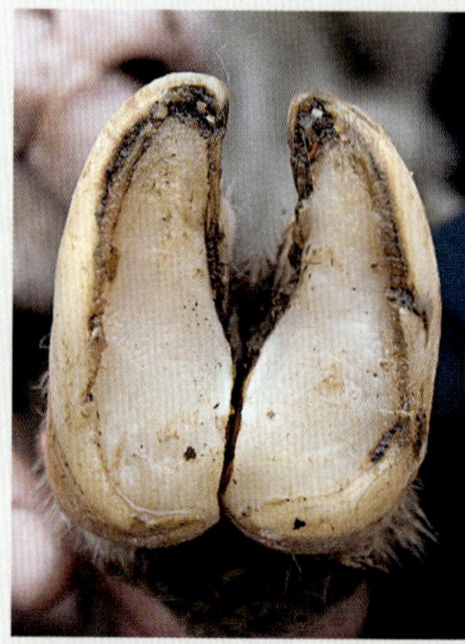

▲ *Domestic Sheep, right front. Wissing, Germany. Raphael Fuchs.*

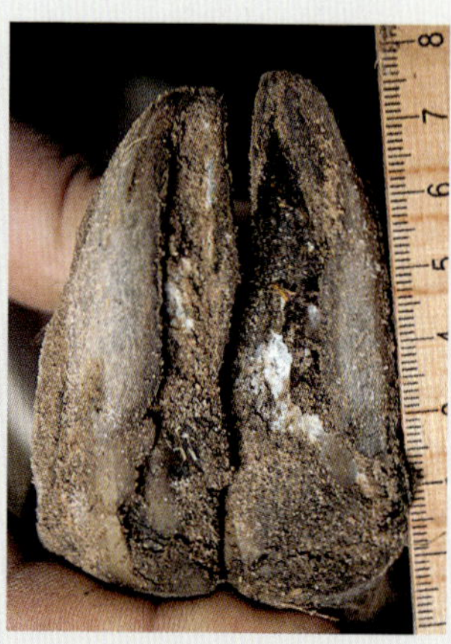

▲ *Domestic Sheep, left hind. Wissing, Germany. Raphael Fuchs.*

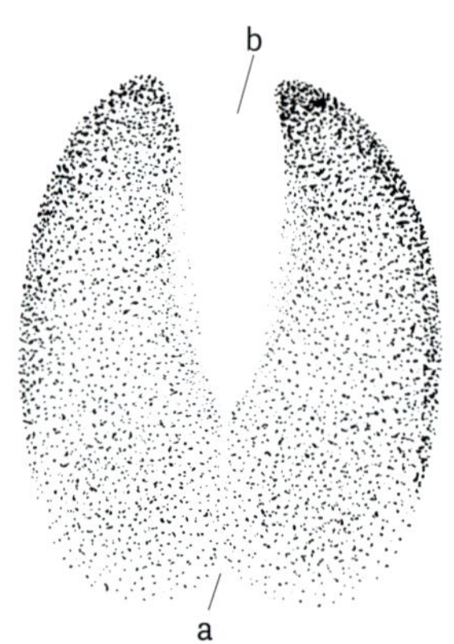

▲ *Domestic Sheep, right front.*

▲ *Domestic Sheep, right front. Wissing, Germany. Raphael Fuchs.*

▲ *Domestic Sheep, left hind. Wissing, Germany. Raphael Fuchs.*

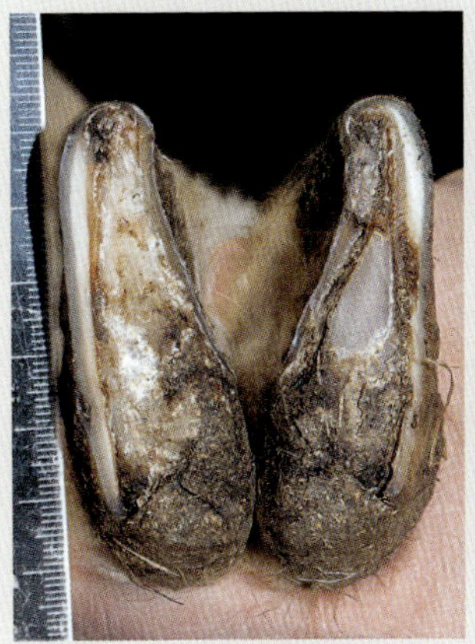

▲ *Domestic Goat, right front. Hemau, Germany. Raphael Fuchs.*

▲ *Domestic Goat, right hind. Hemau, Germany. Raphael Fuchs.*

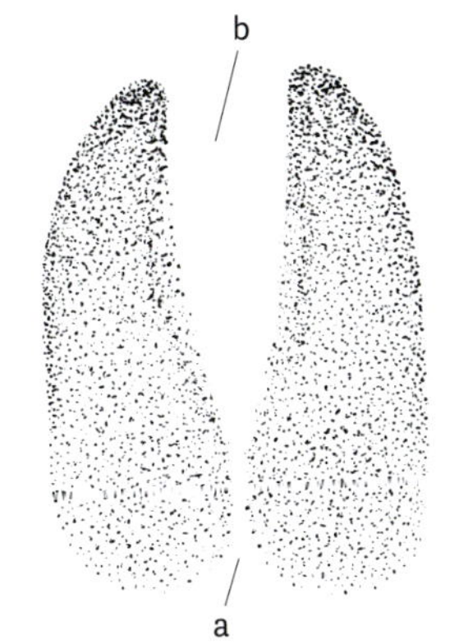

▲ *Domestic Goat, right front.*

suitable distinguishing feature compared to the often similarly sized, but more heart-shaped tracks of Western Roe Deer. The outer edges of the hoof are clearly convex in the front foot, while those of the hind foot are slightly concave. The inner hoof walls tend to be straight to slightly concave. The well-developed pads make the hooves wider and the space between the two hooves much narrower ⓐ. The individual hooves become narrower towards the front. They appear less elongated and are usually more rounded than Mouflon or chamois hooves. The outer edges of the hoof usually leave clear prints. Sheep and goat hooves characteristically spread apart in a wedge shape at the front of the track, something that is noticeable even in slower gaits ⓑ. The front foot print is larger and wider than the hind foot print. The hind foot pads are usually less well-developed.

▲ *Domestic Goat, right front. Hemau, Germany. Raphael Fuchs.*

▲ *Domestic Goat, right hind. Hemau, Germany. Raphael Fuchs.*

	Front		Hind	
White Polled Heath				
Ewe	L 4.9–5.7cm	W 4–4.6cm	L 4.4–5.3cm	W 3.5–4.1cm
Ram	L 5.5–6.4cm	W 4.5–5cm	L 4.9–5.8cm	W 4–4.5cm
Merinolandschaf				
Ewe	L 6–6.5cm	W 5.3–5.8cm	L 5.5–6.8cm	W 4.6–5.5cm
Ram	L 7.5–8.5cm	W 5.8–6.5cm	L 8.4–9.1cm	W 5.5–6.1cm
Pygmy Goat	L 4.5–5.2cm	W 4.1–4.5cm	L 4–4.5cm	W 3–4.2cm
German White Noble	L 4.8–6.5cm	W 4.4–5.2cm	L 4.9–6.2cm	W 3.5–4.5cm

GAITS

Walking is the preferred gait of Domestic Sheep and Goats. As both species usually move in flocks, the large number of similar tracks can be a useful clue to help distinguish them from wild animals. They usually trot to cover longer distances at higher speeds. Domestic Sheep are characterised by a relatively wide trail width in relation to their stride length. The animals flee in bounds and at a gallop when alarmed.

White Polled Heath
Walk
Stride length: 59–82cm
Trail width: 20–30cm

Merinolandschaf
Walk
Stride length: 72–105cm
Trail width: 25–35cm

Pygmy Goat
Walk
Stride length: 56–78cm
Trail width: 20–25cm

Trot
Stride length: 76–91cm
Trail width: 14–19cm
When stride lengths are comparable, goats tend to have narrower trail widths.

Similar tracks
Mouflon and chamois.

◄ *Domestic Goat, overstep walk. Ittelhofen, Germany. Raphael Fuchs.*

▶ *Close-up of bark peeling. Offenbach, Germany. Simone Roters.*

▲ *Domestic Goat bark peeling. Offenbach, Germany. Simone Roters.*

▲ *Excrement of a medium-sized goat breed. Ittelhofen, Germany. Raphael Fuchs.*

BARK PEELING Like Western Roe and Western Red Deer, Domestic Sheep and Goats eat the cambium under the bark of trees. They strip some trees almost completely of their bark, especially in winter and in intensively used areas.

- Max. height of peeling: 1.8m
- Single toothmark width: 0.2–0.4cm

EXCREMENT Egg-shaped or spherical faecal pellets. The individual faecal pellets are often sausage-shaped or clumped together, which can cause them to lose their rounded shape and appear more angular. The colour varies from brownish or greenish black to black. If the animals have eaten new grass, their droppings often clump together or form a lighter-coloured, mushy mass. Chamois droppings are similar, but much larger. Not clearly distinguishable from Mouflon droppings.

- L 0.9–1.9cm D 0.6–1cm (usually ≤ 0.9cm)

BIRDS

BIRDS

A passion for bird tracks

It was a chilly November morning when I arrived at a disused clay pit near the Hoher Fläming Nature Park in Germany with a group of trackers. Paul, an enthusiastic tracker with a passion for birds, had invited me as a CyberTracker Conservation examiner to put him and nine others through their paces. As usual, when the exam began, I asked all the participants not to speak at the exam stations so as not to influence each other. One by one, we showed everyone the stations and asked different questions such as 'what animal made this track?' or 'what gait was this animal moving in?' Everything was as normal until we reached the station with the track that Laura had discovered the day before while scouting: an unusually large, K-shaped footprint. When Paul recognised the print, his face flushed with excitement. He took a second look, looked at me incredulously and then back at the footprint. He began to bounce up and down restlessly on the spot. You could see he was struggling to contain his excitement. But then it burst out of him: 'This is awesome! Man, that's so cool! Guys, I'm going to have to step back for a minute before I give everything away.' Paul walked off and tried to calm down. He had just seen the track of the world's largest owl species (Eurasian Eagle-owl), and I had just seen how much bird tracks can excite people.

Observing and studying birds (Aves) is perhaps the area of natural history that people around the world find most inspiring. Many birds attract attention with their song or splendid plumage. Others are more inconspicuous: they lead a more secretive life and are less well known. The tracks and signs left by birds give us an insight into their diverse world, often even before we hear or see them for the first time. With basic knowledge of the footprints and signs of the birds of Europe, we can learn more about their habitats, way of life and their ecological relationships. While researching for this book, I was first able to identify migratory birds like the Common Redshank in Lusatia by their tracks. Some signs made by birds can be confused with those of mammals. Others are more obviously made by birds, for example woodpecker nest holes, which remain recognisable for years.

FOOT MORPHOLOGY

Looking closely at their feet and footprints can teach us about birds' way of life and evolution. Like mammals, birds' feet have adapted to requirements. All birds are digitigrades whose front limbs have developed into wings. Birds use their feet for many purposes other than walking and they have developed accordingly. A wading bird that lives on the mudflats needs different feet to a swallow that spends most of its life in the air, or a duck that uses its feet for swimming. Different foot structures have developed for different main functions such as swimming, perching, running, gripping and wading and these foot structures leave correspondingly different tracks. Closely related species with similar lifestyles leave similar tracks. There is usually a reliable correlation between the size of a bird and its footprint within a particular taxonomic group. This means that a relatively large, duck-like track is likely to come from a larger duck, like a Mallard, while a relatively small duck track probably comes from one of the smaller duck species, such as Common Teal. This relationship can help to differentiate between species in some cases. This part of the book is intended to give an overview and provide a basis for differentiating between bird groups and some species based on their footprints. It mainly describes typical or common bird tracks. It does not look at the tracks of birds like swifts, which almost never touch the ground, and those of rare species that only appear in the area in exceptional cases.

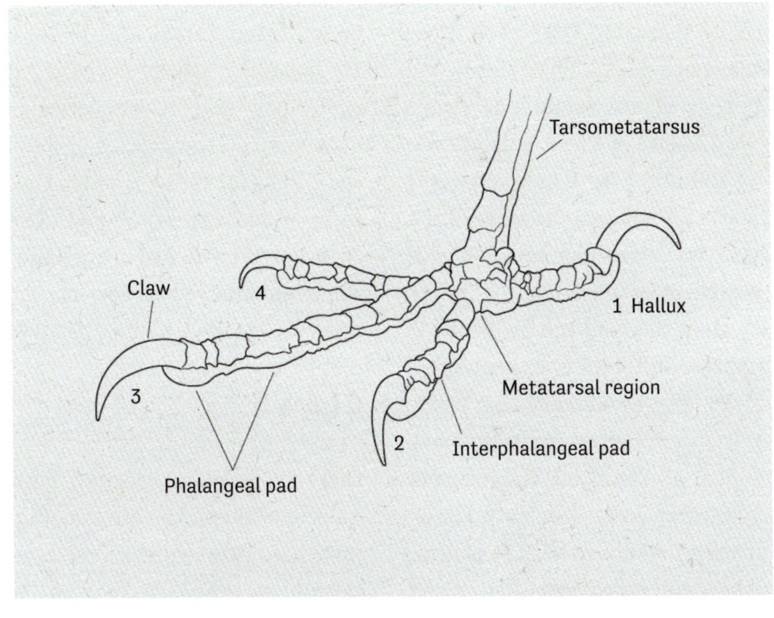

▶ *A typical bird foot.*

STRUCTURE AND FEATURES OF TRACKS

Most bird tracks consist of prints of the toes, the metatarsal region and the claws. Prints of the webbing between the toes of waterbirds is often also visible. Other important features are the size and symmetry of the track.

TOES

The arrangement, length and shape of the toes are important criteria for identifying a species. Shorebirds have long, usually delicate toes, while corvids' toes are sturdier. A toe is formed around several phalanges, with differently shaped pads on the underside. They are called phalangeal pads. The individual phalangeal pads are usually separated from each other by interphalangeal pads. Phalangeal and interphalangeal pads are sometimes visible in clear tracks. The size of these pads or their absence from the print can be an important feature for species identification. Most of the birds discussed here have four toes. As with mammals, we count them from the inside out. Toe 1 is called the hind toe or hallux. It points backwards and has two phalanges. Most passerines have a relatively long toe 1, which can be longer than toe 3 and is mainly used for perching on branches. The hallux is completely absent in some species, such as plovers. In species that spend a lot of time on the ground, toe 1 can be so short and raised

when the bird walks that it often leaves a very faint, indistinct print, or none at all. Toe 2 is called the inner toe. It has three phalanges and is usually the second shortest toe. Toe 3 is called the middle toe, has four phalanges and is usually the longest toe. Toe 4 is called the outer toe. It is the furthest away from the body and, despite having five phalanges, is usually only the second longest toe.

METATARSAL REGION

Unlike mammals, the toes of a bird's foot branch off from the distal end of the tarsometatarsus (furthest from the centre of the body). The tarsometatarsus is made up of fused tarsal and metatarsal bones. Its underside end is covered with pad-like, cushioning skin, which varies in thickness depending on the species and can be seen in the track. This area is called the metatarsal pad or metatarsal region. The presence or absence of a pad print here can be an important clue for species identification.

CLAWS

There is usually a pointed horny covering called a claw on the outermost phalange of each toe. The shape and size of the claws depend on the species. In some birds, the claws are straight and their print is continuous with the toe. In others, the claws are curved and so leave a smaller (but possibly deeper) mark in front of the toe. Unlike mammals, it can sometimes be difficult to distinguish claws from the tips of the toes in bird footprints. We include the claws when measuring bird tracks.

WEBBING

Many swimming birds have flaps of skin between their toes. Webbed feet enable them to propel themselves effectively in water. Webbing can be found between two or more toes, usually between toes 2–4. As in mammals, they can be distal, medial and proximal to the tarsometatarsus. Ducks and geese are examples of birds with distal webbing between toes 2–4, whereas some shorebirds, such as sandpipers, only have proximal webbing between toes 3–4. Only a handful of birds in Europe, like the Pied Avocet, have medial webbing. Lobed flaps of skin on the edge of each individual toe are a special type of webbing. They increase the surface area of the foot, which helps the bird to move in water and also enables it to walk over very soft, muddy ground without sinking. These lobes are shaped differently in different species. For example, grebes have lobes all the way around their toes, coots have 'notched' toe lobes and phalaropes' feet have a combination of notched toe lobes and proximal webbing. The absence of webbing can also be an important clue.

TRACK SIZE AND SYMMETRY

As with mammals, it is often possible to narrow down possible bird species based on track size alone. If animals from similar groups have feet of the same size, track symmetry can also be useful for species identification.

FIVE BASIC TYPES OF BIRD TRACK

The arrangement of the toes and the prints of webbing have a significant influence on the appearance of a bird track. For this reason, the bird species described here are assigned to five basic types of tracks. These basic types are useful for initial classification.

TYPICAL FOOTPRINT, ANISODACTYL

This basic type of footprint is the most common. It is made by the shape of foot that most of us think of when imagining a typical bird footprint. Toes 2–4 point forwards, toe 1 points backwards. This arrangement is called anisodactyl and is useful for perching and gripping. All passerines (Passeriformes) leave prints of this type, as do herons (Ardeidae), storks (Ciconiidae), hawks, eagles and kites (Accipitridae), falcons (Falconidae), doves and pigeons (Columbidae), nightjars (Caprimulgidae), bee-eaters (Meropidae) and hoopoes (Upupidae). Although toes 2 and 3 of kingfishers (Alcedinidae) are fused (syndactyly), they also belong to this category as they make prints in this distinctive shape.

FOOTPRINT WITH SHORT OR ABSENT HIND TOE, TRIDACTYL

Birds that spend a lot of time on the ground may have a correspondingly adapted foot shape where toe 1 is very short or absent. Bird feet with a very short hind toe are technically still anisodactyl and are only referred to as tridactyl when the hind toe is completely absent. However, the tracks of both foot shapes can be clearly distinguished from typical bird footprints because toe 1 leaves only a small print or doesn't leave a print at all. This basic shape occurs in the order of shorebirds (Charadriformes), as well as the families of pheasants and grouse (Phasianidae), cranes (Gruidae), rails (Rallidae) and bustards (Otididae).

FOOTPRINT WITH WEBBING BETWEEN TOES 2–4, PALMATE

Birds that habitually swim often have webbed feet. Their tracks resemble the classic bird footprint with a short or absent hind toe but are clearly distinguishable by the well-developed webbing. The basic shape of these anisodactyl feet with distal

webbing between toes 2–4 is found in the families of ducks and geese (Anatidae), gulls (Laridae), terns (Sternidae) and flamingos (Phoenicopteridae), among others.

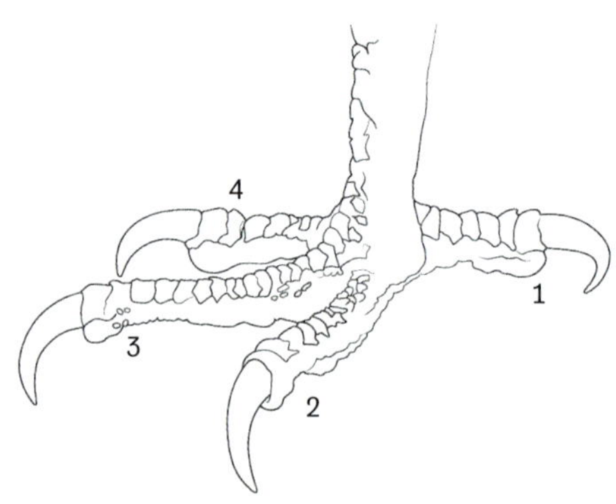

▲ *Anisodactyl foot.*

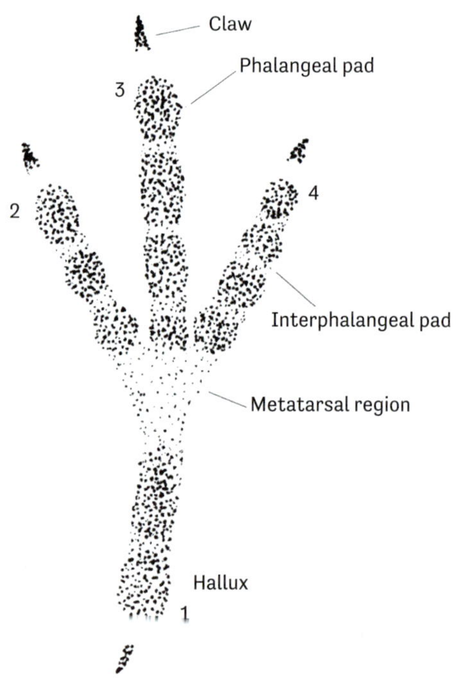

▲ *Anisodactyl track: Toes 2–4 point forwards, toe 1 points backwards.*

FOOTPRINT WITH WEBBING BETWEEN TOES 1–4, TOTIPALMATE

This rarer basic shape differs from the previous one because it has additional webbing between toes 1 and 2. In this kind of track, the webbing between toes 1 and 2, toes 2 and 3 and toes 3 and 4 usually leaves a print. If the prints of the webbing are unclear or difficult to make out, another distinguishing feature is the clearly elongated toe 4 that is longer than toes 2 and 3. In tracks with webbing between toes 2–4, toe 3 is the longest toe. Families with webbing between toes 1–4 include pelicans (Pelecanidae) and cormorants and shags (Phalacrocoracidae).

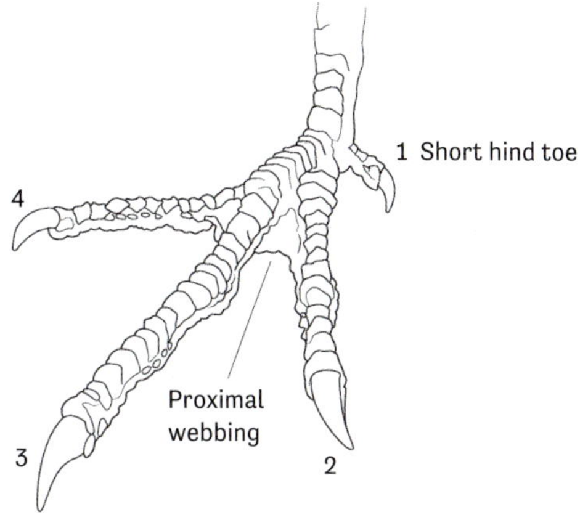

▲ *Anisodactyl foot with a short hind toe.*

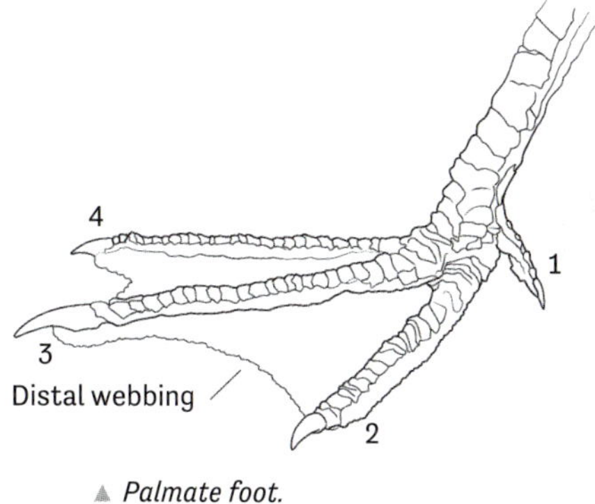

▲ *Palmate foot.*

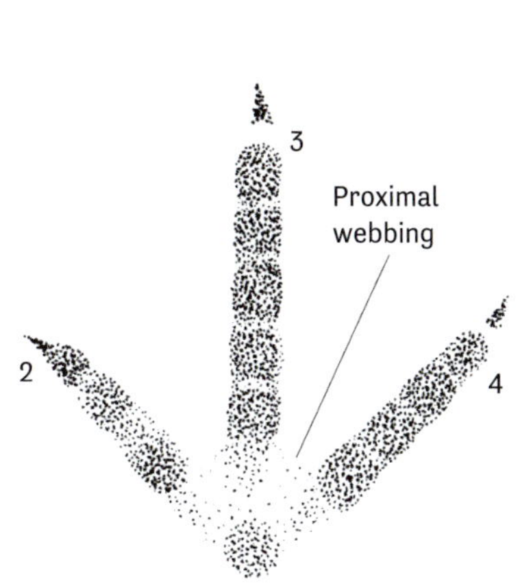

▲ *Track with a short hind toe.*

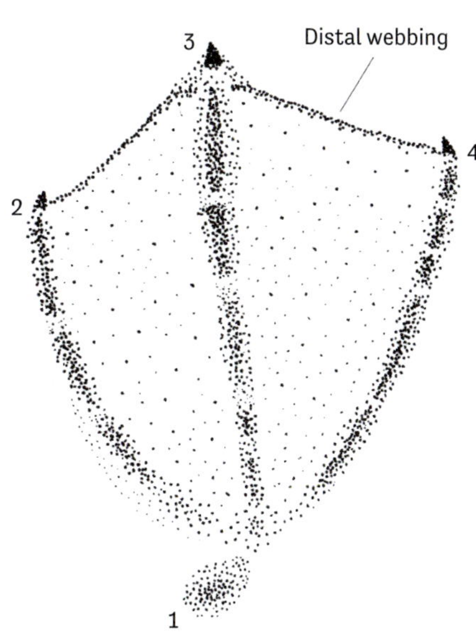

▲ *Palmate track: Clear webbing prints are visible between toes 2–4.*

ZYGODACTYL FOOTPRINT

A bird's foot is called zygodactyl when two toes point forwards and two point backwards. This creates a characteristic X- or K-shaped track. Toes 2 and 3 point forwards and toes 1 and 4 point backwards. This type of toe arrangement is the second most common in bird feet. As most of the species in this category spend little time on the ground, finding tracks is rare. Families with zygodactyl tracks include ospreys (Pandionidae), barn owls (Tytonidae), typical owls (Strigidae), cuckoos (Cuculidae) and woodpeckers (Picidae). Ospreys and owls can turn toe 4 so that it points backwards or sideways.

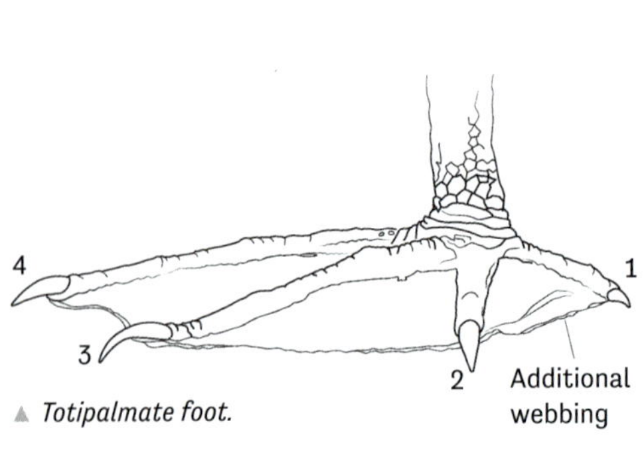

▲ Totipalmate foot.

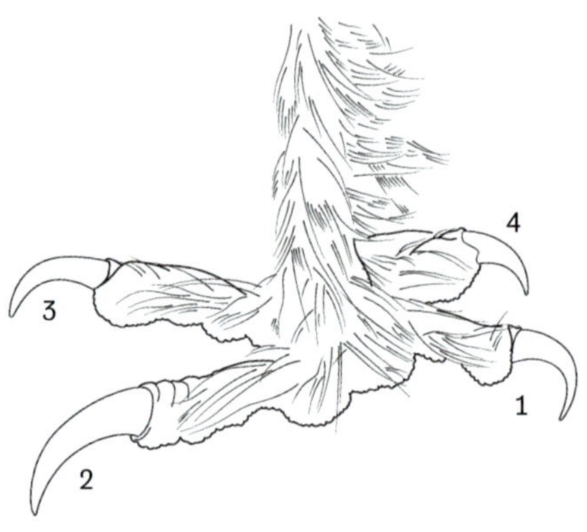

▲ Zygodactyl foot.

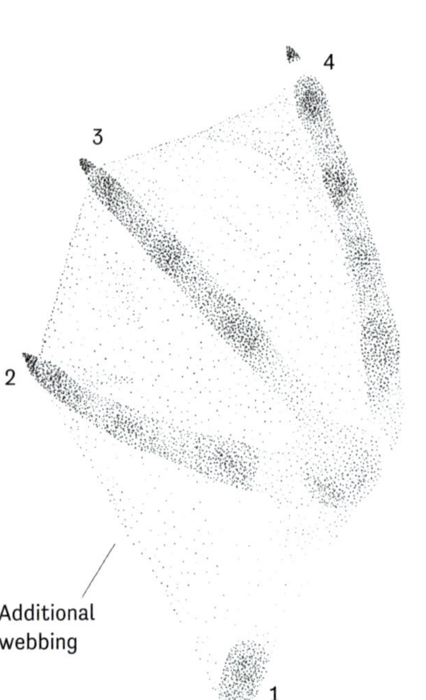

▲ Totipalmate track: Clear webbing prints are recognisable between toes 1–4.

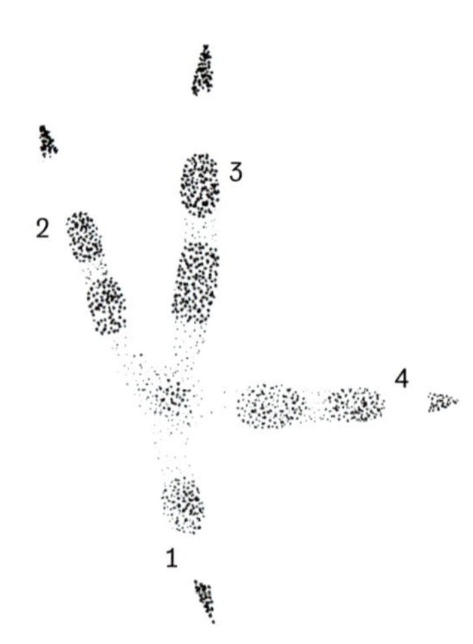

▲ Zygodactyl track: Toes 1 and 4 point backwards and toes 2 and 3 point forwards.

GAITS AND TRACK PATTERNS

Gaits describe how animals move, and track patterns are a sequence of tracks that represent the visible form of the gait on the ground. Being bipedal and with different anatomy, birds' gait and track patterns differ in some respects from those of mammals.

GAITS

Because they only move on the ground on two feet, bird gaits are often easier for us to understand than those of quadrupeds. As with mammals, the gaits of birds can be divided into two groups: symmetrical gaits with continuous track patterns, and asymmetrical gaits with group-forming track patterns.

SYMMETRICAL GAITS WITH CONTINUOUS TRACK PATTERNS

A continuous track pattern consists of two very similar and even rows of consecutive footprints. The two rows are offset in the direction of movement.

Walking and running

When a bird walks or runs, it moves the legs independently of each other. It places the foot of one side of the body down first, followed by the foot of the other side of the body. Each foot is off the ground for the same amount of time, resulting in even distances between the individual footprints. While at least one foot always remains in contact with the ground when walking, both feet temporarily lose contact with the ground when the bird runs, resulting in a moment of suspension. The stride length increases and the trail width decreases. The relationship between stride length and trail width provides information about the speed within a gait, the gait itself (walking or running) and about the species. Short stride lengths with a relatively wide trail width usually indicate birds with shorter legs and a stockier body (e.g., pigeons). Long

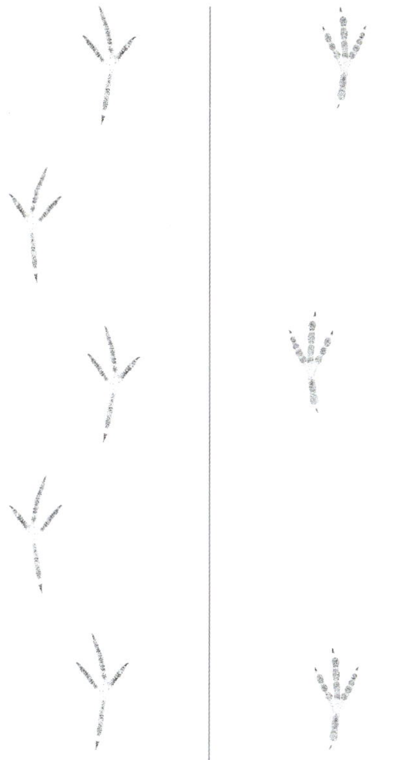

▲ *At a walk, stride lengths tend to be short and trail widths relatively wide.*

▲ *When birds run, there is a moment of suspension, resulting in longer stride lengths and narrower trail widths.*

stride lengths with a relatively narrow trail width tend to indicate birds with long legs and slender bodies (e.g. cranes). Walking and running are gaits commonly used by birds that spend a lot of time on the ground (for example, chickens).

ASYMMETRICAL GAITS WITH GROUP-FORMING TRACK PATTERNS

A group-forming track pattern consists of groups of two tracks that are each made up of the footprints of both sides of the body.

Hop

When a bird hops, it pushes itself off the ground with both legs at the same time. There is a brief moment of suspension before both feet land on the ground again at the same time. This creates a track pattern consisting of adjacent pairs of left and right footprints, as well as large gaps between consecutive pairs of feet in the direction of travel. When the bird hops faster, the stride length increases but the trail width remains more or less the same. Passerines that spend little time on the ground usually hop.

Skip

In contrast to hopping, the legs move independently of each other between moments of suspension and the feet touch the ground at different times when a bird skips. The track pattern shows successive pairs of tracks, each with a left and right footprint, which are arranged out of alignment in the direction of travel. A larger gap is also visible between successive pairs of tracks. This gap represents the moment of suspension, which begins after the last foot to touch the ground lifts off.

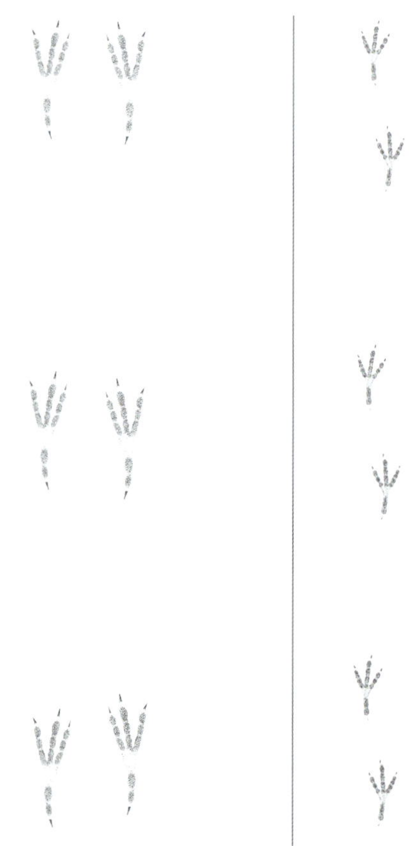

▲ Hopping is characterised by the adjacent footprints of both feet, followed by a gap and then another pair of footprints.

▲ Skipping.

TRACK PATTERNS

As with mammals, the orientation of individual tracks and the appearance and course of a track pattern can help identify the species. Passerine footprints, for example, are usually parallel to the direction of travel or only turn inwards slightly. The tracks of ducks, geese, doves and pigeons, on the other hand, usually clearly turn inwards. Some storks leave footprints that are slightly turned outwards. Linear locomotion is characteristic of galliform birds (gamebirds), whereas wagtails often move in irregular zigzag lines. The relationship between stride length and trail width can also be an important clue for species identification. Ducks often leave a characteristic track pattern with relatively short stride lengths and wide trail widths. Tail or wing marks may be visible in track patterns made during take-off and landing. Most species begin their flight from a standing position. A run-up, as observed in Common Ravens, is rarely required for take-off.

MEASURING TRACKS AND TRACK PATTERNS

TRACKS For tracks with a reliable hind toe print, measure the entire length from the hindmost point of toe 1 to the foremost point of toe 3. Always include visible claw marks in the measurement because the transition from toe to claw is often unclear or not apparent. For tracks with a short hind toe that does not leave prints regularly or does not leave a print at all, measure the length without toe 1. Measure the width at the widest point of the track, at a right angle to the length.

TRACK PATTERNS Measure stride length from any point of a track to the same point of the next track of the same foot. The measurement can be taken from claw tip to claw tip or from metatarsal pad to metatarsal pad. What matters is a consistent approach. Trail width is measured at the widest point from the outer edge of the track of one side of the body to the outer edge of the track on the other side of the body, at a right angle to the stride length.

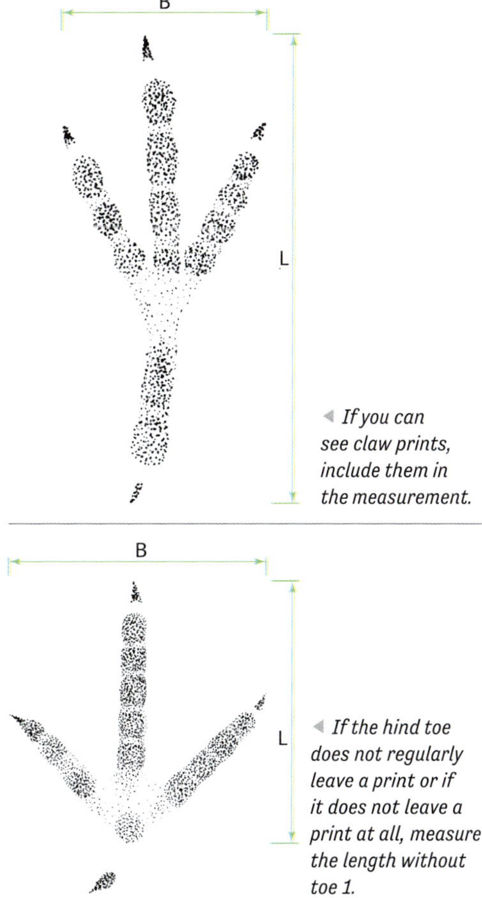

◄ If you can see claw prints, include them in the measurement.

◄ If the hind toe does not regularly leave a print or if it does not leave a print at all, measure the length without toe 1.

GUIDE TO IDENTIFYING BIRD TRACKS

The bird footprints presented are organised according to the five basic shapes described above. The species are summarised in their families and sorted from small to large, within these categories. All track drawings are life-size.

CLASSIC BIRD FOOTPRINT, ANSIODACTYL

Eurasian Wren, *Troglodytes troglodytes*, p.647

House Sparrow, *Passer domesticus*, p.649

Eurasian Blackbird, *Turdus merula*, p.648

Collared Dove, *Streptopelia decaocto*, p.641

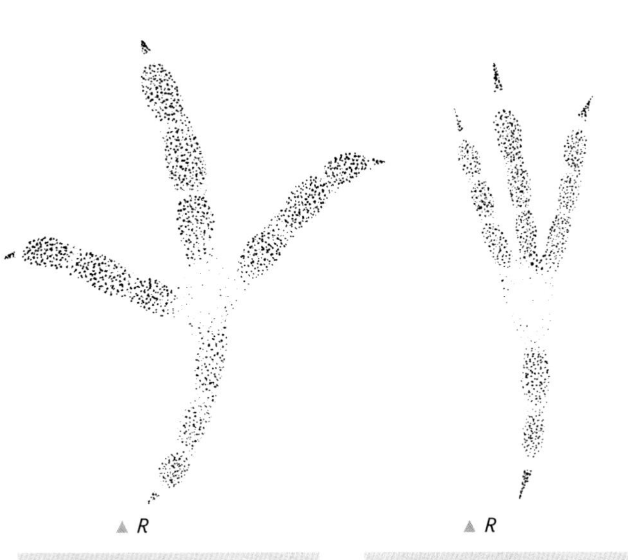

▲ R

Common Woodpigeon,
Columba palumbus, p.639

▲ R

Eurasian Jay,
Garrulus glandarius, p.644

▲ R

Common Magpie,
Pica pica, S. 643

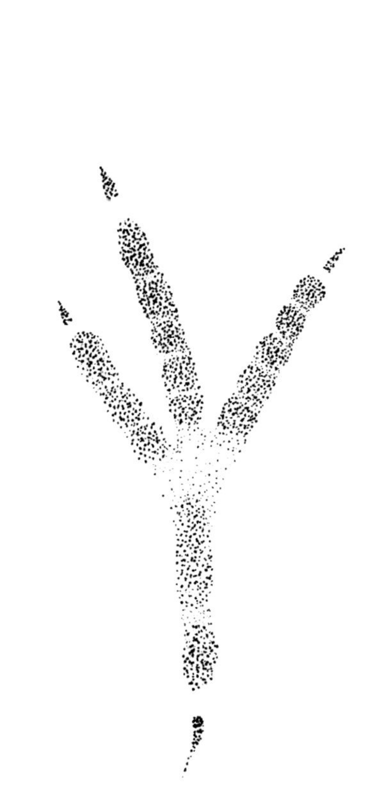

▲ R

Rook, *Corvus frugilegus*, p.645

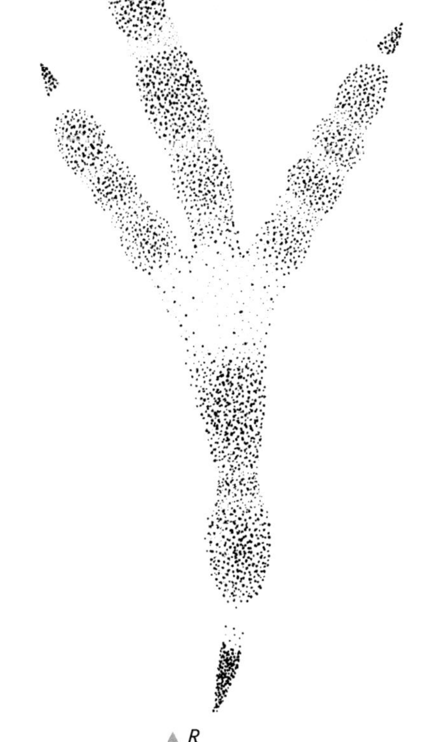

▲ R

Common Raven, *Corvus corax*, p.646

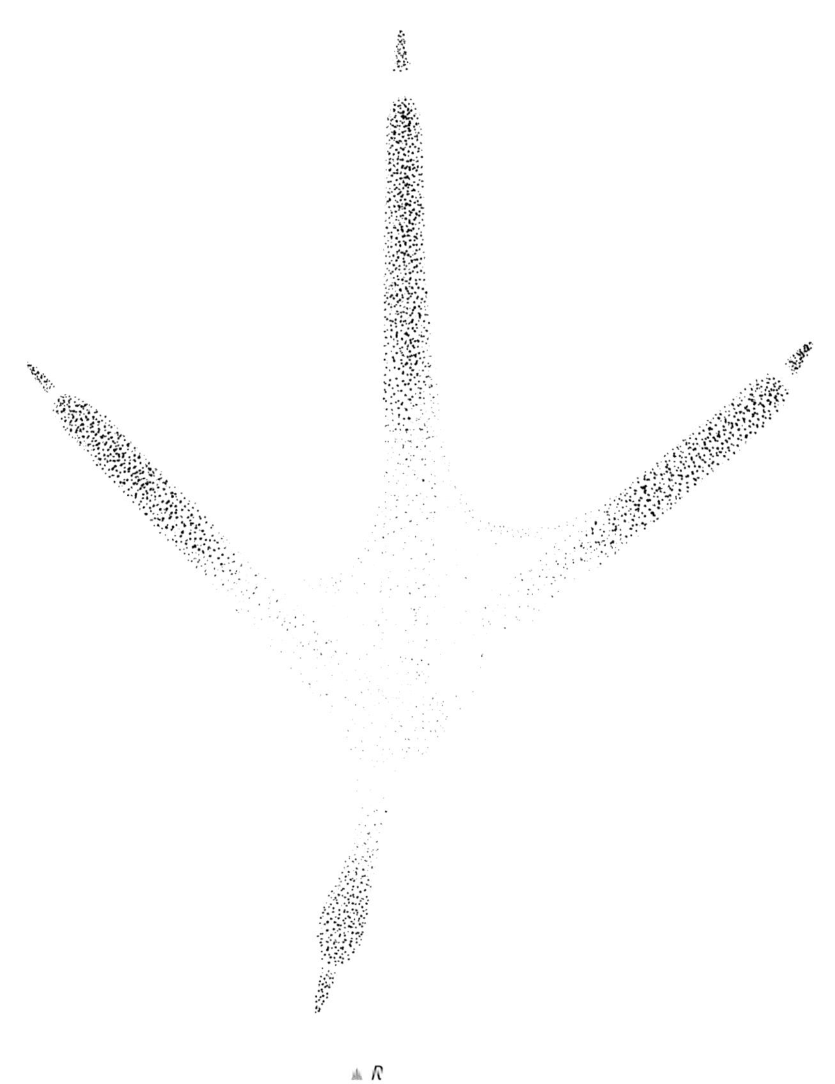

Common Spoonbill, *Platalea leucorodia*, p.635

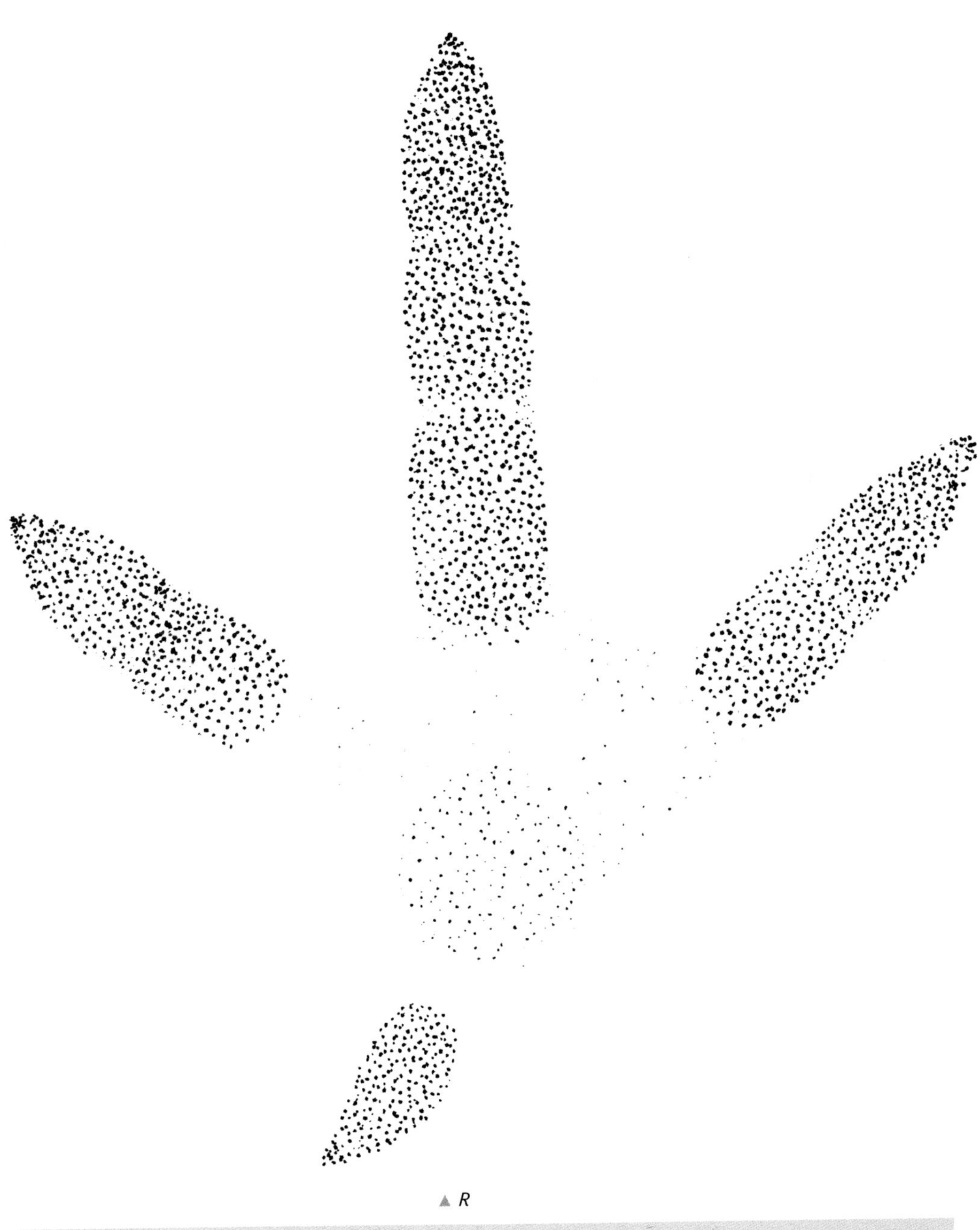

White Stork, *Ciconia ciconia*, p.634

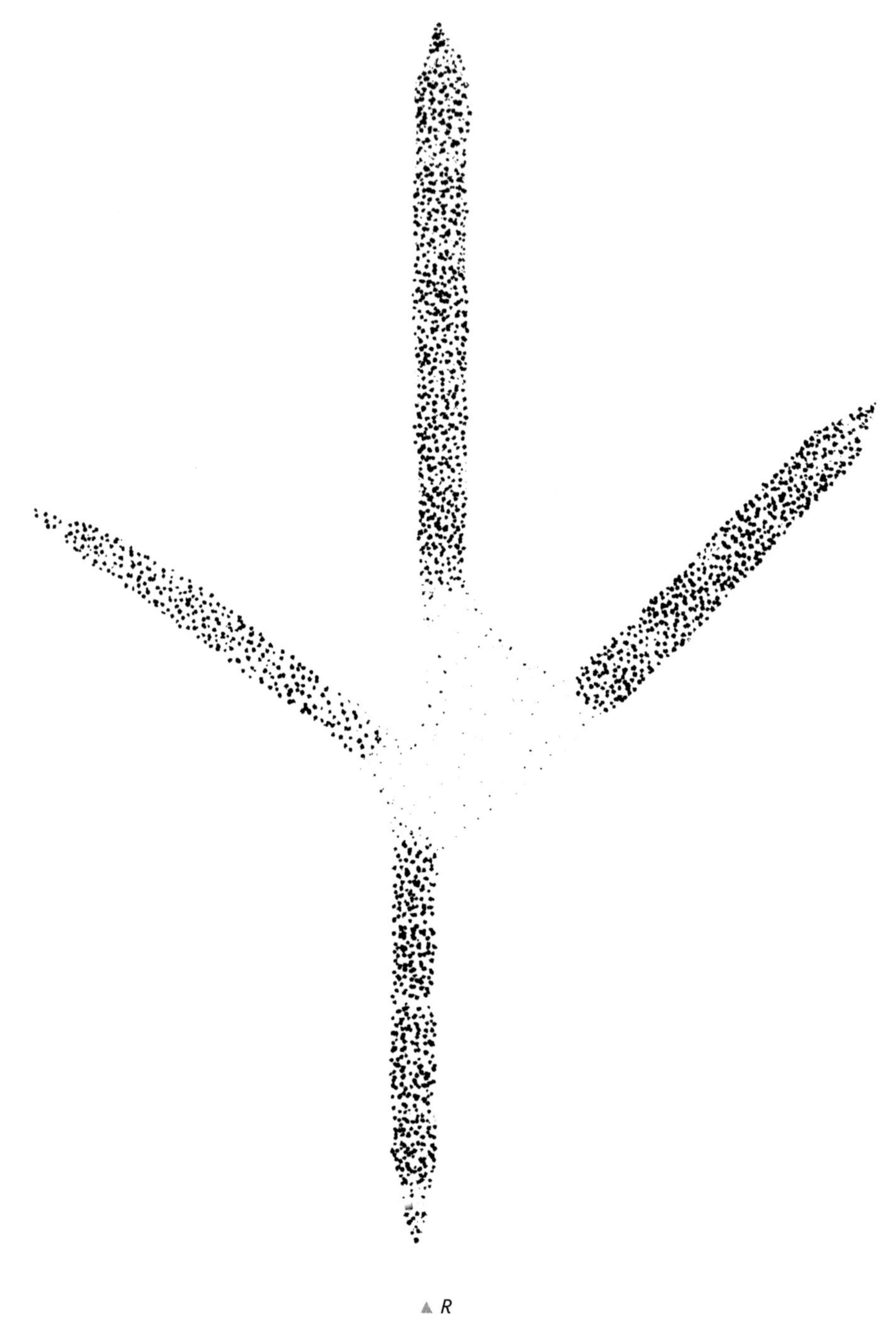

▲ R

Grey Heron, *Ardea cinerea*, p.633

▲ *R*

Common Buzzard, *Buteo buteo,* p.637

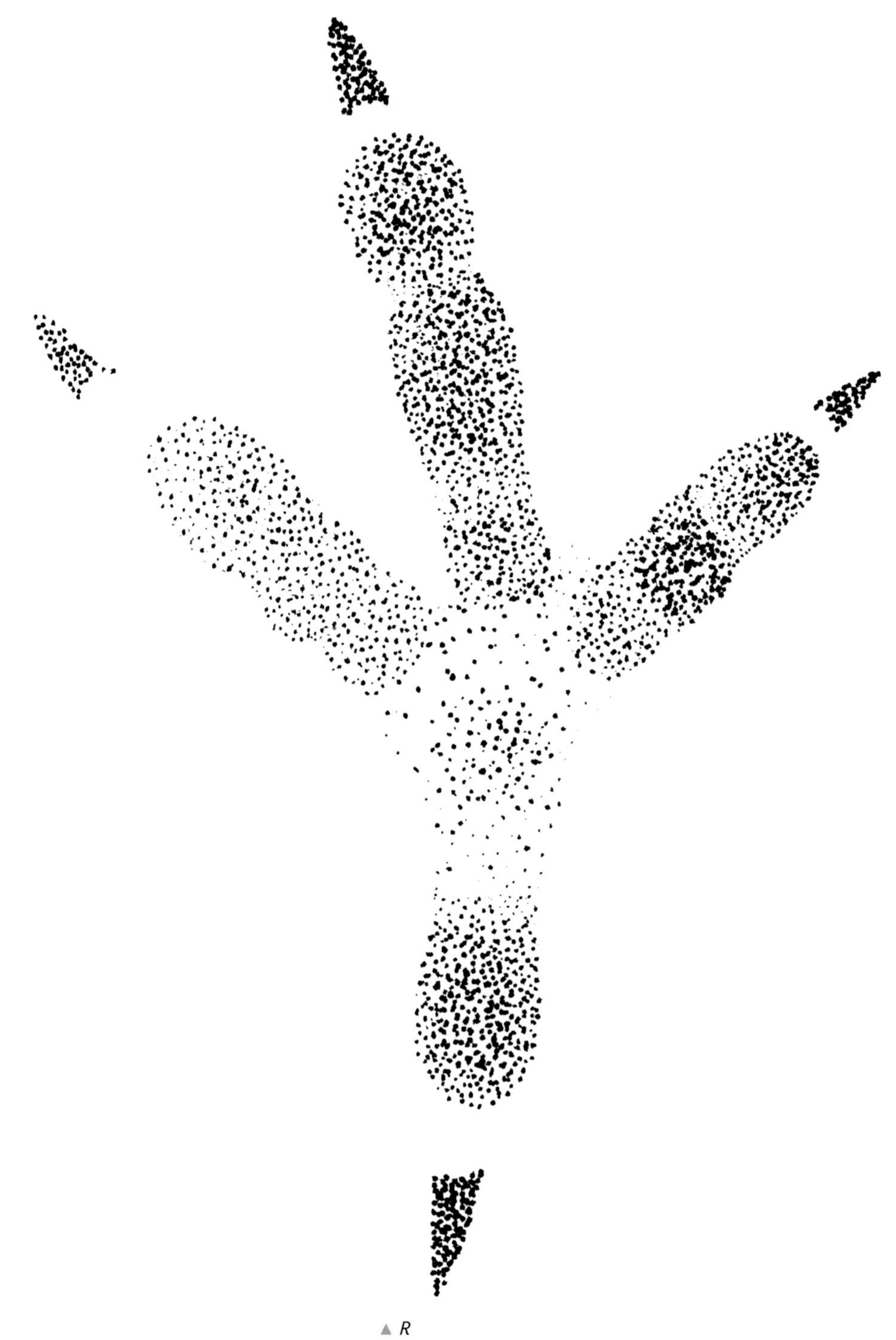

White-tailed Eagle, *Haliaeetus albicilla*, p.638

CLASSIC BIRD FOOTPRINT WITH SHORT OR ABSENT HIND TOE, TRIDACTYL

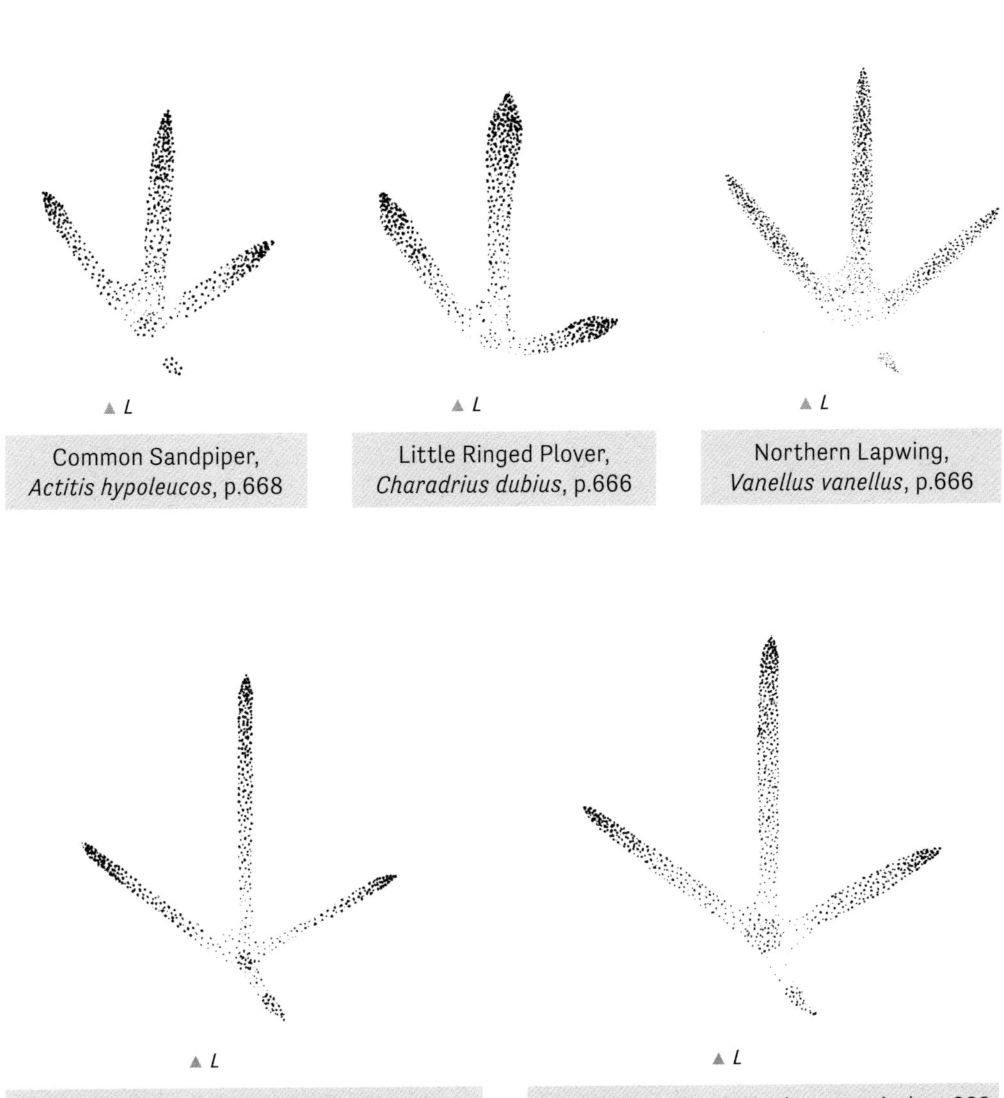

▲ L

Common Sandpiper, *Actitis hypoleucos*, p.668

▲ L

Little Ringed Plover, *Charadrius dubius*, p.666

▲ L

Northern Lapwing, *Vanellus vanellus*, p.666

▲ L

Common Snipe, *Gallinago gallinago*, p.669

▲ L

Eurasian Woodcock, *Scolopax rusticola*, p.668

▲ *L*

Eurasian Curlew,
Numenius arquata, p.671

▲ *L*

Eurasian Oystercatcher,
Haematopus ostralegus, p.664

▲ *L*

Common Moorhen, *Gallinula chloropus*, p.659

Eurasian Coot, *Fulica atra*, p.660

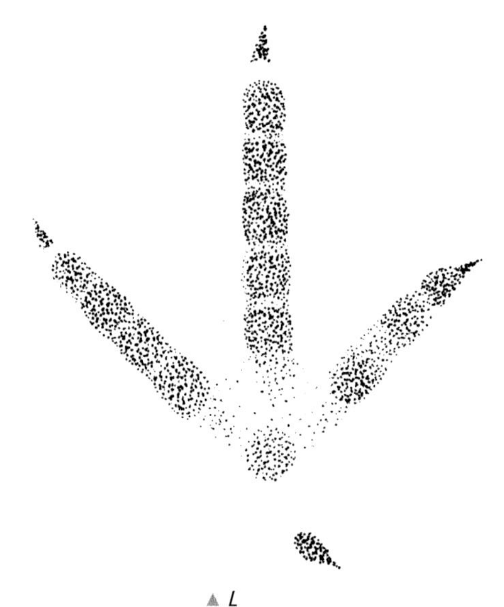

Common Pheasant, *Phasianus colchicus*, p.654

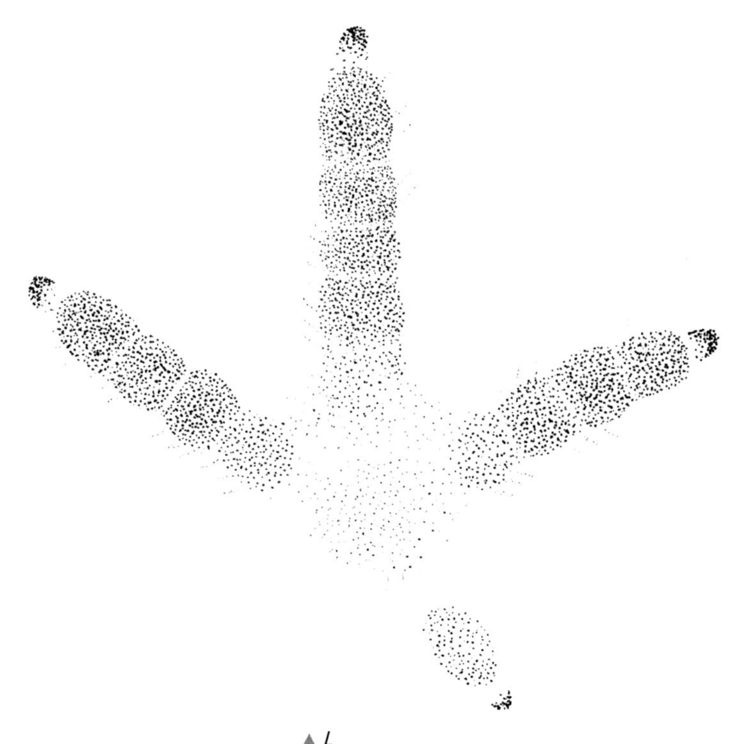

Western Capercaillie, *Tetrao urogallus*, p.656

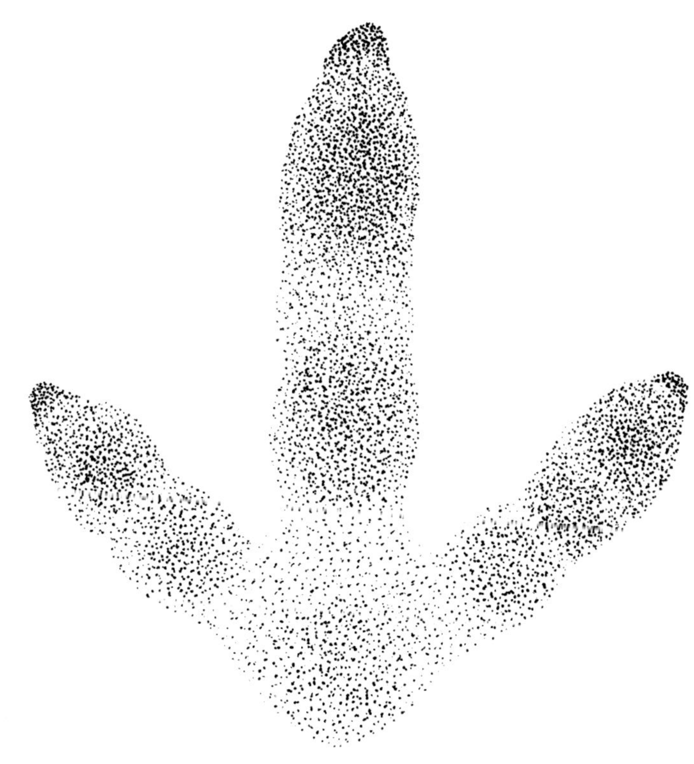

Great Bustard, *Otis tarda*, p.662

▲ R – 50% of original size

Common Crane, *Grus grus*, p.658

BIRD FOOTPRINT WITH WEBBING BETWEEN TOES 2–4, PALMATE

▲ L

Common Tern,
Sterna hirundo, p.682

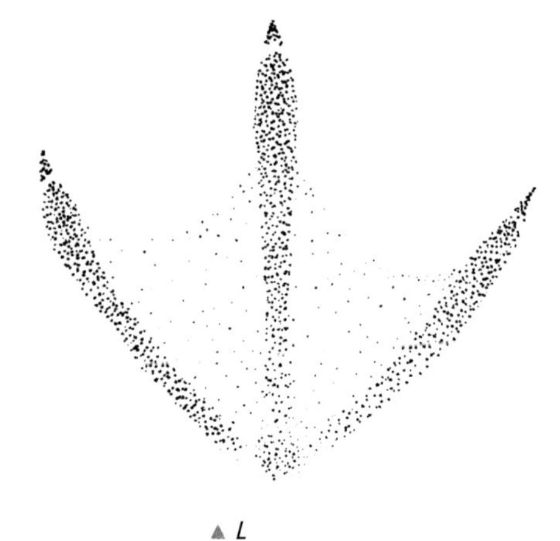

▲ L

Pied Avocet,
Recurvirostra avosetta, p.665

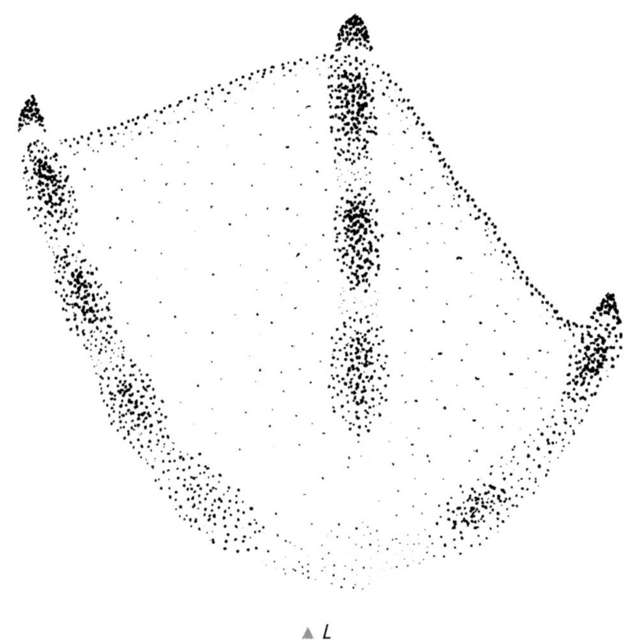

▲ *L*

Great Black-backed Gull, *Larus marinus*, p.680

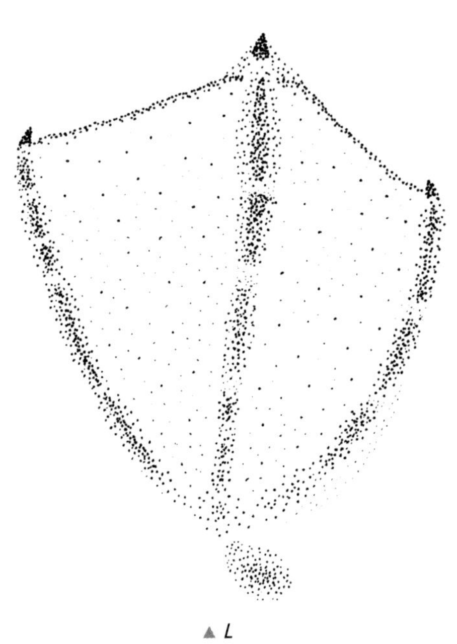

▲ *L*

Goosander, *Mergus merganser*, p.679

Mallard, *Anas platyrhynchos*, p.674

Greylag Goose, *Anser anser*, p.676

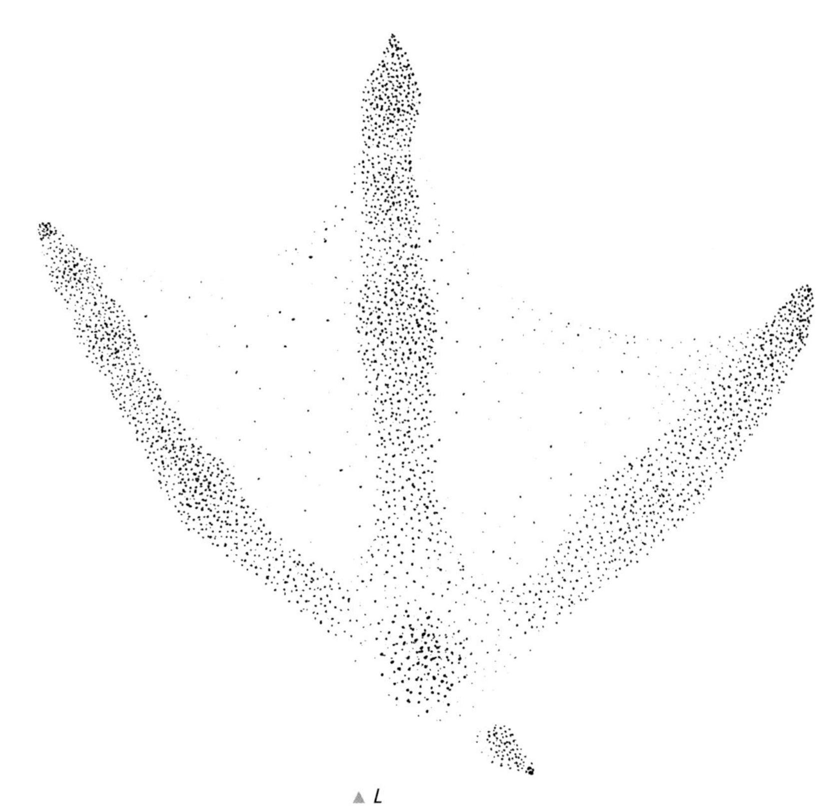

▲ *L*

Greater Flamingo, *Phoenicopterus roseus*, p.672

BIRD FOOTPRINT WITH WEBBING BETWEEN TOES 1–4, TOTIPALMATE

Great Cormorant, *Phalacrocorax carbo*, p.683

ZYGODACTYL BIRD FOOTPRINT

▲ R

Great Spotted Woodpecker, *Dendrocopos major*, p.687

▲ R

Tawny Owl, *Strix aluco*, p.685

ANSIODACTYL

TYPICAL FOOTPRINT

The families covered here are: herons; storks; ibises and spoonbills; hawks, eagles and kites; pigeons and doves; corvids; wrens; thrushes; sparrows and finches.

Herons

HERON TRACKS

Asymmetrical. Slim, elongated toes that remain evenly spaced. Herons (Ardeidae) have a long, straight hind toe, which is almost always recognisable and offset to the inside. Toe 1 is an important identification criterion for heron tracks and gives them their characteristic appearance. Small, proximal webbing between toes 3 and 4 can be visible in clear prints. The metatarsal region is relatively small. Very short, blunt claws that form a continuous print with the toes. Size is the only reliable, known distinguishing feature for differentiating between species like Grey Herons, Little Egrets and Great Egrets; the Little Egret is much smaller than the other two. Although the Great Egret leaves a narrower track with slimmer toes, it cannot always be distinguished from the Grey Heron. As herons spend a lot of time on damp, muddy ground, these details can be lost or a slim toe can appear thicker.

GREY HERON
Ardea cinerea

TRACK

L 13–18.5cm   W 8–9.5cm

GAITS

Walking and running gaits have a very narrow trail width.

Walk
Stride length: 40–65cm

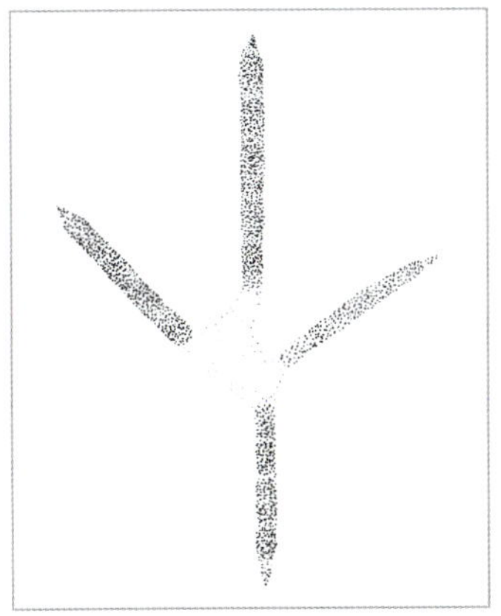

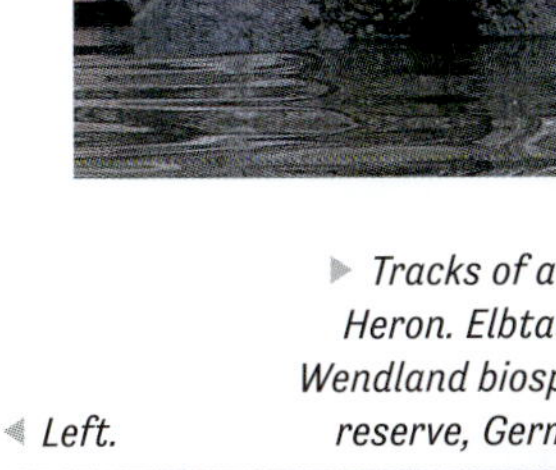

◀ *Left.*

▶ *Tracks of a Grey Heron. Elbtalaue-Wendland biosphere reserve, Germany.*

▲ *Left footprint of a Grey Heron. You can tell which foot is which because the hind toe is offset to the inside. Kreba-Neudorf, Germany.*

▲ *By contrast, the more delicate footprint of a Great Egret. Lausitz, Germany.*

Storks

The White Stork (*Ciconia ciconia*) and Black Stork (*Ciconia nigra*) are the only European representatives of the stork family (Ciconiidae). Their tracks can be found near inland waters, on the shores of reservoirs, or on rubbish dumps where they search for food.

WHITE STORK
Ciconia ciconia

TRACK

L 13–15.5cm W 11–13cm

Asymmetrical. Strong toes with short, blunt claws that leave a continuous print with the toes. Toe 2 often appears slightly shorter than toe 4 in the track. Proximal webbing between toes 3 and 4 may be visible in clear prints. The webbing between toes 2 and 3 is underdeveloped and usually not visible in the print. The Black Stork spends more time in water, so the proximal webbing between toes 3 and 4 is slightly more pronounced. The underdeveloped proximal webbing between toes 2 and 3 can also occasionally be visible. The strong metatarsal pad is only visible in deep prints. The short, broad claws usually do not leave prints. Toe 1 is short, blunt and does not reliably leave a print. Their tracks can be mistaken for heron tracks. However, herons have longer hind toes and storks' toes are much wider.

▲ *Right.*

▲ *Right. Coto de Doñana National Park, Spain.*

GAITS

Walking and running. Comfortable with moving on the ground.

Walk
Stride length: 18–48cm

Run
Stride: over 150cm

Data mainly from the White Stork; Black Stork measurements are comparable.

Ibises and spoonbills

Europe's best-known member of this family (Threskiornithidae) is the Common Spoonbill. Its tracks can be found near shallow waters, such as marshes, and on muddy banks with reeds.

COMMON SPOONBILL
Platalea leucorodia

TRACK

L 12.5–15cm W 10–11.5cm

Asymmetrical. Slim, elongated toes that leave evenly spaced prints. Common Spoonbills have an almost straight hind toe, which is nearly always recognisable and is aligned with toe 3. Proximal webbing between toes 2 and 3 and medial webbing between toes 3 and 4 is visible in clear prints. Toe 4 is longer than toe 2. The metatarsal region is relatively small. Sharp claws that leave a continuous print with the toes. They are much longer than storks' claws.

GAITS

Walking and running with a narrow trail width.

Walk
Stride length: 42–56cm

▲ *Common Spoonbill, right.*

▲ *Right. The webbing between the two forward-pointing toes is a clear distinguishing feature from heron tracks. Vledder, Netherlands. René Nauta.*

Hawks, eagles and kites

TRACKS OF HAWKS, EAGLES AND KITES

Slightly asymmetrical. Hawks, eagles, kites, harriers and buzzards (Accipitridae) have strong toes with clearly recognisable phalangeal and interphalangeal pads. Heron toes are much slimmer in comparison, with less robust phalangeal pads. Unlike herons, accipitrids have no webbing and all the toes fuse at the metatarsal pad. The metatarsal region may or may not leave a print. Long, strong, sharp claws that leave a print separately in front of the toes and are usually clearly visible in the track.

GAITS OF HAWKS, EAGLES AND KITES

Stride length in a walk is much shorter than that of comparably sized herons. Accipitrids also spend less time on the ground, so long, continuous cycles of footfalls are rare. They are usually only attracted to the ground by an animal carcass or other food source.

COMMON BUZZARD
Buteo buteo

TRACK

L 7.5–9.2cm   W 4.8–7cm

Slightly asymmetrical. Strong digital pads. Toe 1 usually leaves a clear print. Toes 2 and 4 are the same length, but toe 2 is wider than toe 4. The angles of the outer toes are usually smaller than those of kites (*Milvus*) or harriers (*Circus*), but the toes can also spread wide apart and then be mistaken for Common Raven tracks. Prints made by the strong claws often vary in length, are usually clearly recognisable and can appear either at a distance ahead of the toes or as a continuous print. The metatarsal region is usually faint or not recognisable at all.

GAITS

Walk
Stride length: 23–29cm

▲ *Right.*
Vledder, Netherlands. René Nauta.

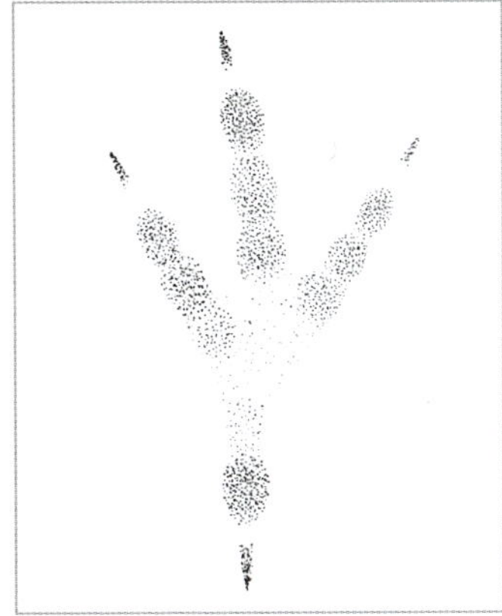

▲ *Right.*

▲ *Right.*
Welzow, Germany. Markus von Hacht.

WHITE-TAILED EAGLE
Haliaeetus albicilla

TRACK

L 15–18.5cm
W 9–13cm

GAITS

Walk
Stride length: 35–50cm

▼ *Right.*

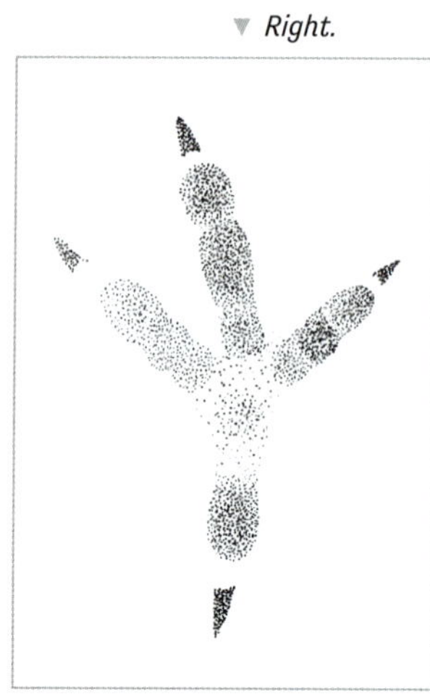

▲ *Right footprint of a White-tailed Eagle. A rare and impressive find. Drehna lake district, Germany.*

Pigeons and doves

Pigeons and doves (Columbidae) leave many footprints, which are easy to recognise due to their characteristic appearance.

PIGEON AND DOVE TRACKS

Asymmetrical. The phalangeal and interphalangeal pads are usually clearly recognisable. Pigeon and dove tracks have a distinctive curved shape. Toe 1 points inwards. Toes 2 and 4 are wide apart and curve slightly backwards. The metatarsal region usually leaves a print. The rather robust claws appear separate from the toes.

GAITS OF PIGEONS AND DOVES

Mainly walking or hopping. Stride length in a walk is relatively short and the trail width relatively wide. The tracks often turn in slightly, resulting in a characteristic track pattern.

▲ *Feral Pigeon, left.*
Vledder, Netherlands. René Nauta.

COMMON WOODPIGEON
Columba palumbus

TRACK

L 5.8–7.4cm W 3.9–5cm

Asymmetrical. The strongest among dove and pigeon tracks with broad toes and strong toe pads. Toe 1 points inwards. Toes 2 and 4 are wide apart and curve slightly backwards. The metatarsal region is strong and usually leaves a print. If the metatarsal region is not recognisable, the track may resemble that of a ground-dwelling bird. Blunt claws that appear separate from the toes.

GAITS

Mainly walking, usually with a conspicuously large trail width and tracks that turn in.

Walk
Stride length: 6–13cm

▼ *The characteristic track of this Common Woodpigeon winds its way from bottom to top. Spessart, Germany.*

▲ *Common Woodpigeon in the snow. The track runs from bottom to top. Märkische Schweiz, Germany.*

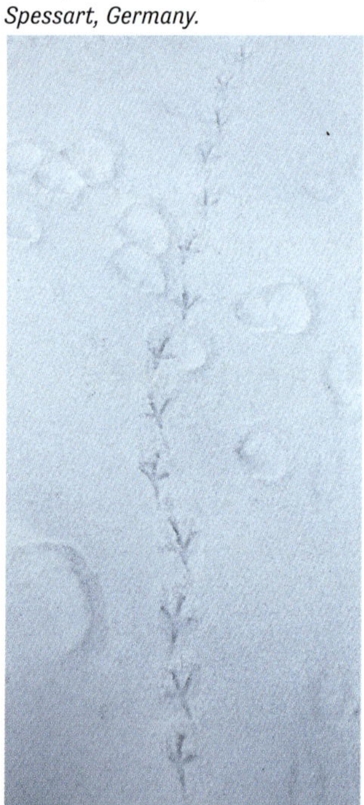

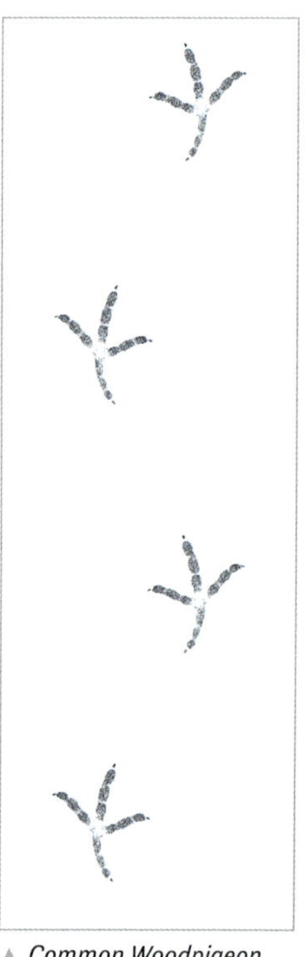

▲ *Common Woodpigeon, walk.*

DIFFERENTIATING BETWEEN PIGEON AND DOVE SPECIES BY THEIR TRACKS

The Common Woodpigeon is Europe's largest pigeon. Stock Dove tracks are slightly smaller. The toes and digital pads appear robust in both tracks. The toes of the Collared Dove and Turtle Dove are much finer and straighter in comparison. Turtle Doves leave the smallest tracks and have the slimmest, straightest toes. Collared Dove tracks appear finer than woodpigeon tracks, but larger and stronger than Turtle Dove tracks.

	Track		Stride length
Woodpigeon	L 5.8–7.4cm	W 3.9–5cm	Walk: 6–13cm
Stock Dove	L 4.5–5.8cm	W 3.5–4.2cm	
Collared Dove	L 4.4–5.4cm	W 3–3.8cm	
Turtle Dove	L 3.5–4.7cm	W 2.4–2.8cm	

▼ *Common Woodpigeon,, left.*

▼ *Collared Dove, left.*

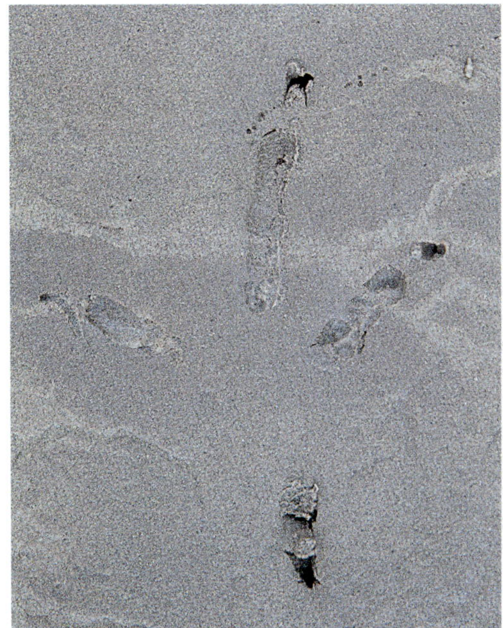

▲ *Common Woodpigeon, left. Pigeon and dove feet usually leave very distinctive tracks. Spessart, Germany.*

▲ *Collared Dove toe prints are much narrower and finer than those of the larger Common Woodpigeon. Coto de Doñana National Park, Spain.*

Corvids

Birds belonging to the crow family (Corvidae) often forage on the ground and leave many footprints. Their characteristic tracks are usually easily identifiable as belonging to this family.

TRACKS OF CORVIDS

Almost symmetrical. Strong toe pads, usually with clearly recognisable phalangeal and interphalangeal pads. The small angle between toes 2 and 3 is a distinctive distinguishing criterion for corvids. These toes are much closer together than toes 3 and 4, resulting in a characteristic appearance ('hugging toe') where the print does not appear clearly symmetrical. The metatarsal region is small and rather indistinct. Small, pointed claws that leave a print separately from the toes. The claw of the hind toe is much longer. Jackdaws make the smallest tracks and Common Ravens the largest.

▼ *From left to right, the right footprint of a Common Raven, Rook, Common Magpie and Eurasian Jay in size comparison. Aaron Tiedemann, Germany.*

▶ *Track of a Rook. Sometimes the claw of toe 1 in corvids leaves distinctively long drag marks. Märkische Schweiz, Germany.*

GAITS OF CORVIDS

Mainly walking or hopping, but often also skipping. The long claw on the hallux often leaves a print in deep, soft ground. It can scrape across the ground, leaving a distinctive long drag mark.

COMMON MAGPIE
Pica pica

TRACK

L 5.8–7.5cm
W 2.6–3.5cm

Typical corvid footprint. A relatively small foot for the size of the bird, with strikingly strong toes. The toes are a key distinguishing feature from the Eurasian Jay. The angle between toes 2 and 3 tends to be slightly larger than in Eurasian Jays or Jackdaws (a).

GAITS

Most members of the crow family spend a lot of time on the ground. Walking, running and skipping are common.

Walk
Stride length: 15–23cm

▶ *Skip.*

◀ *Right track of a Common Magpie.*

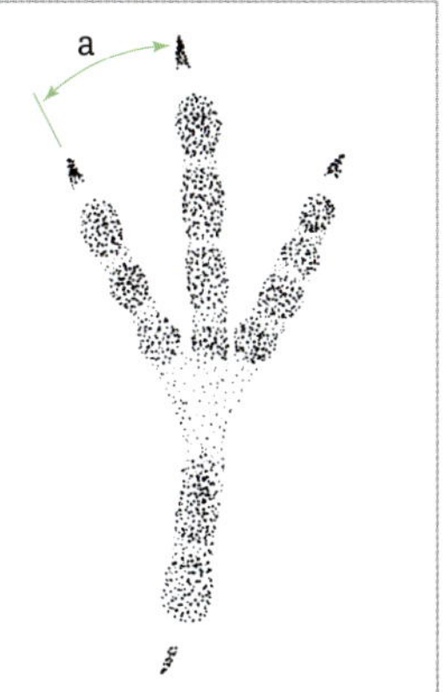

EURASIAN JAY
Garrulus glandarius

TRACK

L 4.3–6.2cm W 1.5–2.5cm

Typical corvid footprint, but with relatively slender toes. Rather small track in relation to the size of the bird. The angle between toes 2 and 4 is small, making the track noticeably narrower than a Common Magpie track (a). Not currently distinguishable from other jay species present in Eurasia.

GAITS

Mainly hopping.

▲ *Right.*

▲ *Hop.*

JACKDAW, CARRION CROW AND HOODED CROW, ROOK

Coloeus monedula, Corvus corone/cornix, Corvus frugilegus

TRACK

Carrion Crow, Rook
L 6.8–8.9cm   W 2.9–4.1cm

Jackdaw
Dimensions comparable to those of the Eurasian Jay.

Typical corvid footprint. Except for the Jackdaw, the tracks are larger than those of Common Magpie and smaller than those of Common Raven. Only the much smaller Jackdaw can be distinguished from the others.

GAITS

Mainly walking, hopping and skipping.

Carrion Crow and Rook in walk
Stride length: 21–41cm

Carrion Crow and Rook hopping
Stride length: 30–55cm

◄ *Carrion Crow tracks near the shore. West Sussex, England.*

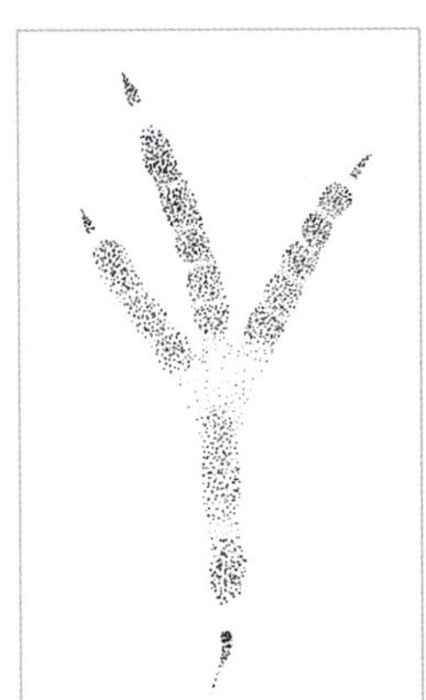

▶ *Rook, right.*

COMMON RAVEN
Corvus corax

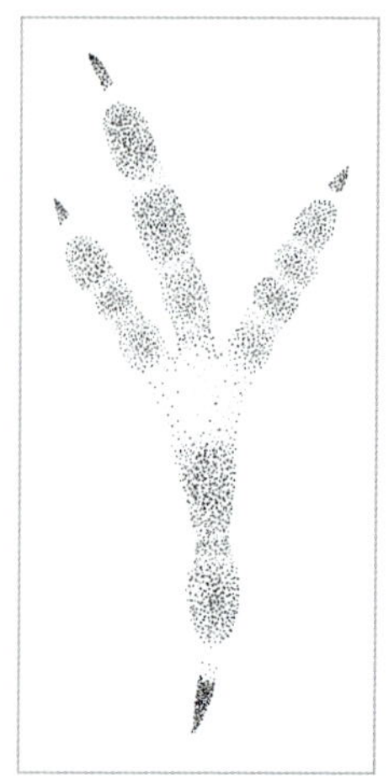

▲ *Right.*

TRACK

L 9.4–13.5cm W 4.5–6.2cm

Typical corvid footprint. Very large with strong toes that normally leave a deep print on soft ground. The strong phalangeal and interphalangeal pads are easily recognisable in clear prints. Size is a clear distinguishing feature from Carrion and Hooded Crows and Rook, as there is no overlap: the smallest footprint of a Common Raven is larger than the largest footprints of the other species. Even if the length is almost the same, Common Raven tracks appear much chunkier in comparison.

GAITS

Mainly walking, hopping and skipping.

Walk
Stride length: 27–50cm

Hop
Stride length: approx. 50–70cm

Passerines

It is difficult or impossible to differentiate between most small passerines (Passeriformes), including tits (Paridae), flycatchers (Muscicapidae), or pipits and wagtails (Motacillidae) by their tracks. Only species that can be reliably identified are presented below.

EURASIAN WREN
Troglodytes troglodytes

TRACK

L 2–3.1cm W 0.9–1.1cm

Symmetrical. Dainty foot, but still quite large in relation to body size. Long, slender toes. Toe 1 appears disproportionately large compared to similarly sized birds. Toes 1 and 3 curve inwards slightly, creating a crescent shape. The metatarsal region usually does not leave a print or only leaves a faint print. Short, sharp claws that leave a continuous print with the toes. Tracks can be found at the edges of watercourses under wooded shrubs and tree trunks.

GAITS

Normally hopping.

Hop
Stride length: 3.5–6cm

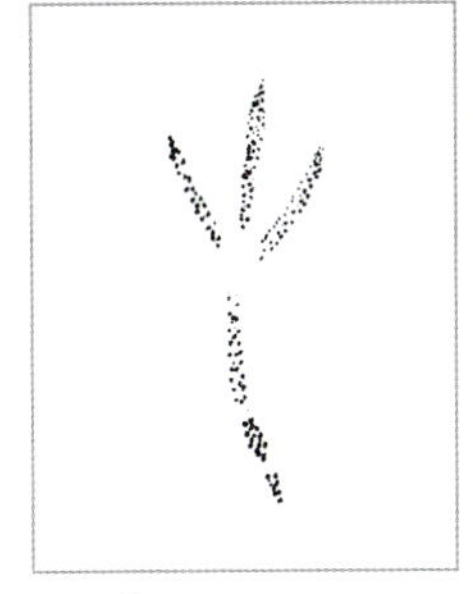

▲ *Left.*

EURASIAN BLACKBIRD
Turdus merula

The Eurasian Blackbird belongs to the thrush family (Turdidae) and leaves one of the most common passerine tracks in Europe.

TRACK

L 4.7–5.6cm W 2.2–2.9cm

Almost symmetrical. Slim, elongated toes. Toe 1 points straight back and has a relatively strong posterior phalangeal pad, which usually leaves a clear print. Toe 1 and toe 3 curve inwards slightly at the tips. If you look at them as a single shape, they form a narrow crescent that curves slightly to the inside. The tips of toes 2 and 4 bend backwards slightly. The metatarsal region usually only leaves a faint print. Very small claws that leave a continuous print with the toes. The claw of the hind toe is longer and leaves a print separately from the toe.

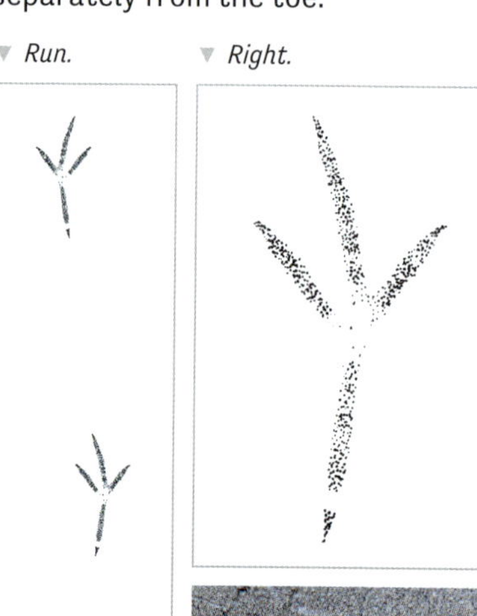

▼ Run.

▼ Right.

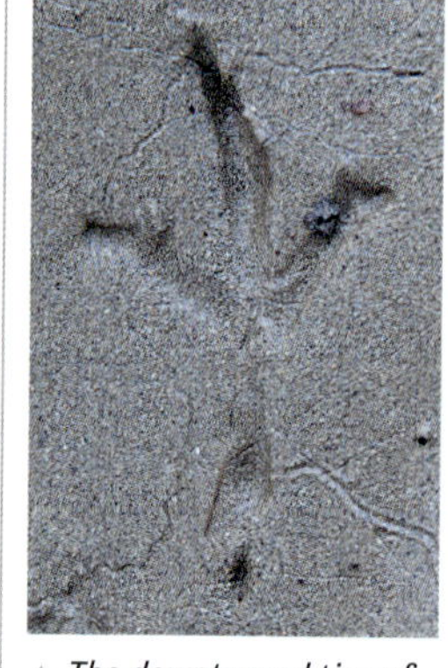

▲ The downturned tips of toes 2 and 4 are a typical feature of this species. Spessart, Germany.

GAITS

Mostly running.

Stride length: 6–10cm

COMMON STARLING
Sturnus vulgaris

TRACK

L 4.9–5.1cm W 2.5cm

Symmetrical. About the size of a Eurasian Blackbird track. Toe 3 is slightly offset towards toe 4 in the metatarsal region, instead of being central. After the breeding season, huge flocks often migrate to fields and meadows in search of food. If you find many tracks of the same kind together, it can point to Common Starlings.

GAITS

Mainly walking.

▲ *Right. Toe 3 is slightly offset towards toe 4 in the metatarsal region, which is a distinguishing criterion from thrushes. Lausitz, Germany.*

HOUSE SPARROW
Passer domesticus

House Sparrow footprints are mainly found on the edges of puddles and muddy areas near settlements.

TRACK

L 2.5–3.5cm W 1.2–1.5cm

Symmetrical. Long, slender toes with short, inconspicuous claws directly in front of them. Very difficult or even impossible to distinguish from other passerines of a similar size. The measurements are intended more as a guide for comparison with other larger or smaller species than for distinguishing House Sparrows from passerines of a similar size.

GAITS

Mainly hopping. Although House Sparrows spend a lot of time on the ground, they rarely leave tracks because they are so light.

Hop
Stride length: 5–11cm

▼ *House Sparrow, right.*

▶ *The House Sparrow's main gait is hopping. Coto de Doñana National Park, Spain.*

▼ *Left and right. Coto de Doñana National Park, Spain.*

COMMON CHAFFINCH
Fringilla coelebs

Because these finches mainly forage on the ground, their footprints can often be found on suitable surfaces. The Brambling (*F. montifringilla*) is a very close relative with similar habits.

TRACK

L 2.5–3.5cm   W 1.2–1.5cm

Symmetrical. Long, slender toes with relatively long claws, which usually leave a clearly recognisable print in front of the toes. The metatarsal region usually leaves a faint print or no print at all.

GAITS

Mainly a fast walk with occasional hopping.

Walk
Stride length: 3.5–6.5cm

▼ In central and southern Europe, millions of Bramblings sometimes gather at roosts. Hilscheid, Germany. Immo Meyer.

◄ A winter flock of Bramblings. Hilscheid, Germany. Immo Meyer.

TRIDACTYL

FOOTPRINT WITH A SHORT OR ABSENT HIND TOE

Members of the pheasant family (partridges and grouse), cranes, rails, bustards, oystercatchers, avocets, plovers, members of the sandpiper family, gulls and terns leave the typical track with an absent or sometimes short hind toe (tridactyl, see page 608).

Pheasants, quails and partridges

Most galliform (gamebird) species in Europe belong to the large pheasant family (Phasianidae). This book covers Common Quail, Grey Partridge, Red-legged Partridge and the non-native Common Pheasant. Phasianidae also includes the subfamily of grouse, from which we will describe the Western Capercaillie in more detail because its foot morphology is representative of other species.

TRACKS OF PHEASANTS

Pheasants have robust, strong toes, with a shape that resembles tracks left by members of the sandpiper family. The hind toe is high up on the leg and does not always leave a print. Proximal webbing between toes 2, 3 and 4 can leave a print in some species. The metatarsal region is strong so it is usually clearly recognisable.

COMMON QUAIL
Coturnix coturnix

Europe's smallest galliform species. Tracks are a rare find.

TRACK

L 3–4.2cm   W 3.2–3.8cm

Asymmetrical. The print of the hind toe is angled inwards, narrow and sometimes not visible. Delicate toes compared to other land fowl. The metatarsal region is strong and usually clearly recognisable. Small, narrow claw marks. Smaller than a partridge.

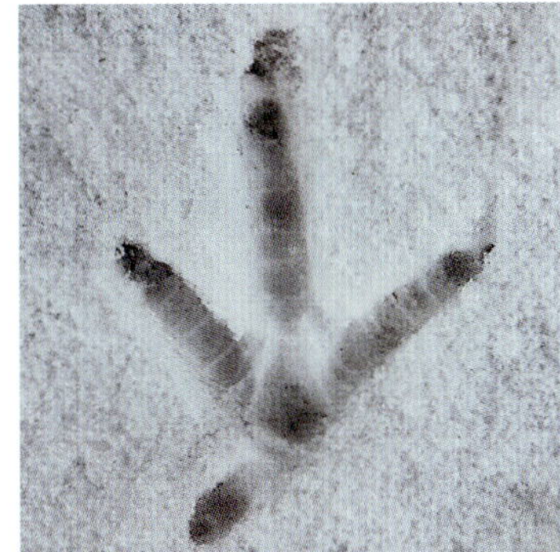

▼ *Domestic Chicken, right footprint. The tracks of galliform birds are generally similar in shape, so the footprint of a Domestic Chicken provides a suitable basis. Extertal, Germany.*

GAITS

Walking and running.

GREY PARTRIDGE AND RED-LEGGED PARTRIDGE

Perdix perdix, Alectoris rufa

TRACK

L 4.2–5.5cm   W 3–4.8cm

Asymmetrical. The print of the hind toe is angled inwards, narrow and sometimes not visible. A small, proximal skin flap between toes 2–4 can be visible in clear prints. The metatarsal region is strong and usually clearly visible. Small, narrow claw marks. The toes are strong compared to those of Common Pheasant.

Similar tracks

The prints of the Red-legged Partridge, a species which is mainly found on the Iberian Peninsula and has been widely introduced for shooting in Britain, are like those of the Grey Partridge.

GAITS

Walking and running. The tracks are striking because they are so close together. Grey Partridges can leave deep furrows in the snow.

◀ Closely spaced footprints are typical of Grey Partridge and Red-legged Partridge (here, Red-legged Partridge). Coto de Doñana National Park, Spain.

▲ Right footprint of a Grey Partridge. Hoher Fläming, Germany.

COMMON PHEASANT
Phasianus colchicus

TRACK

L 5.5–7.5cm W 5–7.2cm

Asymmetrical. The hind toe print is angled inwards and is usually visible. In clear prints, you can make out a small, proximal skin flap between toes 2–4. The metatarsal region is large and normally clearly visible. Large, pointed claw marks, which are usually more clearly recognisable than in Grey Partridges. The track is larger than that of the Grey Partridge, although the toes are proportionately slimmer.

GAITS

Walking and running. The stride length is much longer than that of the Grey Partridge, but the trail width is quite narrow.

Walk
Stride length: 12–45cm

Run
Stride length: 40–95cm

◄ *Left.*
Vledder, Netherlands.
René Nauta.

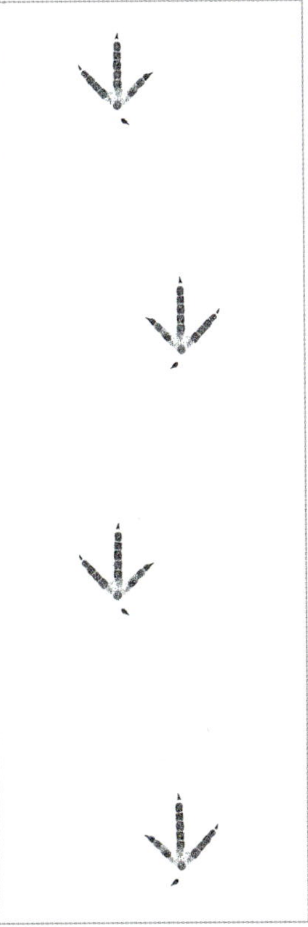

▲ *Walk.*

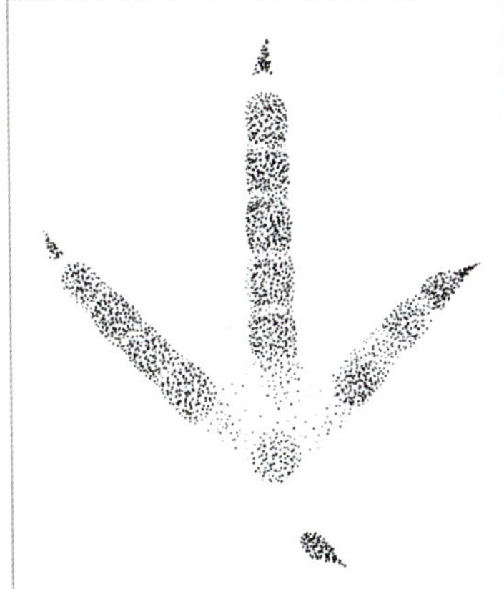

▲ *Left.*

▲ *Left.*
Vledder, Netherlands. René Nauta.

Grouse

The grouse family (Tetraoninae) in Europe includes the Hazel Grouse, Ptarmigan, Black Grouse and Western Capercaillie. Feathered legs down to the toes, which occur mainly in winter, are a characteristic feature of grouse. Unlike members of the partridge family, grouse have skin fringes called pectinations protruding laterally from their toes. These can be visible in clear prints. The footprints of the various grouse species are very similar, so they cannot be distinguished by foot morphology alone. Nevertheless, grouse can often be clearly identified based on their track size, occurrence and habitat.

DIFFERENTIATING BETWEEN GROUSE SPECIES BASED ON TRACK SIZE AND HABITAT

	Track		Habitat
Hazel Grouse	L 4.2–5.2cm	W 4.3–5.1cm	mainly forests with plenty of cover
Ptarmigan	L 5.2–6.3 cm	W 4.5–5.6 cm	mainly alpine, nival zones
Black Grouse	L 5.8–7.5cm	W 5.5–7cm	sparse forests with open areas
Western Capercaillie	L 6–11.7cm	W 7–13cm	fairly sparse, undisturbed forest areas

WESTERN CAPERCAILLIE
Tetrao urogallus

Europe's largest grouse.

TRACK

L 6–11.7cm W 7–13cm
The prints of the cock (male) are much larger and stronger than those of the hen (female). Tracks with a length of more than 10.5cm can be assigned to a cock. Tracks measuring less than 7.5cm in length are made by a hen or young bird.

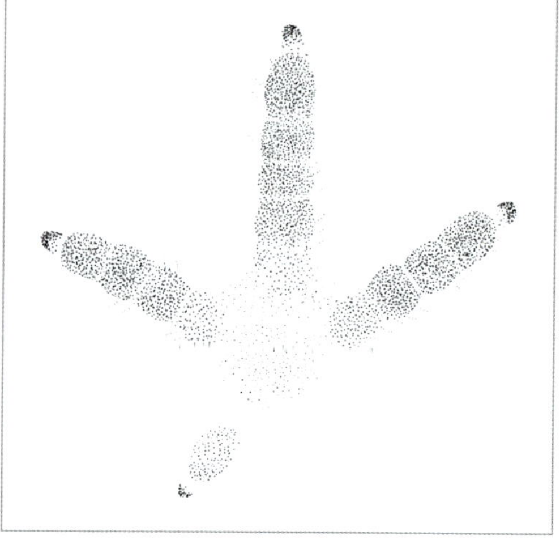

▲ *Right.*

▲ *Right. Jämtland, Sweden. Laura Gärtner.*

Asymmetrical. Typical grouse track. The print of toe 1 is very short, angled inwards and usually clearly recognisable. Toes 2–4 are robust, broad and quite short in relation to the overall size of the track. In winter, comb-like structures called pectinations develop on the toes. This increases the surface area of the track and helps to prevent the bird from sinking into the snow. These horny pins protrude to the sides and can occasionally be visible in very clear prints. The metatarsal region is strong and usually clearly visible. Roundish claw marks that are separate from the toe prints.

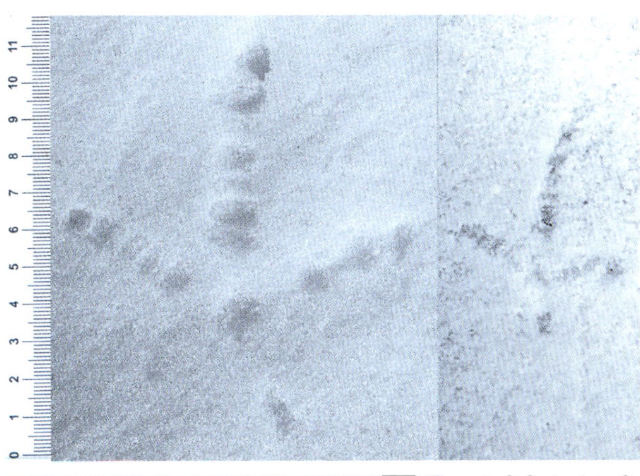

▲ *Left footprint of Western Capercaillie and Ptarmigan in size comparison. Germany. Aaron Tiedemann.*

◄ *The strikingly short stride length in relation to the size of the footprints is typical of the Western Capercaillie. Bavarian Forest, Germany. Lisa Moser.*

◄ *A characteristic track pattern: during courtship, the male Western Capercaillie runs around the mating ground trailing his wings. Jämtland, Sweden. Laura Gärtner.*

GAITS

Walking and running. Stride length in walk is quite short at around 20–30cm. The individual tracks within a track pattern usually turn inwards slightly.

Walk
Stride length: 18–35cm

Cranes

The Common Crane is the only regularly occurring representative of the crane family (Gruidae) in Europe. Its large tracks are often found near wetlands and in meadows and fields.

COMMON CRANE
Grus grus

Along with swans, storks and herons, the Common Crane leaves one of the largest common bird tracks in Europe.

TRACK

L 11–15.5cm   W 15–18cm

Asymmetrical. Long, strong toes. Toe 1 rarely leaves a print and even then, it is only faint. Proximal webbing between toes 3 and 4 can be visible in clear tracks. The metatarsal region is large, well-developed and usually leaves a clear print. Rounded, usually visible claw marks. The backward curving shape of claw 2 can be visible in perfect tracks.

▲ Right.

◀ *The webbing between toes 3 and 4 reveals that this is a right footprint. Lausitz, Germany.*

GAITS

Walking and running. A long-legged bird, it has a correspondingly long stride. Tracks often appear as a distinctive straight line one behind the other.

Walk / run
Stride length: 25–78cm

▶ *Common Crane in walk from bottom to top. Lausitz, Germany.*

◀ *Walk.*

Rails

The rail family (Rallidae) is represented here by the Common Moorhen and the Eurasian Coot. Their tracks can be found on muddy watersides.

COMMON MOORHEN
Gallinula chloropus

TRACK

L 8.9–10.5cm   W 8–10cm

Asymmetrical. Very slender, long and straight toes. Toe 1 is very long and almost always visible in the track, pointing straight back or towards the inside of the path. Small skin flaps on toes 2–4 are sometimes visible. They are much narrower than those of Eurasian Coots as moorhens are less adapted to swimming. The metatarsal region is very small. Narrow, pointed claw marks that usually appear separate from the toe prints.

▲ *Common Moorhen left. Vledder, Netherlands. Aaldrik Pot.*

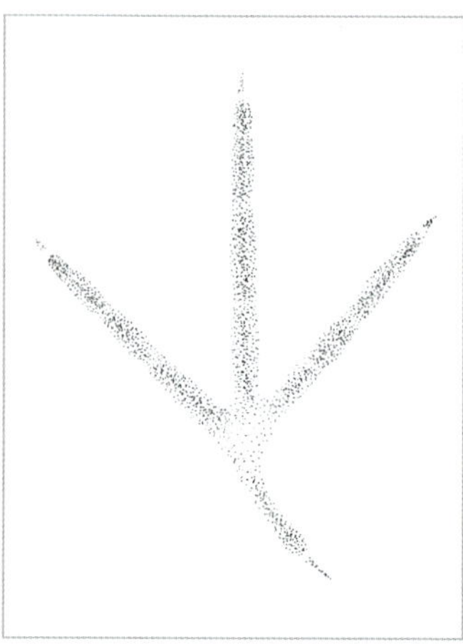

▲ *Left.*

▲ *Left. Note the small lobed flaps of skin on the edge of toes 2–4. Mühlheim, Germany. Simone Roters.*

Strikingly large track in relation to body size.

GAITS

Walking and running, but mainly walking.

Walk
Stride length: 25–30cm

EURASIAN COOT
Fulica atra

TRACK

L 10–13.5cm   W 8–11cm

Asymmetrical. Very long, slender toes. The print of toe 1 is almost always recognisable. It curves towards the inside of the path and has a relatively large claw. The lobed flaps of skin (rather than connected webbing as in ducks) on toes 2–4 can be seen in clear prints. Toe 2 has two distinct lobes and toe 3 has three. The lobes on toe 4 are not clearly divided. The metatarsal region is very small. Long, fine, pointed claw marks. Strikingly large track in relation to body size.

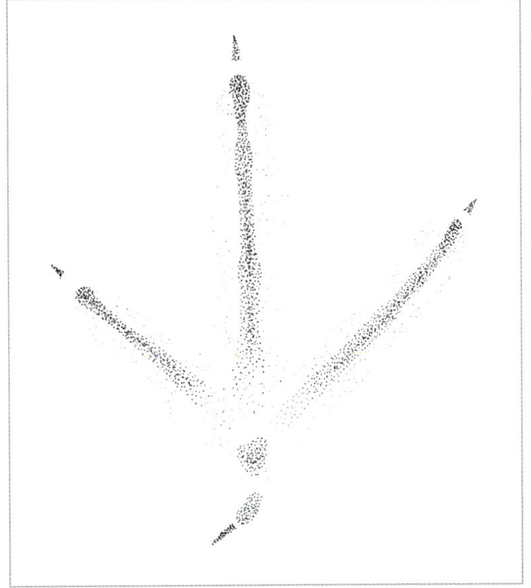

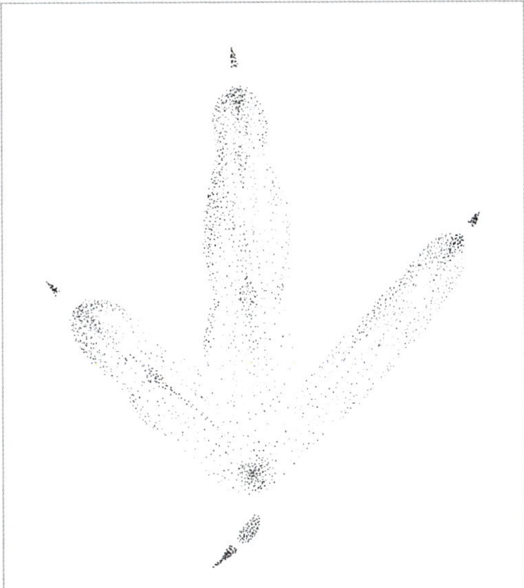

▲ *If the lobes don't leave prints, Eurasian Coot toe prints can appear very narrow. Vledder, Netherlands. René Nauta.*

▲ *Here, the lobes have left clear prints in the snow. Vledder, Netherlands. René Nauta.*

GAITS

Walking and running, but mainly walking. Wavy lines of tracks with short stride lengths are characteristic.

Bustards

Two species from the bustard family (Otididae) occur in Europe. Because both species spend a lot of time on the ground, in fields and other sandy open spaces, their tracks are usually easy to find.

GREAT BUSTARD
Otis tarda

Very pronounced sexual dimorphism. The males, which can weigh up to 15kg, are among the heaviest birds in Europe. Females are about half the size of males and usually weigh less than 5.3kg.

TRACK

♂ L 6.3–7.6cm
♀ L 7.4–9.5cm
♂ W 5.7–6.9cm
♀ W 8.5–9.4cm

Slightly asymmetrical. Toe 1 is absent. Strong footprint, with robust, wide toes that taper towards the tip. The metatarsal region is large and usually clearly visible. On hard ground and at high speed, often only the toes leave a print. Broad claw marks are usually recognisable in front of the toe print.

GAITS

Walking and running. Stride length is quite small in relation to the size of the bird and the trail width is large. The feet turn in slightly when walking.

Walk / run
Stride length: 10–95cm

▶ *Walk.*

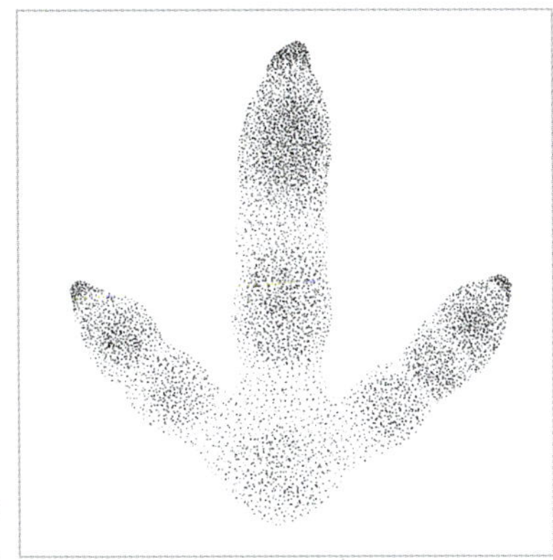

▶ *Left. Great Bustard track.*

LITTLE BUSTARD
Tetrax tetrax

The smallest bustard. Between a partridge and a female Common Pheasant in terms of size.

TRACK

L 3.7–4.5 cm W 3–3.9 cm

Slightly asymmetrical. Toe 1 is absent. Short, strong print with wide toes that usually leave a deep print. Short, blunt claw prints that often merge with the toe prints.

SHOREBIRDS

The large order of shorebirds, or waders, and their relatives (Charadriiformes) includes birds that spend a lot of time on shores, coasts and in different wetlands. They are usually long-legged birds that often forage on the ground in areas with excellent tracking conditions. Many of these habitats are home to a range of different species. A good bird identification book with distribution maps and size information can help to narrow down tracks.

TRACKS OF SHOREBIRDS

Toe 1 is usually very short, sits high up on the leg or is completely absent. Toes 2–4 are slender and toe 2 is shorter than toe 4. In some species, proximal webbing between toes 3 and 4 may leave a print. Depending on the species, the metatarsal region leaves a print often, rarely or not at all.

EURASIAN OYSTERCATCHER
Haematopus ostralegus

TRACK

L 4.2–51cm W 5–6cm

Almost symmetrical. The slightly lobed toe shape makes the toe prints of shorebirds appear relatively broad and short. However, the toes can sometimes look very narrow if the lobed area is not visible on firm ground. Toe 1 is completely absent and toes 2 and 4 are almost the same length. The metatarsal region is very small and often only faintly recognisable or not recognisable at all on firm ground. Small proximal webbing between toes 3 and 4 is occasionally visible in the print. Fine, elongated claw marks may be visible separately from the toe prints.

▲ *Right. The toes, which are quite broad and short for shorebirds, are clearly recognisable. Vledder, Netherlands. René Nauta.*

▲ *Right.*

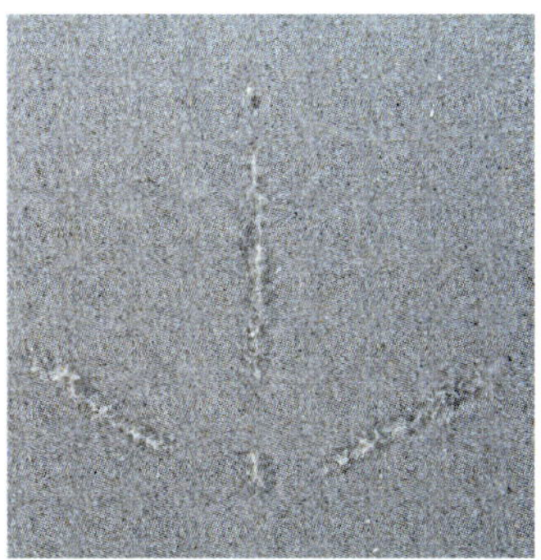

▲ *The broad toes of the Eurasian Oystercatcher can appear long and slender on harder ground, while the prints are relatively short and wide on softer ground. Spiekeroog, Germany. Ulrike Quartier.*

GAITS

Walk
Stride length: 8–20cm

PIED AVOCET
Recurvirostra avosetta

TRACK

L 4.5–5.9cm W 5–6.8cm

Symmetrical. The print of toe 1 is not usually recognisable. Characteristic well-developed medial webbing between toes 2–4.

▼ *Left.*

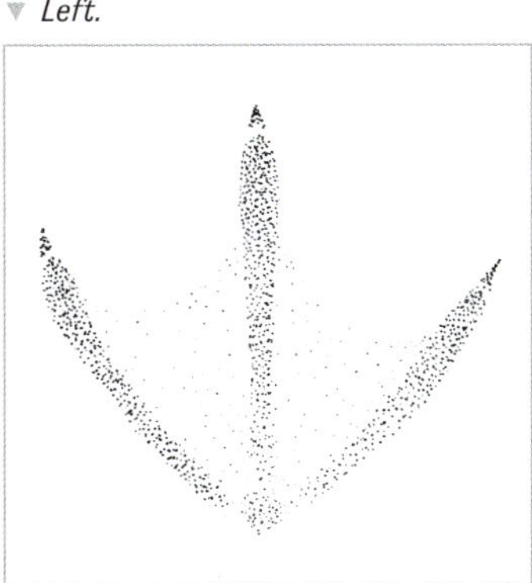

PLOVERS AND LAPWINGS
Charadriidae

Many different plovers can be found on Europe's beaches and shores. We will discuss the most common representatives of this species-rich family here, in ascending order of size: Little Ringed Plover, Ringed Plover, Kentish Plover, Grey Plover, Golden Plover and the robust, pigeon-sized Northern Lapwing. The Little Ringed Plover is the most common plover in Europe. Size is the only reliable way to distinguish between individual species based on their tracks.

TRACK

Little Ringed, Ringed and Kentish Plovers
L 1.9–2.5cm   W 2.3–2.8cm

Grey and Golden Plovers
L 2.6–3.6cm   W 2.7–3.8cm

Northern Lapwing
L 3.5–4.2cm   W 3.5–4.1cm

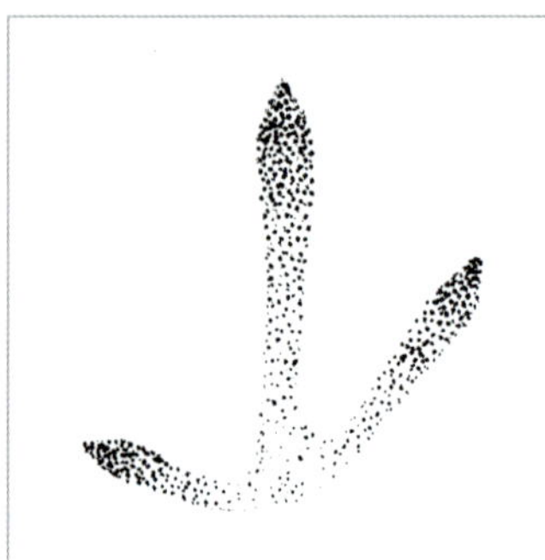

▲ *Little Ringed Plover, right.*

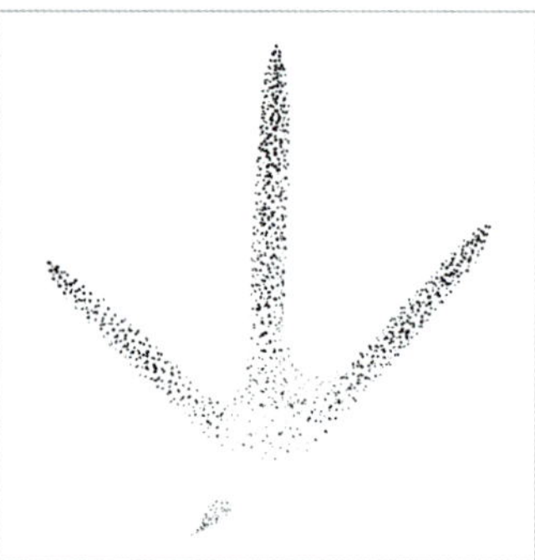

▲ *Northern Lapwing, right.*

▲ *Little Ringed Plover, right. The asymmetrical track is a distinguishing feature from the more symmetrical footprints of the Common Sandpiper. Lausitz, Germany.*

Sliylılly asymmetrical. Toe 1 is very short and situated high up on the leg or absent altogether. It is usually not visible in the track. The relatively wide toe prints can appear cigar-shaped. The angle between toes 3 and 4 is smaller than that between toes 2 and 3. Proximal webbing may leave a print between toes 3 and 4. The metatarsal region is very small and rarely leaves a print. If it does, it is only faint. The claw marks are very short and blunt and there is no gap between them and the toe prints.

GAITS

Walking and running. with a relatively narrow trail width. The tracks usually turn in slightly.

Little Ringed, Ringed and Kentish Plovers
Stride length walking / running: 5–9cm

Grey and Golden Plover
Stride length walking / running: 7–20cm

▲ *Little Ringed Plover tracks are often found close together, for example at this pond in the former opencast mining area of Welzow, Germany.*

Sandpipers

Sandpipers (Scolopacidae) are a large and diverse family of waders. Species covered below include the Common Sandpiper, Eurasian Woodcock, Common Snipe, some members of the *Tringa* genus of waders (shanks) and the Eurasian Curlew.

TRACK

Almost symmetrical to symmetrical. Toe 1 is short and high up on the leg, but leaves a print relatively often (see (a) on page 668). Even a faint hind toe print makes it possible to distinguish the tracks of members of the sandpiper family from those of plovers. Toes 2–4 of birds in the sandpiper family are relatively slender and the same width for their entire length, compared to the cigar-shaped toe prints of plovers. The angles between the toes are more even than in plovers. Proximal webbing can leave a print between toes 3 and 4. The track measurements of the Sanderling (*Calidris alba*) are slightly smaller than those of the Little Ringed Plover. The Sanderling is the only sandpiper without a hind toe, so its tracks can be easily confused with those of small plovers. In this case, identification can be made based on the symmetry of the track, and the toe shape.

GAITS

Walking and running.

COMMON SANDPIPER
Actitis hypoleucos

Europe's most common sandpiper species, whose tracks are mainly found by rocky streams and rivers, lakes and by the coast. It is only slightly larger than the Little Ringed Plover, but with different foot morphology to a plover.

TRACK

L 2.3–3.1cm W 2.4–3.3cm

Classic track of a member of the sandpiper family.

GAITS

Walk / run
Stride length:
6–17cm

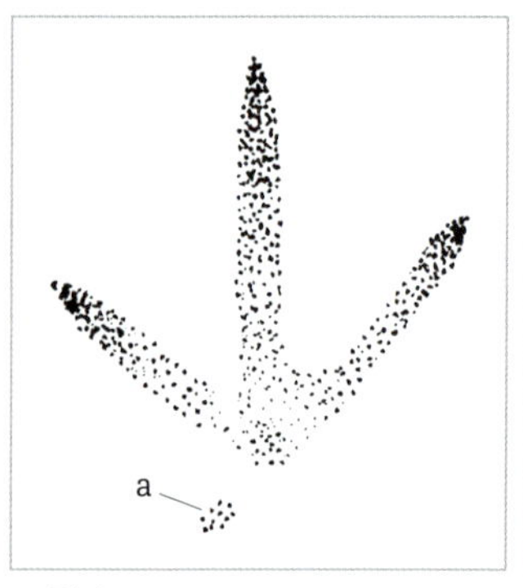

▲ Right.

◀ *Right. Note the narrower, more evenly spaced toes compared to the Little Ringed Plover. Coto de Doñana National Park, Spain.*

EURASIAN WOODCOCK
Scolopax rusticola

Tracks are suitable evidence of this elusive bird. However, it is not always possible to clearly distinguish its footprints from those of the related Common Snipe. With its almost pigeon-sized body, the Eurasian Woodcock is the heavier of the two birds, which is usually reflected in its tracks. In addition to size, habitat also provides an important clue for differentiation.

TRACK

L 4.2–5.1cm   W 4.5–5.3cm

Symmetrical. Exceptionally long, slender toes, but slightly more robust than those of the Common Snipe. Toe 1 often leaves a print. Toes 2 and 4 are almost the same length, which may provide a further clue for differentiation from the Common Snipe. The metatarsal region is narrow. No webbing. Normally fine, clear claw marks.

▲ *Eurasian Woodcock, right.*

COMMON SNIPE
Gallinago gallinago

TRACK

L 3.7–4.8cm   W 4.1–5.2cm

Symmetrical. Exceptionally long, slender toes. Toe 1 often leaves a print. Toe 2 is shorter than toe 4 (a), which may provide a clue for differentiation from the Eurasian Woodcock. The metatarsal region is narrow and protrudes slightly. No webbing. Normally shows clear claw marks. Not always easily distinguishable from Eurasian Woodcock tracks but tend to be smaller and finer.

GAITS

Walk / run   Stride length: 8.5–21cm

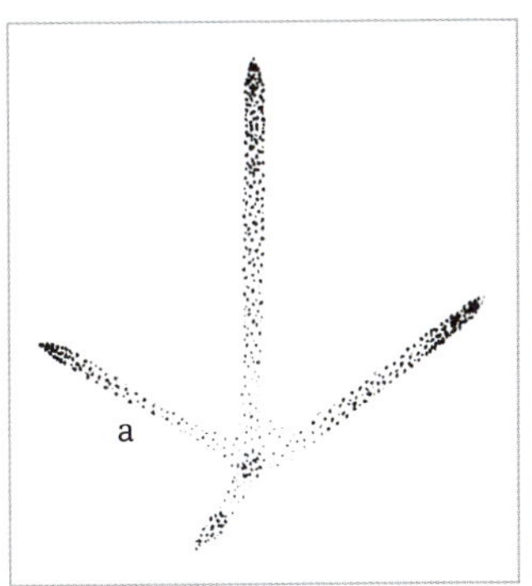

▲ *Right.*

◀ *Right.*
Lausitz, Germany.

SHANKS
Tringa

Around six species are commonly found in Europe. The Wood Sandpiper is the smallest representative. Its track dimensions are almost identical to those of the Common Sandpiper and can easily be confused with this species. The individual tracks tend to be slightly wider and, as the Wood Sandpiper has longer legs, its stride length is usually much longer. The Common Greenshank is our largest shank, with tracks that tend to be in the upper size range. Tracks are indistinguishable apart from size.

TRACK

Wood Sandpiper
L 2.5–3.5cm   W 2.7–4.2cm

Common Redshank, Spotted Redshank, Common Greenshank
L 3.8–5cm   W 3.5–4.5cm

Symmetrical. Long, relatively broad toes for birds in the sandpiper family. They become broader towards the tip and usually leave a deeper print. Toe 1 leaves a print less frequently. The metatarsal region is inconspicuous and rounded. Proximal webbing is occasionally visible between toes 3 and 4. Small claw marks that merge with the toe prints.

GAITS

Mainly walking and running. The stride lengths of these long-legged birds are usually much longer than those of the Common Sandpiper.

Walk / run
Stride length: 8–22cm

EURASIAN CURLEW
Numenius arquata

With its long, curved beak, the Eurasian Curlew is a very distinctive member of the sandpiper family and our largest wader.

TRACK

L 4.8–5.5cm W 4.5–6.7cm

Almost symmetrical. Slim toes that only taper slightly towards the tip. The slightly lobed shape of the toes may be visible in clear prints on soft ground. Toe 1 can leave a print. Toes 2 and 4 are almost the same length. The metatarsal region is small and, on firm ground, is often only faintly recognisable or not recognisable at all. Well-developed proximal webbing between toes 2 and 3, as well as between toes 3 and 4. The latter is often clearly visible. Usually elongated, narrow claw marks that merge with the toe prints.

GAITS

Walk
Stride length: 24–37cm

▼ *Left.*
Vledder, Netherlands. René Nauta.

◄ *Left.*

▶ *Left. The print of toe 1 is a clear distinguishing feature from oystercatchers. West Sussex, England.*

▲ *Eurasian Curlew in walk. The long legs result in a long stride, which can be a distinguishing feature from oystercatchers. West Sussex, England.*

PALMATE

FOOTPRINT WITH WEBBING BETWEEN TOES 2–4

The well-developed webbing is an adaptation to life in the water and excellent for swimming and diving. Tracks with this basic shape are usually found near bodies of water, for example on riverbanks or sandy beaches. However, many species can also be found inland. Families that leave this kind of footprint include flamingos (we will be looking at the Greater Flamingo here), wildfowl (swans, geese, dabbling ducks and diving ducks), gulls and terns.

GREATER FLAMINGO
Phoenicopterus roseus

TRACK

L 8.5–10.2cm   W 9.5–11cm

Asymmetrical. Toe 1 is short, curved inwards and does not reliably leave a print. Toes 2 and 4 clearly curve towards toe 3, but do not point towards it. The metatarsal pad often leaves a print. The claws protrude beyond the webbing. The position of the webbing on toe 3 is characteristic. Flamingos have webbing on toe 3 that is somewhere between the distal form of ducks and geese and the medial form of terns.

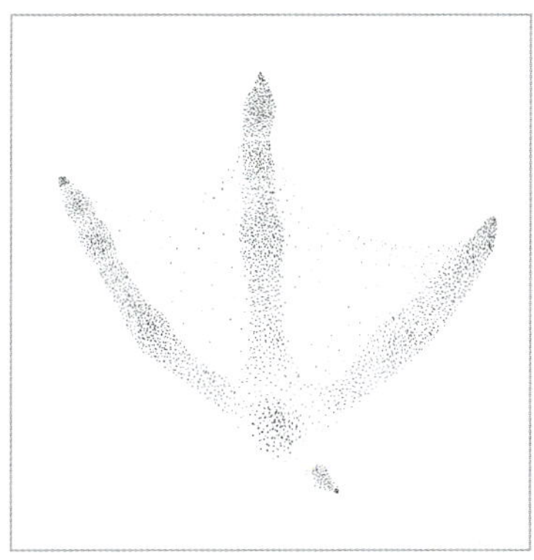

▲ *Left.*

Ducks, geese and swans

The tracks of various ducks, geese and swans (Anatidae; known as wildfowl) can be found on almost every body of water in Europe. You can clearly distinguish between small, medium, large and very large members of this family based on their size. The Common Teal, Mallard, Greylag Goose and Mute Swan are representative of these four sizes. Small to large wildfowl tracks can be confused with those of gulls (Laridae) of a similar size (for differentiation, see page 681).

TRACKS OF DUCKS, GEESE AND SWANS

Asymmetrical. Toe 1 is short and curves inwards. The print of a narrow flap of skin along the outside of toe 2 can be seen on soft surfaces (see (a) in bottom left illustration on page 681). This is characteristic of this family and a clear distinguishing feature from gulls. Toes 2 and 4 are clearly bent towards toe 3. Toe 2 is shorter and wider than toe 4. The metatarsal pad and the hallux often leave a print. The claws protrude beyond the distal webbing.

GAITS OF DUCKS, GEESE AND SWANS

Mainly walking and running. Short stride lengths with relatively large trail widths and footprints that mostly turn inwards (resulting from a wide stance and swaying walk, colloquially known as 'waddling').

MALLARD
Anas platyrhynchos

TRACK

L 5.8–7.5cm
W 5.7–7.5cm

Mallards leave some of the largest tracks of our dabbling ducks. They are indistinguishable from other dabbling ducks of a similar size. Small, clearly recognisable claws.

GAITS

Walk
Stride length: 18–36cm

▲ *Left. Note the narrow flap of skin on the outside of toe 2. Vledder, Netherlands. René Nauta.*

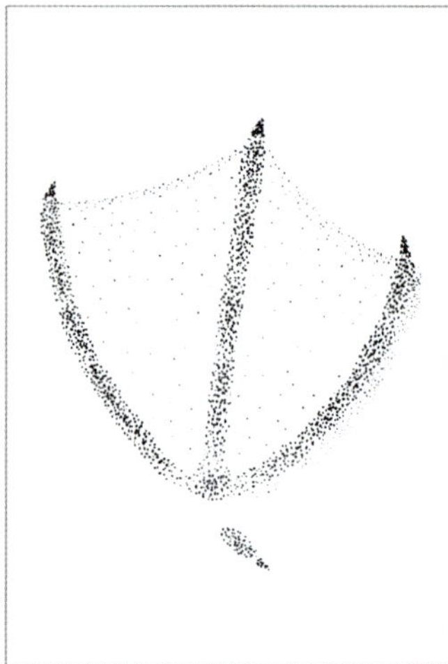

▲ *Left.*

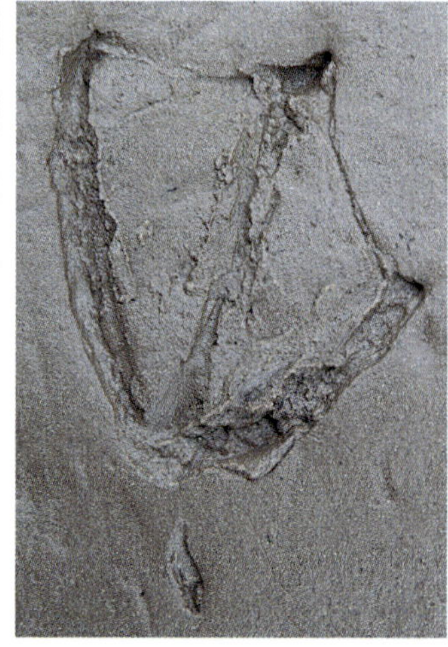

▲ *Left. Note how toes 2 and 4 curve inwards. Elbe near Dresden, Germany.*

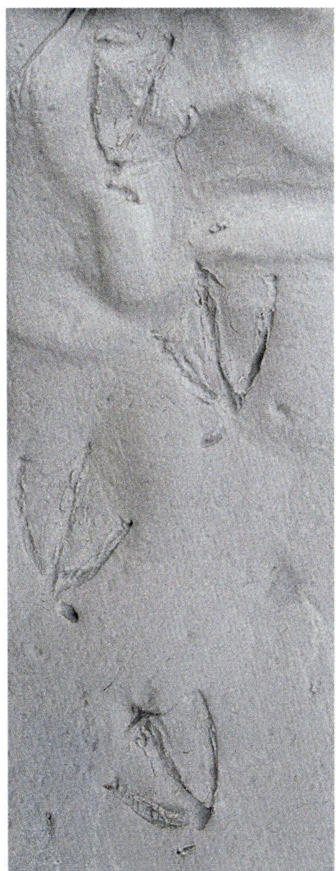

◀ *Mallard in a walk. The short stride lengths and large trail width, combined with the turned in tracks, leave a characteristic track pattern. Elbe near Dresden, Germany.*

▲ *Mallard landing. Perfect tracks can often be found on freshly snow-covered lakes in winter. Märkische Schweiz, Germany.*

COMMON TEAL
Anas crecca

TRACK

L 3.5–4.4cm   W 3.5–4.7cm

Like the Mallard, the Common Teal is a dabbling duck (Anatinae) and the smallest duck in Europe. It can be distinguished from larger ducks like the Mallard by its size. Has slender toes and a very small metatarsal region. Small, roundish claws on the tips of the toes.

GAITS

Walk
Stride length: 15–24cm

GREYLAG GOOSE
Anser anser

TRACK

L 7.5–11.5cm W 7.5–11cm

The tracks are longer and wider than those of ducks. The relatively small metatarsal region and toe 3 usually leave a deeper print than the rest of the foot, giving the track a rounded appearance. Relatively strong, roundish claw marks that merge with the tips of the toes.

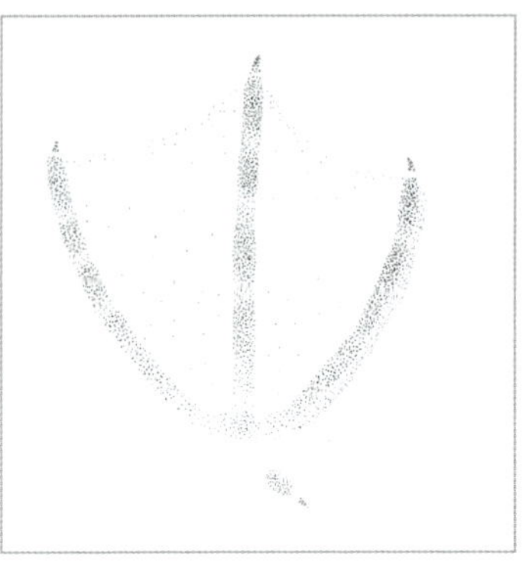

▲ *Left.*

◄ *Left. The metatarsal region and toe 3 characteristically leave a deeper print than the rest of the foot. Lausitz, Germany.*

GAITS

Walk
Stride length: 16–51cm

MUTE SWAN
Cygnus olor

The Mute Swan is the largest species of wildfowl in Europe and can be clearly distinguished from ducks and geese by its size.

TRACK

L 17–20cm
W 14–16cm

GAITS

Walk
Stride length: 26–56cm

▼ *The footprint of a Mute Swan is larger than the hand of an adult human and too large to be confused with the footprints of ducks and geese. Lausitz, Germany.*

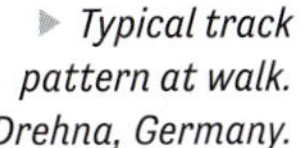
▶ *Typical track pattern at walk. Drehna, Germany.*

DIFFERENTIATING BETWEEN SWANS, GEESE AND DUCKS BY THE SIZE OF THEIR TRACKS

It is normally possible to differentiate between swans, geese and ducks by the size of their tracks. Confusion is possible in the area where large duck species, such as the Mallard, and small goose species, such as the Brent Goose, overlap if both species share a habitat. I am not aware of any reliable distinguishing features apart from size. Every year I see beginners in our tracking courses having difficulty telling the difference, even if the tracks themselves are clear. Looking at different tracks side by side often makes size comparison easier. The following collage gives an overview of the size differences.

▶ *Left foot of the Mallard and right foot of the Mute Swan. Vledder, Netherlands. René Nauta.*

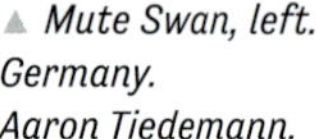

▲ *Mute Swan, left. Germany. Aaron Tiedemann.*

▲ *Greylag Goose, left. Germany. Aaron Tiedemann.*

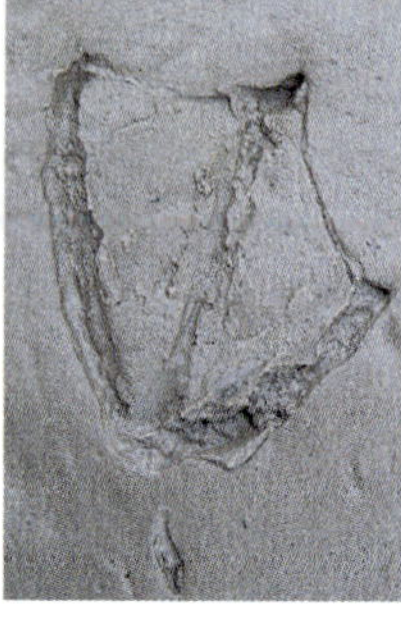

▶ *Mallard, left. Germany. Aaron Tiedemann.*

DIVING DUCKS, SEA DUCKS AND SAWBILLS
Aythyini, Mergini

Unlike dabbling ducks, diving ducks (Aythyini), as well as sea ducks and sawbills (Mergini) spend a lot of time searching for food underwater. For this reason, both groups have feet that are adapted for diving and that are narrower than the relatively wide feet of dabbling ducks. Toe 1 is lobed, which increases the surface area and enables the bird to dive powerfully. If the track is clear enough, this feature can help to distinguish them from dabbling ducks like the Mallard. The feet of diving ducks are also further back on the body, which results in a relatively short stride length. While diving ducks mainly feed on plants, sawbills (*Mergus* and *Mergellus*) are specialists at catching fish. With the narrowest feet of the wildfowl species in Europe, they are highly skilled and fast divers.

DIFFERENTIATING BETWEEN DUCKS BASED ON THE ANGLE BETWEEN TOES 2 AND 4

Brown/Ferguson 1987 use the angle between toes 2 and 4 to differentiate between dabbling ducks, diving ducks and sawbills. They provide the following benchmarks for differentiation: sawbills 25–35°, diving ducks 35–45°, dabbling ducks 65–75°. Based on our practical experience, these values only provide a guide. It is not possible to make a judgment based on the angle between the toes alone, because this angle can vary greatly depending on ground conditions, the angle at which the foot touched the ground and the speed of travel.

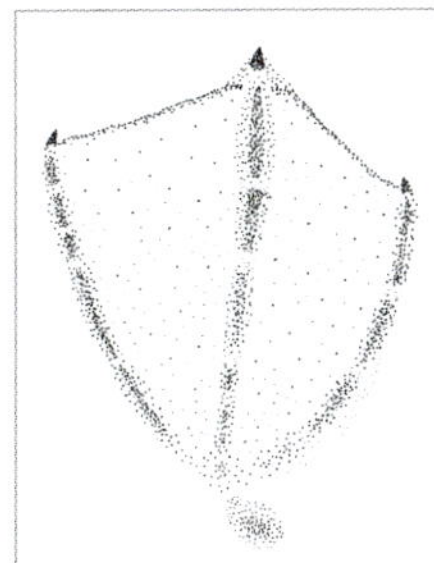

▼ *Common Pochard, left.*

▲ *The narrow angle between toes 2 and 4 is a distinguishing feature from dabbling ducks such as Mallard (page 678). The distribution, habitat and size indicate a Common Pochard. West Sussex, England. René Nauta.*

GULLS
Laridae

Gulls (Laridae) of different sizes occur in Europe. We mainly distinguish between them by the size of their tracks. The Black-headed Gull, Lesser Black-backed Gull and Great Black-backed Gull are described as representatives of small, medium and large gull species.

TRACK

Black-headed Gull
L 3.5–4cm W 4–4.5cm

Lesser Black-backed Gull
L 6–6.5cm W 7–7.5cm

Great Black-backed Gull
L 7–8.5cm W 8–11cm

▲ *Great Black-backed Gull, left. Vledder, Netherlands. René Nauta.*

◀ *Large gull species (Larus spp.), left. Note the deep toe prints and the fainter print of the metatarsal region. There is no hind toe print. Coto de Doñana National Park, Spain.*

Slightly asymmetrical to asymmetrical. Gulls have a short toe 1 that sits further up the leg. Toes 2 and 4 bend towards toe 3, but their tips still point forwards. Toes 2–4 usually leave deep prints. Toe 2 is shorter and wider than toe 4. The metatarsal pad and the hallux often do not leave a print. The claws protrude beyond the webbing and are quite strong compared to duck claws. The tracks of larger gull species are proportionally larger than those of smaller gulls. Apart from their size, the different species cannot be distinguished by other track features. At best, size allows us to differentiate between small, medium and large gulls. The dimensions of Black-headed, Lesser Black-backed and Great Black-backed Gulls provide reference points.

GAITS

Mainly walking and running. The footprints often turn slightly inwards.

Black-headed Gull
Stride length in walk: 6–13cm

Lesser Black-backed Gull
Stride length in walk: 17–45cm

DIFFERENTIATING BETWEEN DUCKS AND GULLS

Ducks and gulls are active in similar habitats and basically have similar foot morphology. Important features for differentiation:

Ducks

a) Print of a narrow skin flap on toe 2.
b) Toes 2 and 4 clearly bend towards toe 3.
c) Toe 1 is stronger, longer and often leaves a print.
d) Claws are often quite sharp and delicate in comparison.
e) Ducks tend to have shorter stride lengths with very turned-in tracks.

Gulls

a) No skin flap on toe 2.
b) Toes 2 and 4 less bent towards toe 3.
c) Toe 1 is finer, shorter and rarely leaves a print.
d) Claws often quite blunt and strong in comparison.
e) Gulls tend to have longer stride lengths and less turned-in tracks.

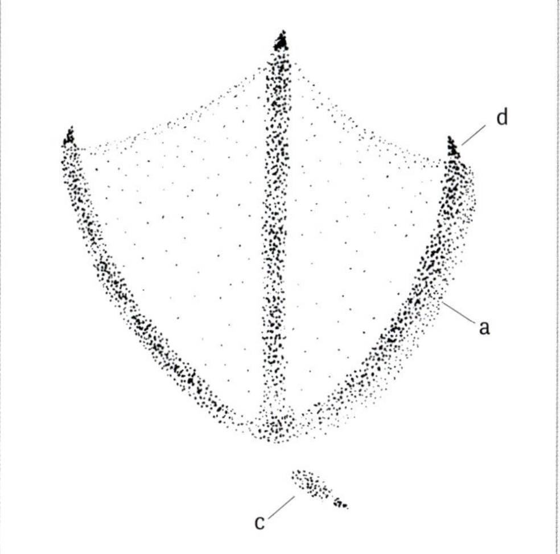

▲ Mallard.

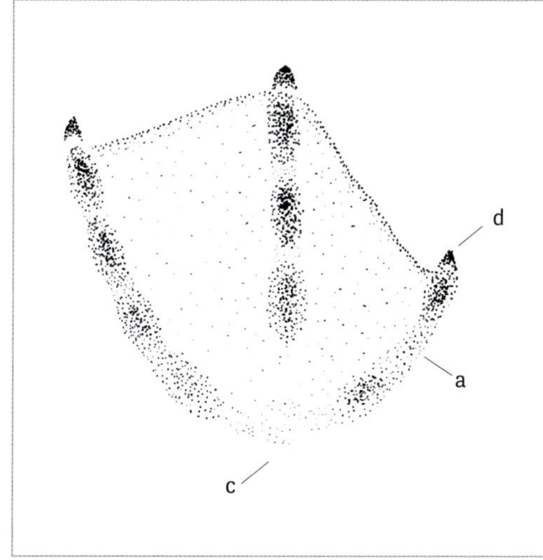

▲ Great Black-backed Gull.

TERNS
Sternidae

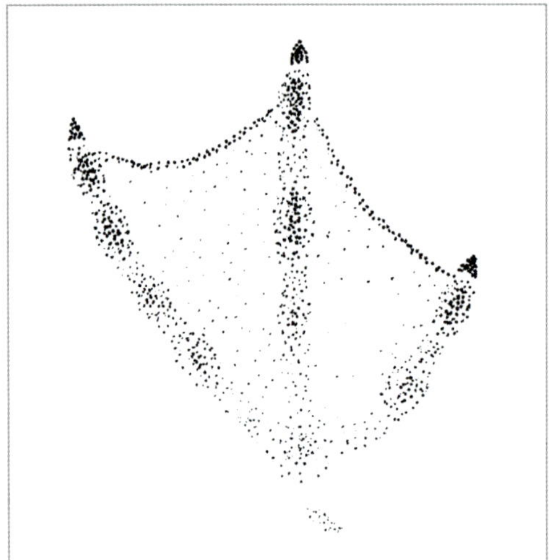

▲ *Tern, left.*

TRACK

L 3–4.8cm W 3.2–4.8cm

Unlike gulls, the foot is only slightly asymmetrical. The webbing connects distally to the tips of toes 2 and 4 and medially to toe 3. In clear prints, you can see that the front 1–1.5cm of toe 3 protrudes in front of the webbing. The claws appear relatively long. Apart from a few gull species, such as the Black-headed Gull and Little Gull, all gull tracks found in Europe are larger than the footprints of our largest terns.

TOTIPALMATE

FOOTPRINT WITH WEBBING BETWEEN TOES 1–4

PELICANS AND CORMORANTS

Pelecanidae, Phalacrocoracidae

The members of these families (we discuss the Great White Pelican and the Great Cormorant here) leave characteristic tracks but finding them is very rare. The Great White Pelican is rare in Europe and the Great Cormorant rarely spends time on the ground.

▲ *Great Cormorant, left.*
Vledder, Netherlands. René Nauta.

▲ *Great Cormorant, left.*

▲ *Great Cormorant, left.*
Vledder, Netherlands. René Nauta.

TRACK

Great White Pelican
L 14–17.5cm   W 9.5–11.5cm

Great Cormorant
L 10–14cm   W 6–8.5cm

Very asymmetrical. Webbed foot: toes 1–4 are connected
with distal webbing and have a relatively long, narrow
shape. Toe 4 is longer than toe 3. Long, pointed claws
protrude beyond the webbing. The tracks of the Common
Shag and Pygmy Cormorant are similar, but smaller or
much smaller. Cormorants and pelicans rarely move on foot
and leave few footprints on the ground.

GAITS

Ferguson *et al.* describe short stride lengths (rarely greater
than 25cm) and a slightly turned-in foot position (Ferguson
et al. 2005, p.67).

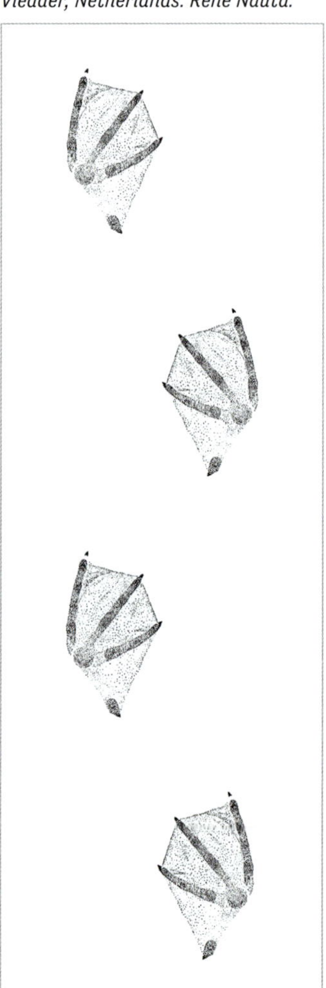

▶ *Walk.*

ZYGODACTYL

ZYGODACTYL FOOTPRINT

Bird families that leave zygodactyl footprints include ospreys, barn owls (Tytonidae), typical owls (Strigidae), cuckoos and woodpeckers. Because of limited data, only barn owls, typical owls and woodpeckers are discussed here.

OWLS

Tytonidae, Strigidae

Owls drink and bathe in puddles, so these are promising places for finding these otherwise rare tracks. We will be looking at tracks of Barn Owl (*Tyto alba*), Little Owl (*Athene noctua*), Long-eared Owl (*Asio otus*), Tawny Owl (*Strix aluco*) and Eurasian Eagle-owl (*Bubo bubo*).

TRACK

Little Owl	Long-eared Owl	Tawny Owl	Eurasian Eagle-owl
L 4.2–5.3cm	L 5.5–6.2cm	L 5.9–7.4cm	L 11–13.7cm
W 3–3.6cm	W 2.7–3.5cm	W 3.4–4.2cm	W 8–10cm

▼ *Barn Owl, right.*
Vledder, Netherlands. René Nauta.

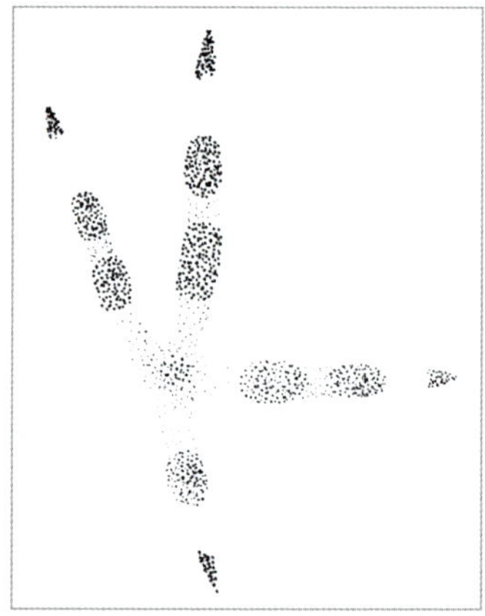

▼ *Tawny Owl, right.*

▼ *An impressive and extremely rare find: the massive right footprint of an Eurasian Eagle-owl. Hoher Fläming, Germany.*

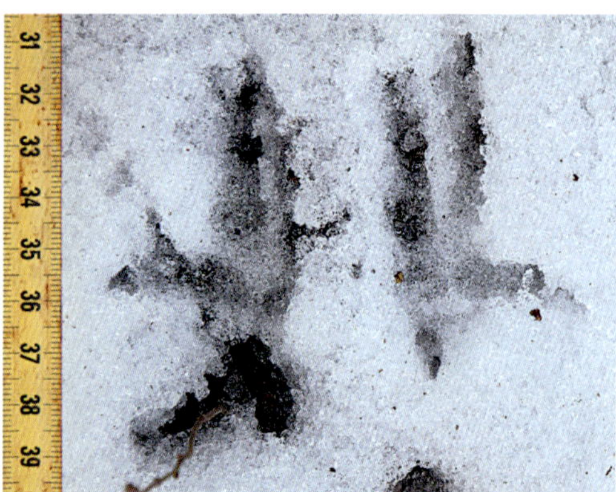

▲ *An owl was sitting here. The K-shape of the track is clearly recognisable. Märkische Schweiz, Germany.*

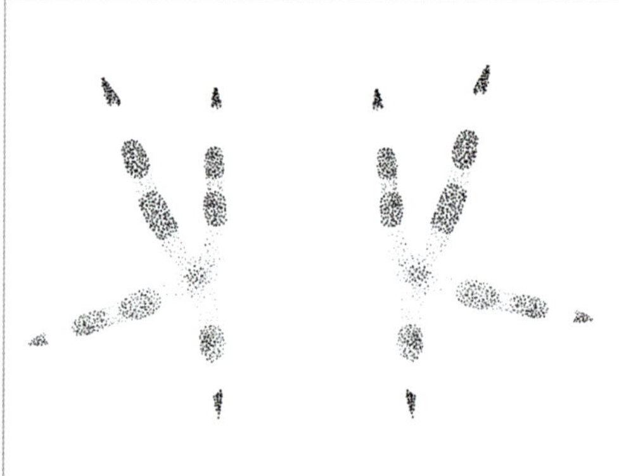

▲ *Tawny Owl, standing.*

Asymmetrical. Toes 1 and 4 point backwards, resulting in a distinctive X- or K-shaped track. Toes 1 and 2 often form a straight line on the inside of the footprint. Toe 3 is usually much longer than toe 2. Toe 4 is movable and can point almost straight backwards or out to the side at a 90° angle. It is usually longer than toe 1. The metatarsal region is pronounced, but usually only leaves a faint print. Long, strong claw marks that are usually clearly recognisable and separate from the toe prints. Barn Owl tracks tend to be slightly smaller (length up to 6.5cm) than those of the Tawny Owl. They have relatively slim toes and prominent pads.

GAITS

Mainly walking. Owls spend little time on the ground, so continuous track patterns are rare. Stride lengths appear short in relation to the size of the animal and trail widths quite large. The tracks are often slightly turned out.

WOODPECKERS
Picidae

Woodpecker tracks are extremely rare. Nevertheless, these birds can often be identified by a variety of conspicuous signs, such as nest holes, feeding marks and droppings (pages 692, 702 and 710). We take the Great Spotted Woodpecker (*Dendrocopos major*) as an example.

TRACK

Great Spotted Woodpecker
L 4.4–6.5cm   W 1.8–2.5cm

Asymmetrical. Toes 1 and 4 point backwards creating a distinctive X- or K-shaped track. Toes 1 and 2 are much shorter than toes 3 and 4. This is typical of all woodpeckers. The toes are more delicate than those of owls and the footprint appears elongated and slender in comparison. Green Woodpeckers land on the ground to feed from anthills. This gives the chance to find tracks. Black Woodpecker footprints can be found next to rotting fallen trees where woodpeckers search for food.

GAITS

As woodpeckers rarely spend time on the ground, prints are rare. I have yet to see a complete track pattern.

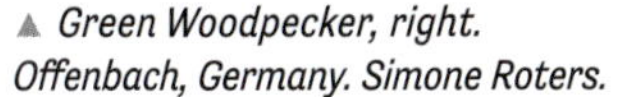

▲ *Green Woodpecker, right. Offenbach, Germany. Simone Roters.*

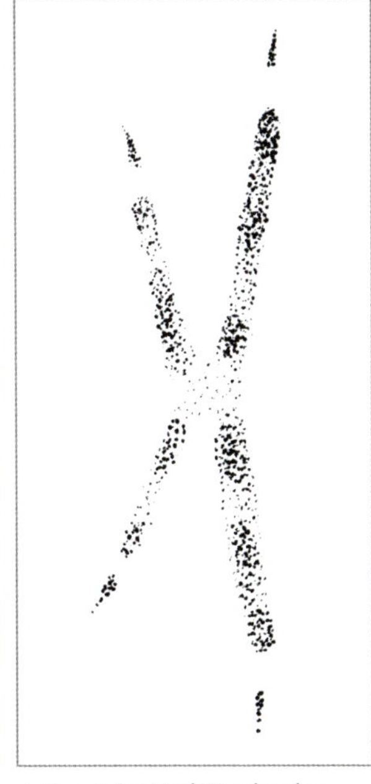

▲ *Great Spotted Woodpecker, right.*

SIGNS

Birds leave many different signs. Some of them are difficult to identify, while others provide clear clues to which species made them. Bird signs can provide information about a particular species even if no footprints are found, and they often tell us something about how birds live.

A dead standing pine with a hole in its trunk made by a Great Spotted Woodpecker, along with beak marks in the wood and a circle of bark at the foot of the tree are all important signs that tell a story. We will be discussing **nests**, **pellets**, **excrement**, **feeding marks** and **other signs** below. Specialist literature exists for identification of feathers and eggs (remember to never disturb an active nest).

NESTS

Many birds build nests to keep their eggs warm and to protect their offspring from predators. Nests come in a huge variety because of the different construction methods and because birds will use virtually any suitable material to build them. Nevertheless, we can identify a reasonably small number of different nest types. Nine common nest types with characteristic examples are described below.

SIMPLE, FLAT PLATFORMS MADE FROM TWIGS AND BRUSHWOOD

Pigeons and doves (Columbidae)

Pigeon and dove nests are usually simple-looking, flat platforms made of loosely put together twigs and brushwood. They are mainly found in trees, bushes or on rocks.

▲ *The makeshift-looking nest of a Common Woodpigeon. Midhurst, England.*

> ### Caution!
>
> Birds are protected during their breeding season. The European directive on the conservation of wild birds prohibits: 'deliberate destruction of, or damage to, their nests and eggs or removal of their nests' and 'deliberate disturbance of these birds particularly during the period of breeding and rearing.' Strict care was taken to abide by this directive and not to disturb the birds in the creation of any of the images shown here. Please make sure that you also follow these rules.

COVERED NESTS WITH A SIDE ENTRANCE

Eurasian Wren (*Troglodytes troglodytes*)

The Eurasian Wren's characteristic spherical nest is usually found close to the ground up to about two metres above ground in dense bushes, in gaps in the root plate of fallen trees or in other sheltered crevices, for example on banks or in old buildings. The nest has a side entrance, very thick walls and a roof. Eurasian Wrens use moss, dry leaves and small branches to build their nests. They line the inside with feathers. Eurasian Wren nests measure around 13–18cm in diameter, with an entrance hole measuring around 2.5cm in diameter.

CUP-SHAPED NESTS IN BUSHES OR TREES

Eurasian Blackbird (*Turdus merula*)

Eurasian Blackbirds build a cup-shaped nest in hedges, bushes and trees, usually around two metres above ground. They use thin twigs, roots, grasses, stalks, dry leaves and moss as building materials and damp earth and bits of plant inside to strengthen the nest. The diameter is approximately 15–18cm.

Song Thrush (*Turdus philomelos*)

A Song Thrush's nest is similar to a Eurasian Blackbird's but the inner cup is lined with mud, which can give the nest a neater appearance. The nest, which is usually close to the trunk, is often found between the forks of branches at a height of around 1.5–4m above the ground. The diameter is approximately 15–18cm.

▲ A spherical nest with a side entrance is typical of a Eurasian Wren. Vledder, Netherlands. René Nauta.

▲ You can differentiate between Eurasian Blackbird and Song Thrush nests because the latter lines the inside with mud. Bielefeld, Germany. Ulrike Quartier.

▲ Eurasian Blackbird nest without mud. Spessart, Germany.

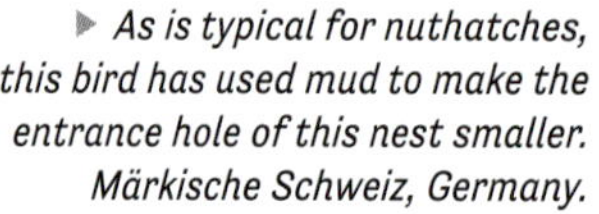

◀ Carrion Crows and some populations of Hooded Crows prefer to nest high up in broad-leaved trees. Gelnhausen, Germany.

▼ Breeding colonies like this one are typical of Rooks. Gelnhausen, Germany.

Rook (*Corvus frugilegus*)

Birds from the crow family build massive, cup-shaped nests from branches and twigs, lined inside with moss, hair, wool or grass. Rooks breed in colonies so several nests are usually found, often in the tops of tall deciduous trees.

CUP-SHAPED NESTS IN TREE HOLLOWS OR CREVICES

Nuthatch (*Sitta* spp.)

Nuthatches usually use existing tree hollows, knotholes or nesting boxes for breeding. They use mud to reduce the entrance hole down to the desired size. Nuthatches can take over Black Woodpecker holes. You can make out fine beak marks in the mud they use for their nests.

▶ As is typical for nuthatches, this bird has used mud to make the entrance hole of this nest smaller. Märkische Schweiz, Germany.

CUP-SHAPED STRUCTURE MADE OF CLAY PELLETS

Swallows and martins (Hirundinidae)

The swallow and martin family is known for its characteristic nests. They build roundish half shells from mud pellets, in sheltered locations high up on walls and under ledges. The entrance is usually located at the top edge. Classic sites are under bridges, on the roofs of stables and churches and on any suitable walls with projections that protect against rainfall.

One exception is the Sand Martin. During breeding season, it digs nesting tunnels in cliffs and steep banks. Classic sites are coastal areas or gravel pits with hard sandy walls, and tunnels may be re-used over several years.

NESTING HOLES IN EARTHY SLOPES

European Bee-eater(*Merops apiaster*)

European Bee-eaters make their nesting holes in dry, earthy slopes. They dig 1–3m deep, mostly horizontal tunnels into the ground. They often live in colonies.

▼ *This former Red-rumped Swallow's nest has been taken over by a Pallid Swift. The large entrance hole has been made by the swift. The entrance 'chimney' on the left is characteristic of a Red-rumped Swallow. Coto de Doñana National Park, Spain.*

▲ *The nesting burrow entrances of Sand Martins in the rock wall of a gravel pit. Märkische Schweiz, Germany.*

▼ *A participant in a Track&Sign evaluation discovered the entrance hole to this European Bee-eater nesting burrow. Coto de Doñana National Park, Spain.*

▼ *Insect remains are often found inside these tunnels. Coto de Doñana National Park, Spain.*

▶ *The slightly horizontal oval entrance hole of a Great Spotted Woodpecker.*

TREE HOLLOWS ABOVE GROUND

Woodpeckers (Picidae)

Members of the woodpecker family carve out hollows in tree trunks or thick branches where they sleep and nest. Due to their shape and size, the entrance holes can occasionally be used for species identification. Entrance holes to Lesser Spotted and Middle Spotted Woodpecker nests measure 3–3.5cm in diameter. The entrance holes of Syrian Woodpecker nests have a diameter of around 3.5–5cm. The Great Spotted Woodpecker usually builds its nest cavities in tree trunks above a height of 2.5m, with an entrance hole diameter of around 4–6cm. Green Woodpecker entrance holes are almost circular and measure around 6–7.5cm in diameter. The Black Woodpecker makes much larger, upright oval entrances measuring around 9–13cm × 7.5–10cm.

▶ *Greylag Goose nest. Vledder, Netherlands. René Nauta.*

▲ *A Mute Swan on its nest. Wissing, Germany. Raphael Fuchs.*

SHALLOW HOLLOWS AND SHAPELESS NEST MOUNDS NEAR WATER

Greylag Goose (*Anser anser*)

Geese scrape shallow hollows in an open location or use existing hollows that they line with down and small feathers. They may also use stalks and other nesting material from the surrounding area.

Mute Swan (*Cygnus olor*)

Large pile of immediately available plant material, often reeds and brushwood, near water. The hollow of the nest is lined with a small amount of down.

LARGE PLATE-SHAPED, INACCESSIBLE NESTS

Accipitriform birds of prey and storks build large, plate-shaped nests, usually in inaccessible locations high up in trees or on rocky ledges, but a few species, such as harriers, nest on the ground. They are built from branches, twigs and stalks. Birds often add leafy branches during the breeding season.

Nests of this type that have been used for years are often reinforced when seeds from the materials used germinate, so they can become very stable. The inside may be lined with animal hair or similar.

Storks (Ciconiidae)

Storks build their large, flat nests from twigs, grasses and earth. They prefer elevated locations near settlements, such as old chimneys, electricity pylons, trees and rocks. They often reuse nests, which can grow into extensive structures after use by several generations of storks.

▲ *Suitable White Stork nesting sites are often used by several generations. Coto de Doñana National Park, Spain.*

▲ *The nest of a Eurasian Goshawk. Offenbach, Germany. Simone Roters.*

EXCRETA

In birds, we distinguish between two types of excreta: pellets and droppings. Pellets can consist of various indigestible prey remains, such as chitin, fish scales, hair or bones. These 'gastric pellets' are regurgitated through the beak. Bird droppings are excreted through the cloaca and consist of a mixture of urine and excrement.

PELLETS

Owl pellets are a familiar sign for many naturalists. That birds such as corvids, gulls, herons, terns, kingfishers, nightjars and storks also produce pellets is less well known. Even small birds like thrushes, flycatchers, robins and wrens regurgitate pellets that are sometimes only the size of a grain of wheat. Many pellets are fragile and quickly disintegrate into their constituent parts. Carefully examining a pellet normally allows us to draw conclusions about what species it came from and can, for example, reveal what prey was eaten. Pellets can provide us with valuable information about other species in a habitat, such as which rodents are present and in what numbers.

Owl pellets are an excellent way of identifying both the bird and its diet, because parts of prey such as bones remain mostly intact, so it is often possible to identify the species and even the number of animals predated. For this reason, we discuss owl pellets in detail below. Most other pellets are more difficult to identify and are only listed with the characteristic features of different groups of birds.

OWL PELLETS

Owls normally regurgitate pellets before eating, to make room for the next meal. Depending on the amount of food ingested, they usually regurgitate one or two pellets per day. For precise identification, you should pay attention to the size, shape, colour, contents and condition of the pellet, as well as the location of the find. The owl pellets listed are arranged in ascending order of relative size.

Differentiating between owl pellets and mammal excrement

Owl pellets can be confused with mammal excrement. However, unlike mammal excrement, owl pellets do not contain any fruit remains and often mainly consist of fur and bones. Carnivore droppings usually have an earthy consistency and, in contrast to the odourless gastric pellets of owls, an unpleasant smell.

▼ *Pellets. From top left to bottom right: Eurasian Pygmy Owl, Little Owl, Northern Hawk-owl, Boreal Owl, Tawny Owl, Ural Owl, Barn Owl, Snowy Owl, Great Grey Owl, Eurasian Eagle-owl. Spessart, Germany.*

SIZE Pellet size is influenced by pronounced sexual dimorphism (females being larger), as well as the type and amount of food eaten. Correspondingly, pellets can vary greatly and size alone is not always sufficient for clear identification. Average dimensions for length are given below. The size of a pellet of a particular species will generally be within these values. Extreme variations in size are shown in brackets afterwards and illustrate the area of overlap with the pellets of other species. The diameter is listed as a range, with the mean value in brackets. Most of the owl pellets shown here come from the Eekholt Wildlife Park in Germany, where all the animals are given the same food. In the wild, the contents of pellets vary according to species, season and habitat. The proportions and shapes of the pellets are given as examples and are intended to provide a basis for differentiating between owl pellets.

Eurasian Pygmy Owl (*Glaucidium passerinum*)

Small. Rare. Lives in seclusion in quiet, older forests with a high proportion of conifers, old growth and dead wood. Depends on the presence of nesting cavities. Pellets can mainly be found in nesting holes. Pellets often contain the bones of thrush-sized and smaller birds, as well as insect, fish and small mammal remains. Heavily bitten bones of prey are characteristic (March 2007).

- L 2–2.8cm (1.5–4.5cm)
- D 1–1.5cm (1.2cm)

Little Owl (*Athene noctua*)

Small. Widespread. The Little Owl has adapted to cultivated landscapes, so its pellets can be found near villages. It nests in suitable hollows in trees and cavities in walls and rocks. It prefers open areas. The slender pellets contain many beetle remains, such as legs, elytra and wings, at times of year when insects are flying.

- L 2.5–3.8cm (1.5–6cm)
- D 1–1.8cm (1.3cm)

Boreal Owl (*Aegolius funereus*)

Small. Rare. Boreal Owls depend on stands of old wood with nesting cavities and adjoining open spaces where they can hunt. The pellets are more bulbous than those of the Little Owl and largely contain the remains of small mammals, especially mice. They almost never eat insects, which is another important difference to the Little Owl.

- L 2.5–3.8cm (1.8–6cm)
- D 1–2cm (1.5cm)

Northern Hawk-owl (*Surnia ulula*)

Medium. Rare. The Northern Hawk-owl is mainly found in the northern taiga and in mountain forests. It is dependent on cavities for nesting, with surrounding open areas where it can hunt. In summer, Northern Hawk-owls almost exclusively eat voles. In winter, the proportion of small mammals in their diet decreases and the number of birds they catch, such as Hazel Grouse and ptarmigans, increases significantly.

- L 3.5–4.8cm (3–7.5cm)
- D 1.5–3.5cm (2.2cm)

Barn Owl (*Tyto alba*)

Medium. Widespread. Barn Owls avoid colder regions and mountainous areas. They have adapted to humans and use farm buildings, attics and church towers as daytime resting or nesting places. Their pellets are relatively

▼ *Eurasian Pygmy Owl. The pellets are often noticeably small and relatively fine. Spessart, Germany.*

▼ *Little Owl. The pellets are often quite slim. Spessart, Germany.*

▲ *Boreal Owl. The pellets are usually more bulbous than Little Owl pellets. Spessart, Germany.*

▲ *A rare Northern Hawk-owl pellet. Spessart, Germany.*

common at these sites. They are different from other owl pellets because of their shiny, smooth coating. The gastric pellets are tightly packed and roundish with rounded, blunt ends. Fresh pellets have a smooth, black surface and almost look as if they have been painted. Older pellets are grey in colour. Barn Owls mainly prey on voles and shrews. Other owl species generally avoid shrews or only eat them in small quantities. Other birds and amphibians also only make up a small part of their diet. When there is a surplus of prey, Barn Owls create food stores, which are also found at the breeding site along with the pellets. A store can contain more than 60 prey items.

- L 3.5–4.5cm (2.5–8cm)
- D 1.8–3.5cm (2.6cm)

Typical location: churches, farms, near villages

Short-eared Owl (*Asio flammeus*)

Medium. Rare. Apart from northern Great Britain, Norway, Sweden and Finland, there are only small populations in Europe. The Short-eared Owl is heavily dependent on vole populations and lives in low-lying, damp and open areas with plenty of cover. These include wet meadows and floodplains, moors and coastal dune landscapes. The pellets are usually slightly longer and much slimmer than Barn Owl pellets. They can be found under suitable perches, such as fence posts, within the habitat.

- L 3.5–5.5cm (3–8.5cm)
- D 1.5–3.5cm (2.2cm)

Tawny Owl (*Strix aluco*)

Medium. Widespread. The most common owl species in Europe. The highly adaptable Tawny Owl can be found in a variety of habitats. It prefers deciduous and mixed forests but can also be found in gardens and urban parks. Nevertheless, Tawny Owl pellets are more difficult to find than Barn Owl pellets, for example, as Tawny Owls often vary their daytime resting places and their individual gastric pellets are usually scattered around a wider area. The pellets usually contain the remains of small mammals, birds, frogs and toads, as well as small twigs, blades of grass and pine needles, so they are often irregularly shaped. Tawny Owls can compensate for a scarcity of rodents as a food source by eating more birds.

- L 3.5–5.5cm (3–8cm)
- D 1.5–3cm (2.4cm)

Typical location: forests and parks

▼ *A dark and shiny coating is characteristic of Barn Owls. Spessart, Germany.*

▼ *Short-eared Owl pellets are usually more elongated than Barn Owl pellets. Spessart, Germany.*

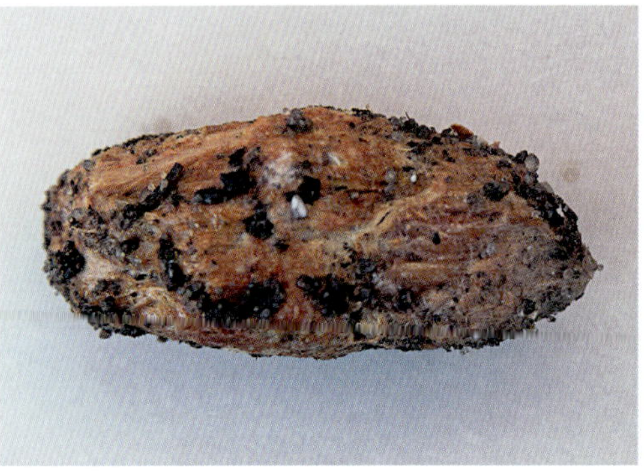

▲ *Tawny Owls frequently change their daytime resting places, so individual pellets can usually be found scattered around an area. Spessart, Germany.*

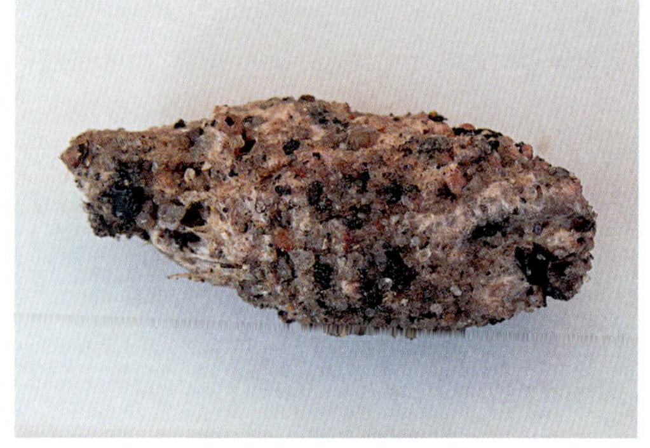

▲ *Ural Owl pellets are usually slightly larger than Tawny Owl pellets. Spessart, Germany.*

▲ *Great Grey Owl pellets mainly contain small mammal remains. Spessart, Germany.*

▲ *Many gastric pellets around the nest of a Snowy Owl are a characteristic sign. Spessart, Germany.*

Ural Owl (*Strix uralensis*)

Large. Rare. The main range is limited to the north-east of Europe. As with Tawny Owls, finding large amounts of pellets is rare, as Ural Owls do not have fixed daytime resting places. The gastric pellets of Ural Owls tend to be slightly larger than those of Tawny Owls. Large Tawny Owl pellets are indistinguishable from Ural Owl pellets.

- L 5.5–6.5cm (3.5–10cm)
- D 1.8–3cm (2.5cm)

Great Grey Owl (*Strix nebulosa*)

Large. Rare. The main range is limited to the coniferous forest taiga of north-eastern Europe. The pellets mainly contain the remains of small mammals. Despite its size, the Great Grey Owl mainly catches and eats vole-sized prey. Birds make up only a very small part of its prey. Outside the breeding season, the number of shrews eaten increases significantly. Barn Owls show similar behaviour.

- L 6–7cm (3.5–11cm)
- D 2–3.8cm (2.9cm)

Snowy Owl (*Bubo scandiacus*)

Very large. Rare. Only found in Sweden, Norway and the north of Finland, where it inhabits mountains or plateaus (Fjell) above the coniferous tree line and the Arctic tundra. The Snowy Owl's main prey are lemmings, so its breeding is closely linked to fluctuations in the lemming population. Snowy Owls lay their eggs in a simple dug out hollow in the ground, around which large quantities of droppings and pellets can often be found. The droppings and pellets fertilise the ground, resulting in a distinctive area of lusher vegetation around the nest. Snowy Owls repeatedly return to favoured spots on the ground to dismember their prey. Mass finds of gastric pellets in these places are characteristic.

- L 7–8cm (4.5–15cm)
- D 2–4cm (2.7cm)

Eurasian Eagle-owl (*Bubo bubo*)

Very large. Rare. Lives in a wide variety of habitats, preferring structured landscapes and rocky terrain for its nesting cavities. The Eurasian Eagle-owl also breeds in quarries or human settlements. It is the largest owl species in the world and complete pellets are usually difficult to confuse with other owl pellets due to their size. White-tailed Eagle gastric pellets are similar in size, but Eurasian Eagle-owl pellets often contain relatively large bones. As the Eurasian Eagle-owl is an opportunistic feeder, its pellets contain the remains of prey that are common in its territory. A significant proportion of prey generally consists of mice and rats. Eurasian Eagle-owls also prey on voles, rabbits, hares and partridges. Fish and seabirds form part of its diet in coastal areas. Eurasian Eagle-owls also eat diurnal birds of prey and other owls. In exceptional cases, entire upper and lower legs of Brown Hare can be found in Eurasian Eagle-owl pellets. Outside the breeding season, their diet mainly consists of smaller prey, especially mice and voles.

- L 7–8.5cm (4.5–16cm)
- D 2–4.5cm (3.4cm)

Typical location: rock walls.

RAPTOR PELLETS

The digestive systems of raptors involve stronger acids than those of owls, so the bones they ingest are usually completely dissolved or at least half digested. Food remains are generally difficult or impossible to identify. It is often only possible to identify a matted mass of mouse hair or the remains

▼ *Because of their size, whole Eurasian Eagle-owl pellets are unlikely to be confused with other owl pellets. This pellet is fresh. Spessart, Germany.*

▼ *Eurasian Eagle-owl pellets are an excellent source of small mammal skulls. This is an old pellet. Märkische Schweiz, Germany.*

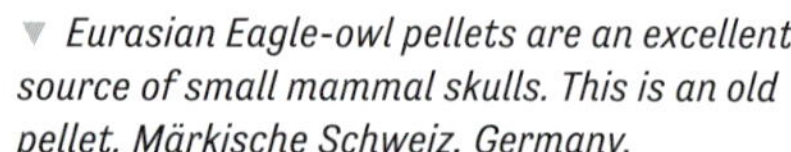

▲ *A gull pellet. Wangerooge, Germany. Ulrike Quartier.*

▲ *Common Raven pellet. The loose egg shape is typical of corvids. Midhurst, England.*

▼ A Grey Heron pellet. Large, dense pellets consisting of small mammal fur and feathers are typical of herons. Maintal, Germany. Simone Roters.

▶ Common Kingfisher regurgitating a pellet. Netherlands. Jeroen Kloppenburg.

of feathers. The location of the find and the relative size can sometimes provide sufficient information for species identification.

GULL PELLETS

Many gulls have opportunistic feeding habits, so the content, shape and colour of their pellets vary greatly. If the gastric pellets mainly consist of insect, mussel, fish, crab or snail remains, they are usually loose and fall apart easily. If they consist mainly of plant food such as husks and grains, the pellets are usually firmer. Pellets are often short and cylindrical or almost spherical in shape and typically yellowish, yellow-brown, white-grey or grey in colour.

CORVID PELLETS

Yellowish or yellowish-brown pellets that contain a variety of elements. Seeds, grass husks, insect remains, feathers and bones are typical. They are characterised by a high proportion of plant remains and a loose egg shape. It is not always possible to reliably distinguish the pellets of different corvids from one another. The location of the find and the relative size can be significant.

HERON PELLETS

These usually tightly compressed pellets mainly consist of small mammal or bird remains. Herons usually completely digest fish remains. Though they sometimes contain small pieces of bone, heron pellets mainly consist of fur and feathers. They are typically found in locations such as riverbanks, fields and under nesting trees.

COMMON KINGFISHER PELLETS

The whitish pellets consist almost exclusively of tiny fish remains. They are between 1–2cm long and around 1cm wide. As they usually fall into the water after regurgitation and quickly disintegrate into their components, they are rarely found. If they are found, then typical locations are shore and bank areas, on slightly elevated places such as stones.

EXCREMENT

Birds excrete urine and droppings together via their cloaca. Because they are excreted together, a white urine coating or a crusty white cap can usually be seen on their droppings. This feature is common to bird and reptile droppings. Below, we differentiate between five main groups of bird droppings and give average measurements for the most common species.

INSECTIVOROUS BIRDS SUCH AS WOODPECKERS, SWALLOWS AND MARTINS, AND WAGTAILS

Woodpecker droppings mainly consist of the indigestible chitin remains of insects. These distinctive droppings are cylindrical or J-shaped and covered with a white membrane.

Green Woodpecker droppings, in particular, contain almost exclusively ant remains and can usually be found near ant colonies. Rotten, dead and lying tree trunks with scattered wood debris and woodpecker strike marks are a classic sign of a Green Woodpecker looking for food. If you look closely, you will often find the bird's droppings in these places. Other insectivorous birds such as **House Martins** and **wagtails** leave smaller, roundish, hard balls of excrement, large quantities of which can often be found around the nest. The colour normally varies between grey, grey-black and black.

HERBIVOROUS WILDFOWL AND GAMEBIRDS

These elongated droppings consist almost exclusively of plant fibres and often have a correspondingly irregular, coarse surface. There is usually a white urine cap at one of the two ends, which distinguishes it from mammal excrement.

MALLARDS, GEESE AND SWANS These birds leave relatively thick, cylindrical droppings with blunt ends. They mainly consist of plant remains but may also contain the remains of invertebrates. Wildfowl also occasionally pass shapeless, mushy, cowpat-like droppings. In both cases, the colour can be green, green-grey, brown or dark grey.

▲ *Green Woodpecker droppings are often J-shaped and covered with a white membrane. Märkische Schweiz, Germany.*

▲ *Opening them up reveals that Green Woodpecker droppings consist almost exclusively of ant remains. Mecklenburgische Seenplatte, Germany.*

◀ *Family of Canada Geese, Jetzendorf, Germany.*

▲ *Excrement of Mallard (top), Greylag Goose (centre) and Mute Swan (bottom). Unlikely to be confused because of their size, but less easy without direct comparison. Germany. Aaron Tiedemann.*

▲ *Geese repeatedly return to their favourite resting places. This can lead to conspicuously trampled areas with piles of droppings and feathers. Lausitz, Germany.*

Mallard

- L 4–6cm   D 0.7–1.2cm

Greylag Goose

- L 4–8.5cm   D 0.8–1.7cm

Mute Swan

- L up to 18cm   D 1.6–2.4cm

GAMEBIRDS produce two types of excrement. The first is soft, thick, mushy caecal pellets, which are relatively rare. They are often only excreted once a day, usually in the morning. The colour varies from olive to dark brown to black. The second type of excrement is usually slightly curved (C-shaped), cylindrical and odourless. It usually consists of tightly packed plant material. A white urine cap can often be seen at one end of fresh droppings. The colour varies from yellowish, yellowish-brown to brown. If the bird has eaten berries, its droppings usually take on their colour. If the diet consists of large quantities of herbaceous ground plants, even the cylindrical droppings can become mushy and shapeless. During the breeding season, hens can excrete larger, mostly fibrous, solid, bulbous droppings. These distinctive droppings can be interpreted as proof that the birds are breeding. Large quantities of the more solid, C-shaped droppings can pile up at nesting and resting sites and under roosting trees. Depending on the species, the number of droppings can provide information about how long the bird or birds stayed there. In winter, Black Grouse defecate approximately every ten minutes and Western Capercaillie approximately every 12 minutes.

GROUSE In winter and spring, **Hazel Grouse droppings** often consist of catkins from birch, alder and hazel trees. This gives the droppings a relatively yellowish, dense and fine appearance. In spring, **Black Grouse droppings** consist mainly of birch buds. **Western Capercaillie droppings** look similar to Black Grouse droppings but are much larger. In winter, Western Capercaillie droppings consist almost exclusively of pine needles.

Hazel Grouse
- L 1.2–2.6cm   D 0.5–0.7cm

Ptarmigan droppings are comparable in size but tend to be larger.

Black Grouse
- L 2–3.8cm   D 0.6–1cm

Western Capercaillie
- L 4–7.5cm   D 0.7–1.4cm

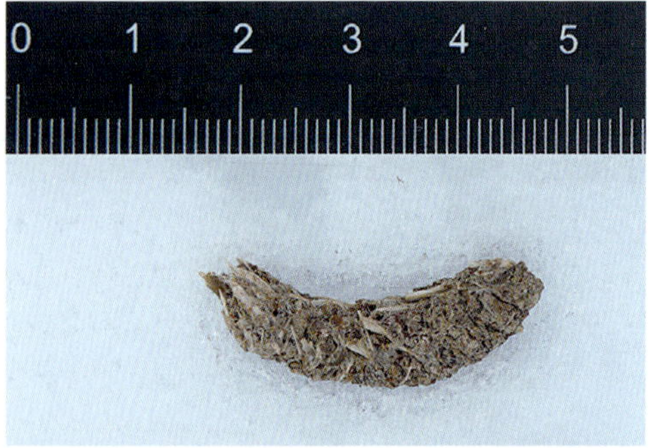

It is often possible to determine the sex of adult Western Capercaillie based on the size of the cylindrical droppings. A diameter of 0.9cm and smaller can be attributed to the hen. A diameter over 1cm comes from the cock (Gjerde 1990).

◄ *Ptarmigan (above) and Western Capercaillie (below) droppings. Aaron Tiedemann.*

▶ *The mushy caecal dropping of a Common Pheasant. Midhurst, England. John Rhyder.*

▶ *The round excrement of a Domestic Chicken, consisting of several cylindrical droppings. Extertal, Germany.*

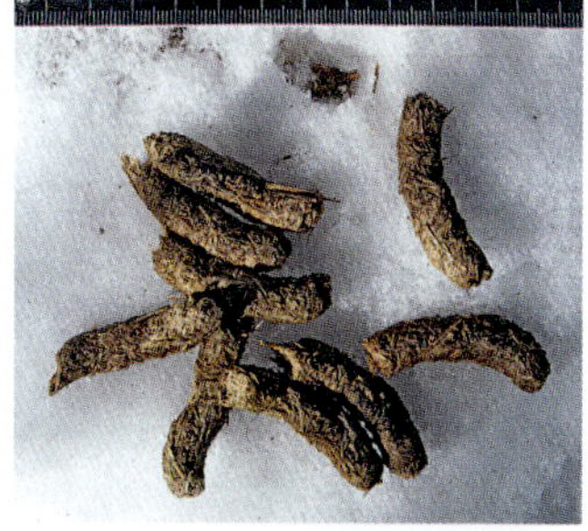

◀ *Large piles of C-shaped, cylindrical Western Capercaillie droppings can be found under roosting trees. Jämtland, Sweden. Laura Gärtner.*

▲ *Common Pheasant droppings can be deposited in a C-shape or in thicker clumps but are generally more irregularly shaped than grouse droppings. Midhurst, England. John Rhyder.*

▶ *The Western Capercaillie's winter droppings usually consist largely of pine needles. They are very large with a rough surface. Värmland, Sweden.*

▼ *Red-legged Partridge droppings consist of a large ball of several thin, cylindrical droppings, with a white urine cap at one end. Coto de Doñana National Park, Spain.*

PHEASANTS AND PARTRIDGES

Grey Partridge droppings are typical of the droppings of birds from this family. In contrast to the C-shape of grouse droppings, Grey Partridge droppings consist of a spherical collection of several thin cylindrical droppings, each with a pointed end.

The part that is excreted first has a more rounded end and usually a white urine cap, while the rest of the excrement is dark in colour. Grey Partridge droppings are like Domestic Chicken droppings, but much smaller. **Common Pheasant droppings** are similar in shape, but about twice the

size of Grey Partridge droppings. Unlike grouse excrement, partridge excrement is largely made up of seed and grass remains. Needle remains are comparatively rare. The droppings have an even composition and shape. The habitat where the droppings were found is also an important distinguishing feature.

Common Pheasant
- L 1.5–2cm D 0.4–0.5cm

The droppings of Common Quail, Partridge, Red Grouse and Grey Partridge are like those of Common Pheasants but only about half as long and slightly thinner.

SEED- AND FRUIT-EATING BIRDS SUCH AS THRUSHES, FINCHES AND PIGEONS

Seedeaters' droppings typically consist of seeds and fruit. They are usually semi-solid, relatively small and contain many undigested seeds. The colour depends on the contents and is often light grey, grey-blue, purple, red or black. If a bird has eaten wild fruits, such as elderberries, droppings will be almost completely dyed this colour. **Eurasian Blackbirds**, which occasionally mainly feed on elderberries, pass a typical blackish, purple-coloured blob of droppings with some solid particles. **Pigeons and doves** often produce thin, tubular droppings coiled into a roundish pile. The white urine spot is often visible in the centre.

▶ *Common Woodpigeons produce circular droppings. Hoher Fläming, Germany. Paul Wernicke.*

▶ *Eurasian Blackbird droppings vary greatly depending on the food consumed. Vledder, Netherlands. René Nauta.*

▼ *Eurasian Blackbird excrement where mainly seeds have been eaten. Vledder, Netherlands. René Nauta.*

▼ *Large quantities of pigeon droppings can accumulate under roosting trees. Midhurst, England.*

▲ *Eurasian Jay droppings vary greatly depending on the food eaten. These droppings show that mainly seeds have been eaten. Spessart, Germany.*

▶ *White Stork droppings are mainly runny and shapeless. Coto de Doñana National Park, Spain.*

▼ *Droppings, tracks and pellets of a gull. Wangerooge, Germany. Ulrike Quartier.*

◀ *This Common Buzzard excrement was sprayed from left to right. Spessart, Germany. Laura Gärtner.*

HERONS, CORMORANTS AND OTHER FISH-EATING BIRDS

When birds feed mainly on fish, they excrete shapeless, whitish-grey blobs that smell of fish. The nesting and roosting trees of herons and cormorants are often spotted white with droppings. The high nutrient input that comes from the excrement of a cormorant colony can cause these trees to gradually die.

BIRDS OF PREY AND CARRION-EATERS

Raptors produce liquid droppings that they excrete in a stream from their nest or resting place. Conspicuous white blobs often collect in the branches and on the ground under the nest. Conspicuous piles of owl droppings can form under popular perches. Unlike raptors, owls do not pass their droppings as a powerful stream, so it is often possible to tell the difference.

FEEDING MARKS

Birds leave a wide variety of feeding marks on fruits and berries, nuts and seeds, as well as on cones, fungi, mussels, snails and other shell animals, and on trees, feathers, carcasses and bones. Birds use their beaks to peck, poke, bore and scratch. This often leaves visible marks. To precisely identify a feeding mark, it is important to distinguish whether the food remains show tooth marks from a mammal or beak marks from a bird, as confusion is possible. Some signs, such as woodpecker feeding marks, are very distinctive and clearly identifiable. Examples of common, distinctive or easily identifiable bird feeding mark types are shown and possible locations are listed below.

SIGNS ON FRUIT AND BERRIES

Many birds eat berries and fruits such as sloes and rose hips, apples, pears and plums. It is rarely possible to identify exact species solely based on feeding marks on fruit. Birds like thrushes, jays, and pigeons and doves generally eat the flesh of larger fruits, often leaving behind only the skin or a 'shell' hollowed out by their beaks. Seedeaters like finches or crossbills peck holes in the flesh of the fruit to reach the seeds.

SIGNS ON NUTS AND SEEDS

Birds like Eurasian Nuthatches, Common Magpies, Jackdaws, jays, woodpeckers, Hawfinches and other finches, and Great Tits feed on seeds and nuts, especially in autumn and winter. Although it is often not possible

▲ *A Eurasian Jay has been feeding on these fallen apples. Spessart, Germany.*

◀ *The many strong beak marks suggest the work of a Eurasian Jay. Lausitz, Germany.*

to identify the species from the feeding marks, the size of the beak marks and the way the beak has been used can occasionally provide a clue. Strong triangular beak marks on hazelnuts, walnuts and acorns are a sign of larger birds, such as Eurasian Jays, whereas neatly arranged small holes are more indicative of smaller birds, like tits.

Woodpeckers and nuthatches wedge hazelnuts into crevices in bark, and then hammer them open with their beaks. Nuthatches tend to make holes with smaller beak marks than those made by woodpeckers.

Remains of smaller nuts and seeds, such as beechnuts or sunflower seeds, are not useful for species identification. You can sometimes find large piles of leftover shells, together with lots of small, white blobs of droppings in and under popular perches and resting places, as well as in trees and bushes.

SIGNS ON CONES

Woodpeckers (*Dendrocopos*)

Woodpeckers of the genus *Dendrocopos* wedge cones into crevices in bark, rocks or tree stumps to eat the conifer seeds. This sign is known as a 'woodpecker anvil' and can be very distinctive. Woodpeckers wedge the cones into a suitable gap with the point facing up and then peck apart the scales with their beak. After eating the seeds on one side, the bird pulls out the cone, turns it around, secures it again and then pecks apart the scales on the other side. It repeats this process until most of the seeds have been eaten. This gives the cones a characteristically dishevelled appearance, with the scales sticking out on different sides

◄ *A woodpecker or nuthatch has repeatedly used the deep crevices in the bark of this oak as somewhere to open hazelnuts. Märkische Schweiz, Germany.*

◄ *Tits have opened these sunflower seeds with their beak to eat the seeds. Spessart, Germany.*

but still intact and the bottom part of the cone hardly touched. Most of the scales are still attached to the axis. This distinguishes them from cones eaten by mice and squirrels (page 164). Large piles of these cones can build up under woodpecker anvils.

Crossbills

Cones eaten by crossbills (*Loxia* spp.) look less dishevelled than ones eaten by woodpeckers. While woodpeckers peck at the cone with their chisel-like beak, crossbills tear open individual cone scales along the length of the cone. This creates a neat longitudinal section with the individual cone scales often divided in the centre. The specialised beak shape that gives crossbills their name makes this possible. Their long, thin upper and lower mandibles cross over at the tips. The bird tilts its head to the side and wedges this special tool between the cone scales. It then powerfully moves its lower jaw sideways to lift the scale, allowing it to grab the seed beneath with its tongue. The characteristic central longitudinal incision is created when

▼ *Great Spotted Woodpeckers leave cones with a characteristically ruffled appearance. Spessart, Germany.*

▲ *A broken pine trunk has been used as a woodpecker anvil. The cones are scattered on the ground. Bieszczady, Poland.*

▼ *Close-up of a Woodpecker anvil. Bieszczady, Poland.*

the bird closes or withdraws its beak. The very thick cone scales of pine cones are an exception because they usually remain intact.

SIGNS ON FUNGI

Various bird species, like **jays**, **starlings** and **thrushes**, eat the flesh of fungi. They often also eat invertebrates living in the fungi. Conspicuous triangular beak marks on the top of the fungus are a clear distinguishing feature from fungi eaten by mammals or snails.

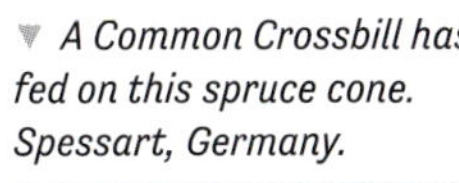

▼ *A Great Spotted Woodpecker has fed on this spruce cone. Spessart, Germany.*

▼ *A Common Crossbill has fed on this spruce cone. Spessart, Germany.*

◄ *Note the conspicuous beak marks on the top of the mushroom. Hunsrück, Germany. Simone Roters.*

SIGNS ON MUSSELS, SNAILS AND OTHER SHELL ANIMALS

Eurasian Oystercatchers mainly eat cockles, mussels, crabs, limpets, winkles and whelks. They wedge softer shells firmly in place with their feet and strike the thinnest part with their beak. The birds open harder shells by repeatedly dropping them onto hard surfaces from a height of several metres. Another technique is to push the tip of their beak into a slightly open mussel shell to cut through the posterior sphincter. They then open the shell by gently twisting and opening their beak. Oystercatchers leave conspicuous piles of shells. **Eiders** and **shelducks** also feed on mussels and other shellfish. Fragments of shells can be found in their habitat. They occur in the Wadden Sea and other coastal areas.

Song Thrushes open the shells of snails (Helicidae) to get at the snail inside. They use a hard, usually slightly raised surface for this, known as a 'thrush anvil'. This is generally a suitable stone, but they may also use the rails of railway tracks or similar. The thrush holds the edge of the shell in its beak and strikes it on the anvil. Large amounts of shell fragments are often scattered around the anvil. Anvils are typically found on the edges of woodland or roads.

SIGNS ON TREES

In winter, **Western Capercaillies** mainly eat the needles and buds of pines, firs and spruces. They often repeatedly feed from

▲ *A classic thrush anvil. Lausitz, Germany.*

▲ *Participants on the one-year wildlife tracking course examine shell fragments from this thrush anvil. Märkische Schweiz, Germany.*

certain trees, resulting in clear signs of browsing. Trees that appear heavily thinned by browsing are a characteristic sign.

Great Spotted Woodpeckers leave well-known and distinctive, as well as less well-known and subtle signs on trees. Conspicuous and common signs include standing dead conifers, usually pines, that have been stripped of their bark. Woodpeckers search for insect larvae living underneath. They remove the bark using a combination of striking, pecking and levering with their beak. A circular pile of bark can often be seen on the ground around the trunk. The wood shavings are normally much finer than those left by the Black Woodpecker.

The often recognisable holes in the wood are caused by the Great Spotted Woodpecker striking with its chisel-like beak to bring out hidden insect larvae. To extract them, it uses its long, sticky tongue which is covered with short barbs.

▶ *Preferred feeding trees of the Western Capercaillie can have a heavily thinned appearance. Sweden. Laura Gärtner.*

▲ The Western Capercaillie breaks off buds and branches with its powerful beak. Sweden. Laura Gärtner

▶ A woodpecker has searched this rotting tree trunk for insect larvae. Spessart, Germany.

Examining exposed deadwood more closely can reveal further signs. A special type of beak mark is created by pecking sideways between the bark and the wood: a fine, noticeably straight and parallel double line in the deadwood, running at a right angle to the grain. The two lines are made by the upper and lower beak.

Claw marks from climbing can also be found on these trees. Claws make simple scratch marks that are usually less straight than beak marks.

Rotting, fallen trees also provide a good food source for woodpeckers. The combination of pecked holes and wood shavings scattered in many directions makes it possible to distinguish them from tree trunks that a European Badger or Brown Bear has torn open with its claws (page 158). Woodpecker droppings are often found at

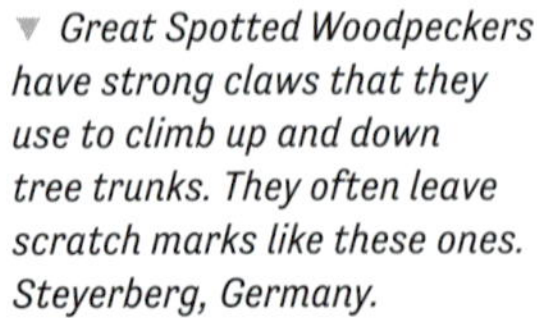

◀ *Debarked, dead standing conifers like this pine tree are a clear sign of the work of a Great Spotted Woodpecker. Spessart, Germany.*

▼ *Great Spotted Woodpeckers have strong claws that they use to climb up and down tree trunks. They often leave scratch marks like these ones. Steyerberg, Germany.*

▲ *These straight, parallel double lines running across the grain are the beak marks of a Great Spotted Woodpecker. Steyerberg, Germany.*

▲ *The holes made by the Great Spotted Woodpecker are much smaller than those made by the Black Woodpecker. Märkische Schweiz, Germany.*

such feeding sites and can be used as further clues for species identification.

Europe's largest woodpecker, the **Black Woodpecker** (*Dryocopus martius*), feeds mainly on wood-dwelling ants during the summer months, such as carpenter ants (*Camponotus*), which build their nest chamber systems in dead or living trees. The Black Woodpecker can locate ants by sound and pecks deep, very large, rectangular holes with its chisel-like, powerful beak to reach its prey. These holes can be up to 50cm long and 10–20cm wide. Large shavings several centimetres wide usually pile up on the ground under these trees.

RINGING is a method used mainly in spring by the **Great Spotted Woodpecker**, **Middle Spotted Woodpecker** and **Eurasian Three-toed Woodpecker** to get at nutritious tree sap. The woodpecker pecks holes in the bark, which then fill with tree sap. When made on

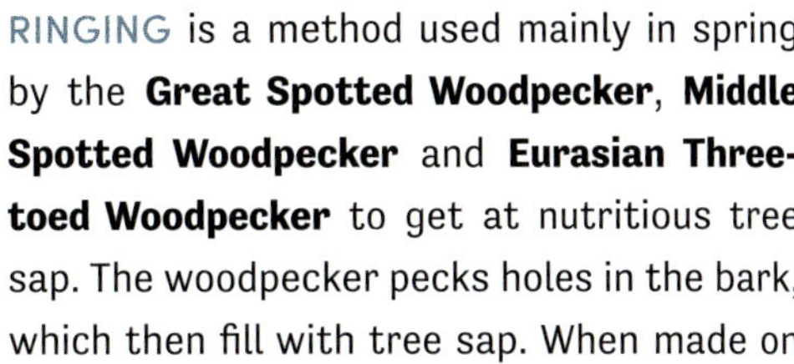
▼ *Pine trees where Black Woodpeckers have fed can become conspicuous features in the landscape. Spessart, Germany.*

▼ *Roundish, usually closely spaced strike marks like these come from a Great Spotted Woodpecker. Midhurst, England.*

◄ *John Rhyder explains Great Spotted Woodpecker ringing during a Cybertracker evaluation. Midhurst, England.*

▲ *A Black Woodpecker has exposed the nest chamber system of these wood-dwelling ants. Bieszczady, Poland.*

branches, these holes are usually in a row, while they often run horizontally around the trunk like a ring.

FEATHERS FROM TEARING OR PLUCKING

When we find a feather, we often wonder where it came from and how it got there. Was it removed by a bird or animal or did it just fall out? Birds lose individual undamaged feathers for various reasons, for example during moulting. Larger quantities of feathers usually indicate that a predator has caught and plucked the bird. If a mammalian predator kills a bird, it will bite off smaller feathers at the quills and pull out larger feathers with its teeth, usually causing severe damage. Tufts of feathers can often be found stuck together by saliva. Fox-sized and larger predators often damage the plumes as well.

A raptor will pluck feathers out of avian prey with its beak. You will often see a beak mark, a notch, a kink or even a break in the quill.

CARCASSES AND BONES

It is not easy to identify which species killed an animal by looking at carcasses and bones. This is often only possible by considering additional information, such as type of prey and habitat. **Owls** usually swallow their prey whole or tear larger animals into pieces. They

▼ *Remains of a Great Spotted Woodpecker killed by a Red Fox. Märkische Schweiz, Germany.*

◀ *Feathers of different bird species. Note the beak marks roughly in the centre of the quill. Offenbach, Germany. Simone Roters.*

▲ *Bitten off quills and tufts of feathers stuck together by saliva are signs of a kill by a mammal. Märkische Schweiz, Germany.*

only pluck out a few feathers and hardly leave behind any remains. Plucking feathers and decapitating their prey is typical of **raptors**, but Red Foxes are also known to do this. Of the remaining body, they usually eat the breast first.

Raptors usually carry their prey to a raised location, such as a tree stump or fence post, to pluck it. Along with the plucked feathers, the beak, legs and wings are often left behind. Prey remains left by a carnivorous mammal are normally more widely scattered and usually show chewed bones with tooth marks.

Eurasian Sparrowhawks and **Eurasian Goshawks** spend most of their time in the forest. Eurasian Sparrowhawks mainly hunt tree-dwelling, small to dove-sized birds, while Eurasian Goshawks often take pigeons and ducks. The plucking sites are very similar, usually under cover and elevated, such as on

a tree stump or mound of earth. Broken bones are left lying around. Signs of beak marks may occasionally be recognisable on the bones.

Common Buzzards usually hunt at the edges of woodland and in open areas such as fields and meadows. Their prey mainly consists of rodents such as mice and squirrels, but they can also catch rabbits, hares, birds, amphibians and reptiles. It is always important to take regional and seasonal features into account. Seasonal fluctuations can occur and in harsh conditions Common Buzzards tend to hunt birds rather than small mammals. Common Buzzards pull the fur off smaller mammals up to the size of squirrels and tear rabbits and hares to shreds. They eat bones whole or break them into pieces. Common Buzzards open up the belly of toads with their beaks and then eat them from the inside out, leaving behind the skin and entrails.

▲ *Raptorsoften eat the head and breast meat of their prey first. Vledder, Netherlands. René Nauta.*

▲ *A Common Buzzard has killed this mole and placed it on a straw bale but was startled shortly afterwards. Offenbach, Germany. Simone Roters.*

Some **shrikes** (Laniidae) store food. They do this by securing their prey, such as mice and small birds, between forks of branches or by impaling insects, amphibians and reptiles on thorns or barbed wire. Classic locations for these shrike 'larders' are the edges of woodland and roadsides with thorn bushes or barbed wire.

The 'bones' of dead **cuttlefish**, known as cuttlebones, are a favourite beak whetstone for many birds and a source of calcium. The triangular beak marks of birds are very different from the toothmarks of small rodents. Cuttlebones are typically found on beaches.

OTHER SIGNS

BEAK MARKS IN THE GROUND

Many waders have long beaks that they use to search for food in loose ground. Depending on the species and beak shape, they can leave distinctive marks. **_Calidris_ sandpipers** cover the ground quickly in search of food. They leave relatively shallow, often zigzagging marks that follow the trail of a running bird. The **Eurasian Curlew** has a powerful beak measuring over 10cm long, which it uses to search for prey living deep in the sand. The bird usually stands still for some time, so a collection of deep boreholes can be found

▲ *The impaled prey of a Red-backed Shrike. Lausitz, Germany.*

▲ *Cuttlebone. Note the triangular beak marks.*

together with tracks where the bird has moved on the spot. Inland, perhaps the best-known marks of this kind are made by **Eurasian Woodcock** and **Common Snipe**. Both have a highly specialised probing beak that can be opened underground like a pair of tweezers to grab prey and pull it out. They search extensive muddy patches so thoroughly that the ground can end up looking like a sieve, with more than 20 holes possible in an area of 5cm². These conspicuous exploratory holes can give clues to the presence of Eurasian Woodcock and Common Snipe, even in the absence of tracks. Their boreholes are smaller and shorter than those of Eurasian Curlew. Classic locations are loose, soft soils, for example beach and shore areas or drained fishponds.

Swallows and martins, and some other birds, like nuthatches, also leave beak marks in the ground when collecting building materials. The birds prefer to pick mud from puddles and then shape them into clay and loam lumps for nest building. These beak marks in the ground are much shallower than exploratory holes.

▼ *Beak marks like these are characteristic of Common Snipe and Eurasian Woodcock. The birds use their long beaks to search for food. Lausitz, Germany.*

▼ *A Eurasian Woodcock has probed the soft ground of a drained fishpond in search of food, leaving distinctive marks. Lausitz, Germany.*

DUST BATH SITES

Galliform birds are known for dust bathing, but many other birds do it too. To make a dust bath, birds create shallow depressions in dry, sandy soils by scratching with their feet or pecking with their beaks. They throw the sand up and over their bodies with their wings, just like water when bathing. This has a cleaning effect, removing dirt and ectoparasites and regulating the oil content of the plumage.

Grains of sand can often be found on the vegetation around these hollows. This can help to distinguish between dust bath sites and digging marks. Feathers are often found in the dust bath, especially in summer and

▲ *A Barn Swallow collects clumps of clay to build its nest. Lausitz, Germany. Markus von Hacht.*

◄ *Shallow beak marks like these are created when the birds collect building materials. Lausitz, Germany. Markus von Hacht.*

autumn. In areas with little suitable sand, birds will also use upturned tree plates of fallen trees, gravel or even ashes from old campfires. Dust bath sites are usually roundish and their size roughly corresponds to the size of the bird.

DEDUCING BIRD SIZE FROM DUST BATH DIAMETER

Dust bath diameter	Bird size
about 6–9cm	e.g. House Sparrow
about 8–12cm	e.g. Eurasian Blackbird
about 13–17cm	e.g. Hazel Grouse
about 10–20cm	e.g. Grey Partridge
about 20–30cm	e.g. Common Pheasant
about 25–60cm	e.g. Western Capercaillie (D > 50cm indicates a male)

▶ *A dust bath, probably from a Eurasian Skylark or Meadow Pipit. Märkische Schweiz, Germany.*

AMPHIBIANS AND REPTILES

AMPHIBIANS AND REPTILES

Toad droppings

I remember how surprised I was the first time I saw toad droppings. Until then, I hadn't thought about the fact that toads produce droppings too. This sparked an interest in the tracks and signs of amphibians and reptiles. The more I learn about them, the more often I find signs of the presence of these fascinating animals. Previously, I had ignored any tracks or signs that I did spot because I couldn't tell which animal made them. As is so often the case, I only noticed certain things once I knew what to look out for. I now find toad droppings often, especially in unexpected places. By broadening our understanding of tracking, we can learn about the biodiversity, ecological relationships and stories of the animals that live in a habitat that would otherwise remain hidden from us.

With around 250 species in Europe, amphibians and reptiles are a diverse group. Many of these animals are highly sensitive to environmental influences. They are known as 'bioindicators' that can provide information about changing environmental conditions. This part of the book presents detection patterns that will help you roughly categorise tracks and signs found in the field. Precise species identification is often not possible or requires specialised literature. Filip Tkaczyk, an American tracks and signs specialist, has written a comprehensive book about the tracks and signs of North American amphibians and reptiles. As far as I know, a comparable work for Europe has not yet been published. Some tracks of amphibians or reptiles can be mistaken for mammal tracks, such as the tracks of toads, which resemble mole footprints.

TRACKS

SALAMANDERS AND NEWTS
Salamandridae

True salamanders and newts (Salamandridae) are a family of tailed amphibians (Caudata) whose members live in aquatic or terrestrial habitats.

TRACK

True salamanders usually have four toes on the front foot and five toes on the hind foot. None of the toes have claws. The hind feet are slightly larger than the front feet. The toes are straight and narrower at the tips than at the base. As many salamanders have similar foot morphology, you should consider distribution maps, habitat and body size as further criteria for precise identification. Salamanders and newts weigh very little and rarely leave clearly recognisable tracks.

GAITS

Salamanders mainly move in an understep walk, where the hind foot touches down behind the front foot on the same side of the body. A direct register walk is also possible. A study from the US showed that salamanders only move in symmetrical gaits such as walk and trot and do not gallop or bound (Petranka 1998). In rare cases, some salamanders may switch to an overstep trot. This energy-intensive gait is only used over short distances in an emergency. In contrast to the turned-in footprints of frogs and toads, the prints of the front and hind feet are usually parallel to the direction of travel. The track pattern is characterised by the combination of small footprints in walk with a large trail width and a usually clear, wavy drag mark made by the tail.

- Track formula: 4f × 5H

▶ *Fire Salamander.*

An Alpine Newt. When viewed from below, we can clearly make out the special shape of the toes. Bielefeld, Germany. Ulrike Quartier.

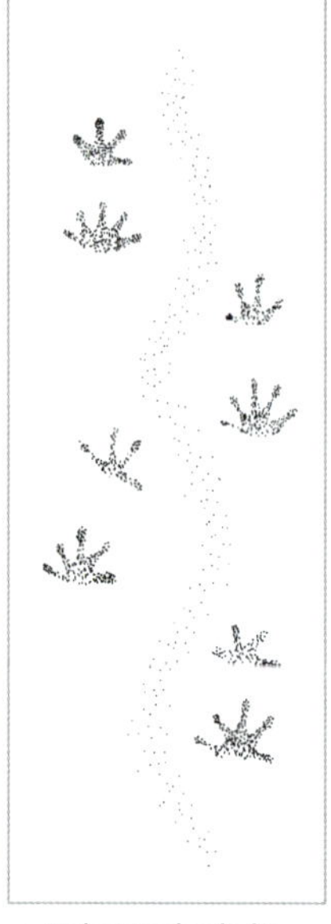

▲ Salamander in its preferred gait, the understep walk.

◄ Fire Salamander in understep walk. Toes that are wider at the base and a large trail width are characteristic of salamanders. Hainburg, Germany. Simone Roters.

FROGS AND TOADS

Anura

Frogs and toads belong to the order of tailless amphibians (Anura), although the names 'frog' and 'toad' have no distinguishing significance taxonomically.

◄ European Green Toad.

TRACK

Frogs and toads have four toes on the front foot and five toes on the hind foot, each without claws. Most species have medial or distal webbing on the hind foot, but this can only be made out in tracks on suitable ground. Because they provide the main propulsion for locomotion, the hind feet are much larger than the front feet.

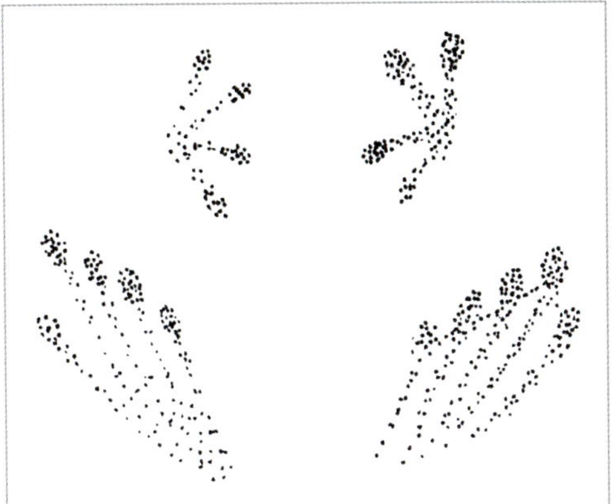

Sitting frog. The bottom tracks are made by the hind feet and are often only recognisable in the field as a diagonal row of four dots.

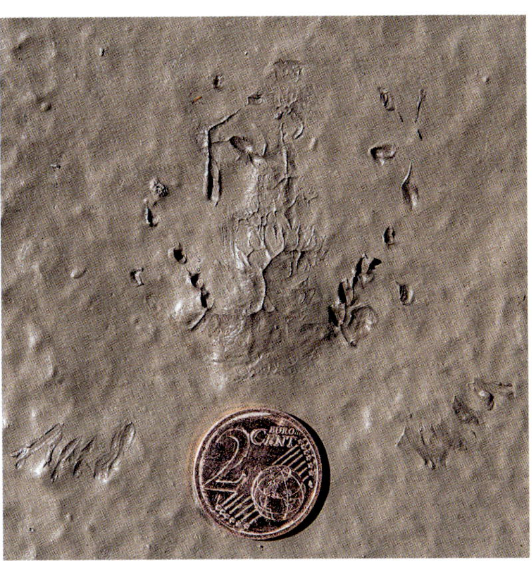

▲ *The almost complete print of the body of a frog. Lausitz, Germany.*

▲ *Left hind foot print (bottom) and left front foot print (top) of a European Green Toad. The turned-in front foot print can resemble a 'K'. Lausitz, Germany.*

Unlike other tailless amphibians, the family of **tree frogs** (Hylidae) have front feet and hind feet of approximately the same size. Their feet have large, suction cup-like toes that help them climb. Both features are important clues for identifying tree frog tracks.

Members of the **European spadefoot toad** family (Pelobatidae) have smooth front feet and hind feet with very pronounced webbing on the hind feet.

They get their name from a strong, keratinised bump (callus internus) located directly beneath the base of the innermost toe (toe 1), which is an adaptation to digging, like a spade. The print of this bump can look like a sixth toe and is a clear distinguishing feature of spadefoot toads.

Members of the family of **true frogs** (Ranidae) have long, tapered toes on the front feet and long, thin toes on the hind feet. During their reproductive period, many male tailless amphibians develop conspicuous nuptial tubercles on the inside of toe 1, which give them a better grip during the mating embrace (amplexus). The nuptial tubercles of males can be visible in perfect prints at least during the mating season and enable us to determine the animal's sex.

GAITS

Frogs and toads either walk or hop. The front feet are turned in so they point towards each other. The hind feet are wider apart and normally point forwards. Like mammals, frogs and toads walk by moving the diagonally opposite front legs and hind legs either simultaneously or individually. The hind feet usually understep so they touch down behind the front feet.

When they hop, the animals push off so powerfully with their hind legs that they land with their front legs first after a moment of suspension. The hind legs touch down close behind the front legs. Depending on the ground conditions, the animal's belly can also leave a print when it hops. Frogs prefer to hop, while toads mainly walk. This can help us tell the difference between frogs and toads, but toads can also hop and frogs can walk. Frogs generally make longer bounds than toads. Extremely long bounds come from frogs. Unlike newt and salamander tracks, there are no tail drag marks.

- Track formula: 4f × 5H

▼ *A toad in an understep walk. The inward-facing K-shape of the front feet, combined with four dots arranged diagonally to the direction of travel behind them, is characteristic of toads.*

▲ *A toad in its preferred gait, the understep walk. The direction of travel is from bottom to top and the sequence of footfalls is: RH, RF, LH, LF, RH, RF, LH, LF. Lausitz, Germany.*

▼ *Swimming tracks of a frog. The left and right hind legs push off alternately. Lausitz, Germany.*

TRUE LIZARDS
Lacertidae

True lizards are usually ground-dwelling animals that mostly live in dry habitats and belong to the order of scaled reptiles (Squamata) together with geckos and snakes, which we also discuss here.

▲ *Algerian Psammodromus (Psammodromus algirus). Note how the outer toe (toe 5) sits further back on the hind foot. Coto de Doñana National Park, Spain.*

TRACK

Lizards have five toes on the front foot and five on the hind foot. Each toe has a claw. All four feet are longer than they are wide, but the hind feet are larger than the front feet. The toes are long, slender and often curved. The toes of the front foot are roughly the same length. The two outer toes usually point to the sides and toes 2–4 point forwards. Toe 4 is the longest toe on the hind foot. The position of toe 5 of the hind foot is striking. It sits further back than all the other toes and points outwards at about 90°. Due to their

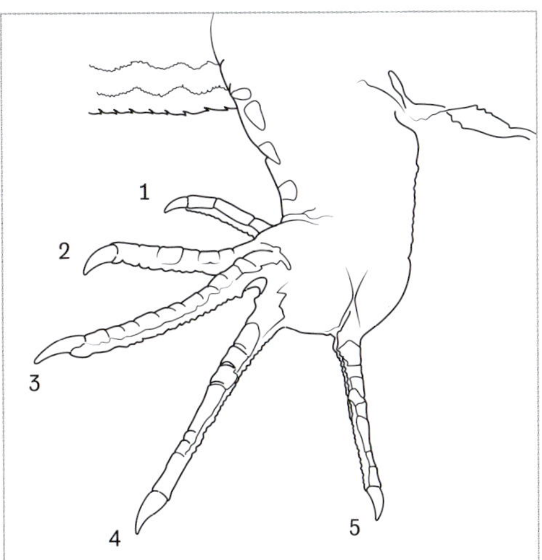

▲ *Lizard, left hind foot.*

low weight, lizards rarely leave clear tracks. The tracks they do leave are most likely to be found in fine, dry sand.

GAITS

Lizards move in the same gaits as mammals. However, their legs splay out to the sides of the body. This means that their spine moves side to side when they are in motion, whereas in four-legged mammals it moves up and down.

Lizards generally move at a walk or trot. Depending on their speed, these gaits can be understep, direct register or overstep. Understep walk is a rare gait for lizards and occurs mainly in cooler temperatures, in a state of relaxation or with the intention of avoiding detection. Lizards typically favour the direct register walk, unless they need to cover longer distances more quickly. The overstep walk is usually used as a transition into trot. It is not clear whether understep trot occurs in this group of animals. Lizards probably trot too quickly for their hind feet to land behind their front feet. I have yet to see a lizard in understep trot. The direct register trot and the overstep trot are the most common gaits observed.

Due to their anatomy, lizards do not gallop. Instead, they have a special form of locomotion known as the 'two-legged run'. In this gait, the hind feet propel the animal forwards so quickly and powerfully that the front feet no longer touch the ground. This is the fastest possible form of locomotion for lizards and is usually only used over short distances. As the animals often occur in dry landscapes, very detailed tracks are a rare find. Often only the track pattern and the gait provide clues as to who the footprints belong to. A fast, sometimes jerky run, often with drag marks from the body and/or tail, are characteristic.

- Track formula: 5f × 5H + C

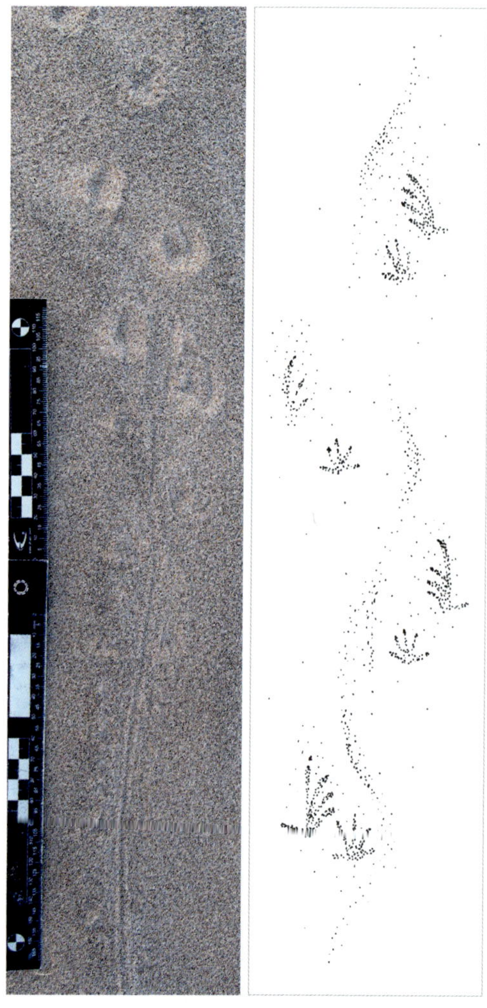

▼ A lizard in its most common gait, the overstep trot.

▷ A lizard moves from bottom to top, accelerating from a direct register walk to a direct register trot. In the bottom half of the image, you can see the drag mark of the tail. Coto de Doñana National Park, Spain.

GECKOS
Gekkonidae

▲ *Common Wall Gecko.*

TRACK

Geckos have five toes on the front foot and five on the hind foot. The toes have claws, but they are too small and short to leave marks in the track. Toe 4 is longer than toe 3. The tips of the toes are wider and covered with sticky lamellae, which enable geckos to grip even on smooth glass surfaces. In clear tracks, these adhesive discs can be seen as small, circular prints on the toe pads. The toe prints of a gecko appear relatively short and broad compared to other lizards. The midfoot pad is small and usually not visible. In rare cases, it is recognisable as a small area between the toes. The front foot is roughly symmetrical and the five toes are arranged in a fan shape.

The hind foot is slightly larger than the front foot, rather asymmetrical and toe 5 points almost 200° backwards.

GAITS

Geckos use the same gaits as true lizards. The hind feet generally land further out than the front feet. Tail drag marks tend to be less common because the animal lifts its tail off the ground when moving. Tail drag marks are mainly seen when the animal comes to a complete stop or changes direction abruptly.

- **Track formula: 5f × 5H + C**

SNAKES
Serpentes

The snake families found in the area include blind snakes (Typhlopidae), boas (Boidae), colubrids (Colubridae) and vipers (Viperidae).

◄ *Common European Adder.*

LOCOMOTION AND TRACK PATTERNS

Snakes are the only vertebrate animals discussed in this book that have no legs and no feet, so their locomotion is interesting and unusual. Europe's snakes can use four different types of locomotion.

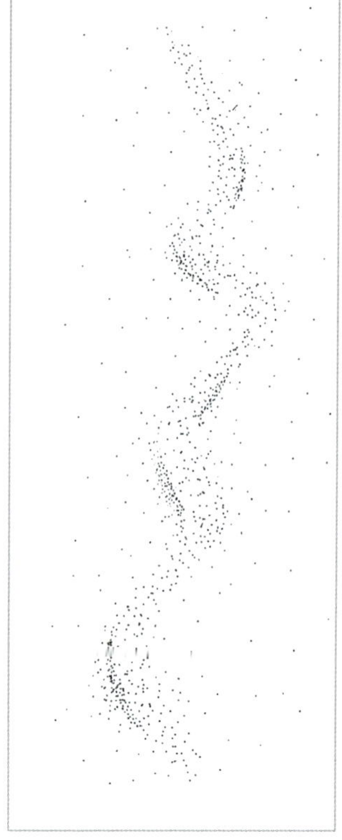

▲ *Winding.*

◄ *Note the outer edge of this winding track. The higher piles indicate the snake's direction of travel. Coto de Doñana National Park, Spain.*

Winding in an S-shape (a horizontal, wave-like movement) is the most common form of locomotion. The snake pushes itself diagonally forwards from both sides, compensating for the lateral forces, and moves forwards. This movement leaves a mark shaped like a sine wave. There can be a raised pile of substrate on the outer edge of individual loops. If there are several piles of substrate, examining them can help to determine the direction of travel.

Snakes can also move in a straight line (**rectilinear locomotion**) by propelling themselves with periodic waves of muscle contractions, and by bracing the edges of the large scales on their abdomen against the ground and pulling their body along in a mainly straight line. This movement only leaves a small amount of conspicuous substrate and its position can be used to determine the direction of travel. Piles of substrate that indicate the direction of travel can mainly be observed at points where the snake changed direction. Rectilinear locomotion is a comparatively slow form of movement and is mainly typical of larger snakes with a heavy body. Long, thin snakes can also crawl straight, especially when slowing down or just before stopping. Rectilinear locomotion is mainly used to slip into cover slowly and without being noticed. Snakes that feel threatened or are agitated usually switch to a faster mode of locomotion.

Sidewinding (sinuous, lateral locomotion) is an unusual form of locomotion where the snake lifts the front part of its body off the ground and puts it back down a little further to the side. At the same time, it makes an S-shaped movement from its head to its tail, causing the body to move forwards diagonally. Only a small part of the body touches the ground. This movement leaves a conspicuous track pattern that looks like a series of more or less parallel lines running diagonally to the direction of travel. The individual lines can resemble a J shape, with the point of the lower loop of the J pointing in the direction of travel.

▶ *A snake leaves a characteristic track in the sand as it sidewinds.*

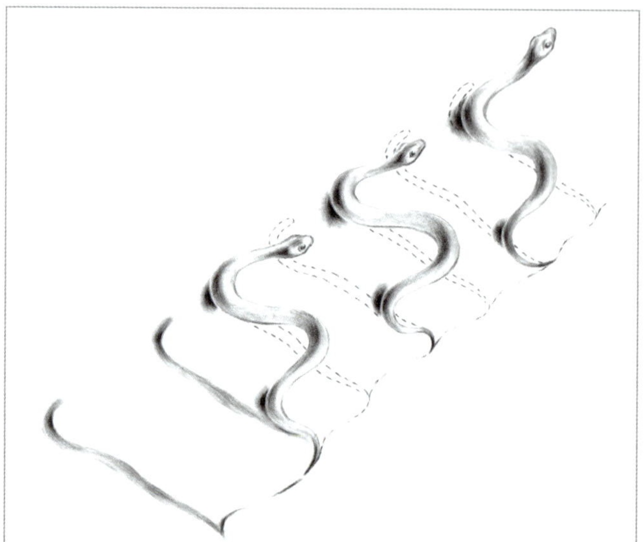

▶ *In sidewinding tracks, the bottom loop of the J points in the direction of travel. The fine lines at the other end are tail drag marks and can only be recognised occasionally.*

Long, thin snakes, in particular, sidewind when they want to move quickly. A fine, narrow line may be recognisable between the otherwise clearly separate, diagonal lines. Desert-dwelling snakes, in particular, seem to use this gait to move quickly over hot, loose sand. A classic example is the venomous Lataste's Viper (*Vipera latastei*), which is found on the Iberian Peninsula. Large snakes with a heavy body are unlikely to move by sidewinding, as it would mean repeatedly lifting large parts of their body off the ground. In parts of Andalusia, for example, this makes it possible to distinguish between venomous and non-venomous snakes. The non-venomous Ladder Snake (*Zamenis scalaris*) is larger and slower than the Lataste's Viper. Although both can move by winding, the Lataste's Viper

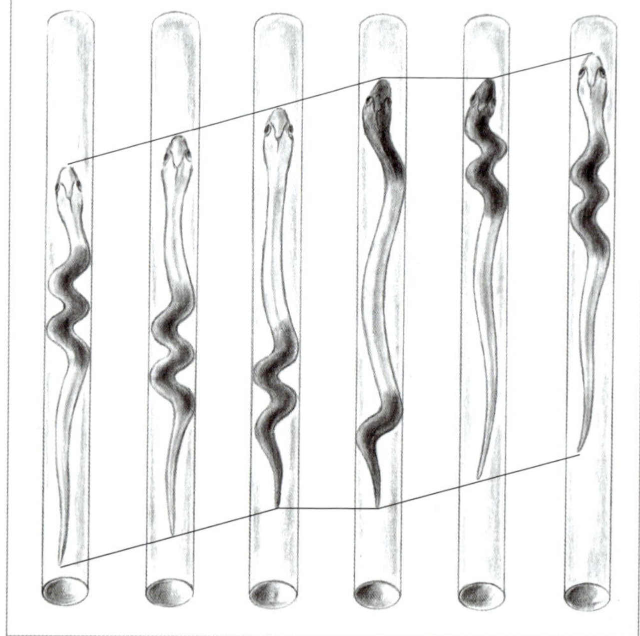

▲ *Concertina movement of a snake.*

favours sidewinding for higher speeds. Since there are no non-venomous snakes in this region that would move by sidewinding, the type of locomotion can give a clue as to whether the track was made by a venomous or non-venomous snake.

Concertina locomotion is a combination of horizontal and linear movement that we rarely encounter in the field. The snake moves the rear part of its body forwards so it is almost parallel to the front part. The front part then shoots forwards and the body stretches out. The snake then uses its rear end to get purchase and the process begins again. This slow form of locomotion is suitable for climbing, for example.

Snakes generally favour certain forms of locomotion according to their anatomy. The surface and the situation also have a major influence on the way they move. Knowing which species occur in an area and which 'gaits' these species favour in which situation can enable precise identification.

TORTOISES AND TURTLES
Testudinata

Members of the following tortoise and turtle families can be found in the area: marine turtles (Cheloniidae), tortoises (Testudinidae), Eurasian pond and river turtles (Geoemydidae) and terrapins (Emydidae).

▶ *Greek Tortoise.*

TRACK

Tortoises have five toes on the front foot and five on the hind foot, usually with strong claws. Often only four toes are visible in the track. Hind foot tracks can appear square. Marine turtles spend almost their entire life in the water and only come ashore to lay their eggs. Their feet have developed into paddle-like flippers. Tortoises spend almost all their time on land. They are adapted to a terrestrial lifestyle, with broad, pillar-like legs and strong, blunt claws. The legs and feet of terrapins have evolved for locomotion on land and in water. They have webbed toes and clearly recognisable claws.

GAITS

Tortoises generally only move on land in an understep walk with a characteristically large trail width. We can distinguish between three different types of understep walk. The first two gait variations, 'full understep' and 'half understep', are forms of extreme understep.

The track pattern of the **full understep walk** can resemble a direct register gait because the hind feet land in the prints of the front feet. Unlike direct register gaits, however, the hind feet step into the penultimate front foot prints, rather than the last ones. The hind legs are so far back that they step into the front foot prints of the previous sequence of footfalls. This is the slowest gait. A tortoise that moves in this way is probably either relaxed or cautious.

The **half understep walk** is slightly faster and can look like an overstep walk. Appearances can be deceptive, however, as the hind legs only overstep the penultimate front foot prints and always remain behind the last front foot prints.

The **understep walk** is the gait most commonly used by tortoises and is comparable with the understep walk of many mammals or even salamanders. The animal places the hind foot directly behind the front foot that it has just put down. A fast understep walk may indicate an escape to water or the nearest cover.

The front feet generally turn inwards slightly, while the hind foot prints tend to be parallel to the direction of travel. The claw marks visible on the tips of the toes can help to determine the direction of travel and recognise changes in direction. Drag marks made by the tail and ventral shell (plastron) are often visible and can provide important clues for precise species identification.

- **Track formula: 5f × 5H + C**

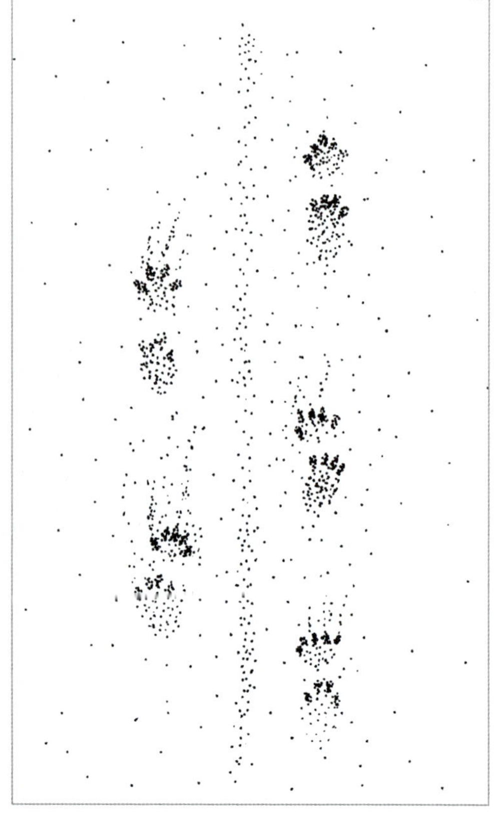

▶ *Tortoise, understep walk.*

SIGNS

Even if you don't find any footprints, signs of amphibians and reptiles can help to identify a species. The spawn, excrement and underground dwellings of selected species are described below. Specialist literature is required to identify eggs.

SPAWN

TOAD SPAWN

Female true toads (Bufonidae) lay gelatinous strings of spawn. Their spawning waters are mainly shallow and still, like puddles, ponds, shallow parts of lakes and slow-flowing waters. The eggs are usually brownish or black in colour and are laid in single or double rows. It is sometimes possible to precisely identify a species by the placement, colour and size of the eggs, the length of the strings and the season.

Strings of eggs laid by the **Common Toad** (*Bufo bufo*) are mainly found from February to April. They are wrapped around branches or aquatic plants and measure 3–5m long and 5–8mm thick. The strings contain around 3,000–8,000 black eggs, usually arranged in a double row.

- Egg diameter: 1.5–2mm

▼ *Spawn strings of a Common Toad, laid in double rows. Hunsrück, Germany. Simone Roters.*

Strings of **European Green Toad** spawn (*Bufo viridis*) can be found between April and June. They are usually 2–4m long and contain similar numbers of eggs to the strings laid by the Common Toad. The eggs are slightly smaller than those of Common Toad and are usually brownish black in colour.

- Egg diameter 1–1.5mm

Natterjack Toads (*Bufo calamita*) lay their 1–2m strings of 2,800-4,000 eggs in single rows from April to August, usually directly on the bottom of the body of water.

- Egg diameter: 1–1.7mm

Female **European spadefoot toads** (Pelobatidae) lay relatively short strings of eggs, almost 1cm thick, between plant stems under water.

Fire-bellied toads (*Bombina*) lay single eggs or up to 10 small clusters of a maximum of 30 eggs. They may be fixed under water or lie on the bottom of the body of water.

FROGSPAWN

Clumps of true frog (Ranidae) spawn can be found from February to May. Unlike toad spawn, frogspawn consists of a ball-shaped clump of transparent jelly blobs, each with a dark, usually black egg at the centre. True frogs lay their spawn in shallow bodies of water, like puddles, ponds and ditches. It is sometimes possible to precisely identify a species by the placement, colour and size of the eggs and clumps and the season. In Europe, the eggs of early-spawning species tend to be dark in colour while those of late-spawning species tend to be light.

The **Common Frog** (*Rana temporaria*) lays a mass of spawn about the size of a fist, consisting of jelly blobs measuring around 1cm. At the centre of each jelly blob is a black egg with a clearly visible white spot on the underside. These clumps of spawn begin to swell and rise to the surface shortly after being laid. The spawn of most other species, however, usually remains under water. Common Frogs form spawning communities, so extensive areas of the water surface can often be covered with clumps of spawn.

- Egg diameter: 1.7–2.8mm

Clumps of spawn produced by the **Moor Frog** (*Rana arvalis*) are like those of the Common Frog. However, the eggs are brownish in colour and the light spot on their underside is less clearly visible than in the Common Frog. On average, the spawn clusters, egg diameter and thickness of the gelatinous layer are smaller than in the Common Frog.

- Egg diameter: 1.5–2mm

▲ *Frogspawn. Frogs lay clumps of spawn while toads lay strings. Lehmkaute, Germany. Simone Roters.*

▲ *Swollen clumps of Common Frog spawn rise to the surface and often cover large areas.*
Offenbach, Germany. Simone Roters.

The spawn clumps of **tree frogs** (Hylidae) are only about the size of a plum and are attached to aquatic plants so they usually remain under water.

STAR JELLY

In wetlands such as marshes and also in areas close to bodies of water, we can find clusters of small spherical formations with whitish lumps of mucus, mainly in spring. These are unfertilised eggs and the swollen fallopian tube of a female frog or toad, regurgitated by a predator. In spring, predators seek out amphibians' spawning grounds for easy prey, and spit out the inedible reproductive organs. This sign is usually left by Common Buzzards, but it can also come from carnivores such as otters, minks or polecats. 'Star jelly' is a vernacular name in some countries, that refers to the fact that the remains appear to have 'fallen from the sky'. And, if they come from a bird of prey, this is actually the case.

▲ *'Star jelly' is found mainly in spring, near the spawning waters of frogs and toads. Jetzendorf, Germany.*

EXCREMENT

TOAD DROPPINGS

Toad droppings mainly consist of the remains of insects like ants or beetles and normally have a correspondingly porous structure with a rough surface. The compact and relatively thick faecal pellets are often surprisingly large and can be mistaken for hedgehog droppings. However, hedgehog droppings are longer and thinner in comparison. An unpleasant odour can also indicate hedgehog droppings. Toad droppings are odourless or have a milder smell. They are normally very light and disintegrate under pressure into many small pieces of insect remains. Unlike frog excrement, which is usually deposited in water, toad droppings can be found far from bodies of water, as toads can tolerate dry environments better. If a toad has fed mainly on ants, the contents may resemble Green Woodpecker droppings, but toad droppings are much thicker and do not have a white urine coating; they are also more compact.

● L 2–4.5cm D 0.5–1.2cm

LIZARD AND GECKO DROPPINGS

Lizard and gecko droppings mainly consist of the exoskeletons of insects and usually have a correspondingly coarse texture with a rough surface. Like bird droppings, the cylindrical faecal pellets have a white urine coating at one end. Urine is normally excreted when the animal begins to deposit excrement. The two parts are often joined together, but they can also be found independently of each other. Lizard droppings tend to be elongated, while gecko droppings are relatively short, thick and slightly tapered at one end. They normally produce very little urine and excrete it at the same time as producing droppings. Gecko droppings are often found clinging to the walls of buildings, while lizard droppings are found on the ground. The size of the droppings mainly depends on the size of the animal. Typical measurements for Sand Lizard (*Lacerta agilis*), Common Lizard (*Zootoca vivipara*), Spiny-footed Lizard (*Acanthodactylus erythrurus*) and Common Wall Geckos (*Tarentola mauritanica*) are between 1–1.9cm.

▼ *Toad droppings are relatively thick and consist of many small insect parts. Märkische Schweiz, Germany.*

▼ *Lizard droppings often have a characteristic white urine coating at one end. Spain. Paloma Troya (SERAFO).*

UNDERGROUND DWELLINGS

Toads dig small holes in the ground to shelter from strong sunlight. These daytime hiding places are generally only slightly larger than the toad itself and are usually elevated on embankments and bank edges as well as near bodies of water. However, they can also be found in sandy, diggable soils such as gravel pits and comparable fallow land. They often have a semi-circular entrance hole.

▶ *Daytime hiding place of a toad. The semi-circular entrance hole is a characteristic feature. Drehna, Germany.*

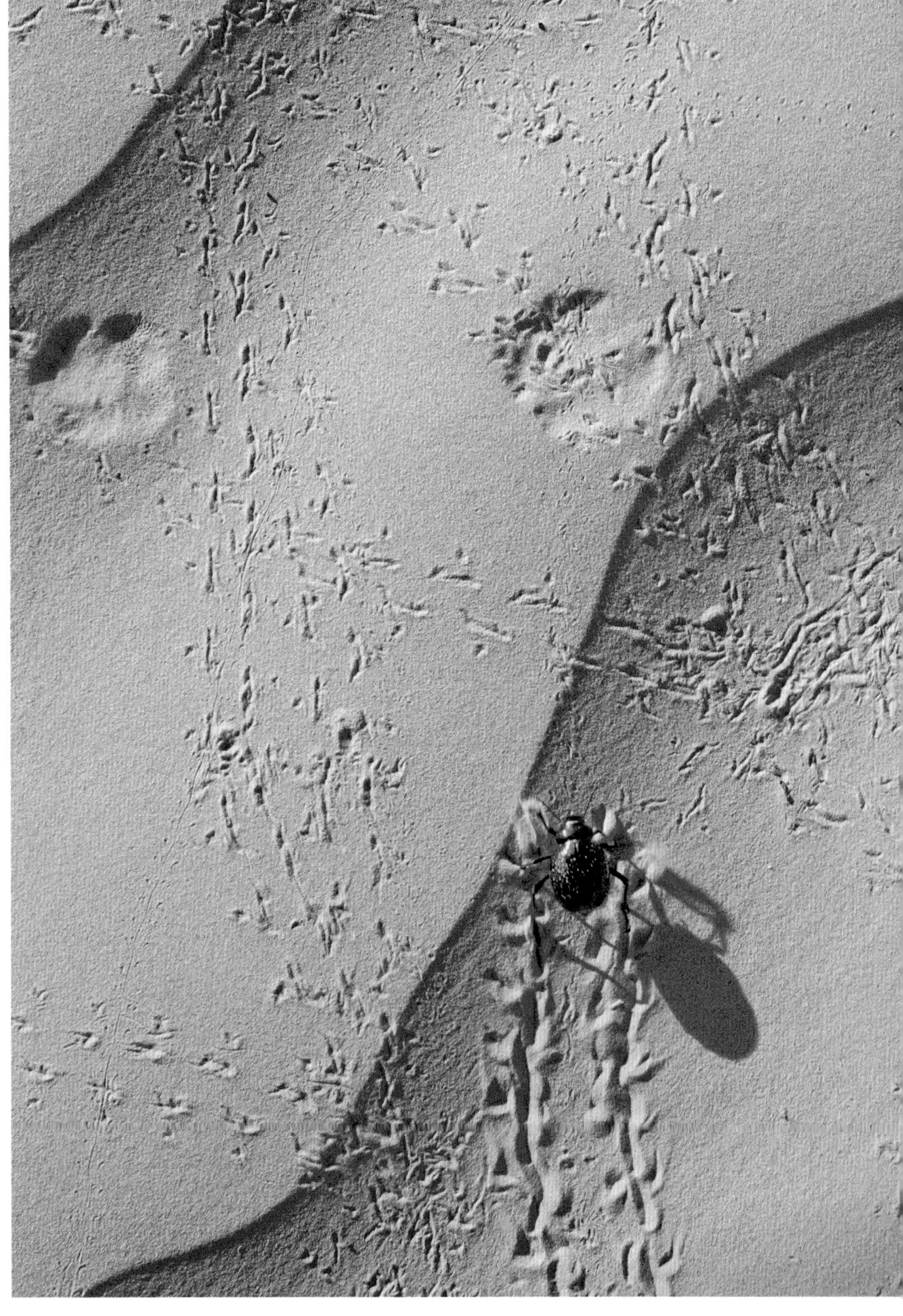

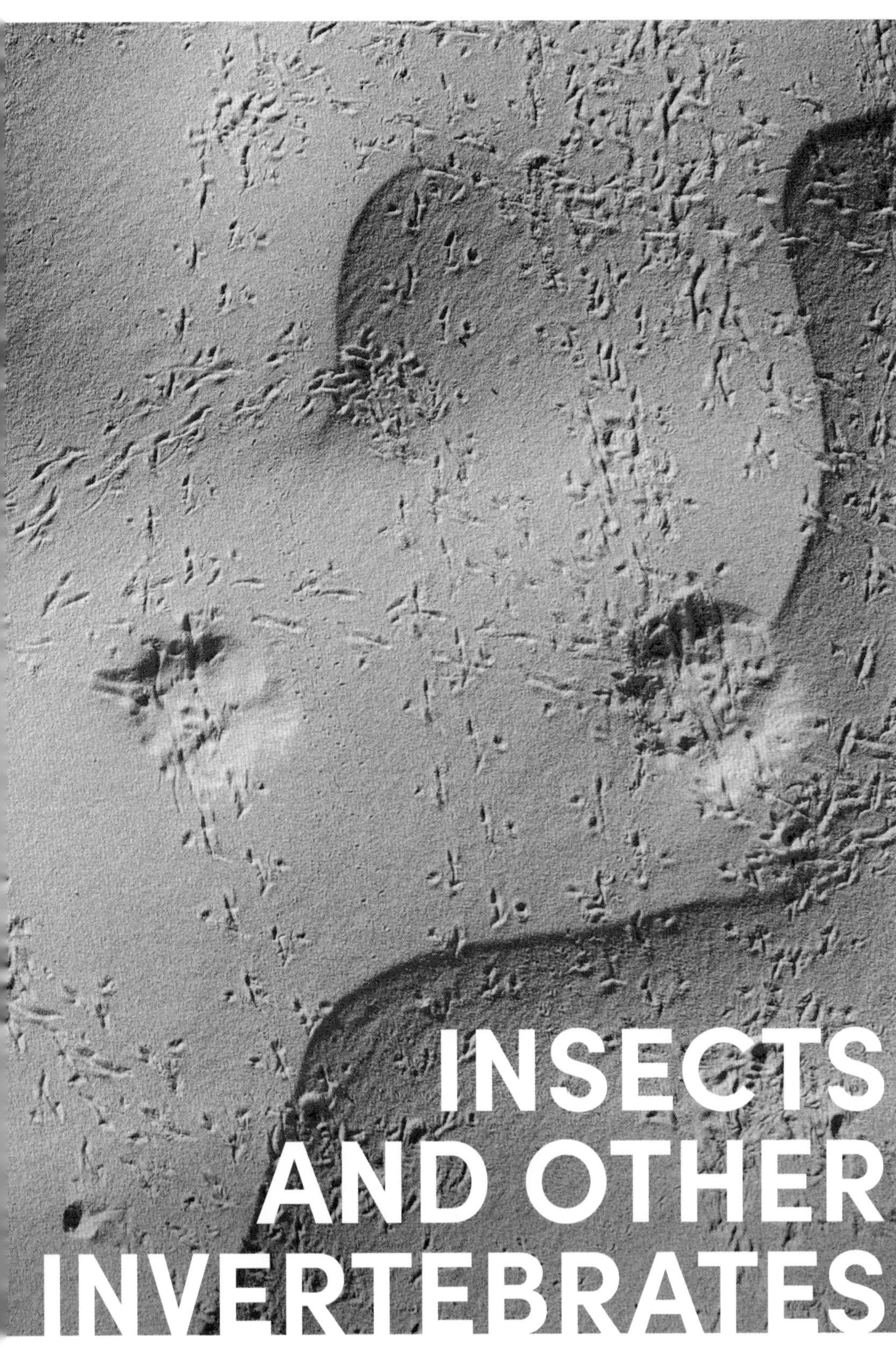
INSECTS
AND OTHER
INVERTEBRATES

INSECTS AND OTHER INVERTEBRATES

'Nothing is so insignificant that it does not leave tracks or should be ignored.'
Mark Elbroch

The world of insects and other invertebrates is highly diverse. Collectively, insects make up most of the animal life on land in terms of biomass and species diversity. Despite this, we know little about their way of life and are constantly discovering new species. Insects and invertebrates are often overlooked or even regarded as pests. They frequently go unnoticed, even though they play a crucial role in keeping habitats in balance. Basic knowledge of their tracks and signs can help us recognise ecological relationships and detect changes within an ecosystem at an early stage. This part of the book describes conspicuous and common traces of insects and other invertebrates. It also mentions signs of species that are important for environmental protection.

TRACKS

BEETLES
Coleoptera

▶ *Dung beetles leave a conspicuous track pattern when at work. The beetle either eats the ball of dung itself or lays eggs in it. Coto de Doñana National Park, Spain.*

Darkling beetles (Tenebrionidae) and **scarabs** (Scarabaeidae) make distinctive tracks that are often found in dry, sandy areas of Europe. Their track pattern consists of a continuous chain of alternating groups of three prints made by the three legs on each side of the body. The prints of the hind legs are furthest back in the group of three. They are usually long and straight and generally point straight back. The prints of the middle legs look similar but are more offset to the side and are the furthest out in the group of three. The prints of the front legs are the shortest and can point in different directions. They are often slightly angled forwards and form the foremost print within a group of three. In fine, dry sand, the prints of the front and hind legs of each side can merge to form a long line, so the shorter front leg prints can be difficult to make out. Sometimes the individual groups of three form an arrow shape, with the arrow pointing in the direction of travel.

- **Width Ø 1.5–3.5cm**

Dung beetles (*Scarabaeus*) leave a special track pattern when rolling a dung ball with their hind legs. Because the beetles move backwards, the prints of the front pairs of legs are on the inside of the track pattern, which is bordered by two parallel outer lines.

◀ The arrow shape of a group of three leg prints can indicate the direction of travel. Here, a darkling beetle ran from right to left. Coto de Doñana National Park, Spain.

▶ Beetle tracks come in a wide variety of sizes. The direction of travel is from right to left.

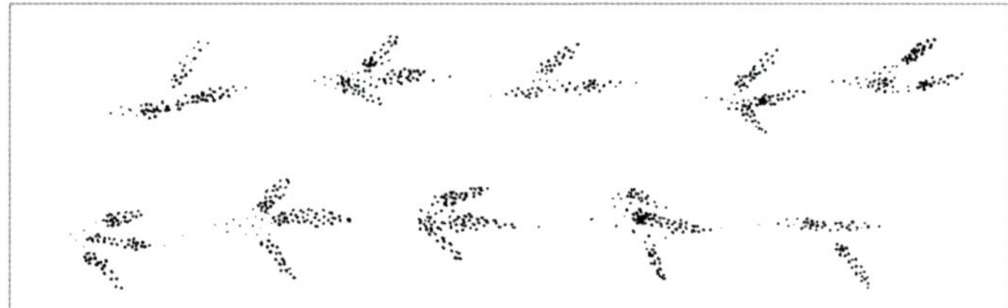

CATERPILLARS
Lepidoptera

▲ Mottled Umber caterpillar.

▶ The prints of the prolegs normally destroy the prints of the true legs.

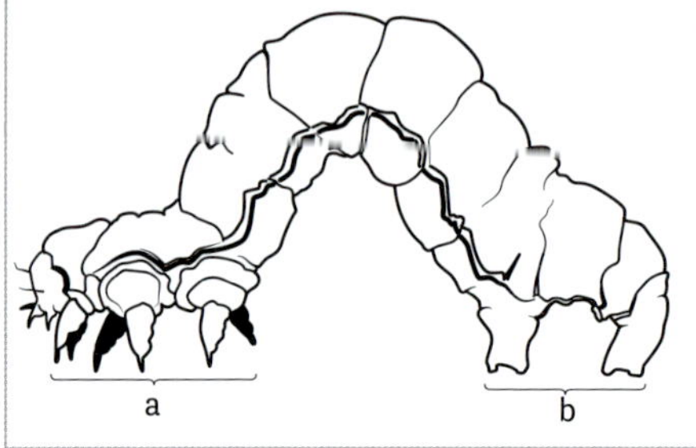

Caterpillars are the larvae of butterflies and moths. They have three pairs of legs (a) in the thorax area of their body. The rear of the body also has several so-called 'prolegs' (b), the number of which varies according to the species. These protrusions are made of skin and widen out towards the bottom like suction cups. They are not true legs because they are not segmented. The prolegs usually occur in pairs in segments 6–9 and are used for locomotion. They push from behind, which gives caterpillars their typical movement pattern. The prolegs destroy the prints of the true legs. This creates a track pattern consisting of a series of prints made by the pairs of prolegs. The stride length is short with a small negative space between the individual prints. The very hairy larvae of tiger moths (Arctiidae) also leave fine traces of their hair on the sides of the path.

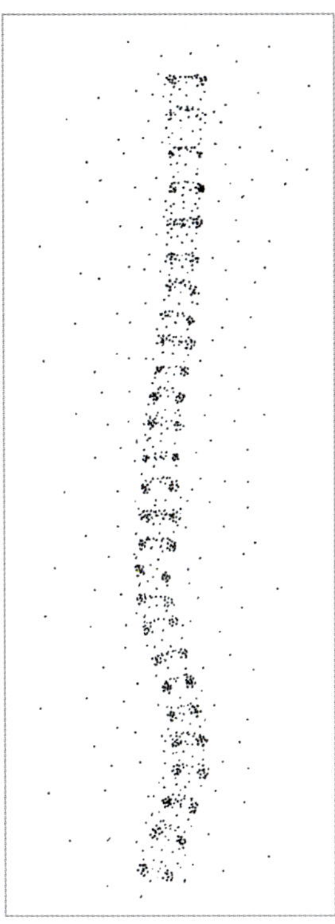

▶ *An edge at the front of the foremost pair of prints is occasionally visible in perfect tracks. This can indicate the direction of travel (here from bottom to top).*

▲ *Note the small negative space between the prints of the prolegs. Caterpillar tracks can look like train tracks. Coto de Doñana National Park, Spain.*

GRASSHOPPERS, LOCUSTS AND CRICKETS
Orthoptera

The tracks of Orthoptera such as short-horned grasshoppers, bush crickets and true crickets are like those of darkling beetles (page 746). However, the prints of the middle legs are often almost perpendicular to the prints of the hind legs and are relatively far from the centre of the path because the legs are longer.

▶ *Meadow Grasshopper.*

This characteristic feature can help us differentiate them from darkling beetles, unless the middle legs are turned further back when they touch the ground and therefore also point diagonally backwards. The prints of the hind legs are normally furthest in and are usually parallel to the path. They can be turned slightly outwards and are much longer than the other footprints. You can often make out prints of the abdomen or drag marks from the hind legs.

Short-horned grasshoppers tend to leave strong footprints and a narrow abdominal drag mark in the centre of the path. This drag mark can be continuous, broken or not recognisable. The negative space between the central drag mark and the footprints is relatively large. The stride length is short in comparison to the trail width, but generally longer than in darkling beetles.

Unlike beetles, members of this group repeatedly make short or long bounds. When they bound, the track pattern consists of two deep prints made by the hind legs and four fainter prints made by the other pairs of legs. On firm ground or in loose sandy soil, you can often only make out a single indentation made by the two hind legs.

▶ Without the abdominal drag mark, grasshopper and cricket tracks can be mistaken for beetle tracks. The almost perpendicular position of the middle pair of legs can be a distinguishing feature.

▲ Landing with the characteristic track of a short-horned grasshopper. The individual footprints are difficult to make out in deep tracks. Coto de Doñana National Park, Spain.

▲ An unknown species of grasshopper or cricket jumping. The prints of its legs have been covered over by the wind. Coto de Doñana National Park, Spain.

STONEFLIES
Plecoptera

Fewer than 150 of the world's 3,000 or so known species occur in Europe. The tracks of these flightless or poorly flying insects can be found near bodies of water. Most stonefly larvae have a forked tail consisting of two appendages (cerci), which drags on the ground as they move, leaving a characteristic trail. The slightly curved, fine prints of the three pairs of legs can often be made out to the left and right of this drag mark. The leftover exoskeletons of the animals can often be found on riverbanks.

▲ *Stonefly.*

MILLIPEDES
Diplopoda

▲ *Millipede tracks consist of two noticeably parallel lines made up of a series of small dots. However, these details can quickly disappear in dry sand. Vledder, Netherlands. René Nauta.*

The body segments of millipedes are almost circular in cross-section. Millipedes are herbivores. Unlike centipedes, they have two pairs of legs per visible body segment. Most move very slowly and often have an endogenous poison that makes them unpalatable as prey for many animals. Millipede tracks can be found at the edges of muddy puddles or in fine, damp sand.

Their track consists of two parallel lines. These lines are made up of a remarkably constant series of small dots, where each dot is the print of one of the many small legs. There is a large negative space between the lines. Our largest millipedes can leave trails up to 1.5cm wide.

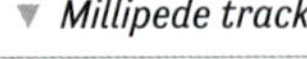

▼ *Millipede track.*

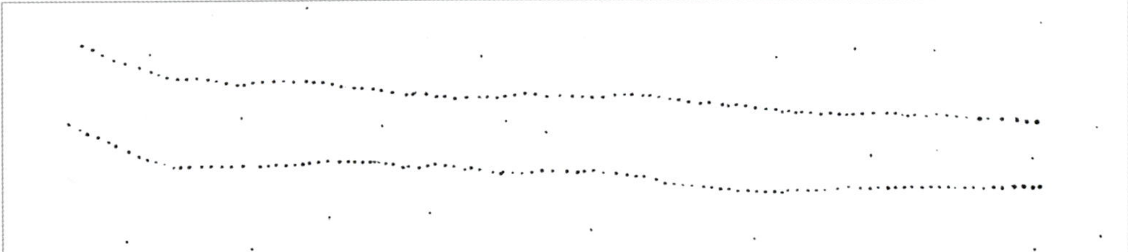

CENTIPEDES
Chilopoda

◀ *Centipedes have long legs, so their trail width is usually quite large. The individual leg prints can have an elongated, undulating appearance. Vledder, Netherlands. René Nauta.*

Centipedes have a long, flat body. They are nocturnal, predatory arthropods that use venom to paralyse their prey. All species that occur in Europe are harmless to humans. Unlike millipedes, centipedes only have one pair of legs per body segment. The legs protrude perpendicular from the body and leave a corresponding track.

SCORPIONS
Scorpiones

Although scorpions have eight legs, often only three footprints are recognisable on each side, so their track patterns can be mistaken for those of beetles. One of the differences is the consistent length of the individual prints, which tend to be oval rather than straight lines. If the feet drag, their prints appear as three or four short, closely spaced, parallel lines. All the prints in a group of three or four are roughly parallel to the path, while beetles and grasshoppers, crickets and locusts leave prints that are more perpendicular.

The track pattern of a scorpion has a large trail width and a short stride length. Groups of footprints appear regular and evenly spaced. The drag mark made by the tail may be continuous, partial or not recognisable. Murie and Elbroch (2005) suggest that the absence of a tail drag mark could indicate a threat or a defensive posture. Eisemann and Charney (2010) question this, as most of the scorpions they followed did not leave a drag mark over long distances. They propose that the absence or presence of drag marks varies by species. I have tracked scorpions with and without tail drag marks over distances of 50–100m. The absence of a drag mark, in conjunction with other data such as trail width, stride length and habitat etc., can presumably tell us something about the species, as well as the condition of the scorpion. However, there is currently not enough data to say for certain.

▲ *Common Yellow Scorpion.*

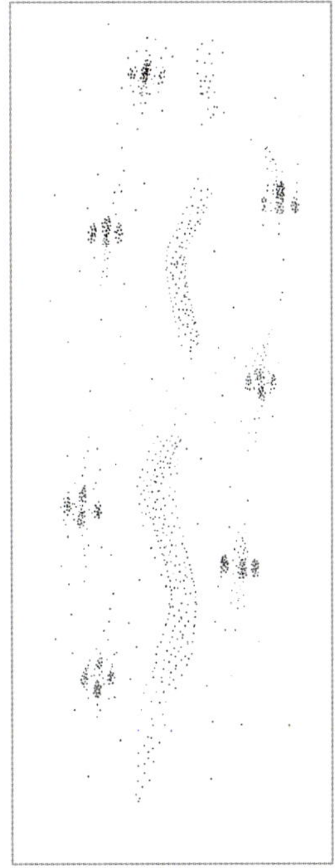

▲ *The characteristic track pattern of a scorpion is three closely spaced, short lines that run parallel to each other.*

▲ *Note the large trail width in relation to the short stride length. Scorpions are known for their regular track pattern. Coto de Doñana National Park, Spain.*

SPIDERS
Araneae

▶ *Spider track patterns can consist of many small dashes and dots. It can be difficult to make out the matching groups of four legs on each side in the field.*

At first glance, spider track patterns can look like a chaotic collection of small dots and dashes. Because spiders have four pairs of legs, assigning the many small prints to the individual legs can be a confusing and laborious task. However, with a little experience and enough time to observe closely, you can often recognise a pattern. The pattern will generally consist of alternating groups of four. In contrast to scorpions, each print can have its own shape and alignment. In the track patterns of funnel weavers (Agelenidae), the front two pairs of legs appear as small dots and the rear two pairs of legs as elongated, curved lines. The tracks of tarantulas (Theraphosidae) normally feature all eight leg marks, which consist of short, broad dashes or single dots. There are often two further, roundish marks closer to the centre of the path that are made by the pedipalps.

CRAYFISH
Astacoidea

Crayfish tracks are easily recognised by the distinctive, fine drag marks made by the outer edges of the tail. The track pattern consists of two parallel lines, on the outside of which are the many prints of the four pairs of legs. Depending on the type of ground, the individual prints of four or just three pairs of legs can appear as single dots, dashes or a combination of dots and dashes. The overall appearance can resemble a scorpion track, as both species have similar anatomy. Prints of the claws of the foremost pair of legs may be recognisable in good ground conditions. Their prints consist of two small, clear holes that are close together.

▲ *Crayfish.*

▲ *On closer inspection, you can make out the two close-set prints of the claws here. Lausitz, Germany.*

▲ *Trackers can learn to create an animal from its tracks. Coto de Doñana National Park, Spain.*

▲ *Many dots arranged in parallel lines indicate the track of a crayfish. Lausitz, Germany.*

COMMON SHORE CRAB
Carcinus maenas

This species of crab has been spread worldwide by humans. It occurs on almost all beaches and along Europe's Atlantic coast and North Sea coast. It is the only animal described here that, like all other crabs, moves sideways. At first glance, the continuous pattern of many small to large dashes and dots is striking. Crabs have five pairs of legs, of which the foremost pair have evolved into pincers. All eight walking legs and the two pincers may leave prints and/

▲ *Shore crabs move sideways. Their tracks consist of dots and dashes, usually without a clear pattern. Portsmouth, England.*

▲ *Tracks and signs tell us stories about the recent past; here a crow has eaten a shore crab. Portsmouth, England.*

or drag marks. The tracks and drag marks are usually very close together in the track pattern. You can often see many different prints, so it can be hard to make out a single print. The continuous pattern of dashes and dots is the most characteristic feature of crab tracks. The tracks often begin or end in a hole in the sand where the crab was or is still buried, waiting for the next tide.

SLUGS AND SNAILS
Arionidae, Helicidae and others

All slugs and snails leave a conspicuous trail of mucus, which is often recognisable even after the animal has gone. Because slugs and snails move in different ways, they leave different mucus trails, which makes it possible to distinguish between the two groups. Slugs leave a more or less continuous trail of slime, while snails leave an interrupted dot-dash pattern.

▲ *Red Slug.*

▶ *Slug slime can become worm-shaped when combined with sand and dry conditions. The outsides of the mucus trail curve inwards, creating an almost closed tunnel. Lausitz, Germany.*

▲ *Continuous mucus trail of a slug. Rodenbach, Germany. Simone Roters.*

▲ *Irregular mucus trail of a Roman Snail. Vledder, Netherlands. René Nauta.*

EARTHWORMS
Lumbricidae

◀ *Earthworm tracks come in different sizes. This is a larger example. Märkische Schweiz, Germany.*

Earthworm tracks can be seen on muddy ground, especially after rain or in the early morning. These tracks are usually less than 5mm wide, rounded at the bottom and more or less straight.

SIGNS
ON TREES AND BARK

Some of the most common signs of wood-dwelling insects are made by beetles (Coleoptera). Bark beetles (Scolytinae), longhorn beetles (Cerambycidae) and jewel beetles (Buprestidae) bore through the bark of living or dead trees to lay their eggs in the cambium or deeper in the sapwood and heartwood.

These beetles can be divided into wood-breeding and bark-breeding species. Once the larvae have hatched, they eat tunnels through the inner bark, cambium or wood of their host. The overall shape of the passageways made by the mother and larvae (maternal and larval galleries) is called the feeding pattern. Large numbers of larvae can kill trees, especially weaker ones. Some species can cause enormous damage to forestry plantations. Warmer, drier summers create favourable conditions for beetles.

▼ *Feeding pattern of a jewel beetle. Lausitz, Germany.*

▲ *The Ribbed Pine Borer (Rhagium inquisitor) is a common longhorn beetle in Europe. It is shown here in its pupal cradle in a pine tree. Heidelberg, Germany. Aaron Tidemann.*

In recent years, evidence has been mounting that global climate change and associated extremes of weather, such as prolonged periods of drought or severe storms, as well as forestry operations, such as planting pure spruce stands, encourage bark beetle infestation. This can lead to mass reproduction and result in the death of of many trees in a plantation. Early detection of beetle infestation can help prevent this and limit damage to forestry.

Boreholes and piles of bore dust at the base of a tree can indicate an infestation. For precise species identification, you should consider the host tree in conjunction with the feeding pattern. It is usually not possible to clearly identify a species by looking at the host tree alone, because it may be infested by different beetle species.

In addition to wood-breeding and bark-breeding beetles, many other insects inhabit trees and leave their feeding patterns on them, such as **cossid miller moths** (Cossidae), **clearwing moths** (Sesiidae), **carpenter ants** (*Camponotus*), **carpenter bees** (*Xylocopa*) and **wood wasps** (Siricidae). In addition to the host tree species, the location, infestation pattern and any excrement or bore dust are important for identification.

To give a better overview, we examine the feeding patterns of beetles together with those of other insects rather than according to their taxonomy. We consider feeding patterns directly beneath the bark, feeding patterns deeper below the bark, and in the sapwood, and feeding patterns in the wood.

FEEDING PATTERNS DIRECTLY BENEATH THE BARK

BARK-BREEDING BARK BEETLES SCOLYTINAE

Shoot and fir engraver beetles are bark-breeding beetles because they mate in a 'mating chamber' directly beneath the bark of their host tree. After mating, the female lays eggs in one or more maternal galleries. The larvae feed on the cambium of their host. Each individual larva creates its own larval gallery that ends in a puparium. Shoot beetles eat little or no cambium, while fir engraver beetles do eat cambium, but often only a few millimetres deep into the sapwood. The larval and maternal galleries are rarely wider than 3mm and are parallel or horizontal to the grain of the wood. Once they have hatched, beetles leave their puparium through small holes in the bark. The resulting feeding patterns are often characteristic of different species, so it is usually possible to precisely identify individual species if you also consider the host tree.

Six-toothed Spruce Bark Beetle
Pityogenes chalcographus

Host tree Mainly spruce, but other coniferous trees like larch, pine, Douglas Fir and fir.

Feeding pattern Starting from the mating chamber, the female creates 3–6 maternal galleries in a star shape, in which to lay her eggs. Each maternal gallery is about 6cm long and 1mm wide. The larvae eat their way at right angles from their maternal gallery, creating 2–4mm long, closely spaced larval galleries. The feeding pattern can resemble a copperplate engraving, hence the alternative common name Spruce Wood Engraver.

European Spruce Bark Beetle
Ips typographus

Host tree Mainly spruce, but other coniferous trees like larch, pine, Douglas Fir and fir. In rare cases, they can also attack deciduous trees, like the Copper Beech.

Feeding pattern Two females mate with one male, who creates the mating chamber. The females create their maternal galleries for laying their eggs in opposite directions, starting from the mating chamber. A double-armed lengthways channel is formed along the grain, with the larval galleries branching off to the sides.

Common and Lesser Pine Shoot Beetle
Tomicus piniperda, Tomicus minor

Host tree Almost exclusively pines, but more rarely spruce and larch.

Feeding pattern One-armed, approximately 10cm long and slightly curved lengthways channel with mating chamber-like enlargements at the ends. The maternal gallery

▲ *Top, the classic feeding pattern of a Six-toothed Spruce Bark Beetle in the bark. Heidelberg, Germany. Aaron Tidemann.*

▼ *Bore dust can be an early indication of Six-toothed Spruce Bark Beetle infestation. Heidelberg, Germany. Aaron Tidemann.*

▲ *Feeding pattern of a Six-toothed Spruce Bark Beetle on a spruce tree. Three to six maternal galleries leading from the mating chamber in a star shape are characteristic. Heidelberg, Germany. Aaron Tidemann.*

▼ *Immature European Spruce Bark Beetles in September. Heidelberg, Germany. Aaron Tidemann.*

▲ *The double-armed lengthways channel with larval galleries at the sides is typical of the European Spruce Bark Beetle. Heidelberg, Germany. Aaron Tidemann.*

usually has a fine resin coating and is found almost exclusively in the inner bark. They make shallow grooves in the sapwood. The larval galleries are long and close together. During the maturation feeding stage, beetles hollow out pine shoots. There will be an entrance borehole with a resin funnel towards the base of the shoot. The pine shoots are hollowed out up to the borehole. The beetles change their feeding site several times during this stage. The hollowed-out shoots stay green, but usually break off during autumn storms and leave a conspicuous covering on the ground.

Ash Bark Beetle
Hylesinus fraxini

Host tree Mainly ash, but in rare cases other deciduous trees such as hazel, beech, maple, locust trees (*Robinia*), walnut and fruit trees.

Feeding pattern Horizontal, double-armed transverse passageway, consisting of two, approximately equal-sized maternal galleries,

▲ *The feeding pattern of the Common Pine Shoot Beetle, looking at the trunk. Heidelberg, Germany. Aaron Tidemann.*

▼ *Centre, the feeding pattern of the Lesser Pine Shoot Beetle with the many characteristic round boreholes. At the sides, the larval galleries of an unidentified jewel beetle in comparison. Heidelberg, Germany. Aaron Tidemann.*

▲ *Feeding pattern of the Common Pine Shoot Beetle. The maternal gallery with a fine resin crust and the mating chamber-like enlargements at the ends are characteristic. Heidelberg, Germany. Aaron Tidemann.*

◄ *Adolescent Lesser Pine Shoot Beetles hollow out pine shoots during maturation feeding. Heidelberg, Germany. Aaron Tidemann.*

leading off a relatively short mating chamber. The larval galleries run more or less vertically, almost at right angles to the maternal galleries. They are usually close together and relatively short at around 4cm in length.

Birch Bark Beetle
Scolytus ratzeburgii

Host tree Mainly diseased, older or weakened birch trees.

Feeding pattern The regularly spaced air holes of the Birch Bark Beetle, which usually run in a vertical line, can even be seen from the outside. These circular holes have a diameter of 2–3mm and run directly above the maternal gallery. The maternal gallery

◄ *Bottom: the short mating chamber of the Ash Bark Beetle points to the left. The two maternal galleries of roughly the same size branch off from it. Heidelberg, Germany. Aaron Tidemann.*

is a one-armed lengthways channel that can reach down to 10cm into the sapwood. The larval galleries that radiate off it are close together, are 15–25cm long and end in a puparium with a large exit hole leading to the outside.

European Oak Bark Beetle
Scolytus intricatus

Host tree Mainly oak, but more rarely other deciduous trees like hazel, beech, poplar, willow, chestnut and elm.

Feeding pattern A relatively short, 1–3cm horizontal, one-armed maternal gallery, from which the larval galleries branch off vertically. Larval and maternal galleries usually leave relatively deep furrows in the sapwood, which can almost completely cover the surface of the branch.

▼ *Here you can see the feeding pattern and maternal gallery of the Birch Bark Beetle in greater detail. Lausitz, Germany.*

▲ *This regular row of holes is a conspicuous and characteristic feature of the Birch Bark Beetle. Neustadt, Germany.*

▲ *The short transverse channel with larval galleries branching off vertically is typical of the feeding pattern of the European Oak Bark Beetle. Quitzdorf, Germany. Aaron Tiedemann.*

Large Elm Bark Beetle
Scolytus scolytus

Host tree   Mostly elms, but in rare cases other deciduous trees such as hornbeam, walnut, Cork Oak, poplar and ash. Large Elm Bark Beetles spread the fungus that causes Dutch elm disease, which primarily threatens Wych Elm populations.

Feeding pattern   Relatively short, vertical and usually 2–5cm long maternal gallery (one-armed lengthways channel). The irregular larval galleries branch off to the sides.

▲ *Feeding pattern of the Large Elm Bark Beetle. Ladenburg, Germany. Aaron Tiedemann.*

LEAF-MINING FLIES
AGROMYZIDAE

This diverse family includes around 900 species in Europe. Almost all of them only reach a few millimetres in length. Adult leaf-mining flies usually feed on sap from green leaves. Some larvae also mine in layers of bark, leaving characteristic feeding galleries in the uppermost layer of wood directly beneath the cambium. These tunnel mines start out as fine, thread-like lines on the branches that are difficult to spot. From there, the leaf-mining flies turn around and feed downwards vertically or in a slightly meandering fashion, parallel to the grain of the trunk. When they reach the bottom, they turn around again and eat their way vertically upwards. This process can be repeated several times before the fully-grown larva leaves the wood to pupate. Brown lengthways traces can be recognisable on the wood of infested tree species, usually birch, alder, poplar and willow. The brown colouring is caused by the excrement of the larvae.

FEEDING PATTERNS DEEPER BELOW THE BARK AND IN THE SAPWOOD

Longhorn beetles, jewel beetles
Cerambycidae, Buprestidae

Unlike bark-breeding beetles, which only eat the cambium layer of their host tree, longhorn beetle and jewel beetle larvae can eat deeper into the sapwood of their host, creating corresponding galleries. It is possible to differentiate between the two groups by the shape of the galleries and exit holes, as well as the ejected bore dust. Jewel beetle larval galleries are broad, flat, oval and usually filled with fine, densely packed bore dust. They usually run in a meandering or zigzag pattern on the surface of the sapwood, directly beneath the bark. The exact feeding patterns depend on the species and can also vary within a species. There is little or no bore dust from the host tree. The exit holes are flat or D-shaped. Longhorn beetles, on the other hand, generally create rounder larval galleries that are filled with coarser and more loosely packed bore dust. Their exit holes are often circular. However, some species create galleries and exit holes that can be as flat as those of jewel beetles, while other longhorn beetle larvae make characteristic boreholes.

Cossid miller moths, clearwing moths
Cossidae, Sesiidae

Caterpillars of cossid miller moths and clearwing moths, which belong to the order Lepidoptera, also create feeding galleries in wood. They often start beneath the bark and then eat their way deeper into the wood, where they usually spend several years as caterpillars before they pupate. Conspicuous piles of tiny caterpillar droppings form on the surface of their host tree (page 783). The wood shavings produced are coarser than the bore dust of most other wood-boring insects. Occasionally, fine silk threads between the droppings and the wood shavings reveal that they were made by a moth caterpillar.

▶ *The broad, flat larval galleries of jewel beetles are usually meandering and filled with fine bore dust. Heidelberg, Germany. Aaron Tidemann.*

▲ *The wood shaving-lined cradle of a longhorn beetle. The coarser bore dust is a distinguishing feature from jewel beetles. Heidelberg, Germany. Aaron Tidemann.*

▲ *Unidentified longhorn beetle larva in an oak tree. Heidelberg, Germany. Aaron Tidemann.*

FEEDING PATTERNS IN THE WOOD

Wood-breeding bark beetles
Scolytinae

Unlike bark-breeding beetles, the larvae of wood-breeding species do not feed on the cambium of their host, but on ambrosia fungi, which the female brings with her in her gastrointestinal tract. The fungus then becomes established in the host tree. The galleries of these beetles, also known as ambrosia beetles, turn black over time due to fungal infestation. The feeding pattern of wood-breeding beetles features an entrance tunnel leading deep into the wood, with maternal galleries extending from it. Galleries created by the parents are long, round and have an even width. Depending on the species, the diameter varies between 0.5–3mm. Unlike bark-breeding beetles, wood-breeding beetles do not create galleries between the wood and bark. The larval galleries are usually short at just 5mm and can resemble the rungs of a single ladder. The parents plug the entrance tunnel with bore dust and create small openings to ensure correct humidity and to dispose of larval excrement and bore dust. When making the entrance tunnel, the beetles produce light-coloured, fine bore dust, which is ejected from the tree. Later, conspicuous and tightly compressed wooden rods from larval feeding can pile up under the tree. The larvae of wood-breeding beetles do not build puparia but leave their host tree as juvenile beetles through the entrance tunnel created by the mother. Ambrosia beetles usually attack deciduous trees. The most common species in Europe include the European Shot-hole Borer (*Xyleborus dispar*), the Fruit-tree Pinhole Borer (*Xyleborus saxesenii*) and the Striped Ambrosia Beetle (*Trypodendron lineatum*), which, as an exception to the rule, only attacks conifers.

Carpenter ants
Camponotus

Carpenter ants are known for building their nests in the wood of trees. The nests consist of a system of tunnels and chambers that run deep into the wood and into the ground beneath the host. Large nesting areas, connected by an underground tunnel system, can extend to encompass several dozen trees. The infested wood has a characteristic spongy appearance. As with an ant's nest in the ground, the many workers transport the wood shavings to the outside, piece by piece. As the wood serves as nesting material and not as food, large piles of wood shavings can build up at the foot of the tree. The Black Woodpecker, which specialises in feeding on carpenter ants, often uncovers these nests

▲ *Beech trunk that has been attacked by an ambrosia beetle* (Xyleborus)*. Heidelberg, Germany. Aaron Tidemann.*

while foraging. Carpenter ants can build their nests in healthy as well as rotten, partially decayed trees or in dead wood. The condition of the infested host tree and the height of the nest can give clues as to the species of ant. The Hercules Ant (*Camponotus herculeanus*) mainly inhabits healthy trees and prefers spruce, but occasionally also colonises pine trees. It is rarely found in deciduous trees and dead wood. Its nests can be built in the trunk up to a height of 6–10m. The Brown-black Carpenter Ant (*Camponotus ligniperda*) prefers conifers and only builds its nests in dead wood. As a rule, it lives in a larger underground section and its nests are no higher than 3m.

Carpenter bees
Xylocopa

Carpenter bees such as the Violet Carpenter Bee (*Xylocopa violacea*) gnaw holes in dead wood. They prefer wood that has been processed by humans, such as roof beams or fence posts. If none of these structures are available, they usually burrow into the underside of fallen tree trunks or in dead branches. The entrance holes can also appear on the vertical side of a suitable substrate.

They are circular and have a diameter of around 1–1.2cm. In contrast to wood wasps and longhorn beetles, the edges of the entrance holes are shaped at an angle and have even bite marks. Insects that bore their way out of a hole leave exit holes with straight edges. When gnawing the entrance

▼ *The striking nest of the Jet-black Ant* (Lasius fuliginosus)*. Heidelberg, Germany. Aaron Tidemann.*

▼ *Carpenter ant.*

holes, the bees leave a small pile of bore dust underneath. This is absent from the exit holes of wood wasps and longhorn beetles.

Wood wasps
Siricidae

Wood wasps are large sawflies that usually lay their eggs in weakened trees or trees with damaged bark using their long, sharp-ended ovipositor. At the same time, they inject a fungus into the host, which causes the wood to decompose and later provides a food source for the growing larvae. The larval galleries begin in the sapwood near the surface and only penetrate deeper into the heartwood after some time. There are no visible galleries directly beneath the bark. Because the larvae grow slowly, the larval galleries also increase in diameter (approx. 0.5–0.8cm). The circular larval galleries are blocked with fine, tightly packed bore dust. After two to four years of development, the adult insect leaves the host through an equally round exit hole that can be confused with that of a longhorn beetle

or carpenter bee. The exit holes of several adults are usually clustered at a certain point on the tree trunk.

FEEDING MARKS ON LEAVES

When mammals or birds leave feeding marks on leaves, the leaf has usually been partly or completely bitten, plucked or torn off. The snails and insects discussed here rasp or cut the leaves with their much finer mouthparts, leaving characteristic signs. Caterpillars, beetles, leaf miner moths, sawflies, true crickets, bush crickets, leafcutter bees and leafcutter ants leave various traces on leaves because they use them as a habitat, food source or as somewhere to lay their eggs. Due to the enormous diversity of species within the class of insects, it is often not possible to precisely identify species, or only possible with the help of specialist literature. However, a little background knowledge and practice enable us to make basic distinctions.

▲ *Carpenter bee.*

▶ *Wood wasp.*

It is therefore important to examine the findings in detail.

PITTING, STARTING FROM THE INSIDE OF THE LEAF

Insects with tiny or very delicate mouthparts may start feeding on leaves from the centre. These marks could be left by small leaf beetles (Chrysomelidae), true weevils (Curculionidae), various beetle and sawfly larvae (Tenthredinidae), small caterpillars, earwigs (Dermaptera), thrips (Thysanoptera) and also slugs and snails (Gastropoda).

PITTING, STARTING FROM THE EDGE OF THE LEAF

The mouthparts of larger caterpillars, large scarabs, true crickets, bush crickets, sawflies, leafcutter bees and leafcutter ants are too large to bite into the centre of the leaf. Instead, they cut or eat holes from the outer edge towards the inside of the leaf.

The feeding marks of beetles and caterpillars tend to be irregular, whereas cuts made by leafcutter bees and leafcutter ants are conspicuously clean and rounded.

Leafcutter bees

Various species of leafcutter bee (*Megachile*) are found in Europe. They build their nests in suitable cavities in cracks in walls, burrows and dead wood or dig cavities in pithy plant

▼ *Animals with small mouthparts eat holes like these from the inside of the leaf. Spessart, Germany.*

▲ *Feeding marks on leaves are a common sign. For initial identification, we distinguish between holes that start from the inside of the leaf and holes that start from the edge of the leaf. Both types are shown here. Spessart, Germany.*

▲ *Irregular feeding marks like these, which start from the edge of the leaf, are typical of beetles and caterpillars whose mouthparts are too large to bite into the centre of the leaf. El Rocío, Spain.*

stems. The females create brood combs in these cavities, which they line with cut pieces of leaf. Each comb contains an egg, and pollen as a food source for the larva when it hatches. They are sealed with more leaf material. Cuts made by leafcutter bees are characteristically circular and meticulously made.

Leafcutter ants

Originally native to the tropics and subtropics, leafcutter ants (*Atta* and *Acromyrmex*) can occasionally be found as exotic species in European gardens and on agricultural land. These insects feed on a fungus that they grow in their anthill. Pieces of leaf that the ants work together to cut, transport and process in their nest provide a culture medium for this fungus. Their work is highly coordinated. The cuts made by leafcutter ants are less circular than those made by leafcutter bees. They tend to cut a wide arc and remove a semi-oval part of the leaf edge.

A leafcutter bee at work.

▲ A leafcutter ant at work.

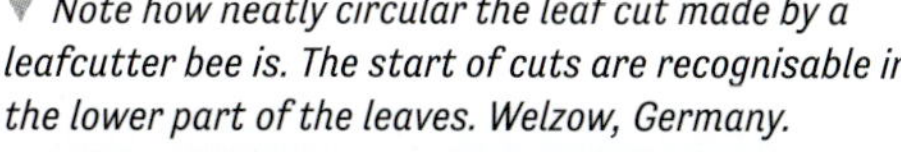
▼ *Note how neatly circular the leaf cut made by a leafcutter bee is. The start of cuts are recognisable in the lower part of the leaves. Welzow, Germany.*

LEAF MINES

Feeding galleries made by leaf miners are called mines or leaf mines. They are channels that run between the upper and lower epidermis of a leaf. The collective term leaf miners mainly includes the larvae of small insects, for example leaf-mining moths (Gracillariidae) or leaf-mining flies (Agromyzidae), but sawflies (Tenthredinidae) and beetle larvae can also create leaf mines. It is often possible to precisely identify species by looking at the host species and the arrangement of the leaf mines. However, describing the hosts and mines of the 900+ known species of leaf-mining flies in Europe is far beyond the scope of this book. There are also numerous species of leaf-mining moths, sawflies and beetles whose host is unknown or whose mining behaviour has not yet been described. For these reasons, we divide them into two basic categories: tunnel mines and blotch mines. These two categories can occur in different variations and combinations.

Tunnel mines

Many linear leaf mines have a narrow and a broad end. If you hold up a leaf like this against a light source, the mines will be clearly visible. The narrow end of a linear mine is where the larva began its development in the leaf tissue. The mine becomes gradually and visibly wider as the larva grows. On closer inspection, you can even see tiny black dots along the centre line inside the mine. These are the droppings of the larva, which it leaves behind as it feeds on the leaf tissue.

Blotch mines

Blotch mines, or 'square mines', appear undefined at first glance so they can be difficult to identify. On closer inspection, however, you can make out differences such as darker areas within the blotch mine and exit holes or patterns.

Horse-chestnut Leaf Miner

From 1990 to the present day, there have been repeated mass reproductions of the Horse-chestnut Leaf Miner (*Cameraria ohridella*) in various parts of Europe. This moth has now become established in most areas. It is mainly found in the Horse Chestnut (*Aesculus hippocastanum*), but occasionally also in the Norway Maple and Sycamore (*Acer platanoides* and *A. pseudoplatanus*). The females lay their eggs on the top of the leaf. As soon as they hatch, the larvae begin to eat

▲ *There are different types of leaf mine. This one shows the incredible aesthetics of nature. Spessart, Germany.*

▼ *A typical tunnel mine. Spessart, Germany.*

▲ *The dark spots inside the mine are the droppings of the larva. Spessart, Germany.*

▲ *The characteristic brown leaf edges and the blotch mines of the Horse-chestnut Leaf Miner. Lausitz, Germany.*

their way into the upper epidermis of the leaf, under the egg membrane. From there, they form blotch mines that grow continuously over the course of several larval stages. The pupae can survive several winters and withstand extremes such as severe cold or dehydration of the leaves before they hatch in spring. Damage caused by the moth causes the leaves to wilt prematurely, but death of trees has so far not been observed. In Germany, France, the Netherlands, Austria and Switzerland, the leaf mines of the Horse-chestnut Leaf Miner are a common sign on Horse Chestnut trees.

OTHER SIGNS ON LEAVES

Leaf rolling

Female leaf-rolling weevils (Attelabidae) lay their eggs in elaborately shaped leaf rolls. They seal up one end of these usually cigar-shaped structures before laying their eggs. The larva develops inside, where it is both protected from predators and supplied with food. Because the insects gnaw or bore into the leaf stalk, the leaf roll begins to wither shortly after the eggs are laid and eventually falls to the ground. This characteristic sign is mainly found from May to August.

Galls

Galls are growth deformities on plants that are caused by another organism. They can be caused by insects such as flies, beetles or wasps, but also by fungi, mites, bacteria or viruses. Insects and mites cause galls to grow by penetrating the tissue of a plant and laying their eggs under the bark or in the leaf tissue. The affected plant produces a gall as a growth reaction. Galls are often very rich in nutrients, so they provide the insect larva inside with a food source as well as protection. Galls rarely cause plants serious damage and small growths can even help them to grow.

▲ *Typical leaf roll on an alder leaf. Bielefeld, Germany. Simone Roters.*

◄ *Various leaf rolls made by a Birch Leaf Roller larva (Deporaus betulae). Märkische Schweiz, Germany.*

▲ *The droplet-shaped leaf gall of a Beech Gall Midge (Mikiola fagi). Spessart, Germany.*

Galls come in a wide variety of shapes, colours and sizes and the appearance of a gall is often unique to the species that caused it. The host plant is also key to identification, as most organisms that cause gall growth specialise in one host. The location of the gall is another clue. Is the gall on a leaf, stem, root or bud? If it is on a leaf, is it on the top or underside, on the edge, in the centre or directly on a vein? More than 1,500 animal species in Europe can cause gall growth. We have presented common or very striking galls here. Excellent specialist literature is also available that often enables precise identification and lets you dive deeper into the topic.

▼ *These distinctively red-coloured, horn-shaped growths are caused by a Sycamore Gall Mite* (Aceria macrorhyncha). *They are found in clusters on the top of maple leaves. Offenbach, Germany. Simone Roters.*

▼ *These characteristic silk button or coin galls come from the gall wasp species* Neuroterus numismalis. *Offenbach, Germany. Simone Roters.*

▲ *This distinctive bramble shoot gall is caused by the gall wasp species* Diastrophus rubi. *Offenbach, Germany. Simone Roters.*

▲ *The roundish oak gall of the Cherry Gall Wasp* (Cynips quercusfolii) *sits on leaf veins on the underside of oak leaves. Bielefeld, Germany. Ulrike Quartier.*

OTHER SIGNS

A selection of common, easily recognisable, strange or rare signs produced by invertebrates are described below. I hope that they will broaden your tracking horizons.

EGGS

Lacewings

After successfully mating in spring, female green lacewings (Chrysopidae) lay their **eggs on long, thin stalks** on tree bark, plant stems or leaves. Depending on the species, the eggs are lined up individually next to each other or laid as a cluster on one stalk.

WEBS AND COCOONS

Bagworm moths

The caterpillars of moths from the bagworm family (Psychidae) have a conspicuous, characteristic **gossamer sac** on their back. It can be covered with a wide variety of materials. They spin the sack onto tree trunks, branches or house walls for pupation.

▼ *Green lacewing eggs have a delicate, graceful structure. They are a rare find. Lausitz, Germany. Simone Roters.*

▲ *Female bagworm moths attract males to their gossamer sac. Lausitz, Germany.*

▲ *The appearance of the larval case of a bagworm moth can vary greatly depending on the species. Coto de Doñana National Park, Spain.*

▲ *The pupal cocoon of a braconid wasp. Offenbach, Germany. Simone Roters.*

Braconid wasps

Parasitic braconid wasps (Braconidae) lay their eggs in hosts, for example butterfly caterpillars, which they paralyse with their ovipositor. The wasp larvae hatch in the host's body, feed on the host and then kill it. The larvae spin a distinctive cocoon for pupation. The outside of the host is often completely covered by this **pupal cocoon**. The cocoon normally starts off whitish and then turns yellowish in colour.

Fairy Lamp Spider

The Fairy Lamp Spider (*Agroeca brunnea*) is known for its beautifully shaped **egg cocoon** that resembles a medieval lamp or lantern and is called a 'fairy lamp'. This sign occurs all year round, but it can unfortunately be difficult to find.

Wasp Spider

The strikingly marked Wasp Spider (*Argiope bruennichi*) is found in large parts of Europe. It prefers sunny locations with large grasshopper, locust or cricket populations. Typical habitats are sandy heaths and dry grasslands, but Wasp Spiders also inhabit wet meadows with low vegetation. The

▼ *Fairy lamps, the egg cocoons that give the Fairy Lamp Spider its name, are rare and very beautiful. Lausitz, Germany. Simone Roters.*

▲ *The basket-like cocoon of a Wasp Spider. Germany. Simone Roters.*

Wasp Spider is known to trackers because of the **basket-like cocoon** in which it lays its eggs. This conspicuous structure is usually attached to grass stems just above the ground and can contain up to 400 eggs. The cocoon is well insulated to help the eggs inside survive the winter.

▲ *The nest of the Northern Pine Processionary Moth mainly consists of pine needles, remains of cocoons and caterpillar droppings. Coto de Doñana National Park, Spain.*

Northern Pine Processionary Moth

The remains of Northern Pine Processionary Moth (*Thaumetopoea pinivora*) cocoons can be found in summer, usually in June and July, and especially in dry, sandy pine forests. The remains of the cocoons can contain large amounts of caterpillar droppings and urticating (stinging) hairs.

> **Caution!**
>
> As with other processionary moths, contact with the caterpillars' urticating hairs can cause an allergic skin reaction (caterpillar dermatitis).

NESTS AND SEALS

Caddisflies

Caddisfly (Trichoptera) larvae have silk glands. They use the silk produced by these glands to weave small objects from their environment into a **protective case** around their abdomen. The cases can be made from a wide variety of materials such as pebbles, duckweed, pieces of wood or fallen leaves or even small live freshwater snails. Caddisflies are considered an indicator of good water quality.

▼ *Caddisfly larva cases can look very different depending on the material used. Bieszczady, Poland.*

▲ *Cuckoo spit is a colloquial term for the nest of a froghopper. Coto de Doñana National Park, Spain.*

Froghoppers

The larvae of most froghoppers (Cercopidae), also known as spittlebugs, live in a **blob of frothy liquid** measuring around 1–2cm in size on plant stems. The larva produces the froth, which protects it against dehydration and predators, by 'pumping' air into a mixture of endogenous substances and plant sap. One or more larvae can live in a ball of froth. This sign can be found almost everywhere in Europe from the beginning of June to the end of September.

Masonry bees

Masonry bees of the genera *Megachile* and *Osmia* are mainly known to trackers because of their fascinating **nests**. They produce a building material from a glandular secretion and parts of leaves or soil, which they use to make cells for their brood. The nests usually consist of several cells arranged one behind the other and are closed at the end with a specially designed nest seal. The nests and their seals are often species-specific and can be found in empty snail shells, hollow plant stems, in the dead wood of hollow branches, in cracks in walls or in the ground. The female lays one egg per cell and then seals them up together with a pollen supply. The larva sometimes feeds on the food supply for several weeks before it spins a cocoon around itself to pupate.

Roman Snail

Some snails, such as the Roman Snail (*Helix pomatia*), can seal up their shell with an attached partition called an **epiphragm**. This air-permeable cover protects against the cold and dehydration and is formed with the help of the body's own secretions and calcium deposits. The snail sheds the seal when it is no longer required.

BOREHOLES AND HOLES

Nut weevils

The females of the very similar-looking nut weevil species *Curculio venosus* and *C. nucum* lay their eggs in unripe acorns or hazelnuts using their long proboscis. The larvae feed on the nuts until they fall to the ground in autumn. They then bore their way out. The often perfectly **circular hole** that can easily be seen with the naked eye is the larva's exit hole.

▼ *Masonry bee nest. Germany. Simone Roters.*

▲ *The epiphragm of a Roman Snail. Germany. Simone Roters.*

▲ *Note the almost circular exit hole of a nut weevil larva. Spessart, Germany.*

▼ *Dung beetle holes are common in Europe. The central, circular entrance is characteristic. Märkische Schweiz, Germany.*

▲ *Beetle hole. The ejected material acts as a self-made track trap where clear prints can often be seen. Coto de Doñana National Park, Spain.*

▼ *The entrance hole of a European Field Cricket. A closer look reveals small black dots, which are the cricket's droppings. Lausitz, Germany.*

◄ *Close-up of the excrement of a European Field Cricket. Lausitz, Germany.*

Beetles

Dung beetles often build their breeding chambers near animal excrement. The perfectly **circular entrance hole** is in a prominent position in the middle of a pile of earth.

When beetles dig a side entrance hole into the soil, they often create a clear path, which can be easily recognised by the fresh ejected material.

European Field Cricket

Small **entrance holes** with a diameter of around 1cm can be found in meadows. If you wait long enough, you might even see a European Field Cricket (*Gryllus campestris*) 'parking' backwards to rest in the safety of its hole. The cricket will often leave droppings in front of the entrance hole, like a latrine.

Wolf spiders

Unlike many other spiders, wolf spiders (Lycosidae) do not catch their prey in a web. Instead, they lurk in silk-lined burrows they have dug themselves. As soon as a prey animal approaches, they dart out to grab it. The burrow can be recognised by an almost **circular opening** in the ground, which is often surrounded by a wall of soil material. It is also criss-crossed with silk that transmits movements near the burrow via vibrations to alert the wolf spider to possible prey. The size of the opening and the wall varies depending on the size of the spider. Impressively large traps can sometimes be found, but all wolf spiders in Europe are harmless to humans.

▶ *The opening of the burrow is criss-crossed by silk. These threads enable the wolf spider to detect potential prey. Coto de Doñana National Park, Spain.*

▲ *A curious wolf spider. Coto de Doñana National Park, Spain.*

▼ *Wolf spider burrow with the corresponding ejected soil in front of it. Coto de Doñana National Park, Spain.*

◀ *Ant trails can be narrow and deep, as here, as well as flat and wide. Hainburg, Germany. Simone Roters.*

TRAILS, FUNNELS AND TUNNELS

Ants

Well trodden-looking **ant trails** are a common find, especially in dry parts of Europe. These 'roads' are the result of the collective labour of countless ants (Formicidae) that have cleared the trail bit by bit. Keeping their trails clear enables efficient transport of raw materials, like a well-maintained motorway network. Ant trails are usually faintly visible paths that stand out from their surroundings due to the different colour of the soil. Most ant trails are up to 1cm wide, but they can also be as much as 20cm wide, depending on the species.

Antlions

The well-known larvae of antlions (Myrmeleontidae) live underground and dig **funnel-shaped pitfall traps** in loose sandy soil. If an ant or other small insect of a similar size steps onto the funnel, it immediately falls to the bottom, where the antlion throws up jets of sand and grabs the prey with its suction pincers. Prey generally do not manage to escape. The larvae move backwards to create a new trap. As they do so, they leave distinctive drag marks in the sand, which differ from those of earthworms due to the frequent changes of direction over a short distance. The size of the trap varies according to the antlion species. There are around 11 different species in Europe. The traps measure 2–5cm in diameter.

▼ *The funnel-shaped pitfall trap of an antlion nymph. The meandering path is typical. Lausitz, Germany.*

▶ *The superficial burrows made by the European Mole Cricket are like those of moles, but much smaller. Lausitz, Germany.*

European Mole Cricket

The **tunnels** of the European Mole Cricket (*Gryllotalpa gryllotalpa*) are like mole tunnels running along the surface, but much smaller. European Mole Crickets live almost exclusively in loose, easily diggable, moist sandy or loamy soil, often near bodies of water. They have shovel-like front legs resembling those of a mole (giving them their name). European Mole Crickets create tunnel systems with a diameter of 0.7–1.5cm.

MOUNDS OF EARTH AND EARTH WALLS

Earthworms

Small piles of worm casts indicate the presence of earthworms (Lumbricidae). Earthworms eat their way through the soil, digest organic material and excrete the remains. The size of the **worm cast** varies depending on the size of the worm. Casts can measure several centimetres in diameter. If you carefully push one of the casts aside, you will usually be able to see the entrance to the burrow in the centre underneath.

▲ *European Mole Cricket burrow. The diameter of the entrance hole is about the thickness of a finger. Lausitz, Germany.*

▶ Earthworms leave piles consisting of excreted soil material. We find them after rain showers or early in the morning. Värmland, Sweden.

◀ Anthills are complex structures that function as air conditioning as well as shelter for the nest, and can reach a considerable size. Värmland, Sweden.

▶ A beetle larva pushes these cylindrical clumps of earth back up through its passageway. Their diameter corresponds approximately to the size of the larva. Coto de Doñana National Park, Spain.

▲ Anthill of the Yellow Meadow Ant, completely overgrown by grass. West Sussex, England.

Beetles

Various beetle larvae make conspicuous **piles of soil** when extending their dwellings. The ejected soil roughly corresponds to the size of the larva. It is a distinctive sign, especially in dry, sandy areas. Studies have revealed that dung beetles use starlight to find their way back to their breeding chamber after searching for food. They create a mental map of the night sky by climbing onto the dung ball, rotating around their own vertical axis and using the positions of the celestial bodies to orientate themselves. This behaviour is called 'dancing'. In hot climates, it also protects against overheating, as the dung ball is cooler than the soil temperature (Byrne 2012). Dung beetles are known to use this type of navigation to move in a straight line towards their breeding chamber, although they usually move backwards.

Ants

The large **anthills** of the Southern Wood Ant (*Formica rufa*) are one of the best-known insect signs in Europe. The anthills, which are usually built from needles, small stones and other surrounding ground material, provide insulation, enabling the ant colony to maintain a constant temperature in their underground nest.

The Yellow Meadow Ant (*Lasius flavus*) is one of the most common ant species in Europe and usually lives in inconspicuous earth nests, for example under stones and roots. However, in semi-natural, undisturbed meadows, it creates distinctive **mounds of earth**. These become completely overgrown with grass and other plants and are not always immediately recognisable as anthills. They can resemble overgrown molehills and are often found in groups.

FEEDING MARKS

Pine Shoot Moth

The Pine Shoot Moth (*Rhyacionia buoliana*) is a small moth whose caterpillars leave distinctive signs. The adult moths swarm from about mid-June. They then lay their eggs individually or in groups on the needles, needle sheaths or shoots of pine trees. The caterpillars spend the winter in the buds. In spring, they hollow out buds and young sprouting shoots from the base to the tip. Depending on the intensity of the feeding, the sprouting buds either die or the young shoots become deformed. This can lead to a characteristic **curvature of the trunk or branches** known as a 'post horn'. The Pine Shoot Moth favours sparse crops, thickets and pole woods, especially in dry locations.

Snail feeding marks on bark and fungi

Snails have a rasping tongue called a radula (Latin for 'scraper'). The radula is a tongue-like band with thousands of microscopic chitinous 'teeth' arranged in rows. Snails eat organic material such as leaf tissue or fungi and rasp algae from stones, tree bark and other surfaces. The feeding mark of a snail is a conspicuous pattern consisting of many individual **scrape marks** made by these small 'teeth'.

When feeding on fungi, small rodents usually leave furrows made by the incisors. Snails leave deep holes in fungi by rasping them from the top. Mammals normally eat from the edge of the fungi.

You can often make out traces of snail slime when a snail has been feeding on a mushroom.

The tiny radula 'teeth' of limpets (Patellidae) are the hardest known biomaterial on earth.

▶ *The caterpillar of the Pine Shoot Moth is responsible for this distinctive 'post horn'. Drehna, Germany. Jonathan Schüppel.*

▼ *Snail feeding mark (right in the image) compared to the feeding mark of a mouse (left in the image). Spessart, Germany.*

▲ *Here a slug has eaten algae from a birch tree with its rasping tongue (radula). Lausitz, Germany.*

EXCREMENT AND REMAINS

Butterflies and moths

Caterpillar excrement is usually easy to recognise because of its strikingly symmetrical structure. The small faecal pellets often have a cylindrical shape with six lengthways grooves and several less pronounced transverse grooves. The shape can resemble parts of a plant. The colour varies from grey-green and green to various brown shades and black. The plant material contained in the faecal pellets is fine and tightly packed. Different caterpillar species may produce their own distinctive variations. Depending on the size of the caterpillar, the faecal pellets can be 2–8mm long.

▶ *Caterpillar droppings. Jetzendorf, Germany. Laura Gärtner.*

▲ *Note the symmetrical structure. The longitudinal and transverse notches are typical of caterpillar droppings. Lausitz, Germany.*

Snails

Snail excrement is thin, sausage shaped, and coiled up like rope. It can be found on leaves, walls and trees and on any other surfaces where snails have crawled. Shiny snail slime is another sign that usually occurs together with the droppings. Depending on the size of the snail, the droppings can be tiny to several centimetres long and up to 4mm thick.

Cuttlefish

The 'bones' of dead cuttlefish (Sepiida) are often found washed ashore. They are known as **cuttlebones**. Cuttlebones are white, mainly made of calcium carbonate and can be shaped like small surfboards. They act as a pressure-resistant buoyancy chamber for the living animal, which can be filled with gas and liquid as required. In the pet trade, cuttlebones are sold as a source of calcium for reptiles or caged birds. Many bird species also use cuttlebones in the wild, so they often have beak marks (see page 718).

◀ *Snail droppings are coiled like rope and can often be found on house walls. Märkische Schweiz, Germany.*

APPENDICES

APPENDICES

LITERATURE (selection)

BOOKS

Abenza, L. (2018): *Aves que dejan Huella.* España, MADbird.

Altum, B. (2016): *Forstzoologie, Insekten.* Nicosia, TP Verone Publishing.

Amman, G. (2006): *Säugetiere und Kaltblüter des Waldes.* Melsungen, Neumann-Neudamm.

Baker, N. (2014): *Fährten lesen und Spuren suchen.* Bern, Haupt Verlag.

Bang, P. and Dahlström, P. (2009): *Tierspuren.* Munich, BLV Verlagsgesellschaft mbH.

Bellmann, H., Spohn, M. and Spohn, R. (2018): *Faszinierende Pflanzengallen.* Wiebelsheim, Quelle & Meyer.

Bergmann, H-H. (2015): *Die Federn der Vögel Mitteleuropas.* Wiebelsheim, AULA-Verlag.

Bergmann, H-H. and Klaus, S. (2016): *Spuren und Zeichen der Vögel Mitteleuropas.* Wiebelsheim, AULA-Verlag.

Bertram, J. F. A. (2016): *Understanding Mammalian Locomotion: Concepts and Applications.* New Jersey, Wiley-Blackwell.

Beutel, R. G. and Leschen, R. A. G. (2016). *Coleoptera, Beetles. Morphology and Systematics.* Berlin, De Gruyter.

Bezzel, E. (2014): *Vogelfedern.* Munich, BLV.

Boback, A. W. (1970): *Unsere Wildenten.* Wittenberg Lutherstadt, Ziemsen Verlag.

Boitani, L. and Mech, D. (2003): *Wolves: Ecology, Behavior and Conservation.* Chicago/London, University of Chicago Press.

Bouchner, M. (1990): *Der große Spurenführer.* Bindlach, Kosmos Verlags-GmbH.

Brauns, A. (1991): *Taschenbuch der Waldinsekten.* Jena/Stuttgart, Gustav Fischer Verlag.

Brown Jr., T (1999): *The Science and Art of Tracking.* New York, Berkley.

Brown, R., Ferguson, J., Lawrence, M. and Lees, D. (2005): *Federn, Spuren und Zeichen der Vögel Europas.* Wiebelsheim, AULA-Verlag.

Brown, R., Ferguson, J., Lawrence, M. and Lees, D. (2009): *Tracks and Signs of the Birds of Britain and Europe.* London, Bloomsbury Publishing PLC.

Brown, R., Lawrence, M. and Pope, J. (1985): *Welches Tier ist das?* Stuttgart, Kosmos Verlags-GmbH.

Busching, W-D. (2005): *Einführung in die Gefieder- und Rupfungskunde.* Wiebelsheim, AULA-Verlag.

Chinery, M. (2012): *Pareys Buch der Insekten.* Stuttgart, Franckh Kosmos Verlag.

Cummins, H. (1929): *The topografic history of the volar pads (walking pads) in the human embryo.* Washington, Elsevier.

Cummins, H. and Midlow, C. (1961): *Finger prints, palms, and soles: an introduction to dermatoglyphics*. New York, Dover Publications.

Curry-Lindahl, K. (1995): *Der Berglemming*. Magdeburg, Ziemsen Verlag.

Davis, A. (2007): *Fährten- und Spurenkunde*. Stuttgart, Franckh Kosmos Verlag.

Diepenbeek, A. (2015): *Veldgids Diersporen*. Zeist, Knnv Uitgeverij.

Dieterlen, F. (2005): Die Säugetiere Baden-Württembergs. In: Braun, M. and Dieterlen, F.: *Die Säugetiere Baden-Württembergs*. Vol. 2, 297-311. Stuttgart, Verlag Eugen Ulmer.

Eiseman, C. and Charney, N. (2010): *Tracks and Sign of Insects*. Mechanicsburg, Stackpole Books.

Elbroch, M. (2003): *Mammal Tracks and Signs*. Mechanicsburg, Stackpole Books.

Elbroch, M. (2006): *Animal Skulls*. Mechanicsburg, Stackpole Books.

Elbroch, M., Kresky, M. and Evans, J. (2012): *Field Guide to Animal Tracks and Scat of California*. Berkeley and Los Angeles, University of California Press.

Elbroch, M. and Marks, E. (2001): *Bird Tracks and Signs*. Mechanicsburg, Stackpole Books.

Elbroch, M. and McFarland, C. (2019): *Mammal Tracks and Signs, Second Edition*. Guilford, Stackpole Books.

Elbroch, M. and Rinehart, K. (2011): *Behavior of North American Mammals*. New York, Houghton Mifflin Harcourt.

Engelmann, W-E. and Obst, F. J. (1981): *Mit gespaltener Zunge. Aus der Biologie und Kulturgeschichte der Schlangen*. Freiburg, Herder Verlag GmbH.

Falkus, H. (1980): *Die Sprache der Tierspuren*. Rüschlikon-Zurich, Verlag Müller Rüschlikon.

Fischer, M. and Lilje. K. E. (2017): *Hunde in Bewegung*. Stuttgart, Franckh Kosmos Verlag.

Fischer, M. and Schumann, H-G. (2008): *Fährten, Spuren und Geläufe*. Melsungen, Neumann-Neudamm.

Galán, J. M. (2012): *Huellas y rastros de la fauna de Doñana*. El, Rocío.

Gibbons, D. (2008): *Stories in Tracks and Signs*. Mechanicsburg, Stackpole Books.

Görner, M. and Hackethal, Dr. sc. H. (1987): *Säugetiere Europas*. Leipzig, Radebeul, Neumann Verlag.

Gotch, A. F. (1997): *Mammals – Their Latin Names Explained*. Dorset, Sterling Pub Co Inc.

Graf, Dr. J. and Wehner, M. (1971): *Der Waldwanderer*. Munich, J. F. Lehmanns.

Grimmberger, E. (2014): *Die Säugetiere Deutschlands*. Wiebelsheim, Quelle & Meyer.

Grimmberger, E. (2017): *Die Säugetiere Mitteleuropas*. Wiebelsheim, Quelle & Meyer.

Grimmberger, E. and Rudolff, K. (2009): *Atlas der Säugetiere Europas, Nordafrikas und Vorderasiens*. Münster, NTV Natur und Tier- Verlag.

Harris, S. and Yalden, D. W. (2008): *Mammals of the British Isles Handbook, 4th Edition*. Southampton, Mammal Society.

Harrison, C. and Castell, P. (2004): *Jungvögel, Eier und Nester der Vögel Europas*. Wiebelsheim, AULA-Verlag.

Hecker, F. (2010): *Welche Tierspur ist das?* Stuttgart, Franckh Kosmos Verlag.

Hering, E. M. (1957): *Bestimmungstabellen der Blattminen von Europa: Einschließlich des*

Mittelmeerbeckens und der Kanarischen Inseln. Berlin/Heidelberg, Springer Verlag.

Hildebrand, M. and Golsow, G. E. (2004): *Vergleichende und funktionelle Anatomie der Wirbeltiere.* Berlin/Heidelberg, Springer Verlag.

Hohmann, M. (2018): *Bewegungsapparat Hund: Funktionelle Anatomie, Biomechanik und Pathophysiologie.* Stuttgart, Sonntag Verlag.

Jedrezejewski. W. and Sidorovich, V. (2010): *The Art of Tracking Animals.* Bialowieza, Mammal Research Institute Polish Academy of Sciences.

Jenrich, J., Löhr, P-W. and Müller, F. (2012): *Bildbestimmungsschlüssel für Kleinsäugerschädel aus Gewöllen.* Wiebelsheim, Quelle & Meyer.

Jenrich, J., Löhr, P-W., Müller, F. and Vierhaus, H. (2016): *Mittel- und Großsäuger: Bildbestimmungsschlüssel anhand von Schädelmerkmalen.* Fulda, Michael Imhof Verlag.

Jenrich, J., Löhr, P-W. and Müller, F. (2010): *Kleinsäuger: Körper- und Schädelmerkmale, Ökologie.* Beiträge zur Naturkunde in Osthessen (ed. Verein für Naturkunde in Osthessen e.V.). Fulda, Michael Imhof Verlag.

Kemp, T. S. (2005): *The Origin and Evolution of Mammals.* Oxford, Oxford University Press.

Klinz, E. (1955): *Die Wildtauben Mitteleuropas.* Wittenberg Lutherstadt, Ziemsen Verlag.

Lang, A. (2004): *Spuren und Fährten unserer Tiere.* Munich, BLV Verlagsgesellschaft mbH.

Laußer, K. (2013): *Tierspuren.* Munich, Gräfe und Unzer Verlag.

Liebenberg, L. (2000): *Tracks and Tracking in Southern Africa.* Cape Town, Photographic Guides.

Liebenberg, L. (1990): *A Field Guide to the Animal Tracks of Southern Africa.* Cape Town/ Johannesburg, David Philip Publishers.

Liebenberg, L. (2000): *Tracks and Tracking in Southern Africa.* Cape Town, Photographic Guides.

Lowery, J. (2013): *Walk with the Animal. A Tracking Methodology.* Frazier Park.

Marchesi, P., Blant, M., Capt, S. (ed.) (2008): *Fauna Helvetica. Säugetiere. Bestimmung.* Neuchatel, CSCF.

Marchesi, P., Mermod, C. and Salzmann, H. C. (2010): *Marder, Iltis, Nerz und Wiesel.* Bern, Haupt Verlag.

März, R. (2007): *Gewöll- und Rupfungskunde.* Wiebelsheim, AULA-Verlag.

McAllister, I (2009): *Wilde Wölfe: die letzten ihrer Art in Kanada.* Munich, Frederking & Thaler.

Mitchell-Jones, A. J. *et al.* (1999): *The Atlas of European Mammals.* London, Bloomsbury Specialist.

Molinari, P., Breitenmoser, U., Molinari-Jobin, A. and Giacometti, M. (2000): *Raubtiere am Werk.* Limena, Rotografica Verlag.

Moskowitz, D. (2010): *Wildlife of the Pacific Northwest.* Portland, Timber Press Inc.

Moskowitz, D. (2013): *Wolves in the Land of Salmon.* Portland, Timber Press Inc.

Müller, F. and Müller D. G. (ed.) (2004): *Wildbiologische Informationen für den Jäger, Band 1 Haarwild.* Remagen, Verlag Kessel.

Müller, F. and Müller D. G. (ed.) (2006): *Wildbiologische Informationen für den Jäger, Band 2 Federwild.* Remagen, Verlag Kessel.

Müller, J. P., Jenny, H., Lutz, M., Mühlenthaler, E. and Briner, T. (2010): *Die Säugetiere Graubündens – eine Übersicht.* Chur, Verlag Desertina.

Müller, P. (2011): *Gelbhalsmaus* (Apodemus flavicollis). In: Broggi, M. F., Camenisch, D., Fasel, M., Güttinger, R., Hoch, S., Müller, J. P., Niederklopfer, P. and Staub, R. Government of the Principality of Liechtenstein (ed.) (2011): Die Säugetiere des Fürstentums Lichtenstein. Vaduz, Amtlicher Lehrmittelverlag.

Nöth, W. (2011): *Origins of Semiosis: Sign Evolution in Nature and Culture.* Berlin, De Gruyter.

Ohnesorge, G. and Scheiba, B. (2012): *Tierspuren und Fährten.* Munich, Bassermann Verlag.

Okarma, H. and Langwald, D. (2002): *Der Wolf: Ökologie, Verhalten, Schutz.* Berlin/ Vienna, Parey Buchverlag.

Olberg, G. (1957): *Tierfährten.* Wittenberg, Ziemsen Verlag.

Olsen, L-H. (2012): *Tierspuren.* Munich, BLV Buchverlag.

Ophoven, E. (2012): *Deutschlands wilde Tiere.* Stuttgart, Franckh Kosmos Verlag.

Pearce, G. (2011): *Badger Behaviour Conservation and Rehabilitation.* Exeter, Pelagic Publishing.

Polly, P. (2007): *Fins into Limbs: Evolution, Development, and Transformation.* Chicago, University of Chicago Press.

Radinger, E. H. (2004): *Wolfsangriffe – Fakt oder Fiktion?* Worpswede, Peter von Döllen Verlag.

Rezendes, P. (1999): *Tracking and the Art of Seeing.* New York, Collins Reference.

Richarz, K. (2006): *Tierspuren.* Stuttgart, Verlag Eugen Ulmer.

Ringleben, H. (1957): *Die Wildgänse Europas.* Wittenberg Lutherstadt, Ziemsen Verlag.

Rohe, W. (2020): *Die Brutbilder der wichtigsten Forstinsekten.* Wiebelsheim, Quelle & Meyer.

Rue, L. L. (1978): *The Deer of North America.* Barrington, Crown Publishers.

Rue, L. L. (2013): *Whitetail Savvy: New Research and Observations about America's Most Popular Big Game Animal.* New York, Skyhorse Publishing.

Schaefer, M. (2017): *Brohmer – Fauna von Deutschland. Ein Bestimmungsbuch unserer heimischen Tierwelt.* Wiebelsheim, Quelle & Meyer.

Schmidt, H-M. and Lanz, U. (2003): *Chirurgische Anatomie der Hand.* Stuttgart/New York, Springer Verlag.

Sidorovich, V. and Vorobej, N. (2013): *Mammal Activity Signs: Atlas, Identification Keys and Research Methods.* Moscow, Veche.

Spitzenberger, F. (2001): *Die Säugetierfauna Österreichs, Band 13.* Graz, Federal Ministry, Agriculture, Forestry, Regions and Water Management.

Stefen C. (2009): Gelbhalsmaus *Apodemus flavicollis.* In: *Atlas der Säugetiere Thüringens.* (Ed: M. Görner) Jena, Verlag AAT.

Svensson, L., Mullarney K., Grant, P. J. and Zetterström, D. (1999): *Der neue Kosmos Vogelführer.* Stuttgart, Franckh Kosmos Verlag.

Tkaczyk, F. (2015): *Tracks and Signs of Reptiles & Amphibians.* Mechanicsburg, Stackpole Books.

Turni, H. (2005): Gelbhalsmaus *Apodemus flavicollis.* In: Braun, M./Dieterlen, F. (ed.): *Die Säugetiere Baden-Württembergs. Volume 2.* Stuttgart, Verlag Eugen Ulmer.

Velasco, F. G. and Santamaría, P. T. (2016): *Guía de Huellas y rastros*. Madrid, Ediciones la Libreria.

Vogel, P. (1995): *Die Säugetiere der Schweiz*. Basel, Birkhäuser Verlag.

Weinberger, I. and Baumgartner, H. (2018): *Der Fischotter*. Bern, Haupt Verlag.

Westheide, W. (2010): *Spezielle Zoologie. Teil 2: Wirbel- oder Schädeltiere*. Berlin/Heidelberg, Springer Verlag.

Wilson, D. E., Lacher, T.E. and Mittermeier, R. A. (2016): *Handbook of the Mammals of the World, Volume 6*: Lagomorphs and Rodents I. Barcelona, Lynx Edicions.

Wilson, D. E., Lacher, T.E. and Mittermeier, R. A. (2017): *Handbook of the Mammals of the World, Volume 7*: Rodents II. Barcelona, Lynx Edicions.

Wilson, D. E. and Mittermeier, R. A. (2009): *Handbook of the Mammals of the World, Volume 1: Carnivores*. Barcelona, Lynx Edicions.

Wilson, D. E. and Mittermeier, R. A. (2011): *Handbook of the Mammals of the World, Volume 2: Hoofed Mammals*. Barcelona, Lynx Edicions.

Wilson, D. E. and Mittermeier, R. A. (2018): *Handbook of the Mammals of the World, Volume 8*: Insectivores, Sloths and Colugos. Barcelona. Lynx Edicions.

Wilson, D. E. and Mittermeier, R. A. (2019): *Handbook of the Mammals of the World, Volume 9*: Bats. Barcelona, Lynx Edicions.

Witt, R. (1994): *Tierspuren, Beobachtungen durch das Jahr*. Munich, Orbis Verlag.

Young, J. and Morgan, T. (2007): *Animal Tracking Basics*. Mechanicsburg, Stackpole Books.

JOURNAL ARTICLES AND INTERNET SOURCES

Afelt, T. *et al.* (1983): Speed control in animal locomotion: Transitions between symmetrical and asymmetrical gaits in the dog. *Acta Neurobiologiae Experemantalis*, 43(4–5), 235–250.

Alexander, R. (1984): The Gaits of Bipedal and Quadrupedal Animals. *The International Journal of Robotics Research*, 3(b), 49–59.

Alkon, P. U. and Saltz, D. (1985): Potatoes and the nutritional ecology of crested porcupines in a desert biome. *The Journal of Applied Ecology*, 22(c), 727–737.

Alonso, J. C. *et al.* (2009): The Most Extreme Sexual Size Dimorphism Among Birds: Allometry, Selection, and Early Juvenile Development in the Great Bustard (*Otis tarda*). *The Auk*, 126 (c), 657–665.

Andrzejewski, R., Babinska-Werka, J., Owadowska, E. and Szacki, J. (2000): Homing and space activity in bank voles Clethrionomys glareolus. *Acta Theriologica*, 35(b), 155–166.

Arbeitskreis Luchs Nordbayern: Risse. Accessible at: https://ak-luchs.de/files/risserkennung.pdf (accessed March 2019)

Barthelmess, E. (2006): Hystrix africaeaustralis, Mammalian Species. *American Society of Mammalogy*, 788, 1–7.

Bertolino, S. *et al.* (2015): Good for management, not for conservation: An overview of research, conservation and management of Italian small mammals. *Hystrix, the Italian Journal of Mammalogy*, 26(a), 25–35.

Bicnevicius, A. R. and Reilly, S. M. (2006): Correlation of Symmetrical Gaits and Whole Body Mechanics: Debunking Myths in Locomotor Biodynamics. *Journal of Experimental Zoology* Part A, 305(11), 923–934.

Blick, T. (2004): Checkliste der Spinnen Mitteleuropas – Checklist of the spiders of Central Europe. Arachnological Society e.V. https://arages.de/fileadmin/Pdf/checklist2004_araneae.pdf (accessed January 2021).

Brown, J. C. and Yalden, D. W. (1973): The description of mammals –2 Limbs and locomotion of terrestrial mammals. *Mammal Review*, 3(d), 107–134.

Bruno, E. and Riccardi, C. (1995): The diet of the crested porcupine *Hystrix cristata L.*. *Mammalian Biology*, 60, 226–236.

Burgin, C. J., Colella, J. P., Kahn, P. L. and Upham, N. S. (2018): How many species of mammals are there? *Journal of Mammalogy*, 99(a), 1–14.

Deer Initiative (2008): Species Ecology Sika Deer. England and Wales best practice guides, www.thedeerinitiative.co.uk (accessed January 2021).

Downing, S. *et al.* (n.d.): Chinese water deer. NNSS, GB non-native species secretariat, www.nonnativespecies.org/index.cfm?sectionid=47

Gipps, J. H. W. (1985): The behaviour of bank voles. *Symposia of the Zoological Society of London*, 55, 61–87.

Gjerde, I. (1990): Determination of sex in capercaillie *Tetrao urogallus* by means of dropping size. *Fauna norvegica*, Series C, Cinclus 13, 91–92.

Griffin, T. G./Main, R. P./Farley, C. T. (2004): Biomechanics of quadrupedal walking: how do four-legged animals achieve inverted pendulum-like movements? *The Journal of Experimental Biology*, 207, 3545–3558.

Harrington, L. A./Harrington, A. L./Macdonald, D. W. (2008): Distinguishing tracks of mink *Mustela vison* and polecat *M. putorius*. *European Journal of Wildlife Research*, 54, 367–371.

Hildebrand, M. (1965): Symmetrical gaits of horses. *Science*, 150 (3697), 701–708.

Hildebrand, M. (1968): Symmetrical gaits of dogs in relation to body build. *Journal of Morphology*, 124, 353–360.

Hildebrand, M. (1980): Analysis of asymmetrical gaits. The adaptive significance of tetrapod gait selection. *American Zoologist*, 20(a), 255–267.

Hildebrand, M. (1989): The quadrupedal gaits of vertebrates. *Bioscience* 39, 766–774.

Horner, A. M. (2011): Crouched Locomotion in Small Mammals: The Effects of Habitat and Aging. Dissertation, College of Arts and Sciences of Ohio University, 121pp. www.semanticscholar.org/paper/Crouched-Locomotion-in-Small-Mammals per cent3A-The-Effects-Horner/d22896fb1de443aeace4e2f632c39968ae3c882f (accessed January 2021).

Ibe, S. C., Salami, S.O. and Ajayi, I. E. (2017): Trunk and paw pad skin morphology of the African giant pouched rat. *European Journal of Anatomy*, 18(c), 175–182.

Jahrl, J. (1995): Historische und aktuelle Situation des Fischotters (*Lutra lutra*) und seines Lebensraumes in der Nationalparkregion Hohe Tauern. Nationalparkinstitut des Hauses der Natur.www.zobodat.at/pdf/

HdN_12_0029-0077.pdf (accessed January 2019)

Kaczensky, P. *et al.* (2011): Wer war es? Spuren und Risse von großen Beutegreifern erkennen und dokumentieren. Ed: Wildland-Stiftung Bayern.

Kaczensky, P. *et al.* (2013): Status, management and distribution of large carnivores – bear, lynx, wolf and wolverine – in Europe. Ed: European Comission. www.researchgate.net/publication/259590863_Status_Management_and_Distribution_of_Large_Carnivores_-_Bear_Lynx_Wolf_and_Wolverine_in_Europe_Part_1 (February 2020)

Kauhala, K. and Winter, M. (2006): *Nyctereutes procyonoides*, Delivering Alien Invasive Species Inventories for Europe (DAISIE). www.europe-aliens.org/pdf/Nyctereutes_procyonoides.pdf (accessed January 2020).

Kimura, S. *et al.* (1999): Palmar and plantar pads and flexion creases of genetic polydactyly mice (Pdn). *Journal of Morphology*, 239(a), 87–96.

Kimura, S. and Kitagara, T. (1986): Embryological development of human palmar, plantar, and digital flexion creases. *The Anatomical Record*, 2016(b), 191–197.

Lapini, L. and Ganlsoßer U. (2012): Der europäische Goldschakal. www.researchgate.net/profile/Luca_Lapini2/publication/272505706_Der_europaische_ Goldschakal/links/54e76d2c0cf277664ffa871c/Der-europaeische-Goldschakal.pdf (last accessed January 2021).

Larivière, S. and Calzada, J. (2001): *Genetta genetta. The American Society of Mammalogists*, 680, 1–6.

Layne, J. L. (1970): Climbing Behavior of *Peromyscus floridanus* and *Peromyscus gossypinus. Journal of Mammalogy* 51(c): 580–591.

Maier, G. (1971): Assessment of the proportion of heavy stem deformation through post-horn pine-shoot borer in Baden. [Rhyacionia], in *Agris* 2013, 188–195. *Allgemeine Forst- und Jagdzeitung* Volume 142 (7).

Michel, F. (1970): Beiträge zur Osteologie der Murmeltiere. *Mitteilungen der Naturforschenden Gesellschaft in Bern*, volume 27. 26–37.

Pasitschniak-Arts, M. and Larivière, S. (1995): *Gulo gulo. The American Society of Mammalogists*, 499, 1–10.

People's Trust for Endangered Species: Your guide to looking for signs of water voles and other riverbank species. https://ptes.org/wp-content/uploads/2015/03/Your-guide-to-looking-for-signs-of-water-voles-and-other-riverbank-species.pdf (accessed March 2020)

Prugh, L.R. and Ritland, C. E. (2005): Molecular testing of observer identification of carnivore faeces in the field. *Wildlife Society Bulletin*, 33(a), 189–194.

Purser, P., Wilson, F. W. and Carden, R. (2009): Deer and forestry in Ireland: a review of current status and management requirements. Woodlands of Ireland. www.woodlandsofireland.com/sites/default/files/DeerStrategy.pdf (accessed November 2019)

Reinhardt, I. and Kluth, G. (2015): Untersuchungen zum Raum-Zeitverhalten und zur Abwanderung von Wölfen in Sachsen-Projekt

„Wanderwolf". LUPUS Institute Germany. www.gzsdw.de/files/endbericht_ projekt_wanderwolf_2012_2014.pdf (accessed March 2020)

Reinhardt, I. and Kluth, G. (2016): Abwanderungs- und Raumnutzungsverhalten von Wölfen (Canis lupus) in Deutschland. *Natur und Landschaft*, 6, 262–271.

Resch, C. and Resch, S. (2020): Waldmaus – *Apodemus sylvaticus*. In: kleinsaeuger. at – Internethandbuch über Kleinsäugerarten im mitteleuropäischen Raum: Körpermerkmale, Ökologie und Verbreitung. *Apodemus* – Priv. Institut f. Wildtierbiologie, Haus im Ennstal. https:// kleinsaeuger.at/apodemus-sylvaticus. html (accessed February 2020)

Robert Koch Institute (2019): Informationen zur Vermeidung von Hantavirus- Infektionen. Berlin, www.rki.de > Infektionskrankheiten A–Z > Hantavirus. www.rki.de/DE/Content/Infekt/EpidBull/ Merkblaetter/Ratgeber_Hantaviren.html (December 2019)

Romani, T., Giannone, C., Mori, E. and Filacorda, S. (2018): Use of track counts and camera traps to estimate the abundance of Western Roe Deer in North- Eastern Italy: are they effective methods? *Mammal Research*, 63(a). 477–484.

Schellhorn, R. (2009): Eine Methode zur Bestimmung fossiler Habitate mittels Huftierlangknochen. Dissertation, Tübingen. https://publikationen.unituebingen.de/ xmlui/bitstream/handle/10900/49294/ pdf/Dissertation_Schellhorn_2009. pdf?sequence=1&isAllowed=y (accessed January 2019)

Stier, N. *et al.* (2010): Untersuchung zur Raumnutzung von Damwild. Final report 1999–2000, Oberste Jagdbehörde Mecklenburg-Vorpommern. https:// refubium.fu-berlin.de/bitstream/handle/ fub188/6832/Diss_E_Gleich_Damwild_ UM.pdf?sequence=1 (accessed October 2019)

Tansley, D (n.d.): Water for wildlife: a guide to water vole ecology and field signs. Essex Wildlife Trust.

Usherwood, J. R. (2010): Inverted pendular running: a novel gait predicted by computer optimization is found between walk and run in birds. *Biology Letters*, 6, pp. 765–768.

Wörner, F. G. (2014): Der Luchs. Heimkehrer auf leisen Pfoten. Tierpark Niederfischbach e.V./Ebertseifen Lebensräume e.V. https:// docplayer.org/21739457-Tierpark- niederfischbach-e-v-ebertseifen- lebensraeume-e-v-dr-frank-g-woerner- der-luchs-heimkehrer-auf-leisen-pfoten. html (accessed January 2019)

Ziekur, I. (2006): Adaptive Differenzierungen bei afrikanischen Muroidea (Rodentia). *Stuttgarter Beiträge zur Naturkunde series A*, edition 689. 1–72. www. zobodat.at/pdf/Stuttgarter-Beitraege- Naturkunde_689_A_0001-0070.pdf (accessed January 2019)

Zschille, J., Heidecke, D. and Stubbe, M. (2004): Verbreitung und Ökologie des Minks – *Mustela vison*. Schreber 1777 (Carnivora, Mustelidae) – in Saxony- Anhalt. *Hercynia N.F.* 37, 103–126.

Zosia, W./Nel, J. A. J. (1976): Climbing Behaviour in Three African Rodent Species. *Zoologica Africana* 11(a): 183–192.

GLOSSARY

Adult
Fully grown, sexually mature animal.

Altricial
Young animals whose senses and physiological characteristics, such as fur or feathers, develop during the initial period after birth (the young are often born blind and/or naked). They are dependent on care and attention from their parents in the nest or dwelling.

Ambush hunt
Hunting method where a predator waits for or carefully stalks prey over a long period of time and then suddenly attacks from a hidden position, taking advantage of the element of surprise. The predator often changes its starting point for the attack.

Arboreal
Tree-dwelling.

Basal
At the bottom, located at the base.

Biotope
Habitat shared by different species.

Breeding season
The time when animals that breed seasonally produce their young.

Cambium
Growth layer, especially in trees, between the sapwood and the bark. It is also found in other plant shoots and roots and is primarily responsible for widthways growth.

Carnivores
Animals that almost exclusively eat other, usually vertebrate animals. Mammalian carnivores often have very pronounced canines for holding onto prey and carnassial teeth with which they can bite through animal tissue with the help of their powerful jaw muscles.

Castor, castoreum
Strong-smelling secretion that beavers use for grooming and territory marking. It is still used today as an ingredient in homeopathic remedies.

Circumpolar
Located around a pole.

Concave
Describes the surface of a three-dimensional body that is curved in or hollow.

Convex
Describes the surface of a three-dimensional body that bulges out or is rounded outwards.

Delayed implantation
Delayed implantation of an embryo. See also Embryonic diapause.

Distal
Description of anatomical position. Away from the centre of the body.

Domestication
Process where humans have selected wild animals or plants and bred them over a

longer period to become domesticated animals or cultivated plants.

Dorsal
Located towards the back.

Ecological niche
Describes the interrelations between a species and the environmental factors relevant to it. The niche is often described as a 'job' or 'position' within the ecological community of different organisms in a biotope.

Ecotypes
Populations of a species that are genetically and physiologically adapted to certain environmental conditions through selection but are not distinct enough to be classified as a separate species.

Embryonic diapause
Phenomenon where fertilised eggs do not immediately start growing into an embryo. The egg implants in the uterus but cell division only begins after a certain time. It allows animals to give birth to and rear young at the most optimal time.

Endemic
Plants or animals that only occur in a specific, clearly defined area.

Fauna
The animal kingdom. Describes all animal species occurring in a defined area.

Fraying
Stripping off the velvet by rubbing antlers against trees or bushes. Seen in deer. Often used as a synonym for marking by striking the antlers on trees, especially in Western Roe Deer.

Gestation period
Period when a fertilised egg grows into an embryo in the uterus of an animal. It lasts until the actual birth and corresponds to pregnancy in humans. The length varies greatly depending on the species, tending to be short in small animals and long in larger ones. The gestation period is not necessarily identical to the period between mating and birth, see also Embryonic diapause.

Granivores
Animals that eat seeds or grain. Specialisation in the consumption of seeds and grains is most frequently seen in birds, but also in small rodents.

Habitat
Where an animal or plant species lives, within a particular biotope.

Heat stress
When animals are exposed to extreme heat or sunlight and must find compensation strategies in order not to exceed the maximum tolerated body temperature. Strategies for avoiding heat stress can include behaviours such as panting, reduced physical activity or seeking out shady spots.

Herbivores
Animals that mainly eat a plant-based diet. Mammalian herbivores often have rather broad incisors (except for rodents) and rather flat molars for grinding plant fibres. They do not have canine teeth.

Home range
Area used by an animal to fulfil essential needs, such as foraging, searching for resting places, mating and rearing young. The home range may include a defended territory, and is usually divided into other different areas that the animal visits at different times for different purposes. Unlike a territory, animals do not defend their entire home range against other members of the same species.

Insectivores
Order of mammals that feed primarily on insects and their larvae, as well as worms. They have many small, pointed teeth that are adapted to their predatory way of life.

Juvenile
Young animal; often not yet sexually mature.

Mammalia
Mammals. Animals that suckle their young with milk. See mammals section.

Maquis
A Mediterranean landscape characterised by dense, evergreen bushes.

Mating season
A specific, time-limited period in the annual cycle of an animal species during which the female is fertile and receptive to mating. Depending on the species, it occurs once or several times a year. Mating season is often accompanied by special behaviours and social structures within a species that are characteristic of this time. In mammals, it often occurs in late summer/autumn.

Medial
Description of anatomical position. Located towards the centre of a body structure.

Molar
Grinding tooth at the back of the mouth of a mammal.

Myxomatosis
Infectious disease caused by smallpox viruses that is transmitted to rabbits by gnats and mosquitoes. Infection led to severe population decline throughout Europe with a mortality rate of 40–60 per cent.

Non-native species
Animal species that have colonised and become established in an area where they were not originally native, following accidental or deliberate introduction by humans.

Oestrus
Phase of the oestrous cycle of female mammals when they are fertile and ready to mate. Also known as being on or in heat.

Omnivores
Animals that eat both plant and animal food and therefore have a very broad and varied diet. Omnivore mammals, such as humans, have incisors, canines and molars that are all relatively equally developed.

Opportunistic
Describes animals that can adapt quickly to different conditions and changes in their environment. Unlike specialists, opportunists can make flexible use of available resources and adapt to different food supplies, habitats or weather conditions, for example.

Pedipalps
The tactile organ on the head segment of arachnids, which is used, among other things, to hold and turn prey. It has various other functions, depending on the species.

Perpendicular
At a 90° angle. Dew claws can splay perpendicularly to the leg during fast movement.

Polygyny
Reproductive strategy where a male mates with several females (the reverse is polyandry).

Population
All animals of a species in a certain area where they live permanently and where they reproduce.

Precocial
Young animals that are already relatively well developed at the time of birth and whose senses and other physiological characteristics, such as fur or muscles, enable them to leave their nest or dwelling very soon after birth and follow their parents.

Predator
An animal that naturally preys on other living animals.

Promiscuity, promiscuous
Sexual behaviour where both sexes mate with several animals. This may be to increase genetic diversity and fertility, but also to avoid inbreeding.

Proximal
Description of anatomical position. Located towards the centre of the body.

Pursuit predation, pursuit predator
Hunting strategy that is often carried out in groups or packs, where the prey is pursued until it becomes exhausted and can be caught by the predator.

Rotating
Turning around an axis, moving in a circle.

Rut
The mating season in ungulates, also the behaviours observed during that time, whereby males fight one another in competition for female attention, and in some cases attract a harem of several females over time.

Sexual dimorphism
Differences in the appearance of sexually mature females and males, for example in body size or colour.

Species
Fundamental unit of biology that describes groups of individuals capable of reproducing with each other and that differ from other species in certain characteristics.

Subterranean
An animal that lives underground.

Superfetation
Double pregnancy or over-fertilisation. Some species can be fertilised again during an existing pregnancy, leading to a high number of offspring per year.

Taiga

Boreal coniferous forest. Refers to a type of geographical landscape with characteristic climate and vegetation that is only found in the northern hemisphere. The taiga is dominated by conifers, the winters are long and cool and the growing season in summer is short.

Taxonomy

Branch of science concerned with classification where, for example, animal species are classified according to their evolutionary relationships.

Territorial behaviour, territorial

Behaviour of animals that defend their territory against members of the same species. Some signs that animals leave, such as fraying on trees or urinating in prominent places, act as markings that other animals of the same species will recognise.

Territory

An area that an animal defends against other members of the same species. Animals often use visual or scent markings to keep other animals, who are competing for food or reproductive rights, at a distance. The size of a territory largely depends on the species in question.

Transverse

Across, diagonally opposite.

Tubercle

Hairless, bony callus, lump or bump found on the soles of the feet, for example in toads. Unless they occur in amphibians, they are referred to here as pads.

Tundra

Geographical landscape type. Characterised by a treeless, barren landscape and a cold, subpolar climate. The tundra is located far to the north and south, bordering the polar zones, and is scarcely inhabited by humans due to its extreme conditions.

Velvet

Skin with blood supply on the antlers of deer, which is rubbed off on trees (fraying) when the antlers stop growing.

Vertebrates

Includes all animals with a backbone or spinal column, as opposed to invertebrates. The subphylum of vertebrates includes the large groups of mammals, birds, amphibians, reptiles and bony and cartilaginous fish.

Vibrissae

Tactile hairs, whiskers. Stiff, often long hairs on the face that animals use to feel their environment or perceive other stimuli from their environment.

DOCUMENTING TRACKS

Clues found on the trail: ....

Droppings: ....

Urine marks: ....

Other: ....

<table>
<tr><td colspan="2">Front foot: drawing and dimensions</td><td colspan="2">Hind foot: drawing and dimensions</td></tr>
<tr><td colspan="2" rowspan="2">Track pattern: drawing and dimensions
(Stride length, trail width, gait):</td><td colspan="2">Time of day: ....

Temperature: ....

Wind direction: ....

Wind strength: ....

Precipitation: ....

Cloud cover (1/8): ....

Moon phase: ....

Ground conditions: ....</td></tr>
<tr><td>Length of the tracks: ....</td><td>Age of the track: ....</td></tr>
</table>

MAPPING

Draw in: trees, woodland, fields, hills, rivers, streams, lakes, sea, paths, roads, buildings, etc.

Nearest village/town: ....

County/state: ....

Topographic map no.:............................. Coordinates: ....

About the author

Joscha Grolms was born in Hildesheim, Germany, in 1983. His fascination with nature began at an early age.

After finishing high school, he spent a year living in the North American wilderness of Wisconsin. It was there that Joscha became captivated by the stories of animals and began reading their tracks. His search for the tracks and signs of mammals, birds, amphibians, reptiles, insects and other invertebrates has taken him to various countries in North America, Africa and across Europe. During his training as a tracker, Joscha has learned both from renowned scientists and indigenous trackers. Inspired by well-known trackers Louis Liebenberg (South Africa) and Mark Elbroch (US), he has spent many years observing, measuring, photographing and researching. He has also passed numerous examinations to obtain certificates in North America, the UK, the Netherlands, Spain and Germany.

Joscha is now an internationally recognised tracker and one of the few specialists in the tracks and signs of animals in Europe. Together with Mark Elbroch, Louis Liebenberg and the experts from CyberTracker Conservation, he works as an evaluator with the aim of establishing a global standard for tracking. Joscha also works as a lecturer and co-director of the wilderness school Wildniswissen, where he offers seminars, excursions and a one-year tracking course. He lives with his partner in Solling, Germany.

www.wildniswissen.de
joscha@wildniswissen.de
www.tierspuren-europas.de

Image sources

This book contains 1,122 photographs and 502 drawings/diagrams.

Photographs:
The photographs are by Joscha Grolms, unless otherwise stated in the captions or listed as follows:
blickwinkel/S. Meyers: 50/51
Hecker, Frank: 188, 293, 464, 469, 633 top, 653 top right, 746 2nd from bottom, 747 bottom, 766 l., 687 top
Bellmann, Heiko/Frank Hecker: 749 top, 767 l.
Körner, Sebastian: 85 (2)
Wetzel, Bernhard: 644 top, 647 top

Mauritius Images:
59, 186, 234, 262, 278, 280, 286, 297, 304, 312, 318, 320, 334, 338, 354, 355, 384, 386, 390, 402, 409, 428, 430, 457, 490, 491, 510, 558, 564, 596, 597, 638 top, 645 top, 648 l., 653 bottom, 660 bottom, 662, 664, 671 top, 674 top, 751 top, 767 right, 793

Nature Picture Library:
Andy Rouse: 192, 411, 769 top; Andy Trowbridge: 586; Bruno D'Amicis: 742; Danny Green/2020VISION: 52; David Pattyn: 659 bottom; David Tipling: 685; Dietmar Nill: 198; Edwin Giesbers: Front cover photo, 722/723, 735; Emanuele Biggi: 733; Erlend Haarberg: 12/13, 526; Espen Bergersen: 16; Fabrice Cahez: 371; Florian Möllers: 360; Folkert Christoffers: 639 bottom; Franco Banfi: 176; George Sanker: 476; Hermann Brehm: 683; Igor Shpilenok: 55, 438; Ingo Arndt: 237; Jasper Doest/Minden: 364; John Waters: 200; Jorma Luhta: 656 top; Jose B. Ruiz: 602/603; Jouan Rius: 34 bottom; Kerstin Hinze: 677 top; Markus Varesvuo: 676 top; Matthew Maran: 53; Mike Wilkes: 344; Paul Hobson: 22; Peter Cairns: 2 (and detail on back cover), 784/785; Philippe Clement: 270, 492; Sergey Gorshkov/Minden: 406, 576; Solvin Zankl: 206; Staffan Widstrand: 731; Stefan Huwiler: 20; Sven Zacek: 518, 802/803; Sylvain Cordier: 256; The Big Picture: 210, 452; Tom Vezo: 484

Shutterstock.com:
aaltair: 666 top; Aklion: 753 top; ArCa: 57; bluebamboo: 718 right; DaniloDjekovic: 668 top; Erni: 220; fernando sanchez: 380; Giedriius: 669 centre right; hfuchs: 649 bottom; Holm94: 732 top; Ihor Hvozdetskyi: 675

Drawings:
Drawings are by Joscha Grolms with the exception of the following:
Laura Gärtner. Even-toed ungulate and bird tracks using templates and
    specifications provided by the author.
Ulrike Quartier. Animal silhouettes on the front endpaper and pp. 61 (2), 63
    (2), 68 (2), 81 (2), 135, 139 (5), 605, 608 top, 609 top left, 609 top right, 610
    top left, 610 top right, 729 bottom, 746 bottom.
Jörn Lies: Gaits in the animal portraits using original drawings and
    specifications by the author.
Siegfried Lokau: Pp. 68, 77, 734 (2) and measurement diagrams on the back
    endpaper.
Helmuth Flubacher: Labelling, measurement and orientation lines in the
    drawings, diagrams and photographs by Joscha Grolms and Ulrike Quartier
    as well as diagram on p. 90 based on a template by Laura Gärtner.

INDEX

Differentiating between:

bound and lope 89
peeling and striking 154
walk and trot 81
Black Rats and Brown Rats 348
Brown Hares and Mountain Hares 224
canids and cats 362
Domestic Dogs and Wolves 395
ducks and gulls 681
European Beavers and Coypus 274
Field Mice, Common Voles and Bank Voles 330
Grey and Red Squirrels 240
grouse species 656
hamsters and rats 315
Muskrats and Coypus 284
pigeon and dove species 640

Red Deer and Common Fallow Deer 536
Red Foxes and Arctic Foxes 415
Sika and Common Fallow Deer 556
Stoats, European Polecats and American Mink 481
swans, geese and ducks 678
Typical mice and Field mice 341
voles by the size of their faecal pellets 279
Water voles and Brown Rats 289
Western Roe Deer and Reeves's Muntjac 561
White-toothed shrews and Long-tailed shrews 196
Wild Boar and Western Red Deer 502
Yellow-necked Mice and Wood Mice 324

Track formula for mammals

The track formula gives trackers important information. It allows taxonomic order and often even family membership to be determined at a glance.

XF × Yh + C

XF   Number of front foot toes normally imprinted (front foot track)

Yh   Number of hind foot toes normally imprinted (hind foot track)

C   Claw prints are distinguishable in both tracks

The capitalisation of the letters 'F' and 'h' indicates the relative size of the front foot and hind foot track.

For example: **4F × 4h + C**

This track formula describes a front foot track with four toe prints and a smaller hind foot track, also with four toe prints. Claw prints are distinguishable in both tracks. The number of toe prints and their relative size in this example are a match for both cats and dogs. Since the claw prints of cats are rarely visible, this is a dog track.

Abbreviations and symbols used

General

D   Diameter

W   Weight

HTL   Head-trunk length

Max.   Maximum

Min.   Minimum

Tl   Tail length

sp.   species (singular)

spp.   species (plural); example '*Tringa* spp.' means 'several species of the genus *Tringa*'

♂   male

♀   female

Tracks and track patterns

W   Width

GL   Group length

Hf   Hind foot

L   Length

LH   Left hind

LF   Left front

RH   Right hind

RF   Right front

TW   Trail width

SL   Stride length

Ff   Front foot

IGL   Inter-group length

Measuring tracks and track patterns

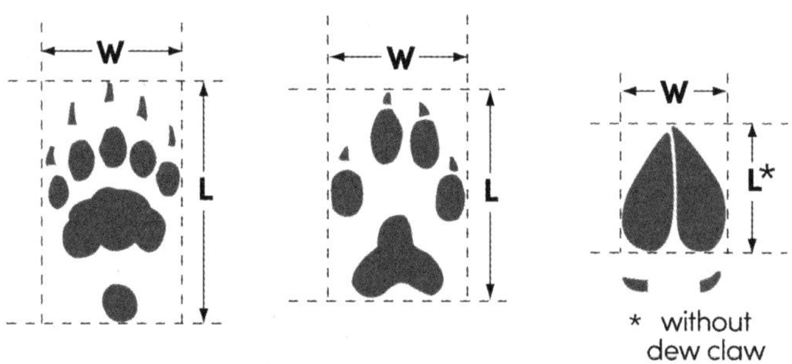

Continuous track patterns

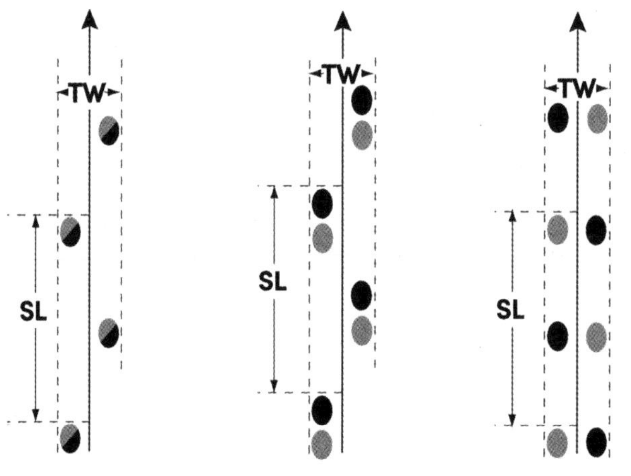

= Front foot   **L** = Length   **SL** = Stride length

= Hind foot   **W** = Width   **TW** = Trail width

= Hind atop front track (direct register)